T0343432

Horizontal Gene Transfer

Horizontal Gene Transfer

Second edition

Edited by

Michael Syvanen

*Department of Medical Microbiology
and Immunology
School of Medicine
University of California
Davis, California, USA*

and

Clarence I. Kado

*Davis Crown Gall Group
University of California
Davis, California, USA*

ACADEMIC PRESS

A Division of Harcourt, Inc.

San Diego San Francisco New York Boston London Sydney Tokyo

This book is printed on acid-free paper.

Copyright © 2002 by ACADEMIC PRESS

All Rights Reserved.
No part of this publication may be reproduced or transmitted in any form or by any means, electronic or mechanical, including photocopy, recording, or any information storage and retrieval system, without permission in writing from the publisher.

Academic Press
A Division of Harcourt, Inc.
Harcourt Place, 32 Jamestown Road, London NW1 7BY, UK
http://www.academicpress.com

Academic Press
A Division of Harcourt, Inc.
525 B Street, Suite 1900, San Diego, California 92101-4495, USA
http://www.academicpress.com

ISBN 0-12-680126-6

Library of Congress Control Number: 2001094563

A catalogue record for this book is available from the British Library

The cover electron micrograph shows the bacterium *Agrobacterium tumefaciens* attached to a pea cell. Courtesy of Dr Martha Hawes, University of Arizona, Tucson.

Transferred to digital print 2007

Printed and bound by CPI Antony Rowe, Eastbourne

Working together to grow
libraries in developing countries

www.elsevier.com | www.bookaid.org | www.sabre.org

ELSEVIER   BOOK AID International   Sabre Foundation

Contents

A color plate section appears between pages 110 and 111

Foreword

The era of high-speed sequencing and computer-based annotating systems have generated enormous data bases from which many novel discoveries are being made. Comparative genomic nucleotide and amino acid sequence analyses have revealed many sequence cassettes that appear highly conserved, thus raising the question whether these sequences were introduced by horizontal gene transfer mechanisms, or whether they were fortuitous occurrences.

The prior recombinant DNA era provided valuable insights on the extent of conserved genetic mechanisms and brought to realization that genes can be moved across species barriers, a finding that was supported 20 years earlier than the recombinant DNA era by bacterial genetic experiments. In fact, a number of elegant bacterial genetic studies had implicated genetic transmission of foreign DNA into bacteria and into plants. The natural transfer of plasmid DNA from *Agrobacterium tumefaciens* to plant cells, resulting in the integration of the foreign plasmid DNA into the chromosome of the plant followed by its expression to generate phenotypic change is the best case of horizontal gene transfer that occurs in nature. Certainly, bacteriophages were well known to mediate horizontal gene transfer long before the *Agrobacterium* story. Although this was a dramatic discovery, horizontal gene transfer among microbes did not have the impact of that mediated between a microbe and a eukaryote.

The genomics era and its ever growing data bases provide vast opportunities to explore potential horizontal gene transfer systems that may exist between microbes and eukaryotes and between lower eukaryotes and higher eukaryotes. The direction of horizontal gene transfer may be dominant in one direction, but may occur in the opposite direction (retrotransfer).

In reviewing the chapters of this book, I have come to the realization that the discovery of horizontal gene transfer among distinct organisms has increased substantially in this new millennium and a "new" scientific vocabulary has been introduced. The terms "cross-species gene transfer", "lateral gene transfer" and "horizontal gene transfer" have been used interchangeably, with the last two terms used most frequently. To avoid future confusion, the term "horizontal gene transfer" should represent the transfer of genes across distinct species, especially when interkingdom gene transfer takes place. The term "lateral gene transfer" could be retained to accommodate gene transfer between distinct species within a kingdom, viz., between prokaryotes, or between eukaryotes.

Clarence I. Kado

Preface

The seminal experiment that illustrated the ability of genetic information to flow between species slipped by largely unnoticed. In 1959, Tomoichiro Akiba and Kunitaro Ochia discovered antibiotic resistance plasmids. The most surprising attribute of this new class of plasmids was that they carried resistance genes to multiple antibiotics and that they moved among different bacterial species, spreading resistance genes, and thereby demonstrating that genetic information can flow from one species to another (Akiba et al., 1960; Ochia et al., 1959). The implications of this finding would have profound effects ranging from the applied field of genetic engineering to the very theory of evolution. Early papers probing the deeper theoretical implications of horizontal gene transfer began to appear in the 1970s, though they were not widely acknowledged or accepted. Fritz Went, in 1971, wrote a review on similar traits that are shared by unrelated flowering plants thereby illustrating many examples of parallel evolution. In addition, he noted that the traits are shared among plants that occupied the same ecosystems. In this context he proposed that these unrelated plants were exchanging genes. He cited bacterial plasmid transfer as a precedent for such events. Krassilov in 1977 arrived at a similar model for flowering plant evolution based on his paleontological studies of the emergence of angiosperms in the fossil record. Anderson in 1970 and Reanney in 1976 suggested that horizontal gene transfer could affect evolution in the animal kingdom, and Hartman, in 1976, suggested that horizontal gene transfer might effect speciation. There were a series of theoretical papers that cited horizontal gene transfer as an explanation for the widespread occurrence of parallelisms in the fossil record (Krassilov, 1977; Erwin and Valentine, 1984; Reanney, 1976; Jeppsson, 1984; Syvanen, 1985).

Meanwhile, genetic engineering experiments began to produce startling results. In 1976, Struhl et al. placed DNA from yeast into a histidine deficient mutant of *Escherichia coli* that resulted in the restoration of histidine biosynthesis. This DNA contained a histidine biosynthesis gene from the yeast genome. What seems commonplace today was difficult to comprehend back in 1976 – genes from a eukaryotic organism artificially introduced into a bacterium could actually function. Davies and Jimenez in 1980 showed that a bacterial neomycin phosphotransferase gene would express aminoglycoside resistance in yeast, showing that a bacterial gene could be expressed in a eukaryote.

As a bacteriologist, I had personally incorporated the findings of Akiba and Ochia into my scientific world-view. I was intrigued by the implication of Struhl's experiment. In the course of a discussion of a review of Crick's book about the unity of the genetic code entitled *Life Itself* (Crick, 1981), it occurred to me that horizontal movement of genes could shed light on this question provided such gene transfer was a factor in major evolutionary transitions. If this conjecture was correct, it could provide an alternative explanation for not only the unity the genetic code, but many other biological unities as well. At this point, I was unaware of the works of Went, Reanney, Krassilov, Hartman and Anderson. I wrote up my ideas in 1982, and they were finally published in 1985. During this period, the field of genetic engineering was exploding. Palmiter et al. in 1983 produced the first transgenic mouse that expressed a foreign gene, the human growth hormone gene. Result after result confirmed that it was possible for genes to cross species boundaries and to express their phenotype. These experiments all demonstrated that genes could be made to cross species boundaries in the laboratory. The fundamental question that remained was whether these events occurred in nature, and whether they occurred at a frequency high enough to effect evolution. Hopefully, this collection of articles will be but one of many which will begin to explore this question.

By the mid-1980s, numerous mechanisms for horizontal gene transfer were firmly established, not only for bacteria but also for metazoans and, in addition, many heretofore difficult to explain biological phenomena were easily handled by a horizontal gene transfer theory. However, there was a paucity of observations giving direct support to these speculations. With the rapid increase in the nucleic acid database over the past decade, the situation has changed. This book covers some of these more recent developments.

Today, researchers in many unrelated areas are making observations related to horizontal gene transfer, which has resulted in the unusual breadth of topics included in this volume. This book does not attempt a comprehensive survey of horizontal gene transfer, but rather attempts to sample various areas with a primary focus on material from active research areas. The chapters in this book deal with three questions.

First, can genes, or more specifically DNA move from one species to an unrelated one? Thus, a section of this book is devoted to the subject of transfer mechanisms, a phenomenon well documented in bacteria but also found in plants and animals. Obviously transfer mechanisms exist, the subsidiary questions are: how widespread are the mechanisms? And, do they operate in natural environments?

Second, what is the evidence that horizontal gene transfer contributes to existing genotypes of species? The primary evidence supporting evolutionary significant horizontal transfers involves phylogenetic reasoning. This is an area where the evidence is accumulating in the gene and protein sequence databases. Two problems

are repeatedly encountered – defining the topology of a gene tree and estimating divergence times following molecular clock assumptions. There are a number of contributions discussing results obtained from phylogenetic analysis and problems associated with this approach.

The third question raised by the central hypothesis is that if the mechanisms exist and events can be documented, does horizontal gene transfer actually play any significant evolutionary role? Or, does a theory that incorporates migrant DNA have utility in explaining more general biological phenomena. To this end, more conjectural papers that directly address macroevolutionary patterns and trends are presented.

REFERENCES

Akiba, T., Koyama, K., Ishiki, Y., Kimura, S. and Fukushima, T. (1960) The mechanism of the development of multiple-drug-resistant clones of *Shigella*. *Jpn J. Microbiol.* **4**: 219.

Anderson, N.G. (1970) Evolutionary significance of virus infection. *Nature* **227**: 1346–1347.

Crick, F. (1981) *Life Itself: Its Origin and Nature*, Simon and Schuster, New York.

Davies, J. and Jimenez, A. (1980) A new selective agent for eukaryotic cloning vectors. *Am. J. Tropical Med. Hygiene* **29**(5 Suppl): 1089–1092.

Erwin, D.H. and Valentine, J.W. (1984) Hopeful monsters, transposons and metazoan radiation. *Proc. Natl Acad. Sci. USA* **81**: 5482–5483.

Hartman, H. (1976) Speculation on viruses, cells and evolution. *Evolution Theory* **3**: 159–163.

Jeppsson, L. (1986) A possible mechanism in convergent evolution. *Paleobiology* **12**: 37–44.

Krassilov, V.A. (1977) The origin of angiosperms. *Bot. Rev.* **43**: 143–176.

Ochia, K., Yamanaka, T., Kimura, K. and Sawada, O. (1959) Inheritance of drug resistance (and its transfer) between *Shigella* strains and between *Shigella* and *E. coli* strains. *Nihon Iji Shimpo* **1861**: 34 [in Japanese].

Palmiter, R.D., Norstedt, G., Gelinas, R.E. et al. (1983) Metallothionein–human GH fusion genes stimulate growth of mice. *Science* **222**(4625): 809–814.

Reanney, D. (1976) Extrachromosomal elements as possible agents of adaptation and development. *Bacteriol. Rev.* **40**: 552–590.

Struhl, K., Cameron, J.R. and Davis, R.W. (1976) Functional genetic expression of eukaryotic DNA in *Escherichia coli*. *Proc. Natl Acad. Sci. USA* **73**(5): 1471–1475.

Syvanen, M. (1985) Cross-species gene transfer; implications for a new theory of evolution. *J. Theor. Biol.* **112**: 333–343.

Went, F.W. (1971) Parallel evolution. *Taxon* **20**: 197–226.

Michael Syvanen

Contributors

Ronald M. Adkins
Biology Department,
University of Massachusetts,
Amherst, Massachusetts, USA

Rita M. P. Avancini
Department of Entomology,
University of Illinois at Urbana-Champaign,
Urbana, Illinois, USA
(present address: Lexington, Massachusetts,
USA)

L. Aravind
National Center for Biotechnology
Information, National Library of Medicine,
National Institutes of Health,
Bethesda, Maryland, USA

William B.N. Berry
Department of Geology and Geophysics,
University of California,
Berkeley, California, USA

Meghan E. Bowser
Genotypes Inc.,
San Francisco, California, USA

James R. Brown
Bioinformatics Department,
GlaxoSmithKline,
Collegeville, Pennsylvania, USA

A. Burmester
Lehrstuhl für Allgemeine Mikrobiologie und
Mikrobengenetik,
Friedrich-Schiller Universität,
Jena, Germany

R. N. Burns
Department of Genetics,
Queens Medical Centre,
University of Nottingham,
Nottingham, UK

Richard Calendar
Department of Molecular and Cell Biology,
University of California,
Berkeley, California, USA

George Chisholm
Genotypes Inc.,
San Francisco, California, USA

Jonathan B. Clark
Department of Zoology,
Weber State University,
Ogden, Utah, USA

Patrice Courvalin
Unité des Agents Antibactériens,
Institut Pasteur,
Paris, France

Pierre Darlu
Inserm U 155,
Kremlin-Bicêtre, France

Martin Day
Cardiff School of Biosciences,
Cardiff University,
Cardiff, UK

Erick Denamur
Inserm U 458,
Hôpital Robert Debré,
Paris, France

Russell F. Doolittle
Center for Molecular Genetics,
University of California,
San Diego, La Jolla, California, USA

Christophe Douady
Bioinformatics Department,
GlaxoSmithKline,
Collegeville, Pennsylvania, USA

Norman C. Ellstrand
Department of Botany and Plant Sciences and
Center for Conservation Biology,
University of California,
Riverside, California, USA

Gayle C. Ferguson
Department of Plant and Microbial Sciences,
University of Canterbury,
Christchurch, New Zealand

Bryant E. Fong
Genotypes Inc.,
San Francisco, California, USA

M. E. Ford
Pittsburgh Bacteriophage Institute,
Department of Biological Sciences,
University of Pittsburgh,
Pittsburgh, Pennsylvania, USA
(present address: Division of Gastroenterology
and Hepatology, University of Pittsburgh
School of Medicine, Pennsylvania, USA)

Lynne M. Giere
Genotypes Inc.,
San Francisco, California, USA

J. Peter Gogarten
Department of Molecular and Cell Biology,
University of Connecticut,
Storrs, Connecticut, USA

Sylvie Goussard
Unité des Agents Antibactériens,
Institut Pasteur,
Paris, France

Catherine Grillot-Courvalin
Unité des Agents Antibactériens,
Institut Pasteur,
Paris, France

Ruth M. Hall
CSIRO Molecular Science,
North Ryde, New South Wales, Australia

James F. Hancock
Department of Horticulture,
Michigan State University,
East Lansing, Michigan, USA

Hyman Hartman
IASB,
Cambridge, Massachusetts, USA

G. F. Hatfull
Pittsburgh Bacteriophage Institute,
Department of Biological Sciences,
University of Pittsburgh,
Pittsburgh, Pennsylvania, USA

Jack A. Heinemann
Department of Plant and Microbial Sciences,
University of Canterbury,
Christchurch, New Zealand

R. W. Hendrix
Pittsburgh Bacteriophage Institute,
Department of Biological Sciences,
University of Pittsburgh,
Pittsburgh, Pennsylvania, USA

Katrin Henze
Institut für Botanik III,
Heinrich-Heine Universität Düsseldorf,
Düsseldorf, Germany

Nathan C. Hitzeman
Genotypes Inc.,
San Francisco, California, USA

Ronald A. Hitzeman
Genotypes Inc.,
San Francisco, California, USA

Susan Hollingshead
Department of Microbiology,
University of Alabama,
South Birmingham, Alabama, USA

Michael J. Italia
Bioinformatics Department,
GlaxoSmithKline,
Collegeville, Pennsylvania, USA

Clarence I. Kado
Davis Crown Gall Group,
University of California,
Davis, California, USA

Margaret G. Kidwell
Department of Ecology and Evolutionary
Biology,
The University of Arizona,
Tucson, Arizona, USA

Eugene V. Koonin
National Center for Biotechnology
Information,
National Library of Medicine,
National Institutes of Health,
Bethesda, Maryland, USA

Valentin A. Krassilov
Paleontological Institute,
Moscow, Russia

David J. Lampe
Department of Entomology,
University of Illinois at Urbana-Champaign,
Urbana, Illinois, USA
(present address: Department of Biological
Sciences, Duquesne University, Pittsburgh,
Pennsylvania, USA)

Jeffrey G. Lawrence
Department of Biological Sciences,
University of Pittsburgh,
Pittsburgh, Pennsylvania, USA

Guillaume Lecointre
Service de Systématique moléculaire (GDR
CNRS 1005),
Muséum National d'Histoire Naturelle,
Paris, France

Wen-Hsiung Li
Ecology and Evolutionary Biology,
University of Chicago,
Chicago, Illinois, USA

Chin Y. Loh
Genotypes Inc.,
San Francisco, California, USA

Eugene L. Madsen
Department of Microbiology,
Cornell University,
Ithaca, New York, USA

Kira S. Makarova
National Center for Biotechnology
Information,
National Library of Medicine,
National Institutes of Health,
Bethesda, Maryland, USA
and
Department of Pathology,
F.E. Hebert School of Medicine,
Uniformed Services University of the Health
Sciences,
Bethesda, Maryland, USA

William Martin
Institut für Botanik III,
Heinrich-Heine Universität Düsseldorf,
Düsseldorf, Germany

Ivan Matic
Inserm E9916,
Faculté de Médecine Necker-Enfants Malades,
Université Paris V,
Paris, France

Robert V. Miller
Department of Microbiology and Molecular
Genetics,
Oklahoma State University,
Stillwater, Oklahama, USA

Gisela Mosig
Department of Molecular Biology,
Vanderbilt University,
Nashville, Tennessee, USA

Lorraine Olendzenski
Department of Molecular and Cell Biology,
University of Connecticut,
Storrs, Connecticut, USA

Honor C. Prentice
Department of Systematic Botany,
Lund University,
Lund, Sweden

Michael D. Purugganan
Department of Genetics,
North Carolina State University,
Raleigh, North Carolina, USA

Alfred Pühler
Department of Genetics,
University of Bielefeld,
Bielefeld, Germany

Miroslav Radman
Inserm E9916,
Faculté de Médecine Necker-Enfants Malades,
Université Paris V,
Paris, France

Loren H. Rieseberg
Department of Biology,
Indiana University,
Bloomington, Indiana, USA

Steven A. Ripp
Center for Environmental Biotechnology,
University of Tennessee,
Knoxville, Tennessee, USA

Hugh M. Robertson
Department of Entomology,
University of Illinois at Urbana-Champaign,
Urbana, Illinois, USA

Claus Schnarrenberger
Institut für Pflanzenphysiologie und
Mikrobiologie der FU Berlin,
Berlin, Germany

K. Schultze
Lehrstuhl für Allgemeine Mikrobiologie und
Mikrobengenetik,
Friedrich-Schiller Universität,
Jena, Germany

Joana C. Silva
National Center for Biotechnology
Information,
National Institutes of Health,
Bethesda, Maryland, USA

M. C. M. Smith
Department of Genetics,
Queens Medical Centre,
University of Nottingham,
Nottingham, UK

Jay V. Solnick
Departments of Internal Medicine and Medical
Microbiology and Immunology,
University of California,
Davis, California, USA

Felipe N. Soto-Adames
Department of Entomology,
University of Illinois at Urbana-Champaign,
Urbana, Illinois, USA
(present address: Department of Biology,
University of Vermont, Burlington, Vermont,
USA)

Michael J. Stanhope
Bioinformatics Department,
GlaxoSmithKline,
Collegeville, Pennsylvania, USA

Michael Syvanen
Department of Medical Microbiology and
Immunology,
School of Medicine,
University of California,
Davis, California, USA

François Taddei
Inserm E9916,
Faculté de Médecine Necker-Enfants Malades,
Université Paris V,
Paris, France

Andreas Tauch
Department of Genetics,
University of Bielefeld,
Bielefeld, Germany

Olivier Tenaillon
Inserm E9916,
Faculté de Médecine Necker-Enfants Malades,
Université Paris V,
Paris, France

K. Voigt
Lehrstuhl für Allgemeine Mikrobiologie und
Mikrobengenetik,
Friedrich-Schiller Universität,
Jena, Germany

Kimberly K. O. Walden
Department of Entomology,
University of Illinois at Urbana-Champaign,
Urbana, Illinois, USA

Carole I. Weaver
Genotypes Inc.,
San Francisco, California, USA

Mark E. Welch
Department of Biology,
Indiana University,
Bloomington, Indiana, USA

Richard J. Weld
Department of Plant and Microbial Sciences,
University of Canterbury,
Christchurch, New Zealand

Donald I. Williamson
Port Erin Marine Laboratory
(University of Liverpool),
Isle of Man, UK

Yuri I. Wolf
National Center for Biotechnology
Information,
National Library of Medicine,
National Institutes of Health,
Bethesda, Maryland, USA

A. Wöstemeyer
Lehrstuhl für Allgemeine Mikrobiologie und
Mikrobengenetik,
Friedrich-Schiller Universität,
Jena, Germany

J. Wöstemeyer
Lehrstuhl für Allgemeine Mikrobiologie und
Mikrobengenetik,
Friedrich-Schiller Universität,
Jena, Germany

Glenn M. Young
Department of Food Science and Technology,
University of California,
Davis, California, USA

Olga Zhaxybayeva
Department of Molecular and Cell Biology,
University of Connecticut,
Storrs, Connecticut, USA

Plasmids and Transfer Mechanisms in Bacteria

Section 1 of this book deals with plasmids and mechanisms of gene transfer in bacteria. We will not be attempting a comprehensive review of plasmid biology, which has been the subject of many excellent reviews and is even well covered in many textbooks. Rather, a potpourri of topics will be sampled that illustrate recent developments and unexpected findings related to plasmid-mediated gene transfers. Plasmids figure prominently in the discussion of horizontal gene transfer because a large number of plasmids will stimulate conjugal transfer of bacterial DNA to cells from an extremely broad range of organisms. These include transfer to unrelated bacteria, yeasts and other fungi and plants. Chapter 1 by Heineman and Chapter 5 by Kado deal with the evolution of conjugal plasmids themselves. These vectors of horizontal transfer experience dual evolutionary pressures of survival in hosts via vertical transmission and the ability to adapt in new environments after horizontal flow.

The remaining chapters in this section describe some of the more recent interesting plasmid-related developments. In Chapter 2, Hall describes integrons, a site-specific recombination system that serves to assemble new antibiotic resistance genes into pre-existing transposable elements. In Chapter 3, Tausch and Pühler describe an unusual antibiotic resistance plasmid that is a mosaic of elements found previously from throughout the bacterial kingdom. This is noteworthy because the genetic rearrangements and gene transfers that gave rise to this plasmid have likely occurred in the past 50 years, while the genes come from a group of distantly related organisms that last shared a common ancestor approximately 1.5–2 billion years ago. In Chapter 6, Weld and Heinemann review protein transfers, a topic that has captured attention in recent years because of its importance in pathogenic mechanisms. Protein transfer is probably also important in ensuring survival of transferred DNA in foreign cells. As is clear, bacteria have numerous and highly adapted mechanisms in place to facilitate the transfer of DNA from donor to recipient cells. These mechanisms do not respect species boundaries.

The question as to whether or not these mechanisms operate in natural populations is the subject of the remaining three chapters in this section. It has been known for many years that conjugal plasmid transfer occurs among bacteria in hospitals, farms and natural environments. Along these lines, Madsen in Chapter 4 has an interesting story that documents the emergence of a plasmid that makes enzymes which degrade coal tars and has spread among different bacterial species in a toxic waste dump. Chapters 7 (by Day) and 8 (by Miller and Ripp) show evidence that the DNA transfer mechanisms of transformation and bacterial virus transduction operate efficiently in natural environments.

CHAPTER 1

Recent History of Trans-kingdom Conjugation

Gayle C. Ferguson and Jack A. Heinemann

Conjugation is a mechanism of horizontal gene transfer (HGT) first observed between bacteria. The conjugative mechanism appears to be analogous, and sometimes homologous, to other means of transferring genes from bacteria to possibly members of every biological kingdom. As such, conjugative mechanisms of DNA transfer are necessary for a host of spectacular phenotypes such as symbiosis, virulence and antibiotic resistance. The conjugative mechanism is also related to the means of translocating and transferring proteins from bacteria to other species. Thus, this nearly generic form of macromolecular transport may move genes and other molecules across species boundaries. Some of these molecules may have immediate effects (e.g. through pathogenesis) and some lasting effects (e.g. through inheritance). There is even evidence that inheritable effects can be caused by transferred proteins. Interest in HGT, previously considered on the fringe, has increased dramatically due to the realization that HGT is not an anomaly but a biological fundamental.

INTRODUCTION

The idea that genes are transferred at any appreciable frequency between species has evolved from one scorned by molecular phylogenists to a mainstream concept. Previously, only frustrated phylogenists would dwell on the odd DNA sequence that could unlace the bootstrap analysis (Gogarten et al., 1999). Whole chromosome sequencing of organisms, however, is beginning to validate the concept that genomes are littered with "carcasses" of DNA from other species – some genes remaining functional and neutral, beneficial, or deleterious to the host, and some slowly fading away into the background average G + C content of the new host.

The extent of horizontal gene transfer (HGT) between organisms is difficult to determine for two main reasons. DNA sequence information is, first, limited by the simplicity of the four letter code and secondly, by the constraints on the sequence when it must reproduce in synchrony with the host (Heinemann, 2000b; Heinemann and Roughan, 2000). Thus, the mechanisms of HGT as well as bioinformatic tools are required to quantify the extent of HGT.

The renaissance in HGT thinking brought about by bioinformatics has a history and origin different from the mechanism studies. These studies identify the means by which genes move between two neighbors that may or may not share a vertical lineage. Studies describing the gene transfer mediated by viruses, plasmids, transposons and transformation are much older than bioinformatics. Mechanism studies did not make HGT a mainstream concept, though, because they were considered "laboratory phenomenon" or "interesting exception to the rule for most genes or most organisms" by many. The mechanism studies did, however, open

Horizontal Gene Transfer
ISBN: 0-12-680126-6

3

Copyright © 2002 by Academic Press.
All rights of reproduction in any form reserved.

imaginations to the potential for HGT and legitimized those who subjected it to serious study.

This review will focus on gene transfer between prokaryotes and eukaryotes by mechanisms that are identical, or similar, to bacterial conjugation. The review will not be a systematic account of all the literature relevant to HGT and conjugation. Instead, it will focus on publications that represent unambiguous conflations of ideas that led to HGT becoming an independent phenomenon for study and established bacterial conjugation as a central, general, mechanism for interkingdom gene transfer (Amábile-Cuevas and Chicurel, 1992; Heinemann, 1992). We begin with an abbreviated history of the merger between HGT and crown gall disease in plants that has developed an inseparable link with bacterial conjugation. Finally, we will discuss bacterial conjugation as a paradigm of interkingdom macromolecular exchange mechanistically connected to pathogenesis.

By the mid-twentieth century, interspecies gene transfer was recognized as an important means by which bacteria acquired antibiotic resistance. Those findings, as indeed most early studies in gene transfer, remained focused on the particular genes or organisms of interest. Our review of the literature suggests to us that a change in thinking about HGT was gaining momentum in the late 1960s. Subsequently, a number of studies examined HGT as a possible phenomenon in its own right, without need of allusion to important organismal adaptations, the success of pathogens (e.g. viruses and *Agrobacterium tumefaciens*), or the exception to the rule that all prokaryotic biology can appear to be to botanists and zoologists!

THE CONVERGENCE OF INTERKINGDOM DNA TRANSFER AND CROWN GALL

A. tumefaciens was clearly linked to crown gall tumors in some plants long before the 1960s (references in Stroun et al., 1970; Nester and Kosuge, 1981; Zhu et al., 2000). However, the seminal clues that the nature of the disease was inseparable from DNA transfer to the host emerged in that

decade. Work by Kerr demonstrated that *A. tumefaciens* virulence characters were transmitted between bacteria, by an unknown mechanism (Kerr, 1969). In the late 1970s, the DNA that caused gall formation, T-DNA, would be identified as a component of a conjugative plasmid, called Ti, in *A. tumefaciens* (Nester and Kosuge, 1981). The T-DNA was subsequently found integrated into plant chromosomes (Thomashow et al., 1980; Yadav et al., 1980; Zambryski et al., 1980).

The search for T-DNA illustrates two different approaches to the study of interkingdom gene transfer operating simultaneously. One group of researchers, which we arbitrarily call the generalists, was dominated by the sense that HGT was a phenomenon independent of the particular biology of the donor and recipient organisms, such as the biology of the phytopathogen *A. tumefaciens* and its potential plant hosts. The other, which we refer to as the specialists, used the power of the causal relationship between *A. tumefaciens* and the gall tissue to discover HGT. The two approaches had complementary strengths and both endured the inevitable false positive and negative results that accumulate whenever techniques are pushed to their extreme limits of sensitivity.

The path to the discovery of the discrete DNA sequences transferred from *A. tumefaciens* to the host, and even to other soil bacteria, was itself a study in the limits of the contemporary molecular techniques. The pioneers at the roots of the crown gall mystery during the 1960s and 1970s were also at the leading edge of molecular biology and biochemistry. From such an edge, there is the risk of accumulating negative results, that is, for example, of not seeing DNA transfer (see below). New techniques also require refinement to distinguish between the noise at their limits of detection and true signals. The results of these early studies were consistently "equivocal, but collectively they suggested that bacterial nucleic acids might play a role in tumorigenesis" (Drlica and Kado, 1975).

Generalists and specialists

Both generalists and specialists were reporting the transfer of bacterial nucleic acids and possibly proteins to eukaryotes by the late 1960s.

The nucleic acids were invariably pursued in bacteria-free tissues by hybridization (references in Drlica and Kado, 1975) or hybridization and density centrifugation (Stroun et al., 1970; Stroun and Anker, 1971, 1973).

The conclusiveness of the hybridization method itself, however, was systematically challenged (Drlica and Kado, 1975). Hybridization methods used to demonstrate the presence of bacterial DNA in eukaryotes were often flawed because a control measurement of hybrid thermal stabilities or dissociation profiles was omitted (Chilton et al., 1974; Drlica and Kado, 1974; Kado and Lurquin, 1976). With improved techniques applied later in the 1970s, *A. tumefaciens* nucleic acids were not detected in tumors (Chilton et al., 1974; Drlica and Kado, 1974). The data of some groups were unable to be reproduced at this experimental standard (for an excellent discussion on the technology of the period, see Drlica and Kado, 1974).

Why did some detect nucleic acids while others did not? One possible explanation is that the sporadic claims of nucleic acid detection were artefacts generated by techniques pushed to their limits. A second possibility is that the practitioners of state-of-the-art techniques are important contributors to detection limits. A third possibility is experimental design. Of course, these three possibilities are not mutually exclusive and cannot be distinguished retrospectively.

With the increase in rigor applied to hybridization experiments came an increase in the precision for calculating the detection limits of the techniques (Drlica and Kado, 1974, 1975; Kado and Lurquin, 1976). Chilton et al.'s DNA–DNA hybridization technique, for example, limited detection to one bacterial genome per three diploid plant genomes and "would not detect single or even multiple copies of a small specific fraction (<5%) of the bacterial ... genome in tumor DNA" (Chilton et al., 1974). Such famous negative results cannot, unfortunately, be directly compared with all reported positive detection of nucleic acids because of differences in determining the sensitivities of the techniques. Thus history cannot distinguish between sporadic artefacts and individual experimenters as explanations for different results from all contemporary experiments.

Some groups monitored the production of bacteria-specific nucleic acids in eukaryotic tissues (Stroun et al., 1970). Although these studies were also not above the criticisms leveled against other hybridization studies and were not consistently reproduced (discussed in Drlica and Kado, 1975), ongoing RNA synthesis potentially provided access to larger quantities of nucleic acids complementary to the probe. In contrast, those groups searching only for transferred bacterial DNA were limited by the small number of copies of those sequences in preparations of eukaryotic genomes. History cannot distinguish between possible sporadic artefacts and differences in experimental design as the explanation for different data from all the different experimenters.

Some generalists introduced further confusion when they reported that DNA transfer occurred from not just *A. tumefaciens*, but also *Escherichia coli*, *Bacillus subtilis* and *Pseudomonas fluorescens* to both plants and animals. Hence, "The relationship of (these observations) to the crown gall disease (was) ambiguous" (Drlica and Kado, 1975). Since only *A. tumefaciens* induced tumors, the mechanism of putative nucleic acid transfers from these other bacteria may have been irrelevant to that conducted by *A. tumefaciens* when it induced tumors.

The generalist view was to be eclipsed by the finding of particular T-DNA sequences in plants and the characterization of a mechanism that could account for its transfer. T-DNA transfer would, for a time, serve as the paradigm of interkingdom gene transfer systems. The generality of HGT would be revived in the 1980s by the finding that bacterial conjugative plasmids and T-DNA were different DNA transferred by the same mechanism (Heinemann, 1991; Sprague, 1991), providing retrospective credence to generalists' claims if not vindication of early experiments.

Critical experimental limits to HGT detection

Until recently, interkingdom DNA transfer has been mostly observed through the isolation of phenotypically recombinant organisms (i.e. gene

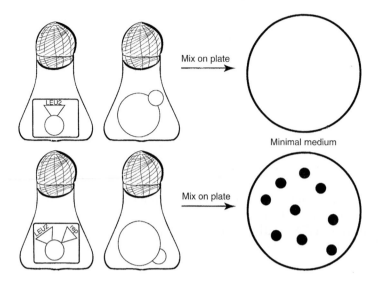

FIGURE 1.1 Illustration of the original experiment demonstrating DNA transfer from bacteria to yeast by conjugation. The rationale for the experiment was that DNA transfer was more generic than could be detected by DNA amplification or the formation of recombinant organisms, which requires DNA transmission (Heinemann and Sprague, 1989). As a test, specially constructed donor bacteria (rectangles) were mixed with genotypically marked recipient yeast (circles with "buds") and plated on medium (large open circles) permissive to the growth of only recombinant yeast. The conjugative plasmids (open circles inside bacteria) were modified to carry either the selectable yeast LEU2 gene or both LEU2 and a DNA sequence that permits replication of extrachromosomal DNA in yeast (rep). Colonies of yeast recombinants (solid black circles) were recovered at a frequency of up to 10% (per donor bacterium) when the plasmid carried yeast-specific replication sequences. Since the DNA introduced into the conjugative plasmids was not responsible for DNA transfer (Bates et al., 1998; Heinemann and Sprague, 1989; Heinemann, 1991), these experiments unequivocally demonstrated that transmission (necessary for detecting recombinants because the DNA is subsequently inherited vertically) was a poor indicator of transfer and the absence of experimentally demonstrated transmission did not imply the absence of DNA transfer.

transmission). DNA transfer can be inferred from any instance in which donor genes are recovered from recipient organisms. This is usually accomplished by selecting recombinant phenotypes. Such phenotypes are the complex product of gene transfer and subsequent stabilization in the germ line of the recipient. Gene transfer is likely not the limiting event in most instances of gene transmission (Heinemann, 1991; Matic et al., 1996). Since inheritable phenotypes or stably maintained DNA sequences remain the easiest way to detect transferred genes, the importance of gene transmission in biasing inferences of the rate and extent of HGT cannot be ignored. In fact, the general reliance on observing recombinant phenotypes or isolating transferred DNA from offspring underestimates HGT (Chilton et al.,

1974; Drlica and Kado, 1974; Heinemann and Roughan, 2000; Heinemann, 2000b).

Several authors over the years have emphasized the importance of distinguishing between gene transfer and transmission to avoid instilling a bias in experimental design and interpretation (reviewed in Heinemann, 1991, 1992). Clark and Warren (1979) made the most systematic justification for the terminology. The first authors to demonstrate the generality of interkingdom conjugation openly acknowledged the influence of that review on their experimental design (Figure 1.1). Confusion between transfer and transmission may have similarly delayed discovery of transfer of DNA from *A. tumefaciens* to plants outside the bacterium's infectious host range (Grimsley et al., 1987).

To illustrate further the importance of distinguishing transfer from transmission, consider the recent report of a DNA virus, that infects animals, evolving via recombination between a DNA virus, that infects plants, and an RNA virus, that infects animals (Gibbs and Weiller, 1999). (Another remarkable intermediate in this chain of events was the likely contribution of a retroviral reverse transcriptase acting on the animal RNA virus to convert an RNA gene into DNA.) The plant virus must have been able to transfer to animals (but caused no obvious phenotype). The many transfer events preceding the evolution of the new variant virus were not detected by selecting or observing a recombinant animal, and likely would not have been detected even with current DNA amplification technologies. The transmission event could be detected, but provides no quantitative information about the frequency of transfers of the original virus to animals.

Furthermore, transferred nucleic acids can be retained by recombination even if whole genes are not inherited (reviewed in Heinemann, 1991; Matic et al., 1996). The extent of this recombination can be masked by the selectivity of homologous recombination enzymes that eliminate long tracts of dissimilar nucleotide sequences better than short tracts (Rayssiguier et al., 1989; Heinemann and Roughan, 2000). Certain environments and mutations that reduce the activity of mismatch repair systems in particular have the effect of reducing selectivity (Matic et al., 1995; Heinemann, 1999b; Vuli'c et al., 1999). Recombination events resulting in the incorporation of short tracts of DNA, even over sequences of extreme genetic divergence, can be difficult or impossible to identify by analysis of DNA sequences (Heinemann and Roughan, 2000).

CONJUGATION AS A PARADIGM SYSTEM OF INTERKINGDOM DNA TRANSFER

The first indication that bacterial conjugation described a general mechanism of interkingdom gene transfer came from the suggestion that certain DNA intermediates observed in *A. tumefaciens* resembled hypothetical DNA intermediates in bacterial conjugation (Stachel et al., 1986). In hindsight, that connection was probably better informed by inspiration than actual data, but nevertheless has withstood significant test.

Conjugation

Bacterial conjugation in its broadest sense has been extensively reviewed, so only a brief description will be provided here (Heinemann, 1992, 1998; Frost, 2000). The focus in this review is on the paradigm conjugative systems defined by the IncP and IncF plasmid groups.

Conjugation mediated by these plasmids requires, at a minimum, a *cis*-acting DNA sequence called the origin of transfer (*oriT*). All other functions (called *tra*) act *in trans* thus allowing plasmids with all *trans*-acting functions also to transfer plasmids with no or a few *trans*-acting functions (Heinemann, 1992). The *trans*-acting gene products are divided further into those involved in DNA metabolism (and are usually specific to a particular *oriT*) and those involved in DNA transport and cell–cell interactions (and thus will interact with a greater range of other plasmids). The conjugative genes specific to DNA metabolism introduce a nick at *oriT* and initiate the unwinding and concomitant transfer of DNA to a recipient cell. Both strands are used as templates for the synthesis of a complementary strand, one in the donor cell and one in the recipient.

Single-stranded plasmid DNA (ssDNA) has been captured in recipient cells, confirming the mechanism of plasmid mobilization. The DNA is recircularized in the recipient. The transport apparatus has not been described biochemically (Heinemann, 2000a), but the genes necessary for forming the apparatus are all plasmid-encoded (Heinemann and Ankenbauer, 1993; Heinemann et al., 1996).

T-DNA is interkingdom conjugation

This uncontroversial model of the conjugative process grounded a model of T-DNA mobilization and transfer proposed by Stachel et al. (1986). Their experiment involved isolating

DNA of the T-DNA region from *A. tumefaciens* (not the plant) after it was induced to prepare the T-DNA for transfer. They provided convincing evidence that linear ssDNA strands defined by the left and right borders of the T-DNA region accumulated in induced bacteria, and that Ti plasmids from induced bacteria had nicks in the border sequences on the strand corresponding to the liberated T-DNA.

It appeared to Stachel et al. that the left and right borders of the T-DNA region, which are characterized as direct repeats, functioned like *oriT* sequences. Nicking and unwinding liberated only the DNA between the nicks, rather than a strand of DNA the length of the Ti plasmid. When the transfer process could not be completed, the T-DNA accumulated in the bacterium.

However, the phenomenology differed from the molecular biology of conjugation in important ways. First, hypothetical ssDNA transfer intermediates *do not* accumulate in bacteria that hold conjugative plasmids even when constitutively induced. Secondly, the conjugative ssDNA was isolated from bacterial recipients; the so-called T-DNA in the Stachel et al. study was never recovered from plants. Thirdly, there existed no evidence at the time that the DNA between tandemly repeated *oriT*s would be liberated during mobilization. Whereas it was shown subsequently that tandem *oriT* repeats do result in mobilization-specific DNA instability in some plasmids (Bhattacharjee et al., 1992; Furuya and Komano, 2000), the repeat of IncP *oriT*s, which are thought to be the closest relatives of the T-DNA borders (Waters et al., 1991; Waters and Guiney, 1993), does not result in mobilization-specific liberation of intervening DNA (Heinemann and Schreiber, personal observation).

Nevertheless, the model has been vindicated by several subsequent genetic tests (Lessl and Lanka, 1994; Christie, 2000). First, T-DNA recombination experiments within plant cells provide evidence that T-DNA is transferred, and enters the nucleus, single-stranded (Tinland et al., 1994). Secondly, the processing reaction between the *cis*-acting border repeat sequences and its putative nick-ase (*virD2*) could be replaced with the *oriT* and its cognate nick-ase

(*mobA*) from the IncQ plasmid RSF1010 (Buchanan-Wollaston et al., 1987). Third, RSF1010 transmission between Agrobacteria was found to be dependent on the other Ti-encoded genes *virA*, *virG*, *virB4*, *virB7* and *virD4* (Beijersbergen et al., 1992). Thus, the *vir* genes, originally identified because they were necessary for virulence, can substitute for *tra* in mediation of conjugative plasmid transfer.

The ability to mix and match genetic requirements of bacterial conjugation and Ti-mediated virulence is consistent with the structural similarities of conjugative and virulence genes (Table 1.1). The *oriT* region of IncP plasmids is homologous to the T-DNA borders (Waters and Guiney, 1993; Frost, 2000), while the *oriT* of the Ti plasmid is homologous to the IncQ *oriT*. Many macromolecular transport systems appear to be composed of gene products homologous to the *tra* functions of conjugative plasmids, including the *vir* genes and type IV protein secretion systems in *Bordetella pertussis*s, *Helicobacter pylori* and *Legionella pneumophila* (Christie, 2000; Frost, 2000) (Tables 1.1 and 1.2).

CONJUGATION IS SUFFICIENT FOR INTERKINGDOM CONJUGATION

A surprise to the crown gall groups was the finding that the transfer of DNA from *A. tumefaciens* to plants was related in part to bacterial conjugation. Meanwhile, yeast studies were soon to show that conjugation could account for interkingdom DNA transfer and that the ability to conjugate with eukaryotic cells is not an evolutionary quirk of *A. tumefaciens*.

In 1989 we crossed bacteria with the yeast *Saccharomyces cerevisiae* using the same plasmids that mediated conjugation between bacteria (Heinemann and Sprague, 1989) (Figure 1.2). *E. coli* transferred a plasmid marked with the *S. cerevisiae* replication origin 2μ and LEU2 gene, to yeast. Recombinant (Leu$^+$) yeast were only formed when the bacteria contained a conjugative plasmid able to mobilize the marker plasmid *in trans*. Formation of Leu$^+$ yeast recombinants was dependent on donor–recipient contact, donor viability, functional

TABLE 1.1 *A. tumefaciens* T-DNA transfer genes that are homologous to genes required for conjugation, protein transfer and virulence in a range of Gram-negative bacteria[a]

	Proposed functions of *vir* genes required for T-DNA transfer from *A. tumefaciens* to plants[b]	*vir* homologues on conjugative plasmids					*vir* homologues involved in protein transfer/virulence					*vir* homologues with as yet unknown function		
		IncF[b]	IncP[c]	pTiC58 (*tra*)[b]	IncW[b]	IncN[b]	*B. pertussis*[b]	*B. abortis*[e] / *B. suis*[d]	*L. pneumophila* (*icm/dot*)[e]	*H. pylori* (*cag*)[b]	*L. pneumophila* (*lvh*)[f]	*R. prowazekii*[f]	*Wolbachia* sp.[g]	*A. actinomycetemcomitans*[h]
virB1	Transglycosylase	*orf169*	*trbN*			*traL*		*virB1*						
virB2	Pilin subunit	*traA*	*trbC*	*trbC*	*trwL*	*traM*	*ptlA*	*virB2*			*lvhB2*			
virB3		*traL*	*trbD*	*trbD*	*trwM*	*traA*	*ptlB*	*virB3*			*lvhB3*			
virB4	ATPase, transport activation	*traC*	*trbE*	*trbE*	*trwK*	*traB*	*ptlC*	*virB4*		*cagE*	*lvhB4*	*virB4*		
virB5	Pilin subunit	*traE*	*trbF*	*trbF*	*trwJ*	*traC*		*virB5*			*lvhB5*			
virB6	Candidate pore former		*trbL*		*trwI*	*traD*	*ptlD*	*virB6*			*lvhB6*			
virB7	Transporter assembly				*trwH*	*traN*	*ptlI*	*virB7*				*rp288*		
virB8	Transporter assembly				*trwG*	*traE*	*ptlE*	*virB8*			*lvhB8*	*rp289*	*virB8*	
virB9	Transporter assembly				*trwF*	*traO*	*ptlF*	*virB9*		*orf15*	*lvhB9*	*rpB9*	*virB9*	
virB10	Coupler of inner and outer membrane subcomplexes	*traB*	*trbI*	*trbI*	*trwE*	*traF*	*ptlG*	*virB10*	*dotG/icmE*	*orf13*	*lvhB10*	*rpB10*	*virB10*	
virB11	ATPase, transport activator		*trbB*	*trbB*	*trwD*	*traG*	*ptlH*	*virB11*	*dotB*	*orf11*	*lvhB11*	*rpB11*	*virB11*	*tadA*
virD4	ATPase, coupler of DNA processing and transport systems	*traD*	*traG*		*trwB*					*orf10*	*lvhD4*	*rpD4*	*virD4*	
virD2	Site-specific single-stranded nicking at the right and left borders		*traI*[i]											
Right and left borders	Site of VirD2 nicking	*oriT*[i]												

Table adapted from Christie (1997a).

[a]Christie (1997a, 2000). [b]Li et al. (1998). [c]O'Callaghan et al. (1999). [d]Frost (2000). [e]Segal et al. (1999). [f]Waters et al. (1991). [g]Masui et al. (2000). [h]Kachlany et al. (2000). [i]Functional homology (Pansegrau et al., 1993). [j]Sieira et al. (2000).

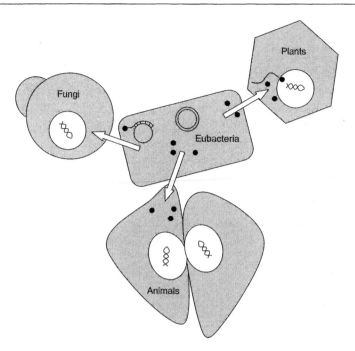

FIGURE 1.2 Bacteria transfer DNA and proteins to plant, animal and fungal cells by similar and related mechanisms. Bacteria transfer DNA (solid lines and large open circles) to both yeast and plant cells by conjugation. Bacterial DNA is integrated into eukaryotic chromosomes (double helices) upon entering the nucleus (white ellipses). Proteins (solid black circles) are transferred to animal cells during pathogenesis. Conjugative plasmids have genes homologous to some genes required for virulence in many bacterial pathogens. Some of those homologous genes are known to be required for DNA or protein transfer.

oriT and *mob* genes, and was independent of exogenous DNAse, indicating that the mechanism of gene transfer was not transformation. *E. coli*–yeast conjugation was subsequently found to be dependent on the same *tra* genes as required for conjugation between *E. coli*, with no additional plasmid-encoded requirements (Heinemann and Sprague, 1991; Bates et al., 1998).

These experiments suggested that DNA transfer from *E. coli* to *S. cerevisiae* occurred by a mechanism analogous to conjugation. The range of yeast able to serve as *E. coli* conjugal recipients has been extended to at least six evolutionary divergent genera (Heinemann, 1991; Hayman and Bolen, 1993; Inomata et al., 1994). Unlike *A. tumefaciens* and plants, *E. coli* and yeast have no known ecological relationship and are not expected to have evolved such a specialized interaction. Therefore interkingdom gene transfer has few, if any, specific requirements evolved within the particular biology of the donor and recipient organism (although

virulence and other phenotypes certainly do have specific requirements).

Interkingdom conjugation is not a species-specific phenomenon

E. coli is not unique in its ability to conjugate with yeast. The T-DNA from *A. tumefaciens* also transferred to *S. cerevisiae*, but by *vir*-dependent conjugation (Bundock et al., 1995). Using *URA3* as a selectable marker with or without the 2μ replication sequence between the T-DNA borders, the frequency of transmission of both replicative and integrative vectors was compared (Bundock et al., 1995). Where transferred T-DNA could replicate autonomously, most transconjugants inherited the vector in its entirety. This was attributed to a failure of VirD2 sometimes to nick the left border, effectively creating a situation where the right border was the only *oriT*. Other transconjugants carried recirularized dsT-DNA molecules.

Interkingdom conjugation is not a plasmid-specific phenomenon

Is the ability to conjugate with eukaryotic cells a particular feature of so-called "broad-host-range" plasmids, such as the IncP family? Bates et al. (1998) compared the ability of conjugation functions from three incompatibility groups to transmit a marked shuttle vector to yeast. IncP plasmids transmitted the shuttle plasmid under conditions where transmission by the narrow-host-range IncF and IncI1 plasmids was not detected (Bates et al., 1998). In contrast, all plasmids were equally capable of transmitting the shuttle plasmid to *E. coli*.

Since recombinants were the only evidence of DNA transfer, it remains formally possible that some aspect of the IncP *tra* system enhances transmission by contributing to the ability of transferred DNA to be inherited. Consistent with this possibility, Heinemann and Sprague did observe F-mediated DNA transmission to yeast using an IncF plasmid derivative instead of mobilizing a shuttle plasmid *in trans* (Heinemann and Sprague, 1989). The higher copy number of their F plasmid derivative may have contributed to the frequency of detectable DNA transmission (Bates et al., 1998).

CONJUGATION AS A CONVERGENCE OF MACROMOLECULAR TRANSPORT SYSTEMS

A. tumefaciens provided an anecdotal link between DNA transfer by conjugation and in pathogenesis. However, in that case, the disease was made possible by the genes transferred but DNA transfer was itself not causing the disease. It has become clear over the past decade that the DNA transport apparatus of conjugation is the ancestor, or at least a sibling (O'Callaghan et al., 1999), of other macromolecular transport systems that are the raison d'être of the disease. As mentioned above, type IV protein secretion genes are homologous to conjugation genes and the transport mechanism for both protein and DNA may be the same (Winans et al., 1996; Christie, 1997a; Kirby and Isberg, 1998; Segal and Shuman, 1998a; Christie and Vogel, 2000).

Bioinformatics

Many homologues of the Ti *virB* genes (B4, B9–11 and sometimes also *virD4*) are found on conjugative plasmids and on chromosomes, as inferred from similarities in sequence and organization. DNA transfer homologues include *tra* of IncN (Pohlman et al., 1994) and Ti (Li et al., 1998), *trb* of IncP and *trw* of IncW (Kado, 1994; Christie, 1997a) plasmids. The *virB* genes have homologues in the pertussis toxin secretion system, *ptl* of *B. pertussis* (Covacci and Rappuoli, 1993; Shirasu and Kado, 1993; Weiss et al., 1993; Farizo et al., 1996). The *cag* pathogenicity island of *Helicobacter pylori*, implicated in contact-mediated secretion of proteins into epithelial cells, is homologous to *virB* (Tummuru et al., 1995; Censini et al., 1996; Christie, 1997b; Covacci et al., 1997). *virB* homologues have also been found in the chromosome of the obligate intracellular parasite *Rickettsia prowazekii* (Andersson et al., 1998), the arthropod intracellular pathogen *Wolbachia* sp. (Masui et al., 2000), the human pathogen *Actinobacillus actinomycetem-comitans* (Kachlany et al., 2000) and are essential for virulence in the intracellular pathogens *Brucella abortus* and *Brucella suis* (O'Callaghan et al., 1999; Sieira et al., 2000).

Relations between protein and DNA secretion systems is not restricted to *vir*. The *icm/dot* genes, essential for *L. pneumophila* survival and replication inside human alveolar macrophages, are homologous to conjugation genes from various plasmids (Segal and Shuman, 1997, 1999; Purcell and Shuman, 1998; Segal et al., 1998; Vogel et al., 1998) (Table 1.2). Fourteen of the *icm/dot* genes are similar, both in sequence and in structural organization, to the *tra* region of IncI plasmid Col1b-P9 (Segal and Shuman, 1999), and *icmE* is homologous to *trbI* of IncP plasmid RK2.

Mechanism

The link between protein and DNA secretory systems is also suggested by mechanistic

TABLE 1.2 *tra* genes homologous to *icm/dot* genes[a]

L. pneumophila *icm/dot*	CollIb-P9 (IncI1)	RK2 (IncP)
icmT	*traK*	
icmS		
icmP	*trbA*	
icmO	*trbC*	
icmI	*traM*	
icmK	*traN*	
icmE		*trbI*
icmG	*traP*	
icmC	*traQ*	
icmD	*traR*	
icmJ	*traT*	
icmB	*traU*	
dotA	*traY*	
dotB	*traJ*	*trbB*
dotC	*traI*	
dotG	*traH*	

[a]Adapted from Segal and Shuman (1999).

studies. For example, a radiolabeled DNA primase (Rees and Wilkins, 1989, 1990) and *E. coli*'s RecA protein (Heinemann, 1999a) were transferred to recipients during bacterial conjugation. In these cases, protein and DNA transfer were associated but the possibility remains that the protein and DNA need not be associated for transfer (Heinemann, 1999a).

Likewise, the decreased stability of T-DNA transferred from *virE2* mutant bacterial donors is complemented by *in planta* expression of VirE2 protein (Rossi et al., 1996) and extracellularly by *virE2*[+] bacteria (Christie et al., 1988; Citovsky et al., 1992), suggesting that VirE2 is also transferred into plants independently of T-DNA. In fact, VirE2, VirD2 and VirF may be transported to plants independently of both T-DNA and the *virB* genes, although tumorigenic *virB*-independent transfer has not been demonstrated (Chen et al., 2000). Intriguingly, tumorigenicity is significantly inhibited when *A. tumefaciens* also carries the mobilizable RSF1010 plasmid (Binns et al., 1995; Stahl et al., 1998). Similarly, RSF1010 attenuates the virulence of *L. pneumophila* (Segal and Shuman, 1998b). In these two cases, the RSF1010:protein mobilization complex and the substrate of the

virulence transport systems are thought to compete (Figure 1.3).

That mutations in *mobA* suppress the effect of RSF1010 on *L. pneumophila* virulence is consistent with this hypothesis (Segal and Shuman, 1998b). The *icm/dot* genes substitute for *tra* supplied *in trans* to transmit RSF1010 to recipient *L. pneumophila* by conjugation, indicating that the RSF1010:MobA complex is a substrate for the secretory system encoded by *icm/dot* (Segal and Shuman, 1998b; Segal et al., 1998; Vogel et al., 1998). The effect of RSF1010 on virulence could be failure to transport efficiently, as yet unidentified, effector proteins that alter vesicle targeting within the macrophage because they are displaced by the RSF1010:MobA complex (Segal and Shuman, 1998a). The *virB* homologue *lvh* does not complement the effect of *icmE/dotB* mutations on virulence, but it did complement the effect of *icmE/dotB* mutations on conjugation (Segal et al., 1999). Thus, the physical requirements for translocating the RSF1010:MobA complex and putative effector protein are not identical.

The effects of RSF1010 on *A. tumefaciens* tumorigenicity are suppressed by over-expression of virB9, *virB10* and *virB11* (Ward et al., 1991), whose products are located in the cell membrane and form the putative conjugation pore (Christie, 1997a). Again, it has been suggested that an RSF1010:MobA complex may displace the T-DNA complex from the translocation apparatus due to the former's higher copy number, the constitutive presence of its processed form, greater affinity for the translocation complex or slow passage through the translocation pore (Binns et al., 1995; Stahl et al., 1998).

The IncW plasmid pSa is an even stronger suppressor of tumorigenicity than RSF1010. Several lines of genetic evidence suggest that the *osa* gene product of pSa blocks protein VirE2 translocation (Chen and Kado, 1994, 1996; Lee et al., 1999). *osa* was first identified as the gene sufficient to cause pSa abolition of oncogencity (Chen and Kado, 1994). The specific effect on VirE2 rather than a protein–DNA complex is supported by the observation that *osa* did not inhibit the conjugative transmission of the Ti plasmid.

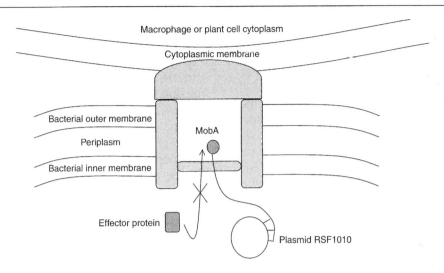

FIGURE 1.3 The mobilizable IncQ plasmid RSF1010 inhibits transmission of T-DNA from *A. tumefaciens* to plant cells. Furthermore, RSF1010 inhibits the ability of *L. pneumophila* to evade fusion of its phagosome with lysosomes inside the macrophage. The *icm/dot* genes that are required to prevent lysozome fusion are also necessary for conjugation of RSF1010. It has been proposed that *icm/dot* is a system that mediates secretion of proteins into the macrophage cytoplasm or phagosome during phagocytosis. The mobilized form of RSF1010 may inhibit virulence by competing with the natural substrate of these protein secretion systems. (Adapted from Segal and Shuman, 1998a.)

The *osa* product also does not inhibit T-DNA transfer. *osa* did not suppress oncogenicity when expressed in *virE2* mutants as long as VirE2 was supplied by separate donors through extracellular complementation, or else it was produced by the recipient plant cell (Lee et al., 1999). The interesting ability for *virE2* mutants to be complemented extracellularly by separate VirE2 donors was suppressed, however, when *osa* was expressed in the protein donor (Lee et al., 1999). Thus, the *osa* product specifically affects VirE2 translocation or function prior to T-DNA entry into the plant cell.

The effects of pSa and RSF1010 on oncogenicity are similar but not identical. First, RSF1010 inhibits both VirE2 translocation and possibly T-DNA transfer, whereas pSa only prevents VirE2 translocation. Secondly, an RSF1010-protein complex is necessary for oncogenic suppression but only the *osa* gene product of pSa is required for suppression (Lee et al., 1999). Thirdly, over-expression of VirB9, VirB10 and VirB11 suppresses the RSF1010 effect on tumorigenicity but not the *osa* effect. These apparent dissimilarities may reflect only quantitative differences in the RSF1010 and pSa mechanisms, since RSF1010 partially inhibits oncogenicity and pSa completely abolishes tumor formation (Lee et al., 1999).

However, the RSF1010 and pSa effects may have different mechanistic explanations. As discussed above, VirE2, VirD2 and VirF proteins are transported across the inner membrane by a *virB24*- and *virD4*-independent mechanism. The *osa* product, but not RSF1010, prevented VirE2, VirF and VirD2 from achieving normal periplasmic levels (Chen et al., 2000). This suggests that the *osa* product and MobA–RSF1010 could inhibit VirE2 translocation at different steps. While MobA–RSF1010 may inhibit the directed translocation of proteins through the putative outer membrane pore, the *osa* product may inhibit translocation across the inner membrane. Such a model is consistent with both the inner membrane localization of Osa (Chen and Kado, 1996) and the observation that VirB10, VirB11 and VirB12 over-expression did not restore tumor formation by *A. tumefaciens* carrying pSa (Lee et al., 1999).

What came first, protein or DNA transfer?

DNA and proteins are probably transferred between species by similar mechanisms. The effects of transferring non-nucleic acid molecules may sometimes be similar too; macromolecules, e.g. prions, other than nucleic acids possess gene-like qualities (Campbell, 1998; Heinemann and Roughan, 2000). Some proteins are not genes, but can influence epigenes that establish heritable phenotypes many generations after the protein has disappeared (Heinemann, 1999a). So conjugation may be a manifestation of protein secretion and, sometimes, protein secretion is another type of HGT.

CONCLUSION

HGT has established itself as a legitimate topic of study independently of the effects of the genes transferred on the biology of donor and recipient organisms. Nevertheless, the study of pathogens like *A. tumefaciens* and *L. pneumophila*, symbionts like *Rhizobium meliloti*, and phenotypes like antibiotic resistance and crown gall, have each contributed to the richness of the evidence supporting the notion that genes are less restricted by our notions of species sanctity than we have previously thought. In particular, the studies of bacterial conjugation, crown gall disease and protein secretion have provided extensive mechanistic insight into how DNA is exchanged between kingdoms, species and siblings.

Extensive similarities between genes identified as either virulence or conjugation determinants provided an early hint that macromolecular transport was a general phenomenon. Those early hints have been vindicated by demonstrations of genetic interchangeability between some determinants (complementation studies) and genetic conflict between others.

DNA is not special cargo but one of a number of molecules that might be transported by the same basic macromolecular transport systems. The ability to move molecules intercellularly has obvious implications for both single and multicellular organisms. Of immediate relevance are the diseases and recombinants that could arise from this nearly generic transport mechanism.

But what of the molecules being transferred? Plasmids and viruses, for example, make excellent evolutionary livings transferring between organisms, even evolving despite their effects on the host. Transfer alone might explain their existence (Cooper and Heinemann, 2000). Did these genetic entities evolve a means to replicate by HGT, or was the existence of macromolecular transport enough for such semi-autonomous entities to evolve? Other kinds of molecules could transmit genetic information (Heinemann and Roughan, 2000). Could HGT be a mechanism for the evolution of genetic entities that are not nucleic acids?

ACKNOWLEDGMENTS

We thank A. Harker for critical reading of the manuscript and C.F. Delwichie for encouraging comments on our contribution to the first edition. JAH acknowledges M. Stroun and P. Anker for their support, reprints and valuable insights. This work was supported in part by the Marsden Fund (Grant M1042 to JAH) and a University of Canterbury Roper Scholarship (to GCF).

REFERENCES

Amábile-Cuevas, C.F. and Chicurel, M.E. (1992) Bacterial plasmids and gene flux. *Cell* **70**: 189–199.

Andersson, S.G.E., Zomorodipour, A., Andersson, J.O. et al. (1998) The genome sequence of *Rickettsia prowazekii* and the origin of mitochondria. *Nature* **396**: 133–140.

Bates, S., Cashmore, A.M. and Wilkins, B.M. (1998) IncP plasmids are unusually effective in mediating conjugation of *Escherichia coli* and *Saccharomyces cerevisiae*: involvement of the Tra2 mating system. *J. Bacteriol.* **180**: 6538–6543.

Beijersbergen, A., den Dulk-Ras, A., Schilperoort, R.A. and Hooykaas, P.J.J. (1992) Conjugative transfer by the virulence system of *Agrobacterium tumefaciens*. *Science* **256**: 1324–1327.

Bhattacharjee, M., Rao, X.-M. and Meyer, R.J. (1992) Role of the origin of transfer in termination of strand transfer during bacterial conjugation. *J. Bacteriol.* **174**: 6659–6665.

Binns, A.N., Beaupré, C.E. and Dale, E.M. (1995) Inhibition of VirB-mediated transfer of diverse substrates from *Agrobacterium tumefaciens* by the IncQ plasmid RSF1010. *J. Bacteriol.* **177**: 4890–4899.

Buchanan-Wollaston, V., Passiatore, J.E. and Cannon, F. (1987) The *mob* and *oriT* mobilization functions of a bacterial plasmid promote its transfer to plants. *Nature* **328**: 172–175.

Bundock, P., den Dulk-Ras, A., Beijersbergen, A. and Hooykaas, P.J.J. (1995) Trans-kingdom T-DNA transfer from *Agrobacterium tumefaciens* to *Saccharomyces cerevisiae*. *EMBO J.* **14**: 3206–3214.

Campbell, A.M. (1998) Prions as examples of epigenetic inheritance. *ASM News* **64**: 314–315.

Censini, S., Lange, C., Xiang, Z. et al. (1996) *cag*, a pathogenicity island of *Helicobacter pylori*, encodes type I-specific and disease-associated virulence factors. *Proc. Natl Acad. Sci. USA* **93**: 14648–14653.

Chen, C.-Y. and Kado, C.I. (1994) Inhibition of *Agrobacterium tumefaciens* oncogenicity by the *osa* gene of pSa. *J. Bacteriol.* **176**: 5697–5703.

Chen, C.-Y. and Kado, C.I. (1996) Osa protein encoded by plasmid pSa is located at the inner membrane but does not inhibit membrane association of VirB and VirD virulence proteins in *Agrobacterium tumefaciens*. *FEMS Microbiol. Lett.* **135**: 85–92.

Chen, L., Li, C.M. and Nester, E.W. (2000) Transferred DNA (T-DNA)-associated proteins of *Agrobacterium tumefaciens* are exported independently of virB. *Proc. Natl Acad. Sci. USA* **97**: 7545–7550.

Chilton, M.-D., Currier, T. C., Farrand, S. K. et al. (1974) *Agrobacterium tumefaciens* DNA and PS8 DNA not detected in crown gall tumors. *Proc. Natl Acad. Sci. USA* **71**: 3672–3676.

Christie, P.J. (1997a) *Agrobacterium tumefaciens* T-complex transport apparatus: a paradigm for a new family of multifunctional transporters in Eubacteria. *J. Bacteriol.* **179**: 3085–3094.

Christie, P.J. (1997b) The *cag* pathogenicity island: mechanistic insights. *Trends Microbiol.* **5**: 264–265.

Christie, P.J. (2000) *Agrobacterium*. In *Encyclopedia of Microbiology* (ed., J. Lederberg), pp. 86–103, Academic Press, San Diego, CA.

Christie, P.J. and Vogel, J.P. (2000) Bacterial type IV secretion: conjugation systems adapted to deliver effector molecules to host cells. *Trends Microbiol.* **8**: 354–360.

Christie, P.J., Ward, J.E., Winans, S.C. and Nester, E.W. (1988) The *Agrobacterium tumefaciens* virE2 gene product is a single-stranded-DNA-binding protein that associates with T-DNA. *J. Bacteriol.* **170**: 2659–2667.

Citovsky, V., Zupan, J., Warnick, D. and Zambryski, P. (1992) Nuclear localization of *Agrobacterium* VirE2 protein in plant cells. *Science* **256**: 1802–1805.

Clark, A.J. and Warren, G.J. (1979) Conjugal transmission of plasmids. *Annu. Rev. Genet.* **13**: 99–125.

Cooper, T.F. and Heinemann, J.A. (2000) Postsegregational killing does not increase plasmid stability but acts to mediate the exclusion of competing plasmids. *Proc. Natl Acad. Sci. USA* **97**: 12 643–12 648.

Covacci, A. and Rappuoli, R. (1993) Did the inheritance of a pathogenicity island modify the virulence of *Helicobacter pylori*? *Mol. Microbiol.* **8**: 429–434.

Covacci, A., Falkow, S., Berg, D.E. and Rappuoli, R. (1997) Pertussis toxin export requires accessory genes located downstream from the pertussis toxin operon. *Trends Microbiol.* **5**: 205–208.

Drlica, K.A. and Kado, C.I. (1974) Quantitative estimation of *Agrobacterium tumefaciens* DNA in crown gall tumor cells. *Proc. Natl Acad. Sci. USA* **71**: 3677–3681.

Drlica, K.A. and Kado, C.I. (1975) Crown gall tumors: are bacterial nucleic acids involved? *Bacteriol. Rev.* **39**: 186–196.

Farizo, K.M., Cafarella, T.G. and Burns, D.L. (1996) Evidence for a ninth gene, *ptlI*, in the locus encoding the pertussis toxin secretion system of Bordetella pertussis and formation of a PtlI–PtlF complex. *J. Biol. Chem.* **271**: 31643–31649.

Frost, L.S. (2000) Conjugation, bacterial. In *Encyclopedia of Microbiology* (ed., J. Lederberg), pp. 847–862, Academic Press, San Diego, CA.

Furuya, N. and Komano, T. (2000) Initiation and termination of DNA transfer during conjugation of IncI1 plasmid R64: roles of two sets of inverted repeat sequences within oriT in termination of R64 transfer. *J. Bacteriol.* **182**: 3191–3196.

Gibbs, M.J. and Weiller, G.F. (1999) Evidence that a plant virus switched hosts to infect a vertebrate and then recombined with a vertebrate-infecting virus. *Proc. Natl Acad. Sci. USA* **96**: 8022–8027.

Gogarten, J.P., Murphey, R.D. and Olendzenski, L. (1999) Horizontal gene transfer: pitfalls and promises. *Biological Bulletin* **196**: 359–362.

Grimsley, N., Hohn, B., Davies, J.W. and Hohn, B. (1987) *Agrobacterium*-mediated delivery of infectious maize streak virus into maize plants. *Nature* **325**: 177–179.

Hayman, G.T. and Bolen, P.L. (1993) Movement of shuttle plasmids from *Escherichia coli* into yeasts other than *Saccharomyces cerevisiae* using trans-kingdom conjugation *Plasmid* **30**: 251–257.

Heinemann, J.A. (1991) Genetics of gene transfer between species. *Trends Genet.* **7**: 181–185.

Heinemann, J.A. (1992) Conjugation, genetics. In *Encyclopedia of Microbiology* (ed., J. Lederberg), pp. 547–558, Academic Press, San Diego, CA.

Heinemann, J.A. (1998) Looking sideways at the evolution of replicons. In *Horizontal Gene Transfer* (eds, C. Kado and M. Syvanen), pp. 11–24, International Thomson Publishing, London.

Heinemann, J.A. (1999a) Genetic evidence for protein transfer during bacterial conjugation. *Plasmid* **41**: 240–247.

Heinemann, J.A. (1999b) How antibiotics cause antibiotic resistance. *Drug Discovery Today* **4**: 72–79.

Heinemann, J.A. (2000a) Complex effects of DNA gyrase inhibitors on bacterial conjugation. *J. Biochem. Mol. Biol. Biophys.* **4**: 165–177.

Heinemann, J.A. (2000b) Horizontal transfer of genes between microorganisms. In *Encyclopedia of Microbiology* (ed., J. Lederberg), pp. 698–706, Academic Press, San Diego, CA.

Heinemann, J.A. and Ankenbauer, R.G. (1993) Retrotransfer of IncP plasmid R751 from *Escherichia coli* maxicells: evidence for the genetic sufficiency of self-transferable plasmids for bacterial conjugation. *Mol. Microbiol.* **10**: 57–62.

Heinemann, J.A. and Roughan, P.D. (2000) New hypotheses on the material nature of horizontally transferred genes. *Ann. NY Acad. Sci.* **906**: 169–186.

Heinemann, J.A., Scott, H.E. and Williams, M. (1996) Doing the conjugative two-step: evidence of recipient autonomy in retrotransfer. *Genetics* **143**: 1425–1435.

Heinemann, J.A. and Sprague, G.F., Jr (1991) Transmission of plasmid DNA to yeast by conjugation with bacteria. *Methods Enzymol.* **194**: 187–195.

Heinemann, J.A. and Sprague, G.F., Jr (1989) Bacterial conjugative plasmids mobilize DNA transfer between bacteria and yeast. *Nature* **340**: 205–209.

Inomata, K., Nishikawa, M. and Yoshida, K. (1994) The yeast *Saccharomyces kluveri* as a recipient eukaryote in transkingdom conjugation: behaviour of transmitted plasmids in transconjugants. *J. Bacteriol.* **176**: 4770–4773.

Kachlany, S.C., Planet, P.J., Bhattacharjee, M.K. et al. (2000) Nonspecific adherence by *Actinobacillus actinomycetem-comitans* requires genes widespread in Bacteria and Archae. *J. Bacteriol.* **182**: 6169–6176.

Kado, C.I. (1994) Promiscuous DNA transfer system of *Agrobacterium tumefaciens*: role of the virB operon in sex pilus assembly and synthesis. *Mol. Microbiol.* **12**: 17–22.

Kado, C.I. and Lurquin, P.F. (1976) Studies on *Agrobacterium tumefaciens*. V. Fate of exogenously added bacterial DNA in *Nicotiana tabacum*. *Physiol. Plant Pathol.* **8**: 73–82.

Kerr, A. (1969) Transfer of virulence between isolates of *Agrobacterium*. *Nature* **223**: 1175–1176.

Kirby, J.E. and Isberg, R.R. (1998) Legionnaires' disease: the pore macrophage and the legion of terror within. *Trends Microbiol.* **6**: 256–258.

Lee, L.-Y., Glevin, S.B. and Kado, C.I. (1999) pSa causes oncogenic suppression of *Agrobacterium* by inhibiting VirE2 protein export. *J. Bacteriol.* **181**: 186–196.

Lessl, M. and Lanka, E. (1994) Common mechanisms in bacterial conjugation and Ti-mediated T-DNA transfer to plant cells. *Cell* **77**: 321–324.

Li, P.-L., Everhart, D.M. and Farrand, S.K. (1998) Genetic and sequence analysis of the pTiC58 *trb* locus, encoding a mating-pair formation system related to members of the type IV secretion family. *J. Bacteriol.* **180**: 6164–6172.

Masui, S., Sasaki, T. and Ishikawa, H. (2000) Genes for the type IV secretion system in an intracellular symbiont, *Wolbachia*, a causative agent of various sexual alterations in arthropods. *J. Bacteriol.* **182**: 6529–6531.

Matic, I., Rayssiguier, C. and Radman, M. (1995) Interspecies gene exchange in bacteria: the role of SOS and mismatch repair systems in evolution of species. *Cell* **80**: 507–515.

Matic, I., Taddei, F. and Radman, M. (1996) Genetic barriers among bacteria. *Trends Microbiol.* **4**: 69–73.

Nester, E.W. and Kosuge, T. (1981) Plasmids specifying plant hyperplasias. *Annu. Rev. Microbiol.* **35**: 531–565.

O'Callaghan, D., Cazevieille, C., Allardet-Servent, A. et al. (1999 A homologue of the *Agrobacterium tumefaciens* VirB and *Bordetella pertussis* Ptl type IV secretion systems is essential for intracellular survival of *Brucella suis*.) *Mol. Microbiol.* **33**: 1210–1220.

Pansegrau, W., Schoumacher, F., Hohn, B. and Lanka, E. (1993) Site-specific cleavage and joining of single-stranded DNA by VirD2 protein of *Agrobacterium*

tumefaciens Ti plasmids: analogy to bacterial conjugation. *Proc. Natl Acad. Sci. USA* **90**: 11538–11542.

Pohlman, R.F., Genetti, H.D. and Winans, S.C. (1994) Common ancestry between IncN conjugal transfer genes and macromolecular export systems of plant and animal pathogens. *Mol. Microbiol.* **14**: 655–668.

Purcell, M. and Shuman, H.A. (1998) The *Legionella pneumophila* icmGCDJBF genes are required for killing human macrophages. *Infect. Immunol.* **66**: 2245–2255.

Rayssiguier, C., Thaler, D.S. and Radman, M. (1989) The barrier to recombination between *Escherichia coli* and *Salmonella typhimurium* is disrupted in mismatch-repair mutants. *Nature* **342**: 396–401.

Rees, C.E.D. and Wilkins, B.M. (1989) Transfer of *tra* proteins into the recipient cell during bacterial conjugation mediated by plasmid ColIb-P9. *J. Bacteriol.* **171**: 3152–3157.

Rees, C.E.D. and Wilkins, B.M. (1990) Protein transfer into the recipient cell during bacterial conjugation: studies with F and RP4. *Mol. Microbiol.* **4**: 1199–1205.

Rossi, L., Hohn, B. and Tinland, B. (1996) Integration of complete T-DNA units is dependent on the activity of VirE2 protein of *Agrobacterium tumefaciens*. *Proc. Natl Acad. Sci. USA* **93**: 126–130.

Segal, G. and Shuman, H.A. (1997) Characterization of a new region required for macrophage killing by *Legionella pneumophila*. *Infect. Immunol.* **65**: 5057–5066.

Segal, G. and Shuman, H.A. (1998a) How is the intracellular fate of *Legionella pneumophila* phagosome determined? *Trends Microbiol.* **6**: 253–255.

Segal, G. and Shuman, H.A. (1998b) Intracellular multiplication and human macrophage killing by *Legionella pneumophila* are inhibited by conjugal components of the IncQ plasmid RSF1010. *Mol. Microbiol.* **30**: 197–208.

Segal, G. and Shuman, H.A. (1999) Possible origin of the *Legionella pneumophila* virulence genes and their relation to *Coxiella burneti*. *Mol. Microbiol.* **33**: 667–672.

Segal, G., Purcell, M. and Shuman, H.A. (1998) Host cell killing and bacterial conjugation require overlapping sets of genes within a 22-kb region of the *Legionella pneumophila* genome. *Proc. Natl Acad. Sci. USA* **95**: 1669–1674.

Segal, G., Russo, J.J. and Shuman, H.A. (1999) Relationships between a new type IV secretion system and the icm/dot virulence system of *Legionella pneumophila*. *Mol. Microbiol.* **34**: 799–809.

Shirasu, K. and Kado, C.I. (1993) The virB operon of the *Agrobacterium tumefaciens* virulence regulon has sequence similarities to B, C and D open reading frames downstream of the pertussis toxin-operon and to the DNA transfer-operons of broad-host-range conjugative plasmids. *Nucl. Acids Res.* **21**: 353–354.

Sieira, R., Comerci, D.J., Sánchez, D.O. and Ugalde, R.A. (2000) A homologue of an operon required for DNA transfer in *Agrobacterium* is required in *Brucella abortus* for virulence and intracellular multiplication. *J. Bacteriol.* **182**: 4849–4855.

Sprague, G.F., Jr (1991) Genetic exchange between kingdoms. *Curr. Opin. Genet. Dev.* **1**: 530–533.

Stachel, S.E., Timmerman, B. and Zambryski, P. (1986) Generation of single-stranded T-DNA molecules during the initial stages of T-DNA transfer from *Agrobacterium tumefaciens* to plant cells. *Nature* **322**: 706–712.

Stahl, L.E., Jacobs, A. and Binns, A.N. (1998) The conjugal intermediate of plasmid RSF1010 inhibits *Agrobacterium tumefaciens* virulence and VirB-dependent export of VirE2. *J. Bacteriol.* **180**: 3933–3939.

Stroun, M. and Anker, P. (1971) Bacterial nucleic acid synthesis in plants following bacterial contact. *Mol. General Genet.* **113**: 92–98.

Stroun, M. and Anker, P. (1973) Transcription of spontaneously released bacterial deoxyribonucleic acid in frog auricles. *J. Bacteriol.* **114**: 114–120.

Stroun, M., Anker, P. and Auderset, G. (1970) Natural release of nucleic acids from bacteria into plant cells. *Nature* **227**: 607–608.

Thomashow, M.F., Nutter, R., Postle, K. et al. (1980) Recombination between higher plant DNA and the Ti plasmid of *Agrobacterium tumefaciens*. *Proc. Natl Acad. Sci. USA* **77**: 6448–6452.

Tinland, B., Hohn, B. and Puchta, H. (1994) *Agrobacterium tumefaciens* transfers single-stranded transferred DNA (T-DNA) into the plant cell nucleus. *Proc. Natl Acad. Sci. USA* **91**: 8000–8004.

Tummuru, M.K.R., Sharma, S.A. and Blaser, M.J. (1995) *Helicobacter pylori* picB, a homologue of the *Bordetella pertussis* toxin secretion protein, is required for induction of IL-8 in gastric epithelial cells. *Mol. Microbiol.* **18**: 867–876.

Vogel, J.P., Andrews, H.L., Wong, S.K. and Isberg, R.R. (1998) Conjugative transfer by the virulence system of *Legionella pneumophila*. *Science* **279**: 873–876.

Vuli'c, M., Lenski, R.E. and Radman, M. (1999) Mutation, recombination, and incipient speciation of bacteria in the laboratory. *Proc. Natl Acad. Sci. USA* **96**: 7348–7351.

Ward, J.E., Jr, Dale, E.M. and Binns, A.N. (1991) Activity of the *Agrobacterium* T-DNA transfer machinery is affected by virB gene products. *Proc. Natl Acad. Sci. USA* **88**: 9350–9354.

Waters, V.L. and Guiney, D.G. (1993) Processes at the nick region link conjugation, T-DNA transfer and rolling circle replication. *Mol. Microbiol.* **9**: 1123–1130.

Waters, V.L., Hirata, K.H., Pansegrau, W. et al. (1991) Sequence identity in the nick regions of IncP plasmid transfer origins and T-DNA borders of *Agrobacterium* Ti plasmids. *Proc. Natl Acad. Sci. USA* **88**: 1456–1460.

Weiss, A.A., Johnson, F.D. and Burns, D.L. (1993) Molecular characterization of an operon required for pertussis toxin secretion. *Proc. Natl Acad. Sci. USA* **90**: 2970–2974.

Winans, S.C., Burns, D.L. and Christie, P.J. (1996) Adaptation of a conjugal transfer system for the export of pathogenic macromolecules. *Trends Microbiol.* **4**: 64–68.

Yadav, N.S., Postle, K., Saiki, R.K. et al. (1980) T-DNA of a crown gall teratoma is covalently joined to host plant DNA. *Nature* **287**: 458–461.

Zambryski, P., Holsters, M., Kruger, K. et al. (1980) Tumor DNA structure in plant cells transformed by A. tumefaciens. *Science* **209**: 1385–1391.

Zhu, J., Oger, P.M., Schrammeijer, B. et al. (2000) The bases of crown gall tumorigenesis. *J. Bacteriol.* **182**: 3885–3895.

CHAPTER 2

Gene Cassettes and Integrons: Moving Single Genes

Ruth M. Hall

Gene cassettes are very simple, small mobile elements that generally include only a single complete gene (or open reading frame) or occasionally two genes and a recombination site called a 59-be that enables them to be mobilized. Movement of cassettes is achieved by site-specific recombination with the reaction catalyzed by members of the IntI-type DNA integrase family (tyrosine site-specific recombinases) that are encoded by integrons. Most commonly, cassettes are incorporated into a specific site, an *attI* site, that is found in the integron adjacent to the *intI* gene. By repeated rounds of recombination between a 59-be and the *attI* site, the IntI integrase can incorporate more than one cassette into the same integron leading to the formation of short or long arrays of gene cassettes. However, because the integron-encoded IntI integrases can also recognize secondary recombination sites, cassettes can also be incorporated at many other locations, allowing them to be widely disseminated. Several classes of integron that are differentiated by differences in the sequences of the IntI integrases have been found and many more are likely to be found in the future. However, the gene cassettes are shared. Among the integron classes, some are part of mobile elements, also known as inte-grons, while others are located on bacterial chromosomes. The integrons that are themselves mobile are most important in spreading gene cassettes from strain to strain and species to species. The integrons that are an integral part of a

bacterial chromosome may act as storehouses of genes for emergencies that are added to, as well as sampled and spread, by the mobile integrons.

INTRODUCTION

When the processes of horizontal gene transfer move DNA from one organism to another there is little impact unless the incoming DNA is stably maintained and expressed in the recipient organism. Several different processes can achieve this outcome and examples include the stable maintenance of a plasmid or incorporation of the incoming DNA into the bacterial chromosome or into plasmids already resident in that cell. Incorporation of new DNA into an existing chromosome (bacterial or plasmid) can occur by homologous recombination, but only if homologous regions are present in both DNA species. However, other specific processes such as transposition or site-specific recombination can also lead to stable incorporation of parts or all of an incoming DNA molecule. The regions that can move in this way are generally discrete genetic elements and the ability of such mobile elements to shift their location enables them to move into entities such as plasmids and conjugative transposons that are able to move from cell to cell with ease. Because of this, translocatable elements are an important force in horizontal gene transfer in the bacterial world and any associated gene or group of genes is

19

Copyright © 2002 by Academic Press.
All rights of reproduction in any form reserved.

ultimately able to gain access to many different organisms and species. In this chapter, the family of mobile elements known as gene cassettes and their host elements, the integrons, are described. Integrons which are themselves mobile are also described. Further information on specific aspects can be found in recent reviews (Hall and Collis, 1995, 1998; Recchia and Hall, 1995a, 1997; Rowe-Magnus et al., 1999).

FUNCTIONAL DEFINITION OF GENE CASSETTES AND INTEGRONS

The definitions of gene cassettes and integrons are to a considerable extent interdependent as integrons are first and foremost defined by their ability to capture gene cassettes. Indeed, this is how they were found. Gene cassettes were initially identified by virtue of the fact that many

different gene cassettes, each containing a different antibiotic resistance gene, can be found in the same sequence context (Cameron et al., 1986; Hall and Vockler, 1987; Ouellette et al., 1987; Wiedemann et al., 1987; Sundström et al., 1988; Hall et al., 1991). This situation allowed the identification of the approximate boundaries of gene cassettes and also of the conserved backbone sequence into which the cassettes slot (Hall and Vockler, 1987; Sundström et al., 1988; Stokes and Hall, 1989). Within this conserved backbone, a gene was found whose product bore a significant resemblance to integrases (tyrosine site-specific recombinases) that are harbored by the genomes of some temperate phage (Hall and Vockler, 1987; Ouellette and Roy, 1987; Sundström et al., 1988; Stokes and Hall, 1989). Experimental evidence that this integrase (IntI1) was active was soon forthcoming (Martinez and de la Cruz, 1988, 1990). The term integron was originally coined to

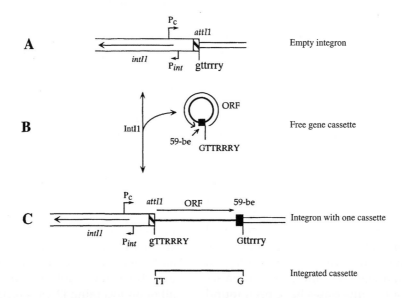

FIGURE 2.1 Insertion of a circular gene cassette into an integron. (A) An empty integron, showing the three distinctive features: an *intI* gene that encodes the IntI integrase, an adjacent recombination site, *attI* (hatched box), and promoters P_c and P_{int}. (B) A gene cassette in its circular form consisting of a gene or open reading frame (ORF) and a 59-be recombination site (filled box). (C) An integron containing one gene cassette, with the boundaries of the integrated cassette shown below. IntI-catalyzed recombination between *attI* in the integron and the 59-be in the circular cassette results in insertion of the cassette into the integron. The ORF in the inserted cassette is now transcribed from P_c. gttrrry (A) and GTTRRRY (B) represent the 7 bp core sites surrounding the recombination crossover point in the *attI* site of the integron and in the 59-be of the circular cassette respectively. The configuration of these bases after incorporation of the cassette is shown in C. Further cassettes may be similarly inserted at *attI*, resulting in the accumulation of arrays of cassettes.

A. Free circular cassette

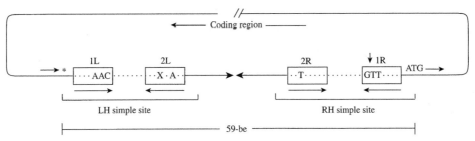

B. Integrated linear cassette

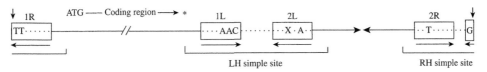

FIGURE 2.2 Structure of gene cassettes. (A) A cassette in its free, circular form. (B) A cassette in integrated, linear form showing the coding region of the gene and the 59-be recombination site. The extent of the coding region (not to scale) is marked by start (ATG) and stop (*) codons. Boxes surround 7 bp core site sequences, related to the consensus GTTRRRY, which lie within putative IntI binding sites that include the core sites as well as flanking bases. Left hand (LH) and right hand (RH) simple sites consist of pairs of binding sites (1L and 2L; 2R and 1R respectively). The relative orientations of the core sites are indicated with arrows. Bases found in all 59-be are shown, while other bases that conform to the core site consensus are represented by dots, and an extra base in 2L is marked by an X. In any individual 59-be, 1L and 1R sites are closely related, as are 2L and 2R, but the bases between them are not. An inverted repeat, represented by a pair of arrows, lies between 2L and 2R and has a variable sequence and length. The recombination crossover point is indicated by a vertical arrow. When the circular cassette is linearized by insertion into the *attI* site of an integron (B), the last six bases of 1R in the circular cassette become the first six bases of the integrated cassette.

describe this group of elements (Stokes and Hall, 1989) but, as further types of integrons have since been found, they are now designated class 1 integrons.

In 1991, all of the sequence information that was available was drawn together and the general and current definition of a gene cassette as a single gene or open reading frame coupled with a downstream 59-be recombination site emerged (Hall et al., 1991). This study also gave rise to a model for the integration and excision of gene cassettes that predicted the existence of a circular form of cassettes (Figure 2.1). Experimental studies of cassette movement, using the IntI1 integrase, were subsequently reported. Precise excision of gene cassettes was demonstrated, thus defining them experimentally (Collis and Hall, 1992a) and the circular form of cassettes was also isolated (Collis and Hall, 1992b). Demonstration

of the incorporation of circular gene cassettes into an integron completed the picture (Collis et al., 1993). An interesting aspect of the latter study is that cassettes are preferentially incorporated at the *attI* site of the integron to become the first cassette in an array of gene cassettes.

Finally, in 1995, a known gene cassette was found in a plasmid that does not contain an integron (Recchia and Hall, 1995b). This demonstrated that cassettes can also move into almost any location, where they become fixed, and further cassettes that have been incorporated at such secondary sites have since been reported.

GENE CASSETTES

Gene cassettes are the smallest of the known types of mobile elements. Generally, they consist of a single gene or ORF together with a downstream IntI-specific recombination site known as a 59-be (59-base element). Occasionally, two ORFs are present in a single cassette. Each cassette is a discrete mobile element that can exist in either a free, closed-circular form or an integrated linear form (Figures 2.1 and 2.2). As cassettes contain no replication functions, the closed circular form cannot replicate and its role is limited to that of an intermediate in cassette movement (Collis and Hall, 1992b). The boundaries of the linear gene cassette form have been precisely located (Stokes et al., 1997) and each cassette commences with a TT doublet and ends with a G residue. This is because the strand exchange that occurs during cassette integration occurs between the G and TT of a completely conserved triplet that forms part of the 7 bp core site (GTTaggc or GTTRRRY) found at one end of 59-be (Stokes et al., 1997) and also in *attI* sites at the point of cassette integration (Hansson et al., 1997). The sequence of any specific 59-be is found by joining the sequence at the right-hand (RH) end of the cassette to that at the left-hand (LH) end, to recreate the circular configuration (Figure 2.2). Because each cassette contains a unique 59-be (see below) the 59-be are named after the gene in the cassette.

Cassettes are compactly organized (Figure 2.2). In the integrated form, as few as 7 bp separate the beginning of the cassette from the initiation codon of the gene, leaving no space for a promoter and little space for a ribosome binding site. Commonly, the termination codon of the gene is very close to, or even within, the region ascribed to the 59-be recombination site. However, occasionally a promoter is present within the cassette and translational attenuation regulatory signals have also been found in the *cmlA1* cassette and other related *cmlA* cassettes (Stokes and Hall, 1991).

Although the cassettes that contain antibiotic resistance genes were the first to be identified, the diverse range of enzymatic functions encoded by the genes that confer resistance indicated that any gene could potentially become associated with a 59-be to form a functional gene cassette. There are now over 60 identified gene cassettes that confer resistance to antibiotics. These include

genes for inner-membrane transporters (efflux proteins), β-lactamases belonging to classes A, B and D, acetyl-, adenylyl-, phosphoryl- and ribosyltransferases with a variety of antibiotic substrate specificities, and dihydrofolate reductases (Recchia and Hall, 1995a; Hall and Collis, 1998; Mazel and Davies, 1999). Among the gene cassettes found in the small chromosome of the *Vibrio cholerae* strain that has been completely sequenced, there are also a few that contain antibiotic resistance genes or potential resistance genes (Heidelberg et al., 2000). However, some of the gene cassettes found in *V. cholerae* or *V. mimicus* strains determine other functions. Gene products with known functions include a heat-stable toxin that is found in relatively few *V. cholerae* strains (Ogawa and Takeda, 1993) and a mannose-fucose-resistant haemagglutinin (Barker et al., 1994) and alipase (Rowe-Magnus et al., 2001). Potential functions for further genes have been proposed on the basis of relationships to known proteins but the majority of ORF remain unidentified (Rowe-Magnus et al., 1999; Clark et al., 2000).

Among the gene cassettes that contain known antibiotic resistance genes, only one includes two open reading frames, the *aacA1*–orfG cassette (Accession. No. AF047479). However there are several cassettes that include two ORF found both in class 1 integrons (Accession No. AF047479) and in the *Vibrio cholerae* collection (Clark et al., 2000; Heidelberg et al., 2000). One of these determines the mannose-fucose-resistant haemagglutinin, but whether both ORF are needed for this activity is not known (Barker et al., 1994). A cassette from a *Xanthomonas campestris* isolate includes both the XbaI restriction and modification genes (Accession No. AF051092). There are also some cassettes where the orientation of the gene (or genes) is reversed. Although these exceptions remain to be adequately accounted for, a possible explanation for cassettes containing genes in the opposite to normal orientation is that the genes are not essential to *V. cholerae* and the sequence has drifted such that the original ORF is no longer detectable.

A very large number of 59-be sites have been found (Stokes et al., 1997). Indeed, each of the cassettes found in class 1 integrons contains a unique 59-be. Initially, 59-be were identified as a consensus sequence of 59 bp found in a few cassettes (Cameron et al., 1986) and this alignment led to the subsequent identification, in other gene cassettes, of related but more diverged sequence elements, some of which had different lengths (Hall et al., 1991). It is now known that 59-be can range in size from 57 to 141 bp (Recchia et al., 1995a). However, the term 59-be has found common currency despite the size variation and has been retained for all members of this family. The VCR found in *V. cholerae* cassettes form a more homogeneous group (Barker et al., 1994) but are also members of the 59-be family (Recchia and Hall, 1997). Indeed 59-be that are closely related to the VCR sequence are found among the 59-be associated with antibiotic resistance genes.

INTEGRONS

As described above, integrons include two distinctive features, an *intI* gene and an adjacent *attI* site, that enable them to capture gene cassettes. They do not necessarily include a gene cassette and class 1 integrons that contain no cassettes have been found in the wild and created experimentally (Bissonnette and Roy, 1992; Collis and Hall, 1992a; Rosser and Young, 1999). However, integrons generally do contain one cassette or an array of two or more gene cassettes. The cassette array can be very long as is the case for the *V. cholerae* chromosomal integrons, where the sequenced strain has more than 170 cassettes (Heidelberg et al., 2000) and other strains are estimated to include at least 100 cassettes (Clark et al., 2000).

As the vast majority of the known gene cassettes contain a gene but not a promoter (Hall et al., 1991; Recchia and Hall, 1995a), an upstream promoter is required for the expression of the genes contained in cassettes. This promoter, P_c, is supplied by the integron and is the third distinctive feature of an integron (Figure 2.1). In the case of the class 1 integrons, which are the commonest type of integrons found in antibiotic-resistant clinical isolates, the P_c promoter is located just inside the beginning of the *intI1* gene (Hall and Vockler, 1987; Stokes and Hall,

1989). All transcripts of the array of gene cassettes start at P_c (Collis and Hall, 1995). The fact that the integron supplies the promoter imposes an orientation constraint on gene cassettes if their genes are to be expressed. In all cases where the gene function is known and expression can be monitored, the orientation of the cassette-associated gene that is found is the one that permits expression. This orientation is achieved only if the 59-be is located downstream of the gene.

MANY CLASSES OF INTEGRON

That there were, in addition to the class 1 integrons, other classes of integrons containing the same gene cassettes, but a different conserved backbone and thus a potentially a distinct IntI gene, was known from the earliest studies (Cameron et al., 1986; Hall and Vockler, 1987; Wiedemann et al., 1987). In fact, the first complete gene cassettes to be sequenced were the *dfrA1* and *aadA1* cassettes that are responsible, respectively, for the resistance to trimethoprim and to spectinomycin and streptomycin conferred by the transposon Tn7. However, these two cassettes were identified as cassettes only after they were also subsequently found in class 1 integrons (Cameron et al., 1986; Sundström and Sköld, 1990; Hall et al., 1991). Indeed, the relationship of the predicted products of the *intI1* gene and the partially sequenced *intI2* gene in Tn7 was identified before the nature of the gene products was known (Hall and Vockler, 1987). However, the Tn7 *intI2* gene (*intI2**) is defective, because it includes an in-frame stop codon that precludes production of a functional protein and this may explain why so few examples of class 2 integrons with different arrays of gene cassettes have been reported thus far (Recchia and Hall, 1995a). As the level of identity between IntI1 and IntI2* was only 45%, it was obvious that these two classes of integrons (Class 1 and 2) were likely to be representative of a vast family of integrons, each encoding a related, but distinct, IntI, but retaining the potential to share the same gene cassettes. Subsequently, a third class of integron was found in clinical isolates of antibiotic resistant *Serratia marcescens* in Japan (Arakawa et al., 1995) and again the cassettes are ones that had been found in class 1 integrons.

INTEGRONS IN BACTERIAL CHROMOSOMES

Although much is known about the impact of gene cassettes on the emergence of multiply antibiotic-resistant strains of Gram-negative bacteria, their provenance is not restricted to antibiotic resistance genes. The small chromosome of *Vibrio cholerae* has recently been shown to include an *intI* gene, *intI4*, adjacent to a long array of gene cassettes (Mazel et al., 1998; Clark et al., 1997, 2000; Heidelberg et al., 2000), and thus contains an integron. Different strains contain different cassette arrays (Clark et al., 2000). The gene cassettes were found first (Ogawa and Takeda, 1993; Barker et al., 1994) but, as in the case of Tn7, they were not initially recognized as such. A repetitive sequence element called a VCR was identified but its similarity to 59-be was recognized later (Recchia and Hall, 1997). However, VCR were shown to be present in the chromosomes of several further *Vibrio* species such as *V. mimicus*, *V. anguillarum*, *V. hollisae* and *V. metschnikvii* but not in others (Ogawa and Takeda, 1993; Barker et al., 1994; Mazel et al., 1998; Clark et al., 2000). The result of these studies are not always consistent, however, they do indicate that an integron and gene cassettes are also likely to be present in many, but not necessarily all, *Vibrio* species. The partially sequenced *intI5* gene from *V. mimicus* has diverged from *intI4* to the same degree as other known chromosomal genes, providing evidence that an integron was a feature of the genome of the common ancestor (Clark et al., 2000). This conclusion has recently been confirmed for other *Vibrio* species (Rowe-Magnus et al., 2001).

MORE INTEGRONS, MORE CASSETTES

It is likely that *Vibrio* species represent only the first case where an integron is found in the

bacterial genome and, therefore, that further examples of gene cassettes and new types of integrons will come to light as more bacterial genomes are sequenced. Further classes of integron may also be found on plasmids recovered from different environments. In fact, some genes whose products are clearly related to the IntI integrases can be found amongst the sequences available in the partially sequenced genomes of *Shewanella putrefaciens*, *Treponema denticola*, *Geobacter sulfurreducens* (Rowe-Magnus et al., 1999, 2001; Nield et al., 2001). However, whether these genes indicate the presence of an integron must await the identification of gene cassettes to go with them. Recently, three new *intI* genes (*intI6, 7* and *8*) have been recovered from environmental soil samples and in two cases a potential adjacent gene cassette has been recovered, together with the *intI* gene (Nield *et al.*, 2001). This confirms that integrons and gene cassettes are likely to be common in the bacterial world.

SOME INTEGRONS CAN MOVE

In the gene cassette/integron system, it is the cassettes that are the mobile elements. However, in the context of their contribution to horizontal gene transfer, it is obvious that integrons and gene cassettes can have a substantial impact, as is the case with respect to antibiotic resistance, only if the integron can gain wide access to a variety of bacterial species. This occurs readily when it is located on a plasmid and this is best achieved if the integron can also translocate. Indeed, class 1 integrons are found in many different genetic contexts, mainly on different plasmids (Hall and Vockler, 1987; Stokes and Hall, 1989; Hall et al., 1994). Class 1 integrons are in fact either transposable elements as exemplified by Tn402 (Rådström et al., 1994) or, more often, defective derivatives of them (Brown et al., 1996; Rådström et al., 1994; Liebert et al., 1999; Partridge et al., 2001a,b). The latter can obviously be moved so long as both outer ends are intact and a set of suitable transposition genes are present in the same cell. Often they are found flanked by a 5 bp duplication of the target site, indicating that they have reached their current location by transposition (Brown et al., 1996; Partridge et al., 2001a,b). In a few cases, class 1 integrons that are unable to mobilize themselves have moved into, and are now found within, another transposon. This is the case for In2 which is found within Tn21 (Liebert et al., 1999), for In4, which is found in Tn1696 (Partridge et al., 2001a), and for In28, which is found in Tn1403 (Partridge et al., 2001b).

The exemplar of a class 2 integron is the transposon Tn7 which contains three gene cassettes *dfrA1–sat2–aadA1* and three other known members of this group are transposons that differ in the identity of the first cassette in the array or have lost this cassette (Recchia and Hall, 1995a). The class 3 integron provides a contemporary example of the rapid spread of resistance genes carried by self-transmissible plasmids. It was first isolated in Japan, in 1993, and has already spread to several other bacterial species (Senda et al., 1996). Our preliminary evidence and that of others (Shibata et al., 1999) indicates that the class 3 integron is, not surprisingly, also a mobile element. In all of these cases, the integron is defined as the complete structure bounded by the terminal inverted repeats or that part of such a structure that remains.

THE RECOMBINATION SYSTEM

The components of the recombination system that effect cassette movement are the IntI-type integrases, the integron-associated *attI* sites and the cassette-associated 59-be sites.

The known IntI integrases form a family of related proteins that share highly significant levels of identity (35–94%). They also share certain features with other members of the integrase or tyrosine recombinase super-family, but pairwise identities between these other integrases and IntI integrases is generally less than 25%. The most obvious of the shared features are the two conserved domains or boxes that are normally used to identify members of this super-family (Ouellette and Roy, 1987; Sundström et al., 1988; Stokes and Hall, 1989). However a recent alignment of the C-terminal catalytic domain of all known members of the

TABLE 2.1 Recombination events catalyzed by IntI1

Participating sites	Integration	Excision
attI1 × 59-be	+	+
59-be × 59-be	+	+
attI1 × *attI1*	+	ND[a]
59-be × 2^0rs	+	–[b]
attI1 × 2^0rs	+	–[b]

[a]Not determined.
[b]Precise excision of a cassette located at a 2^0rs is in most cases unlikely to be possible.

tyrosine recombinase superfamily has revealed further shared "patches" (Nunes-Düby et al., 1998). Other members of the IntI family are known (Nield et al., 2001; Rowe-Magnus et al., 2001), but whether all of them are associated with gene cassettes and are thus part of an integron remains to be established. Only the reactions catalyzed by the IntI1 integrase have been studied in detail and these are described briefly below. However, IntI3 and IntI4 have also been shown to be active (Hall et al., 1999; Rowe-Magnus et al., 2001; Collis and Hall, unpublished).

In addition to recombination between a 59-be and *attI1* site, which occurs when a cassette is incorporated into an integron, IntI1-catalyzed recombination between two 59-be or two *attI1* sites can also occur (Martinez and de la Cruz, 1988, 1990; Hall et al., 1991; Recchia et al., 1994; Hansson et al., 1997; Stokes et al., 1997; Hall et al., 1999; Partridge et al., 2000; Collis et al., 2001). The reactions that have been reported are listed in Table 2.1. The efficiencies of the integration reactions have also been compared (Collis et al., 2001). Recombination between two 59-be sites occurs at high frequency but is less efficient than recombination between a 59-be and the *attI1* site. However, excisive recombination involving two 59-be is important because it can lead to excision of the downstream cassettes in an array. Recombination between two *attI* sites occurs at a much lower frequency than the other reactions (Hansson et al., 1997; Partridge et al., 2000) and is unlikely to be an important event in cassette movement. Recombination between a 59-be and a secondary site (2^0rs) also occurs at low frequency (Francia et al., 1993; Recchia et al., 1994) and is an important

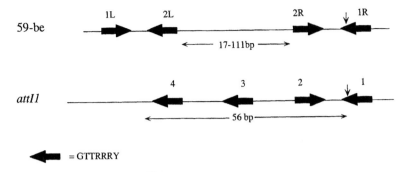

FIGURE 2.3 Recombination sites. The arrangement of 7 bp core sites (arrows) that form part of the larger IntI binding domains is shown. For simplicity the individual core sites are numbered (above the arrows). The vertical arrows show the position of the crossover.

reaction that permits gene cassettes to be integrated at almost any position. The role of recombination between *attI* sites and 2^0rs (Hansson et al., 1997; Collis et al., 2001) is less obvious.

The *attI* sites are part of the integron backbone and are distinguished by the fact that they are the sites into which cassettes are incorporated. In known integrons, they are located upstream of the *intI* gene. However the sequences of the regions adjacent to the first gene cassette in each of the four well-established integron classes are not highly conserved. The extent of the *attI* site has been established experimentally (Recchia et al., 1994; Hansson et al., 1997; Partridge et al., 2000). The complete *attI* site (Figure 2.3), which is required for recombination with a 59-be partner, includes 56 bp from the left side of the crossover point and at least a further 9 bp to the right of the crossover (Hall et al., 1999; Partridge et al., 2000). Within this region, four binding sites for IntI1 have been found and delineated using foot-printing techniques (Collis et al., 1998; Gravel et al., 1998). A single molecule of IntI1 bound to the strongest binding domain protects a total of 14 bp which includes the 7 bp core site regions (Collis et al., 1998). Two of the IntI1 binding domains (1 and 2 in Figure 2.3A) are inversely oriented with respect to one another and form a simple site equivalent to those recognized by other tyrosine recombinases. The additional IntI1-binding domains (3 and 4 in Figure 2.3A) considerably enhance the activity of *attI1* in recombination with a 59-be (Recchia et al., 1994; Hall et al., 1999; Partridge et al., 2000) but are not required for

recombination with a complete *attI1* partner (Hansson et al., 1997; Partridge et al., 2000). They appear to bind IntI1 more strongly than sites 1 and 2 (Collis et al., 1998), and may play a role in retaining the newly synthesized molecules of IntI1 in the proximity of *attI1*.

Simple sites can also be found in the expected positions in *attI2*, 3 and 4 (Collis et al., 1998), but whether these sites resemble *attI1* in binding four molecules of their cognate integrase (IntI2, 3 or 4) remains to be established experimentally. Preliminary data suggest that each IntI preferentially recognizes its cognate *attI* site (Hall et al., 1999), but this also remains to be established rigorously.

The 59-be have a different architecture from that of *attI* sites. All 59-be comprise two regions of 25 bp that are each related to a consensus sequence (Collis and Hall, 1992a; Stokes et al., 1997). In any individual 59-be the consensus regions are imperfect inverted repeats of one another and are separated by a region of highly variable sequence and length that is, in most instances, also an inverted repeat. 59-be are rather unusual in that they include two simple sites, only one of which is the site of strand exchange (Stokes et al., 1997). The LH and RH consensus regions correspond to the bulk of these simple sites but, based on foot-printing data from *attI1* (Collis et al., 1998), the IntI1-binding regions are likely to extend further (Figure 2.3) and a weak consensus is found for some of the bases in the extensions (Collis and Hall, 1992b; Stokes et al., 1997). A striking feature of this family of recombination sites is that the relationship between the sequences of the LH and RH consensus regions is generally retained in preference to adherence to the consensus sequence (Hall et al., 1991; Stokes et al., 1997). This feature is yet to be adequately explained in terms of the activity of these sites. The inverted repeatedness is imperfect in the simple site regions; the 1L and 1R core sites and the 2L and 2R core sites mirror one another but the bases that separate them do not and there is an extra base in 2L (Figure 2.2). The distance between 1L and 2L is 5 bp, but either 5 or 6 bp separate 2R from 1R. It remains to be established which of the differences between the LH and RH simple sites are important in ensuring that the recombination crossover occurs at 1R.

CONCLUSIONS

The role of gene cassettes and integrons in the emergence and spread of antibiotic resistance is well established. The same system has now been implicated in the evolution of bacterial genomes, and the extent of this involvement already appears to be quite significant. Many more new classes of integrons and new gene cassettes will undoubtedly be found in the not too distant future. How gene cassettes are created and how the different integrases recognize the same cassettes as well as other intricacies of the site-specific recombination system are important, but complex, questions that remain to be examined.

REFERENCES

Arakawa, Y., Murakami, M., Suzuki, K. et al. (1995) A novel integron-like element carrying the metallo β-lactamase gene bla_{IMP}. *Antimicrob. Agents Chemother.* **39**: 1612–1615.

Barker, A., Clark, C.A. and Manning, P.A. (1994) Identification of VCR, a repeated sequence associated with a locus encoding a haemagglutinin in *Vibrio cholerae* O1. *J. Bacteriol.* **176**: 5450–5458.

Bissonnette, L. and Roy, P.H. (1992) Characterization of In0 of *Pseudomonas aeruginosa* plasmid pVS1, an ancestor of integrons of multiresistance plasmids and transposons of gram-negative bacteria. *J. Bacteriol.* **174**: 1248–1257.

Brown, H.J., Stokes, H.W. and Hall, R.M. (1996) The integrons In0, In2 and In5 are defective transposon derivatives. *J. Bacteriol.* **178**: 4429–4437.

Cameron, F.H., Groot Obbink, D.J., Ackerman, V.P. et al. (1986) Nucleotide sequence of the AAD(2″) aminoglycoside adenylyltransferase determinant *aadB*. Evolutionary relationship of this region with those surrounding *aadA* in R538–1 and *dhfrII* in R388. *Nucleic Acids Res.* **14**: 8625–8635.

Clark, C.A., Purins, L., Kaewrakon, P. and Manning, P. (1997) VCR repetitive sequence elements in the *Vibrio cholerae* chromosome constitute a mega-integron. *Mol. Microbiol.* **26**: 1137–1138.

Clark, C.A., Purins, L., Kaewrakon, P. et al. (2000) The *Vibrio cholerae* O1 chromosomal integron. *Microbiology* **146**: 2605–2612.

Collis, C.M. and Hall, R.M. (1992a) Site-specific deletion and rearrangement of integron insert genes catalysed by the integron DNA integrase. *J. Bacteriol.* **174**, 1574–1585.

Collis, C.M. and Hall, R.M. (1992b) Gene cassettes from the insert region of integrons are excised as covalently closed circles. *Mol. Microbiol.* **6**: 2875–2885.

Collis, C.M. and Hall, R.M. (1995) Expression of antibiotic resistance genes in the integrated cassettes of integrons. *Antimicrob. Agents Chemother.* **39**: 155–162.

Collis, C.M., Grammaticopoulos, G., Briton, J., et al. (1993) Site-specific insertion of gene cassettes into integrons. *Mol. Microbiol.* **9**: 41–52.

Collis, C.M., Kim, M.-J., Stokes, H.W. et al. (1998) Binding of the purified integron DNA integrase IntI1 to integron- and cassette-associated recombination sites. *Mol. Microbiol.* **29**: 477–490.

Collis, C.M., Recchia, G.D., Kim, M.-J. et al. (2001) Efficiency of recombination reactions catalysed by the class 1 integron integrase IntI1. *J. Bacteriol.* **183**: 2535–2542.

Francia, M.V., de la Cruz, F. and García Lobo, M. (1993) Secondary sites for integration mediated by the Tn21 integrase. *Mol. Microbiol.* **10**: 823–828.

Gravel, A., Fournier, B. and Roy, P.H. (1998) DNA complexes obtained with the integron integrase IntI1 at the *attI1* site. *Nucleic Acids Res.* **26**: 4347–4355.

Hall, R.M. and Collis, C.M. (1995) Mobile gene cassettes and integrons: capture and spread of genes by site-specific recombination. *Mol. Microbiol.* **15**: 593–600.

Hall, R.M. and Collis, C.M. (1998) Antibiotic resistance in gram-negative bacteria: the role of gene cassettes and integrons. *Drug Resist. Updates* **1**:109–119.

Hall, R.M. and Vockler, C. (1987) The region of the IncN plasmid R46 coding for resistance to β-lactam antibiotics, streptomycin/spectinomycin and sulphonamides is closely related to antibiotic resistance segments found in IncW plasmids and in Tn21-like transposons. *Nucleic Acids Res* **15**: 7491–7501.

Hall, R.M., Brookes, D.E. and Stokes, H.W. (1991) Site-specific insertion of genes into integrons: role of the 59-base element and determination of the recombination cross-over point. *Mol. Microbiol.* **5**: 1941–1959

Hall, R.M., Brown, H.J., Brookes, D.E. and Stokes, H.W. (1994) Integrons found in different locations have identical 5′ ends but variable 3′ ends. *J. Bacteriol.* **176**: 6286–6294.

Hall, R.M., Collis, C.M., Kim, M.-J. et al. (1999) Mobile gene cassettes in evolution. *Ann. NY Acad. Sci.* **87**: 68–80.

Hansson, K., Sköld, O. and Sundström, L. (1997) Non-palindromic *attI* sites of integrons are capable of site-specific recombination with one another and with secondary targets. *Mol. Microbiol.* **26**: 441–453.

Heidelberg, J.F., Elsen, J.A., Nelson, W.C. et al. (2000) DNA sequence of both chromosomes of the cholera pathogen *Vibrio cholerae*. *Nature* **406**: 477–483.

Liebert, C.A., Hall, R.M and Summers, A.O. (1999) Transposon Tn*21*, flagship of the floating genome. *Microb. Mol. Biol. Rev.* **63**: 507–522.

Martinez, E., and de la Cruz, F. (1988) Transposon Tn*21* encodes a RecA-independent site-specific integration system. *Mol. Gen. Genet.* **211**: 320–325.

Martinez, E. and de la Cruz, F. (1990) Genetic elements involved in Tn*21* site-specific integration, a novel mechanism for the dissemination of antibiotic resistance genes. *EMBO J.* **9**: 1275–1281.

Mazel, D. and Davies, J. (1999) Antibiotic resistance in microbes. *Cell. Mol. Life Sci.* **56**: 742–754.

Mazel, D., Dychinco, B., Webb, V.A. and Davies, J. (1998) A distinctive class of integron in the *Vibrio cholerae* genome. *Science* **280**: 605–608.

Nield, S.B., Holmes, A.J., Gillings, M.R. et al. (2001) Recovery of new integron classes from environmental DNA. *FEMS Microbiol. Lett.* Submitted.

Nunes-Düby, S.E., Kwon, H.J., Tirumalai, R. et al. (1998) Similarities and differences among 105 members of the Int family of site-specific recombinases. *Nucleic Acids Res.* **26**: 391–406.

Ogawa, A. and Takeda, T. (1993) The gene encoding the heat-stable enterotoxin of *Vibrio cholerae* is flanked by 123 base pair direct repeats. *Microbiol. Immunol.* **37**: 607–616.

Ouellette, M. and Roy, P.H. (1987) Homology of ORFs from Tn*2603* and from R46 to site-specific recombinases. *Nucleic Acids Res.* **15**: 10055.

Ouellette, M., Bissonnette, L. and Roy, P.H. (1987) Precise insertion of antibiotic resistance determinants into Tn*21*-like transposons: nucleotide sequence of the OXA-1 β-lactamase gene. *Proc. Natl Acad. Sci. USA* **84**: 7378–7382.

Partridge, S.R., Recchia, G.D., Scaramuzzi, C. et al. (2000) Definition of the *attI1* site of class 1 integrons. *Microbiology* **146**: 2855–2864.

Partridge, S.R., Brown, H.J., Stokes, H.W. and Hall, R.M. (2001a) Transposons Tn*1696* and Tn*21* and their integrons In4 and In2 have independent origins. *Antimicrob. Agents Chemother.* **45**: 1263–1270.

Partridge, S.R., Recchia, G.D., Stokes, H.W. and Hall, R.M. (2001b) A family of class 1 integrons related to In4 from Tn*1696*. *Antimicrob. Agents Chemother.* In press.

Rådström, P., Sköld, O., Swedberg, G. et al. (1994) Transposon Tn*5090* of plasmid R751, which carries an integron, is related to Tn7, Mu, and the retroelements. *J. Bacteriol.* **176**: 3257–3268.

Recchia, G.D., Stokes, H.W. and Hall, R.M. (1994) Characterisation of specific and secondary recombination sites recognised by the integron DNA integrase. *Nucleic Acids Res.* **22**: 2071–2078.

Recchia, G.D. and Hall, R.M. (1995a) Mobile gene cassettes: a new class of mobile element. *Microbiol.* **141**: 3015–3027.

Recchia, G.D. and Hall, R.M. (1995b) Plasmid evolution by acquisition of mobile gene cassettes: plasmid pIE723 contains the *aadB* gene cassette precisely inserted at a secondary site in the IncQ plasmid RSF1010. *Mol. Microbiol.* **15**: 179–187.

Recchia, G.D. and Hall, R. M. (1997) Origins of the mobile gene cassettes found in integrons. *Trends Microbiol.* **389**: 389–394.

Rowe-Magnus, D.A., Guérout, A.-M. and Mazel, D. (1999) Super-integrons. *Res. Microbiol.* **150**: 641–651.

Rowe-Magnus, D.A., Guérout, A.-M., Ploncard, P., Dychinoco, B., Davies, J. and Mazel, D. (2001). The evolutionary history of chromosomal super-integrons provides an ancestry for multiresistant integrons. *Proc. Natl Acad. Sci. USA* **98**: 652–657.

Rosser, S.J. and Young, H.-K. (1999) Identification and characterization of class 1 integrons in bacteria from an aquatic environment. *J. Antimicrob. Chemother.* **44**: 11–18.

Senda, K., Arakawa, Y., Ichiyama, S. et al. (1996) PCR detection of metallo-β-lactamase (*bla*$_{IMP}$) in gram-negative rods resistant to broad-spectrum β-lactams. *J. Clinical Microbiol.* **34**: 2909–2913.

Shibata, N., Kurokawa, H., Yagi, T. and Arakawa, Y. (1999) A class 3 integron carrying the IMP-1 metallo-β-lactamase gene found in Japan. 39th Interscience Conference on Antimicrobial Agents and Chemotherapy, San Francisco.

Stokes, H.W. and Hall, R.M. (1989) A novel family of potentially mobile DNA elements encoding site-specific gene-integration functions: integrons. *Mol. Microbiol.* **3**: 1669–1683.

Stokes, H.W. and Hall, R.M. (1991) Sequence analysis of the inducible chloramphenicol resistance determinant in the Tn*1696* integron suggests regulation by translational attenuation. *Plasmid* **26**: 10–19.

Stokes, H.W., O'Gorman, D.B., Recchia, G.D. et al. (1997) Structure and function of 59-base element recombination sites associated with mobile gene cassettes. *Mol. Microbiol.* **26**: 731–745.

Sundström, L. and Sköld, O. (1990) The *dhfrI* trimethoprim resistance gene of Tn7 can be found at specific sites in other genetic surroundings. *Antimicrob. Agents Chemother.* **34**: 642–650.

Sundström, L., Rådström, P., Swedberg, G. and Sköld, O. (1988) Site-specific recombination promotes linkage between trimethoprim- and sulfonamide resistance genes. Sequence characterization of *dhfrV* and *sulI* and a recombination active locus of Tn*21*. *Mol. Gen. Genet.* **213**: 191–201

Wiedemann, B., Meyer, J.F. and Zühlsdorf, M.T. (1987) Insertions of resistance genes into Tn*21*-like transposons. *J. Antimicrob. Chemother.* **18**: 85–92.

CHAPTER 3

A *Corynebacterium* Plasmid Composed of Elements from Throughout the Eubacteria Kingdom

Andreas Tauch and Alfred Pühler

Multiple antimicrobial resistance in human pathogens is a global medical problem. Especially, Gram-positive microorganisms show alarming increases in antibiotic resistances. The complete DNA sequence of the 51-kb multiresistance plasmid pTP10 from the Gram-positive human pathogen *Corynebacterium striatum* M82B provides genetic information about acquired resistance mechanisms to antimicrobial agents in this species. Analysis of the genetic organization of pTP10 suggests that the plasmid is composed of a mosaic structure comprising eight DNA segments the boundaries of which are represented by horizontal mobile elements. The DNA segments of pTP10 turned out to be virtually identical to a plasmid-encoded macrolide and lincosamide resistance region from the human pathogen *Corynebacterium diphtheriae*, a chromosomal DNA region from *Mycobacterium tuberculosis*, a mobile chloramphenicol resistance region from the soil bacterium *Corynebacterium glutamicum*, several transposable elements from Gram-negative phytopathogenic *Pseudomonas*, *Xanthomonas* and *Erwinia* species, and to a transposable aminoglycoside resistance region from the Gram-negative animal pathogen *Pasteurella piscicida*. This provides molecular evidence that natural routes exist by which antibiotic resistance genes from bacteria of different habitats and geographical origin can be assembled in a human pathogen. This shows that highly diverged species that last shared a common ancestor about 2 billion years ago can still exchange genes. Consequently, horizontal gene transfer of antibiotic resistance genes is an important mechanism which limits the successful use of antimicrobials in the clinical treatment of human infections.

INTRODUCTION

Drug resistance in human pathogens is the result of overuse of antimicrobials in medicine and agriculture (Witte, 1998). This overuse of antibiotics has led to the rapid evolution of bacteria that are resistant to multiple drugs such that even vancomycin and teicoplanin, the drugs of last resort, are no longer effective against some bacterial isolates. Generally, a microorganism can have either intrinsic or acquired resistance to antibiotics (Tan et al., 2000). Intrinsic resistance is a stable genetic property, arising from mutation in the chromosomal DNA. As each mutation confers only a slight alteration in susceptibility, microorganisms need to accumulate several mutations to become intrinsically resistant to antibiotics. Alternatively, resistance is acquired by genetic exchange with another microorganism from the same or

Copyright © 2002 by Academic Press.
All rights of reproduction in any form reserved.

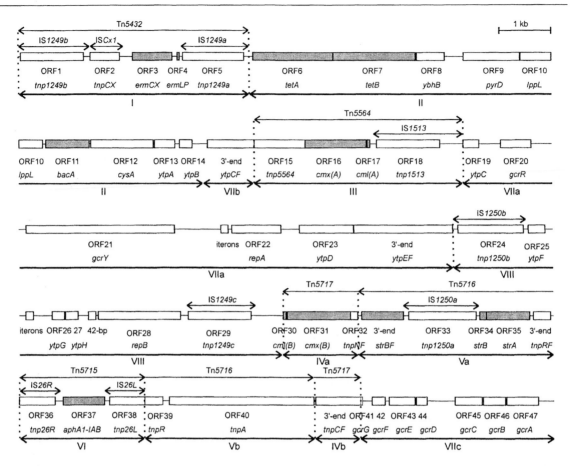

FIGURE 3.1 Genetic organization of the 51409-bp multiresistance plasmid pTP10 identified in *C. striatum* M82B. The organization of the open reading frames (ORFs) and the position of transposons and insertion sequences is presented. The eight DNA segments of pTP10 are specifically marked (I–VIII). Dotted lines represent segment boundaries corresponding to the insertion of mobile elements. The acquired antibiotic resistance genes are marked by filled boxed. Details on the DNA sequence of pTP10 are available from GenBank accession number AF139896.

another genus (Tan et al., 2000). Resistance genes can be transferred among bacteria and can be integrated into the bacterial chromosome to be stably inherited from one generation to the next. Additionally, antibiotic resistance genes can be maintained in an extra-chromosomal state on a bacterial plasmid. Plasmids that can transfer DNA to adjacent bacteria are known as conjugative plasmids. As conjugation can occur in a broad range of species it is one of the main mechanisms through which resistance genes are transferred between bacteria (Dröge et al., 1998, 1999). Therefore, such plasmids can be classed among the horizontal mobile elements

that also include phages, integrons, transposons, and insertion sequences. Accordingly, the transfer of horizontal mobile elements occurs via conjugation, transduction, transposition, and transformation.

Today, integrated genome research offers the chance to analyze large resistance plasmids (R plasmids) and acquired antibiotic resistances at the nucleotide level. Highly automated sequencing machines and processes have been developed to allow large-scale DNA sequencing and subsequent DNA sequence interpretation by bioinformatics. The information obtained within the scope of such projects not only sheds new light on acquired antibiotic resistance mechanisms, but also on horizontal gene transfer and plasmid evolution (Perretten et al., 1997; Tauch et al., 2000).

The genetic data described below focus on a plasmid project which dealt with the DNA sequence analysis of the large multiresistance plasmid pTP10 from the human pathogen *Corynebacterium striatum* M82B (Tauch et al., 2000). The complete DNA sequence revealed insights into how pTP10 is genetically organized and, in particular, how a multiresistance plasmid from a human clinical source has evolved over time. Virtually identical DNA segments have been identified in a soil bacterium and in plant, animal and human pathogens. This finding implies that horizontal gene transfer has played a central role in the evolutionary history of pTP10.

ANALYSIS OF AN ANTIBIOTIC RESISTANCE PLASMID ISOLATED FROM *CORYNEBACTERIUM STRIATUM* BY INTEGRATED GENOME RESEARCH

In recent years, the Gram-positive bacterium *C. striatum* has been recognized with increasing frequency as an important opportunistic human pathogen, especially in immunocompromised patients and in patients under intensive care. In clinical diagnostics, numerous isolates of *C. striatum* were found to be highly resistant to the majority of clinically relevant antibiotics which more and more resulted in the failure of antibiotic treatment of *C. striatum*-mediated human infections. *C. striatum* M82B was initially isolated from the bacterial flora of an otitis media patient in a Japanese hospital. It was shown to carry the R plasmid pTP10 encoding resistances to the antibiotics chloramphenicol, erythromycin, kanamycin, and tetracycline (Tauch et al., 1995a). Since the plasmid genome of pTP10 was determined to be only 51 kb in size, it represents an ideal model system for the analysis of acquired antibiotic resistance by integrated genome research. The total DNA sequence of the R plasmid pTP10 was

determined and subsequently annotated by means of an automated genome interpretation system. In such a way the complete gene structure of pTP10 was identified. The pTP10 sequence contains 47 open reading frames (ORFs) and an additional five incomplete coding regions. Moreover, pTP10 harbors five transposons (Tauch et al., 1995b, 1998) and two additional insertion sequences. Figure 3.1 presents a detailed map of the ORFs found on pTP10 as well as other relevant structural features.

Based on the DNA sequence data, further experiments concentrated on the antibiotic resistance genes of pTP10, some of which are integral parts of horizontal mobile elements (Figure 3.1). Besides the known resistance determinants of pTP10 to chloramphenicol (*cml(A)*, *cmx(A)*), erythromycin (*ermLP*, *ermCX*), kanamycin (*aphA1-IAB*), and tetracycline (*tetAB*), DNA sequence analysis revealed the presence of a duplicated chloramphenicol resistance region (*cml(B)* and *cmx(B)*) and genes probably involved in bacitracin (*bacA*) and streptomycin resistance (*strAB*). However, antibiotic susceptibility studies clearly demonstrated that the bacitracin and streptomycin resistance determinants are inactive on pTP10 (Tauch et al., 2000). In addition, DNA sequence interpretation made it possible to deduce the respective resistance mechanisms encoded on pTP10 and to identify virtually identical resistance genes in Gram-positive and Gram-negative bacteria of different habitats. Most interestingly, for tetracycline a new resistance mechanism could be proposed that is based on heterodimerization of two ABC transporters resulting in an oxytetracycline and oxacillin cross-resistance (Tauch et al., 1999). The deduced data concerning the resistance determinants of pTP10 are summarized in Table 3.1.

HORIZONTAL GENE TRANSFER AND ACQUIRED ANTIBIOTIC RESISTANCE IN *CORYNEBACTERIUM STRIATUM* M82B

Taking into account the DNA sequence annotation and the structural data, the pTP10 plasmid was subdivided into eight genetically distinct DNA segments, six of which are involved in antibiotic resistance. One of the DNA segments present on pTP10 comprises the composite resistance transposon Tn5432 that consists of two IS1249 sequences flanking the erythromycin resistance gene region (Figure 3.1; I). The central part of Tn5432 was found to be virtually identical at the nucleotide level to the antibiotic resistance gene region of plasmid pNG2 from the human pathogen *Corynebacterium diphtheriae* (Figure 3.2A) encoding an inducible resistance

to macrolide and lincosamide antibiotics. In contrast to the genetic arrangement on pTP10, the sequenced resistance region of pNG2 is not part of a Tn5432-like mobile element (Figure 3.2A).

The second DNA segment of pTP10 which is involved in antibiotic resistance is located downstream of Tn5432 and comprises a DNA region with a high G + C content (Figure 3.1; II). The gene products of the respective coding regions showed the highest similarity to proteins encoded by the *Mycobacterium tuberculosis*

TABLE 3.1 Acquired antibiotic resistance gene regions of pTP10 from *C. striatum* M82B and its closest relatives

pTP10 gene region	Resistance genes	Resistance mechanism	Closest relative/ microorganism
Erythromycin resistance	*ermLP, ermCX*	23S rRNA methylation	*Corynebacterium diphtheriae* (Gram-positive human pathogen)
Tetracycline resistance	*tetA, tetB*	Efflux via heterodimerization of ABC transporters	No similarity found
Bacitracin resistance	*bacA*	Phosphorylation of undecaprenol (inactive in pTP10)	*Mycobacterium tuberculosis* (Gram-positive human pathogen)
Chloramphenicol resistance copy A and copy B	*cml(A), cmx(A)* *cml(B), cmx(B)*	Antibiotic export	*Corynebacterium glutamicum* (Gram-positive soil bacterium)
Streptomycin resistance	*strA, strB*	Phosphorylation (inactive in pTP10)	*Erwinia amylovora, Pseudomonas syringae, Xanthomonas campestris* (Gram-negative plant pathogens)
Kanamycin resistance	*aphA1-IAB*	Phosphorylation	*Pasteurella piscicida, Klebsiella pneumoniae* (Gram-negative animal and human pathogens)

FIGURE 3.2 (Opposite.) Comparison of pTP10 DNA segments carrying acquired antibiotic resistance gene regions with virtually identical DNA regions from Gram-positive and Gram-negative bacteria of different habitats. (A) Comparison of the macrolide and lincosamide resistance region of pTP10 (segment I) with a DNA fragment of plasmid pNG2 from the human pathogen *C. diphtheriae*. (B) Comparison of the chloramphenicol resistance transposon Tn5564 (segment III) with transposon Tn45 from the soil isolate *C. glutamicum* 1014. Both elements are virtually identical with the exception of IS1513 which inserted between *cml(A)* and the inverted repeat (IR) of Tn5564. (C) Comparison of the Tn3-type transposon Tn5716 present on pTP10 (segment V) with the streptomycin resistance transposon Tn5393 from plasmid pEa34 of the plant pathogen *E. amylovora*. (D) Comparison between the *aphA1-IAB* aminoglycoside resistance regions from *C. striatum* M82B (segment VI) and the Gram-negative fish pathogen *P. piscicida*. The *aphA1-IAB* gene is part of IS26-based composite transposons which differ in the orientation of the insertion sequences (arrows) and in the length of the spacer sequences.

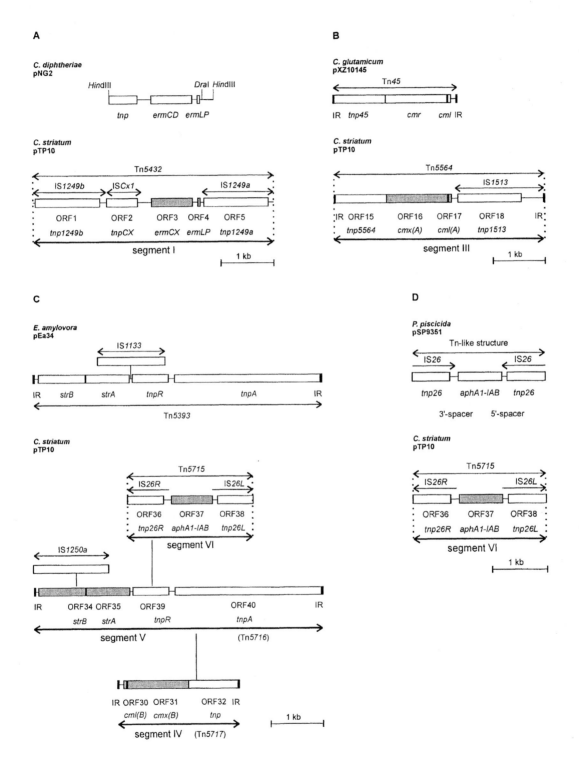

chromosome. Furthermore, it is noteworthy that the genetic organization of these genes on pTP10 is almost identical to the gene arrangement found in the *M. tuberculosis* chromosome. The high G + C region of pTP10 also comprises the *tetAB* genes (Figure 3.1) the deduced proteins of which are similar to chromosomally encoded ATP-binding cassette transporters. These data strongly suggest that the DNA segment derived from the chromosome of a microorganism belonging to the high G + C branch of Gram-positive bacteria.

The third DNA segment of pTP10 comprises the chloramphenicol resistance region that is part of the mobile element Tn*5564* (Figure 3.1; III). The basic molecular structure of this transposon, comprising the resistance gene *cmx*(A) and the transposase gene, is nearly identical at the nucleotide level to the chloramphenicol resistance region of plasmid pXZ10145 from the Chinese soil isolate *Corynebacterium glutamicum* 1014. In contrast to the basic structure present in *C. glutamicum*, Tn*5564* carries the additional insertion sequence IS*1513*, located between the putative leader sequence of *cmx*(A) and the left inverted repeat (Figure 3.2B).

Interestingly, DNA segment IV of pTP10 is represented by a second identical copy of the basic structure of Tn*5564* (Figure 3.1; IVa and IVb). This copy is disrupted by the Tn*3*-type mobile element Tn*5716* (Figure 3.1; Va and Vb). Transposon Tn*5716* contains the genetic information for a transposase, a resolvase and the linked StrAB streptomycin resistance proteins and is identical at the nucleotide level to the basic structure of Tn*5393* from the *Erwinia amylovora* plasmid pEa34 (Figure 3.2C). Tn*5393*-like transposons are structurally very similar and can be distinguished by additional insertions of mobile elements. Although the *strAB* tandem pair of streptomycin resistance genes is widespread among Gram-negative bacteria, it has not been identified in a Gram-positive bacterium to date. Moreover, the association of the *strAB* genes with a transposable element was exclusively found in the Gram-negative phytopathogenic bacteria *E. amylovora*, *Pseudomonas syringae*, and *Xanthomonas campestris* isolated from American agricultural habitats where streptomycin was utilized as bactericide

(Sundin and Bender, 1996). Therefore, it is most likely that the Tn*5393*-like DNA segment of pTP10 derived from a Gram-negative host organism and was transferred to pTP10 by horizontal gene transfer. Interestingly, the Tn*5393* variants occur on large conjugative plasmids in the three phytopathogenic genera (Sundin and Bender, 1996).

The aminoglycoside resistance region of pTP10 (Figure 3.1; VI) is part of the composite transposon Tn*5715* consisting of two IS*26* insertion sequences and a resistance gene encoding an aminoglycoside-3',5''-phosphotransferase. Both IS*26* elements of Tn*5715* were found to be identical to previously sequenced IS*26* elements from the Gram-negative species *Salmonella ordonez*, *Klebsiella pneumoniae*, and *Pasteurella piscicida*. The aminoglycoside resistance protein encoded by Tn*5715* is identical to the AphA1-IAB protein identified in a clinical isolate of *K. pneumoniae* from Chile (Lee et al., 1991) and in the animal pathogen *P. piscicida* SP9351 from a Japanese marine fish farm (Kim and Aoki, 1994). The protein carries four amino acid substitutions when compared with the widely distributed Aph(3')-Ia protein from Tn*903* and was associated with a significantly higher turnover of the aminoglycosides kanamycin and neomycin in *K. pneumoniae* (Lee et al., 1991). The 5'-spacer of Tn*5715*, located between the IS*26L* element and the translational start of the *aphA1-IAB* resistance gene, and the 3'-spacer of Tn*5715* are characteristic for the *aphA1-IAB* gene region of *P. piscicida* (Figure 3.2D). This strongly indicates that both resistance regions derived from a common ancestor molecule. Due to the nucleotide sequence identity and the apparent low G + C content of the *aphA1-IAB* gene region (44.1%), it is obvious that DNA segment VI was transferred from a Gram-negative bacterium to pTP10. Interestingly, the *aphA1-IAB* resistance genes from *K. pneumoniae* and *P. piscicida* are located on transferable R plasmids which might enable the rapid dissemination of this specific type of aminoglycoside resistance across species boundaries.

Furthermore, the automated genome interpretation revealed that the pTP10 plasmid encodes two replication proteins with amino

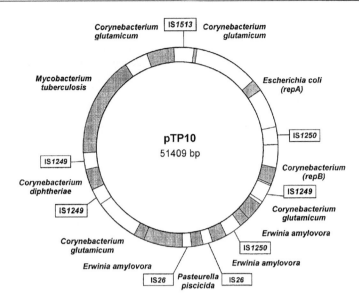

FIGURE 3.3 "Evolutionary map" of the multiresistance plasmid pTP10 from *C. striatum* M82B. Similarities of the pTP10 DNA segments to DNA regions initially identified in bacteria of different origins are shown. The location of insertion sequences on the pTP10 plasmid is indicated.

acid sequence similarities to broad-host-range replicases (Figure 3.1). The *repA* gene region showed the highest similarity to replicases from Gram-negative bacteria whereas *repB* is similar to replicases from Gram-positive bacteria. Due to the presence of two different types of broad-host-range replicases, pTP10 is a naturally occurring shuttle plasmid. This capacity to replicate in various bacterial species might also be responsible for the collection of antibiotic resistance genes from transient host organisms.

Evidently, the pTP10 plasmid represents a "hot spot" in evolution, since it is composed of eight DNA segments (Figure 3.1; I–VIII) from different bacterial origins including soil bacteria and animal, plant, and human pathogens. These DNA modules are combined by insertion sequences and transposons which may have served as mobilization tool for antibiotic resistance determinants (Figure 3.3). Consequently, the evolution of the pTP10 plasmid resulted in a DNA mosaic structure characterized by a collection of distinct DNA segments conferring a multiresistant phenotype.

HORIZONTAL GENE TRANSFER AND ACQUIRED ANTIBIOTIC RESISTANCE IN *LACTOCOCCUS* AND *STAPHYLOCOCCUS* STRAINS

Lactic acid bacteria are important microorganisms in parts of the human and animal body and in environments where spontaneous fermentations of carbohydrate containing substrates occur. The composite 29.9-kb resistance plasmid pK214 is resident in *Lactococcus lactis* subsp. *lactis* isolated from raw milk soft cheese (Perretten et al., 1997; Teuber et al., 1999). It contains the tetracycline resistant determinant *tetS* previously described in *Listeria monocytogenes* and chloramphenicol (*cat*) and streptomycin (*str*) resistance genes from *Staphylococcus aureus*. Three out of five insertion sequences on pK214 are virtually identical to IS*1216* present in the *vanA* resistance transposon Tn*1546*. Comparative sequence analysis of the entire plasmid DNA demonstrates that pK214 is composed of distinct DNA segments which originate in species like *Listeria*, *Staphylococcus* and *Enterococcus* (Teuber et al., 1999). The heterologous segment acquisition may have been driven by the mobile elements of pK214. These data provided molecular evidence on the horizontal gene transfer of antibiotic resistance determinants into fermented and other food.

S. aureus and, especially, coagulase-negative staphylococci are significant nosocomial pathogens. The 46.4-kb plasmid pSK41 represents a prototypical multiresistance conjugative horizontal mobile element in this species (Berg et al., 1998). pSK41 family plasmids typically confer resistance to several aminoglycosides, as well as multidrug resistance to antiseptics and disinfectants. The complete nucleotide sequence of pSK41 revealed that several of the resistance determinants carried by pSK41-related plasmids were found to be located on smaller cointegrated plasmids. Thus, pSK41-like plasmids represent a consolidation of antimicrobial resistance functions, collected via transposition and cointegrative capture of other plasmids (Berg et al., 1998). The capacity of conjugative plasmids to collect resistance activities as they move horizontally through bacterial populations highlights the importance of non-pathogenic microorganisms that act as reservoirs for antimicrobial resistances.

REFERENCES

Berg, T., Firth, N., Apisiridej, S. et al. (1998) Complete nucleotide sequence of pSK1: Evolution of staphylococcal conjugative multiresistance plasmids. *J. Bacteriol.* **180**: 4350–4359.

Dröge, M., Pühler, A. and Selbitschka, W. (1998) Horizontal gene transfer as a biosafety issue: a natural phenomenon of public concern. *J. Biotechnol.* **64**: 75–90.

Dröge, M., Pühler, A. and Selbitschka, W. (1999) Horizontal gene transfer among bacteria in terrestrial and aquatic habitats as assessed by microcosm and field studies. *Biol. Fertil. Soils* **29**: 229–245.

Kim, E. and Aoki, T. (1994) The transposon-like structure of IS26-tetracycline and kanamycin resistant determinant derived from transferable R plasmid of fish pathogen *Pasteurella piscicida. Microbiol. Immunol.* **38**: 31–38.

Lee, K.-Y., Hopkins, J.D. and Syvanen, M. (1991) Evolved neomycin phosphotransferase from an isolate of *Klebsiella pneumoniae. Mol. Microbiol.* **5**: 2039–2046.

Perretten, V., Schwarz, F., Cresta, L. et al. (1997) Antibiotic resistance spread in food. *Nature* **389**: 801–802.

Sundin, G.W. and Bender, C.L. (1996) Dissemination of the *strA-strB* streptomycin-resistance genes among commensal and pathogenic bacteria from humans, animals, and plants. *Mol. Ecol.* **5**: 133–143.

Tan, Y.-T., Tillett, D.J. and McKay, I.A. (2000) Molecular strategies for overcoming antibiotic resistance in bacteria. *Mol. Medicine Today* **6**: 309–314.

Tauch, A., Kassing, F., Kalinowski, J. and Pühler, A. (1995a) The erythromycin resistance gene of the *Corynebacterium xerosis* R-plasmid pTP10 also carrying chloramphenicol, kanamycin and tetracycline resistances is capable of transposition in *Corynebacterium glutamicum. Plasmid* **33**: 168–179.

Tauch, A., Kassing, F., Kalinowski, J. and Pühler, A. (1995b) The *Corynebacterium xerosis* composite transposon Tn5432 consists of two identical insertion sequences, desgnated IS1249, flanking the erythromycin resistance gene *ermCX. Plasmid* **34**: 119–134.

Tauch, A., Zheng, Z., Pühler, A. and Kalinowski, J. (1998) *Corynebacterium striatum* chloramphenicol resistance transposon Tn5564: genetic organization and transposition in *Corynebacterium glutamicum. Plasmid* **40**: 126–139.

Tauch A., Krieft, S., Pühler, A. and Kalinowski, J. (1999) The *tetAB* genes of the *Corynebacterium striatum* R-plasmid pTP10 encode an ABC transporter and confer tetracycline, oxytetracycline and oxacillin resistance in *Corynebacterium glutamicum. FEMS Microbiol. Lett.* **173**: 203–209.

Tauch, A., Krieft, S., Kalinowski, J. and Pühler, A. (2000) The 51409-bp R-plasmid pTP10 from the multiresistant clinical isolate *Corynebacterium striatum* M82B is composed of DNA segments initially identified in soil bacteria and in plant, animal, and human pathogens. *Mol. Gen. Genet.* **263**: 1–11.

Teuber, M., Meile, L. and Schwarz, F. (1999) Acquired antibiotic resistance in lactic acid bacteria from food. *Antonie van Leeuwenhoek* **76**: 115–137.

Witte, W. (1998) Medical consequences of antibiotic use in agriculture. *Science* **279**: 996–997.

CHAPTER 4

Horizontal Transfer of Naphthalene Catabolic Genes in a Toxic Waste Site

Eugene L. Madsen

Coal-tar waste is a naphthalene-rich, tarry material that is a widely distributed subsurface contaminant produced by the manufactured gas plant industry. This paper summarizes recent investigations examining how the naturally-occurring groundwater microbial community in a site in South Glens Falls, New York, has genetically adapted to exposure to coal-tar waste and naphthalene, in particular. After isolation of naphthalene-degrading bacteria from the site and sequencing both their key naphthalene catabolic (*nahAc*) and 16S rRNA genes, incongruent gene phylogeny suggested that horizontal gene transfer (HGT) had occurred. The *nahAc* gene was found to reside on large (70–88 kb) plasmids within the site-derived isolates. RFLP and Southern analysis of the plasmids showed that they were very similar to the archetypal naphthalene catabolic plasmid pDTG1. Laboratory mating assays with cured, rifampin-resistant recipient strains proved the plasmids were self-transmissible. Laboratory mating assays also proved that the native soil microbial community carried donor strains similar to pDTG1. Field mating assays in site sediments and waters indicated that significant barriers to HGT occur in real-world settings. Plans are described for forthcoming studies that will use a combination of *lacZ*-transcriptional fusions and field experiments to discover both molecular and ecological cues for HGT.

INTRODUCTION

Horizontal gene transfer (HGT), the transmission of DNA fragments between organisms of distinctive ancestral lineage, is an increasingly significant area of science because of its evolutionary, biotechnological, and medical implications (Ochman et al., 2000; Worning et al., 2000; Schmidt and Harikein, 1996; Rivera et al., 1998; Jain et al., 1999). The need to pursue knowledge of horizontal gene transfer within the soil and other environments also mirrors broad societal concerns: it has bearing on our understanding of the development and proliferation of antibiotic resistance and other traits (Trevors et al., 1987; Levy and Miller, 1989; Stotzky, 1989; Klingmuller et al., 1990; Lorenz and Wackernagel, 1993; Cutting and Youngman, 1994; Davies, 1994, 1996; Nikolich et al., 1994; Provence and Curtiss, 1994; Syvanen, 1994; Zhou and Tiedje, 1995; Heinemann, 1998; Kroer et al., 1998; Normander et al., 1998; Syvanen and Kado, 1998; Wiener et al., 1998; Davison, 1999; Sobecky, 1999), on mechanisms of plant pathogenesis (Heath et al., 1995; Zhu et al., 2000), and on the adaptation of naturally-occurring microbial communities to environmental contaminants (see below).

BACKGROUND

Concern over the environmental fate and ecological impact of environmental pollutants is long

Copyright © 2002 by Academic Press.
All rights of reproduction in any form reserved.

standing (Alexander, 1981, 1999; Dagley, 1983; Gibson, 1984; Young and Cerniglia, 1995; Atlas and Untermann, 1999). A large body of information about organic environmental pollutant metabolism by pure cultures and environmental samples has accrued (Madsen, 1997; Wackett and Hershberger, 2001). While such studies contribute substantially to our understanding of how soil microorganisms interact with, respond to, and detoxify pollutant compounds, there is little known about the mechanisms which allow naturally-occurring soil microbial populations to adapt to pollutant exposure. Such adaptation, based in fundamentals of population growth, mutation, or gene exchange (Leahy and Colwell, 1990), has very real implications for predicting the environmental fate of organic environmental contaminants. Although rearrangement and horizontal transfer of biodegradation genes between soil microorganisms have been retrospectively proposed as a means of adaptation to pollutants, especially the herbicide, 2,4-D (or other chemical) exposure (van der Meer et al., 1991, 1992; Tiedje, 1994; Fulthorpe et al., 1995; Ka and Matheson et al., 1996), only recently have nucleic acid-based biotechnology methods become available to document relationships between horizontal transfer of biodegradation genes and *in situ* environmental selective pressure for that genetic exchange (Nusslein et al., 1992; Herrick et al., 1993, 1997; Herrick, 1995; Smets et al., 1995; Christensen et al., 1996; Blatny et al., 1997; Normander et al., 1998; Stuart-Kiel et al., 1998; van der Meer et al., 1998; Schenk and Decker, 1999; Hohnstock et al., 2000).

Ongoing research in Madsen's laboratory strives to use physiological, biochemical, and molecular knowledge obtained from individual laboratory isolates to understand the activity of microorganisms in field sites. For the past several years, projects have been aimed at understanding the behavior of naphthalene in coal tar waste-contaminated groundwater and sediments (EPRI, 1996). Reports have focused on *in situ* metabolism of naphthalene as indicated by adaptation and protozoan predators (Madsen et al., 1991), by a field-detected transient metabolic intermediate (Wilson and Madsen, 1996), and by field detection of mRNA transcripts of naphthalene dioxygenase (Wilson et al., 1999). Additional field studies have explained the

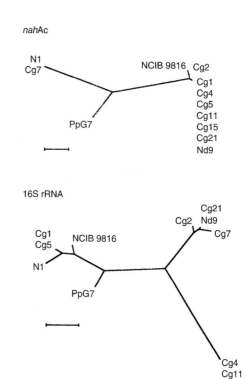

FIGURE 4.1 Phylogenetic trees of *nahAc* (top) and 16S rRNA (bottom) gene sequences from bacteria isolated from the contaimated study site. Horizontal bars represent a distance equivalent to approximately 1% dissimilarity. (From Herrick et al., 1997.)

anomalous persistence of naphthalene in site sediments (Madsen et al., 1996a) and used passive sampling devices to assess the mobility of both contaminants and contaminant-degrading microorganisms (Madsen et al., 1996b). The present paper reviews recently completed and planned investigations examining *in situ* genetic exchange that has almost certainly contributed toward metabolic adaptation by naturally-occurring microorganisms to naphthalene contamination at a coal tar waste-contaminated field site in South Glens Falls, New York.

RETROSPECTIVE PHYLOGENETIC EVIDENCE FOR HGT

Through a series of procedures involving isolation of site-derived bacteria and characterization of their catabolic genes (designing oligonucleo-tide primers to amplify targeted genes; purifying and sequencing the PCR amplification products; analyzing the nucleotide sequences; and performing phylogenetic analyses; Herrick et al. (1997)), we have obtained retrospective evidence for horizontal *in situ* transfer of *nahAc*, a catabolic gene that encodes a key component of the naphthalene dioxygenase enzyme complex. Figure 4.1 shows the result of a phylogenetic analysis in which lack of congruence between the two (16S rRNA and catabolic) genes carried by a group of site bacteria suggests *in situ* horizontal gene transfer. The identical naphthalene catabolic gene (Figure 4.1, top) is shared by seven hosts whose taxonomic identities are distributed broadly across the 16S rRNA phylogenetic tree (Figure 4.1, bottom). Widely divergent bacteria from our contaminated study site carried highly conserved contaminant-catabolizing genes (Herrick et al., 1997).

CHARACTERIZATION OF THE MOBILE GENETIC ELEMENT (PLASMID)

Localization of the catabolic genes

Because naphthalene catabolic genes are generally found on self-transmissible plasmids, we hypothesized that plasmid transfer, whether by conjugation or plasmid transformation has played a role in *nah* gene exchange (Herrick et al., 1997). With an alkaline lysis protocol and standard gel electrophoresis, plasmids comparable in size to the NAH7 plasmid of PpG7 (~80 kb) were detected in all nine site-derived naphthalene-degrading bacteria whose partial *nahAc* sequences were shown in Figure 4.1. To estimate plasmid sizes accurately, PFGE in combination with Southern hybridization was performed on partially purified, S1 nuclease-linearized plasmid preparations from a group of site-derived bacteria and on the archetypical strains NCIB-9816-4 and PpG7 (Yen and Gunsalus, 1982; Serdar and Gibson, 1989; Simon et al., 1993; Goyal and Zylstra, 1997). The 407 bp *nahAc* probe derived from site strain Cg1 hybridized to all *nahAc* homologues under the

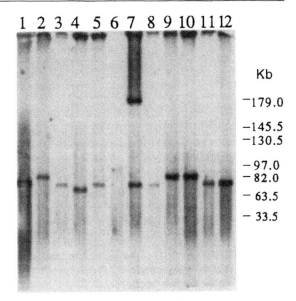

FIGURE 4.2 Southern hybridization to *nahAc* of partially purified plasmids from naphthalene-degrading bacteria separated by CHEF-PFGE. The sizes from molecular markers are indicated in kilobases on the side. Plasmids were from the following reference and site-derived (Cg) organisms: lane 1, NCIB 9816-4; lane 2, Cg1; lane 3, Cg2; lane 4, Cg4 lane 5, Cg5; lane 6, Cg7; lane 7, Cg 8; lane 8, Cg9; lane 9, Cg11; lane 10, Cg15; lane 11, Cg16; and lane 12, Cg21. (From Herrick et al., 1997.)

conditions chosen and the plasmids were found to vary in size from 70 to 88 kb (Figure 4.2).

In order to confirm the location (plasmid versus chromosome) of *nahAc* in the isolates, total genomic DNA from each strain was separated using PFGE and subsequently Southern hybridized using the *nahAc* probe (Herrick et al., 1997). Hybridization occurred with plasmids carried by most site strains, but also with the chromosomal DNA of others. The occurrence of *nahAc* on chromosomes and on different-sized plasmids of naphthalene-degrading bacteria native to our contaminated study site suggests a variety of possible gene transfer mechanisms. A single mobile plasmid may have been transferred between many hosts and subsequently undergone insertions or deletions. In addition, transposons, conjugative transposons, and/or plasmid integration into the chromosome may have operated in positioning *nahAc* on the chromosome of some strains.

Plasmid RFLP analysis

To characterize the naphthalene plasmids, we isolated and purified them from each of their hosts. Restriction enzyme digestion of the naphthalene catabolic plasmids revealed a pattern of restriction fragment lengths common to plasmids from all of our site-derived bacteria and pDTG1. This common pattern was unlike the restriction fragment length polymorphism (RFLP) pattern for the canonical naphthalene catabolic plasmid NAH7. Some of the plasmids contained one or two restriction fragments that were slightly altered (shifted, missing, or additional) from the RFLP pattern shared among the naphthalene plasmids (Stuart-Keil et al., 1998). These alterations in restriction fragment lengths correlated well with differences in the overall size of the plasmids (Stuart-Keil et al., 1998).

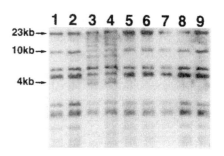

FIGURE 4.3 Southern hybridization of pDTG1 to naphthalene catabolic plasmids retrieved from sediment microorganisms after filter matings with recipient site-derived strain Cg9.CR. The plasmids were cut with *Xho*I and run on a 0.7% agarose gel. Plasmids were isolated from parent strain Cg9 (lane 1) and transconjugant colonies (lanes 2 to 9). (From Stuart-Keil et al., 1998.)

LABORATORY AND FIELD EXPERIMENTS EXAMINING HGT

Although we are aware that transduction and transformation are competing gene transfer mechanisms, all data gathered thus far from our site-specific collection of naphthalene degrading isolates suggest that the mobile element is one or more self-transmissible plasmids. The HGT procedures that have been implemented are analogous to the "fishing expeditions" reported by Bale et al. (1988) and Top et al. (1995, 1996) in which genes from unknown soil microorganisms donated genes to a recipient bacterium that was retrievable from soil. In the laboratory studies by Top et al. (1995, 1996), antibiotic resistant recipient strains were mated with suspensions of soil in microcosms. The mating conditions were on 10% LB agar plates at 30°C overnight. The transconjugants were isolated based on both antibiotic resistance and the catabolic gene (for Top et al. (1995, 1996), this was involved in 2,4-D biodegradation) received from unknown donors in the soil community. Once retrieved, not only was gene exchange

documented but also the mobile element itself, could be characterized further.

Thus, documenting HGT requires retrievable recipient strains. To develop these, all naphthalene-degrading bacteria were repeatedly transferred in LB broth for several months. During this time, most site-derived seep strains, as well as the archetypal naphthalene graders NCIB 9816-4 and PpG7 (Serdar and Gibson, 1989; Yen and Gunsalus, 1982), were cured for their large plasmids (Stuart-Keil et al., 1998). Loss of plasmid correlated with a loss in ability to grow with naphthalene as the sole carbon source. Concurrent loss of the *nahAc* gene was verified by lack of PCR amplification of a 407-bp fragment of *nahAc* in the cured strains, while the parent naphthalene-degrading strains all contained the amplifiable *nahAc* gene. By successively growing spontaneous resistant mutants on increasing concentrations of rifampin, strains that were resistant to rifampin (MIC of greater than 500 µg/ml) were obtained (Stuart Keil et al., 1998). The rifampin-resistant cured strains were then used as recipients in mating experiments. Filter matings between many naphthalene-degrading isolates and their respective cured, rifampin-resistant progeny yielded transconjugant colonies which were able to grow on naphthalene-media amended with rifampin. Plasmid profiles for the donor, recipient, and transconjugant strains clearly demonstrated that the naphthalene catabolic plasmids were donated from donor to recipients and, thus, were self-transmissible (Stuart-Keil et al., 1998).

Microorganisms indigenous to sediment from our Glens Falls, New York, study site were obtained by homogenizing the sediment in a blender with 0.1% $Na_4P_2O_7$. These cells from the sediment were used as donors in filter matings with cured, rifampin-resistant, seep-derived recipient strains. Potential transconjugants were screened to detect the activity of a dioxygenase gene in transforming indole to indigo. Only when one of the strains (Cg9, cured and rif^+) served as a recipient were any indigo-colored putative transconjugant colonies observed. These were picked, restreaked, and verified as transconjugants carrying the *nahAc* gene on a large plasmid. We digested total plasmid extractions from the eight transconjugant strains with *Xho*I, separated the fragments by electrophoresis, and performed Southern analysis with the pDTG1 probe (Figure 4.3). All eight exogenously isolated plasmids (Figure 4.3, lanes 2 to 9) shared the same pattern of nine restriction fragments found in pCg9 (lane 1). Tranconjugant naphthalene-catabolic plasmids in lanes 3 and 4 (Figure 4.3) contained additional restriction fragments. None of the naphthalene-catabolic plasmids previously isolated from our site-derived strains have the same *Xho*I RFLP pattern seen in the transconjugant naphthalene-catabolic plasmids in lanes 3 and 4 (Figure 4.3). Thus, in addition to retrieving the same naphthalene-catabolic plasmid previously isolated from our site-derived strains, our exogenous plasmid isolation procedure retrieved a novel variant of this plasmid. Data in Figure 4.3 clearly show that the microorganisms indigenous to our coal tar-contaminated site possess the self-transmissible naphthalene-catabolic plasmid previously found in site-derived isolates and that its transmission to site-derived recipient strains can be rapidly induced under laboratory conditions (Stuart-Keil et al., 1998).

Eager to extend the laboratory assays of Stuart-Keil et al. (1998) to true *in situ* field experiments, Hohnstock et al. (2000) carried out three field-release mating experiments in which the cured, rifampin-resistant recipient strains were deployed to site sediments enclosed in retrievable open glass cylinders and site groundwater in retrievable membrane-enclosed cartridges. Despite amendments of naphthalene and/or LB broth that accompanied the bacteria inoculated into the study site and substantial survival of the recipients in the field, spontaneous transfer of naphthalene catabolic genes from the native microbial community to added recipient strains could not be detected. However, when both donor (at sufficient titer, $10^7/g$) and recipient cells were mixed in site sediments along with naphthalene, a single transconjugant was obtained in a field experiment (Hohnstock et al., 2000). The results obtained by Hohnstock et al. (2000) underscore the lessons learned by many other investigators (Trevors and van Elsas et al., 1997; Clerc and Simonet, 1998). A wide variety of factors may explain lack of conjugation between the native microbial community and recipient strains under field conditions. We hypothesize that four key interacting factors may prove crucial: (i) type and duration of cell–cell contact, (ii) physiological status of donors and recipients (especially as governed by growth phase carbon amendments), (iii) biochemical/genetic cues for plasmid mobilization, and (iv) post-transfer DNA modification mechanisms of recipients. The means by which naphthalene stimulated HGT (described above) awaits explanation: among the intriguing possibilities are chemotaxis-enhanced cell–cell contact and substrate-activated plasmid conjugation.

FUTURE PROSPECTS: REPORTER CONSTRUCTS TO ASSESS TRANSCRIPTIONAL ACTIVATION OF TRA GENES

There is an immense amount of basic knowledge about genetic exchange that can be used to discover what controls HGT at our specific field site. The broad plan for the future is to continue to move iteratively between laboratory and field experiments to identify two types of cues that govern horizontal gene transfer: intracellular (or genetic) ones and real-world environmental factors. The approach for generating data will implement experiments with archetypal naphthalene-degrading bacteria and site-derived isolates. *lacZ* transcriptional fusions that directly assay activation of the NAH plasmid's transfer function will be prepared.

Within the last year, Dr G. Zylstra (Rutgers University) has sequenced and annotated the entire naphthalene catabolic plasmid (pDTG1) carried by the model naphthalene degrading bacterium *P. putida* NCIB 9816-4. pDTG1 is the plasmid found strongly to resemble (by RFLP analyses and Southern blotting) the plasmids carried by isolates from the coal tar-contaminated study site (see above). The new sequences offer another independent avenue of inquiry. Within the 80 kb sequence of pDTG1, searches for open reading frames and BLAST comparisons have allowed approximate assignments of the function of various genes. There is a cluster of genes with high similarity to the *Agrobacterium* plasmid transfer genes (*tra*) that have been previously characterized (Zhu et al., 2000). The *tra* gene cluster in *Agrobacterium* is carried on the Ti plasmid and codes for proteins critical for initiating the injection of the portion of the Ti plasmid into plant root cells during establishment of crown gall disease (Zhu et al., 2000). Recent discovery of this *tra* gene cluster in pDTG1 is exciting news because it offers a direct way to prepare reporter gene constructs that can probe how plasmid transfer – the basic mechanism of HGT – is regulated in our study site. Using the sequence information available, we have identified a ~2 kb promoter region at the beginning of the *tra* homologue promoter region (coding for plasmid transfer) in pDTG1.

Reporter constructs derived from the naphthalene-degrading archetype strains and our site-derived strains (and, perhaps, nucleic acids extracted from site soils) will allow direct assessment of genetic control of HGT. After completing the reporter assays on a wide variety of constructs, we will investigate many possible environmental, nutritional, and other (quorum sensing?; Dunny and Winans, 1999) factors that regulate the gene transfer operon of bacteria native to our site. The variety of potential key cues for activating gene transfer may include: intermediary metabolites of naphthalene catabolism, O_2, extractable soil factors, naphthalene, cell density, extractable factors from cell cultures, temperature, starvation, carbon source, pilus formation, and cell binding to surfaces. These and others that arise will be tested on mixed microbial communities native to our site. The information obtained from these reporter gene fusion assays will be used to guide field mating assays. By pursuing this two-pronged approach we expect to be able to test a variety of hypotheses about HGT of naphthalene catabolic genes at our study site. In this way, we hope to achieve significant insights into the molecular and ecological mechanisms by which the naturally occurring microbial communities adapt to organic contaminants.

ACKNOWLEDGMENTS

During preparation of this manuscript, the author received support from the National Science Foundation (MCB-0084175), from the New York State Agriculture Experiment Station (189434) and CEFIC LRI Programme. Expert manuscript preparation from Patti Lisk is gratefully appreciated.

REFERENCES

Alexander, M. (1981) Biodegradation of chemicals of environmental concern. *Science* 211: 132–138.

Alexander, M. (1999) *Biodegradation and Bioremediation*, 2nd edn, Academic Press, San Diego, CA.

Atlas, R. and Unterman, R. (1999) Bioremediation. In *Manual of Industrial Microbiology* (eds, A.L. Demain and J.E. Davies), 2nd edn, pp. 668–681. American Society for Microbiology Press, Washington, DC.

Bale, M.J., Fry, J.C. and Day, M.J. (1988) Transfer and occurrence of large Hg resistance plasmids in river epilithon. *Appl. Environ. Microbiol.* 54: 972–978.

Blatny, J.M., Brautaset, T, Winther, L.H.C. et al. (1997) Construction and use of a versatile set of broad-host-range cloning and expression vectors based on the RK2 replicon. *Appl. Environ. Microbiol.* 63: 370–379.

Christensen, B.B., Sterberg, C. and Molin, S. (1996) Bacterial plasmid conjugation on semi-solid surfaces monitored with the green fluorescent protein (GFP) from *Aequorea victoria* as a marker. *Gene* 173: 59–65.

Clerc, S. and Simonet, P. (1998) A review of available systems to investigate transfer of DNA to indigenous soil bacteria. *Antonie van Leeuwenhoek* 73: 15–23.

Cutting, S.M. and Youngman, P. (1994) Gene transfer in Gram-positive bacteria. In *Methods for General and Molecular Bacteriology* (ed., P. Gerhardt), pp. 348–364, American Society for Microbiology, Washington, DC.

Dagley, S. (1983) Biodegradation and biotransformation of pesticide is the earth's carbon cycle. In *Residue Reviews* (eds, F.A. Gunther and J.D. Gunther), vol. 85, pp. 127–137, Springer-Verlag, New York, NY.

Davies, J. (1994) Inactivation of antibiotics and the dissemination of resistance genes. *Science* **264**: 375–382.

Davies, J. (1996) Bacteria on the rampage. *Nature* **383**: 219–220.

Davison, J. (1999) Genetic exchange between bacteria in the environment. *Plasmid* **42**: 73–91.

Dunny, G.M. and Winans, S.C. (1999) *Cell–cell Signaling in Bacteria*. American Society for Microbiology Press, Washington, DC.

EPRI (1996) Characterization and monitoring before and after source removal at a former manufactured gas plant (MGP) disposal site. Electric Power Research Institute, Pleasant Hill, CA. Publ. No. TR-105921.

Fulthorpe, R.R., McGowan, C. , Maltseva, O.V. et al. (1995) 2,4-dichlorophenoxyacetic acid-degrading bacteria contain mosaics of catabolic genes. *Appl. Environ. Microbiol.* **61**: 3274–3281.

Gibson, D.T. (ed.) (1984) *Microbial Degradation of Organic Compounds*. Marcel Dekker, New York, NY.

Goyal, S.M. and Zylstra, G. (1997) Genetics of naphthalene and phenanthrene degradation by *Comamonas testosteroni*. *J. Ind. Microbiol. Biotechnol.* **19**: 401–407.

Heath, J.D., Charles, T.C. and Nester, E.W. (1995) Ti plasmid and chromosomally encoded two-component systems in plant cell transformation by *Agrobacterium* species. In *Two-Component Signal Transduction* (eds, J.A. Hoch and T.J. Silhavy), pp. 367–385, American Society for Washington, Washington, DC.

Heinemann, J. (1998) Looking sideways at the evolution of replicons. In *Horizontal Gene Transfer* (eds, M. Syvanen and C.I. Kado), pp. 11–23, Chapman and Hall, London.

Herrick, J.B. (1995) Detection, divergence, and phylogeny of a naphthalene dioxygenase gene and naphthalene-degrading bacteria native to a coal tar-contaminated site. Ph.D. Thesis. Cornell University.

Herrick, J.B., Madsen, E.L., Batt, C.A. and Ghiorse, W.C. (1993) Polymerase chain reaction amplification of naphthalene catabolic and 16S rRNA gene sequences from indigenous sediment bacteria. *Appl. Environ. Microbiol.* **59**: 687–694.

Herrick, J.B., Stuart-Keil, K.G., Ghiorse, W.C. and Madsen, E.L. (1997) Natural horizontal transfer of a naphthalene dioxygenase gene between bacteria native a contaminated field site. *Appl. Environ. Microbiol.* **63**: 2330–2337.

Hohnstock, A.M., Stuart-Keil, K.G., Kull, E.E. and Madsen, E.L. (2000) Naphthalene and donor cell density influence field conjugation of naphthalene catabolic plasmids. *Appl. Environ. Microbiol.* **66**: 3088–3092.

Jain, R., Rivera, M.C. and Lake, J.A. (1999) Horizontal gene transfer among genomes: The complexity hypothesis. *Proc. Natl Acad. Sci. USA* **96**: 3801–3806.

Ka, J.O. and Tiedje, J.M. (1994) Integration and excision of a 2,4-dichlorophenoxyacetic acid-degradative plasmid in *Alcaligenes paradoxus* and evidence of its natural intergeneric transfer. *J. Bacteriol.* **176**: 5284–5289.

Klingmuller, W., Dally, A., Fentner, C. and Steinlein, M. (1990) Plasmid transfer between soil bacteria. In *Bacterial Genetics in Natural Environments* (eds, J.C. Fry and M.J. Day), pp. 133–151, Chapman and Hall, London.

Kroer, N., Barkay, T., Sorensen, S. and Weber, D. (1998) Effects of root exudates and bacterial metabolic activity on conjugal gene transfer in the rhizosphere of a marsh plant. *FEMS Microbiol. Ecol.* **25**: 375–384.

Leahy, J.G. and Colwell, R.R. (1990) Microbial degradation of hydrocarbons in the environment. *Microbiol. Rev.* **54**: 305–315.

Levy, S.B. and Miller, R.V. (1989) *Gene Transfer in the Environment*. McGraw-Hill, New York, NY.

Lorenz, M.G. and Wackernagel, W. (1993) Bacterial gene transfer in the environment. In *Transgenic Organisms-risk Assessment of Deliberate Release* (eds, W. Wöhrmann and J. Tomiuk), pp. 43–64, Birkhauser-Verlag, Basel.

Madsen, E.L. (1997) Methods for determining biodegradability. In *Manual of Environmental Microbiology* (eds, C.J. Hurst et al.), pp. 709–720, American Society for Microbiology, Washington, DC.

Madsen, E.L., Sinclair, J.L. and Ghiorse, W.C. (1991) *In situ* biodegradation: microbiological patterns in a contaminated aquifer. *Science* **252**: 830–833.

Madsen, E.L., Mann, C.L. and Bilotta, S. (1996a) Oxygen limitations and aging as explanation for the persistence of naphthalene in coal-tar contaminated surface sediments. *Environ. Toxicol. Chem.* **15**: 1876–1882.

Madsen, E.L., Thomas, C.T., Wilson, M.S. et al. (1996b) *In situ* dynamics of aromatic hydrocarbons (AHs) and bacteria capable of AH metabolism in a coal tar waste-contaminated field site. *Environ. Sci. Technol.* **30**: 2412–2416.

Matheson, V.G., Forney, L.J., Suwa, Y. et al. (1996) Evidence for acquisition in nature of a chromosomal 2,4-dichlorophenoxyacetic acid/α-ketoglutarate dioxygenase gene by different *Burkholderia* spp. *Appl. Environ. Microbiol.* **62**: 2457–2463.

Nikolich, M.P., Hong, G., Shoemaker, N.B. and Salyers, A.A. (1994) Evidence for natural horizontal transfer of *tetQ* between bacteria that normally colonize humans and bacteria that normally colonize livestock. *Appl. Environ. Microbiol.* **60**: 3255–3260.

Normander, B., Christensen, B.B., Molin, S. and Kroer, N. (1998) Effect of bacterial distribution and activity on conjugal gene transfer on the phylloplane of the bushbean (*Phaseolus vulgaris*). *Appl. Environ. Microbiol.* **64**: 1902–1909.

Nusslein, K., Maris, D., Timmis, K. and Dwyer, D.F. (1992) Expression and transfer of engineered catabolic pathways harbored by *Pseudomonas* spp. introduced into activated sludge microcosms. *Appl. Environ. Microbiol.* **58**: 3380–3386.

Ochman, H., Lawrence, J.G. and Groisman, E.A. (2000) Lateral gene transfer and the nature of bacterial innovation. *Nature (London)* **405**: 299–304.

Provence, D.L. and Curtiss, R. (1994) Gene transfer in Gram-negative bacteria. In *Methods for General and Molecular Bacteriology* (ed., P. Gerhardt), pp. 317–347, American Society for Microbiology, Washington, DC.

Rivera, M.C., Jain, R., Moore, J.E. and Lake, J.A. (1998) Genomic evidence for two functionally distinct gene classes. *Proc. Natl Acad. Sci. USA* **95**: 6239–6244.

Schenk, S. and Decker, K. (1999) Horizontal gene transfer involved in the convergent evolution of the plasmid-encoded enantioselective 6-hydroxynicotine oxidases. *J. Molec. Evol.* **48**: 178–186.

Schmidt, E.R. and Harikein, T. (eds) (1996) *Transgenic Organisms and Biosafety: Horizontal Gene Transfer, Stability of DNA, and Expression of Transgenes.* Springer-Verlag, Berlin.

Serdar, C.M. and Gibson, D.T. (1989) Isolation and characterization of altered plasmids in mutant strains of *Pseudomonas putida* NCIB 9816. *Biochem. Biophys. Res. Commun.* **164**: 764–771

Simon, M.J., Osslund, T.D., Saunders, R. et al. (1993) Sequences of genes encoding naphthalene dioxygenase in *Pseudomonas putida* strains G7 and NCIB 9816–4. *Gene* **127**: 31–37.

Smets, B.F., Rittmann, G.E. and Stahl, D.A. (1995) Quantification of the effect of substrate concentration on the conjugal transfer rate of the TOL plasmid in short-term batch mating experiments. *Lett. Appl. Microbiol.* **21**: 167–172.

Sobecky, P.A. (1999) Plasmid ecology of marine sediment microbial communities. *Hydrobiologia* **401**: 9–18.

Stotzky, G. (1989) Gene transfer among bacteria in soil. In *Gene Transfer in the Environment* (eds, S.B. Levy and R.V. Miller), pp. 165–222, McGraw-Hill, New York, NY.

Stuart-Keil, K.G., Hohnstock, A.M., Drees, K.P. et al. (1998) Plasmids responsible for horizontal transfer of naphthalene catabolic genes between bacteria at a coal tar-contaminated site are homologous to pDTG1 from *Pseudomonas putida* 9816–4. *Appl. Environ. Microbiol.* **64**: 3633–3640.

Syvanen, M. (1994) Horizontal gene transfer: Evidence and possible consequences. *Annu. Rev. Genet.* **28**: 237–261.

Syvanen, M. and Kado, C.I. (eds) (1998) *Horizontal Gene Transfer.* Chapman and Hall, London.

Top, E.M., Holben, W.E. and Forney, L.J. (1995) Characterization of diverse 2,4-dichlorophenoxyacetic acid degradative plasmids isolated from soil by complementation. *Appl. Environ. Microbiol.* **61**: 1691–1698.

Top, E.M., Malltseva, O.V. and Forney, L.J. (1996) Capture of a catabolic plasmid that encodes only 2,4-dichloro-phenoxyacetic acid: α-ketoglutaric acid dixoygenase (TfdA) by genetic complementation. *Appl. Environ. Microbiol.* **62**: 2470–2476.

Trevors, J.T. and van Elsas, J.D. (1997) Quantification of gene transfer in soil and the rhizosphere. In *Manual of Environmental Microbiology* (eds, C.J. Hurst, G.R. Knudsen, M.J. McInerney et al.), pp. 500–508, ASM Press, Washington, DC.

Trevors, J.T., Barkay, T. and Bourquin, A.W. (1987) Gene transfer among bacteria in soil and aquatic environments: A review. *Can. J. Microbiol.* **33**: 191–198.

van der Meer, J.R., Zehnder, A.J.B. and deVos, W.M. (1991) Identification of a novel composite transposable element, Tn5280, carrying chlorobenzene dioxygenase genes of *Pseudomonas* sp. Strain P51. *J. Bacteriol.* **173**: 7077–7083.

van der Meer, J.R., deVos, W.M., Harayama, S. and Zehnder, A.J.B. (1992) Molecular mechanisms of genetic adaptation to xenobiotic compounds. *Microbiol. Rev.* **56**: 677–694.

van der Meer, J.R., Werlen, C., Nishino, S.F. and Spain, J.C. (1998) Evolution of a pathway for chlorobenzene metabolism leads to natural attenuation in con-taminated groundwater. *Appl. Environ. Microbiol.* **64**: 4185–4193.

Wackett, L.P. and Hershberger, C.D. (2001) *Biocatalysis and Biodegradation.* ASM Press, Washington, DC.

Warren, L. and Koprowski, H. (eds) (1991) *New Perspectives in Evolution.* Wiley-Liss, Inc., New York, NY.

Wiener, P., Egan, S. and Wellington, E.M.H. (1998) Evidence for transfer of antibiotic-resistance genes in soil populations of streptomycetes. *Molec. Ecol.* **7**: 1205–1216.

Wilson, M.S. and Madsen, E.L. (1996) Field extraction of a unique intermediary metabolite indicative of real time *In situ* pollutant biodegradation. *Environ. Sci. Technol.* **30**: 2099–2103.

Wilson, M.S., Bakermans, C. and Madsen, E.L. (1999) *In situ*, real-time catabolic gene expression: Extraction and characterization of naphthalene dioxygenase mRNA transcripts from groundwater. *Appl. Environ. Microbiol.* **65**: 80–87.

Worning, P., Jensen, L.J., Nelsun, K.E. et al. (2000) Structural analysis of DNA sequence: evidence for lateral gene transfer in *Thermotoga maritima. Nucl. Acids Res.* **28**: 706–709.

Yen, K.M. and Gunsalus, I.C. (1982) Plasmid gene organization: naphthalene/salicylate oxidation. *Proc. Natl Acad. Sci. USA* **79**: 874–878.

Young, L. and Cerniglia, R.S. (eds) (1995) *Microbial Transformation and Degradation of Toxic Organic Chemicals.* John Wiley and Sons, New York, NY.

Zhou, J.-Z. and Tiedje, J.M. (1995) Gene transfer from a bacterium injected into an aquifer to an indigenous bacterium. *Molec. Ecol.* **4**: 613–618.

Zhu, J., Oger, P.M., Schrammeijer, B. et al. (2000) The bases of crown gall tumerogenesis. *J. Bacteriol.* **182**: 3885–3895.

CHAPTER 5

Horizontal Transmission of Genes by *Agrobacterium* Species

Clarence I. Kado

Agrobacterium tumefaciens and *A. rhizogenes* represent real-life examples of organisms that horizontally transfer genes to eukaryotes using a promiscuous gene delivery system. The gene transfer machinery reminds us of the way mosquitoes, while feeding fortuitously, pick up genes of other organisms and transmit them into others. The T-DNA transport apparatus is composed of VirB proteins whose amino acid sequence remains highly conserved among various pathogenic bacteria. T-DNA also requires the T-pilus to transform host cells, like the mosquito the T-pilus may serve as the proboscis to deliver the T-DNA and perhaps acquire novel host genes during the process. Retrotransfer of genes by *Agrobacterium* species could account for the presence of *rol* and *ros* genes on the Ti plasmid and on the chromosome of this bacterial species. Based on amino acid sequence motif of the C2H2 zinc finger in Ros regulatory protein, the *ros* gene might have originated from a marine organism rather than from a plant.

INTRODUCTION

The transmission of foreign genes between organisms has been long known. Many instances of the transfer of microorganisms such as bacteria, protozoans and worms into higher organisms is an established phenomenon occurring continuously in nature. Depending on the biological system that has evolved, the transmitted microbe may or may not survive in the foreign environment of the recipient cell. DNA released by cell lysis is usually one circumstance detrimental to the in-coming microbe but results in the opportunity for its DNA, if not degraded, to become established in the genome of the host cell. In each case of microbe transfer, a specialized machinery to deliver the microorganism is required. One classic example is the mosquito that is equipped with a needle-like proboscis to facilitate access to blood vessels by first injecting anti-clotting proteins before sipping up blood for its meal (Figure 5.1 (see color plates)). Infected mosquitoes serve as the carrier as well as the transmitter of the microorganism and viruses. Establishment of infection will depend on the degree of host specificity of the microorganism transmitted. Dissemination of a given microorganism into non-host provides opportunities for the released DNA to become integrated into the genome of the recipient. Although purely speculative, prolong exposure by these opportunities could result in the integration and eventual employment of novel genes.

The above example deals with higher organisms transmitting microorganisms into cells of primarily eukaryotes. The following example, which is often overlooked, is a well established case of horizontal gene transfer between prokaryote and eukaryote occurring presently in nature. *Agrobacterium tumefaciens* and *A.*

Copyright © 2002 by Academic Press.
All rights of reproduction in any form reserved.

rhizogenes represent two root-associated (rhizo-plane) bacterial residents that transfer and integrate their genes into higher organisms such as plants, fungi and actinomycetes. This chapter focuses briefly on the mechanism used by *A. tumefaciens* to deliver DNA into plants, the nature of the genes constituting the DNA transport system, and the identity of genes transferred and acquired by *A. tumefaciens* through evolution. The chapter will begin with a reflection on a previous hypothesis that the large Ti and Ri plasmids resident in *A. tumefaciens* and in *A. rhizogenes*, respectively, represent selfish DNA that originated from a simple origin of DNA replication into a macromolecule containing numerous genes that confer properties essential for plasmid–host survival (Kado, 1998).

THE TI PLASMID IS A SELFISH DNA

Ti (tumor-inducing) and Ri (root-inducing) plasmids are large (95–280 kilobase pairs long) covalently-closed replicons occurring in all virulent strains of *A. tumefaciens* and *A. rhizogenes*, respectively. Ti plasmids also exist in other *Agrobacterium* species such as *A. vitis*. Ti and Ri plasmids confer on the *Agrobacterium* the molecular machinery to transfer and integrate oncogenes into the genome of their hosts. The products of the oncogenes catalyze the synthesis of auxin (indole-3-acetic acid) and cytokinin, plant growth hormones that stimulate hypertrophy and hyperplasia of transformed cells, culminating in the formation of large tumors and prolific growth of roots known as hairy root. Oncogenes are located in the T-DNA (transferred DNA), which are specific sectors on the Ti and Ri plasmids. Besides oncogenes, opine synthesizing genes are also located on the T-DNA. Opines are amino acid-organic acid conjugates and phospho-disaccharides that are mainly utilized by *Agrobacterium* species. The tumors and hairy roots abundantly produce opines and thus provide a specific environment for *A. tumefaciens* and *A. rhizogenes*. The opine catabolic genes on the Ti plasmid confer chemotaxis to their cognate substrates (Kim and Farrand, 1998).

The survival of *A. tumefaciens* and *A. rhizogenes* ensures the survival of the Ti and Ri plasmids, respectively. These plasmids reside as single copy macromolecules that replicate with the bacterial chromosome. Adverse environmental effects leading to cell death promotes the conjugative transfer of the Ti and Ri plasmids into recipient bacteria. This reaction could be considered a valuable escape mechanism for Ti/Ri plasmid survival. Both plasmids contain all of the genetic information for conjugation and delivery of the entire plasmid molecule into the recipient cell.

DEVELOPMENT OF A DNA DELIVERY SYSTEM

The transfer of DNA between bacteria has long been thought to require the close pairing of donor and recipient cells by "specific pair formation," whereby donor and recipient are resistant to separation upon dilution, leading to "effective pair formation," whereby a conjugal bridge is established (Curtiss, 1969), the combination of which has been later termed "mating pair formation" (Mpf) (Lessl et al., 1993; Anthony et al., 1999). The exact makeup of the Mpf is not clear, but a number of plasmid-borne genes are needed for this function. Basically, specific pairing occurs via the generation and attachment of a conjugative pilus, followed by pilus retraction, presumably leaving an exit pore for the DNA to travel from the donor and into the recipient cell. Exactly how the DNA is moved and what type of the system is used to transfer the DNA into the recipient cell has never been clearly resolved and has remain speculative even though several models have been proposed, basically using *E. coli* plasmids, in particular F plasmid (Curtiss, 1969; Willetts and Skurray, 1987; Wilkins and Lanka, 1993).

In the case of T-DNA transmission, the DNA transfer machinery is likely to be distinct from F-mediated conjugative transfer that occurs in *E. coli* and related species. As with the F- and R-mediated plasmid systems, a pilus is generated in the T-DNA transfer system. T-DNA transfer requires a pilus, called the T-pilus (Lai and

Kado, 1998) that is generated when *virB* genes are induced by plant signal molecules. There are eleven genes that comprise the *virB* operon and these genes are needed for T-pilus biogenesis as well as T-DNA transfer. Five of the eleven VirB proteins encoded by the *virB* genes make up the DNA transfer apparatus while VirB4 and VirB11 are ATPases and VirB2 is the propilin protein that is processed into a cyclic T-pilin subunit (Eisenbrandt et al., 1999). VirB1 bearing a transglycosylase-like sequence, VirB3 and VirB5 may be part of the chaperone/usher proteins to stabilize the VirB6-10 proteins escorted to the respective membrane location. As recently predicted (Kado, 2000), VirB4 and VirB11 may form pentameric rings within the transport pore of T-DNA transporter.

Interestingly, the VirB proteins display amino acid sequence homologies to a number of proteins used in the formation of the protein/nucleic acid secretion systems in various pathogenic bacteria, including Rickettsiae (Christie and Vogel, 2000; Lai and Kado, 2000). The proteins secreted are generally involved in facilitating infection, while the secretion of nucleic acid is likely to establish residency of the nucleic acid itself in a new cell and to expand its territory. The *virB* gene homology extends beyond sequence similarities to similarities in gene order, suggesting that the array of genes involved in coding the protein/nucleic acid secretion system were derived as a cluster from a common ancestor. Subsequent small mutations were necessary for effective adaptation of the organism to its environment; hence, some *virB* genes were less conserved than others in various bacterial species that interact with a host (Figure 5.2 (see color plates)).

Plasmid-mediated transfer of DNA requires a conjugative pilus. For example, in the absence of F-pili, conjugative transfer of DNA does not occur. Likewise, T-DNA transfer requires the T-pilus since, in the absence of T-pili, *A. tumefaciens* is avirulent. It is therefore apparent that conjugative pili are essential for DNA transfer. However, it remains unclear on whether the conjugative pili serve as a pipeline for DNA to transit from the donor into the recipient cell. Electron microscopic examination of F- and T-pili has revealed the presence of a lumen

that is wide enough to accommodate the single-stranded (ss) DNA strand complexed with a pilot protein. This tubular filament might serve as the conduit for transferring DNA. However, direct evidence is needed to show the presence of DNA along the length of the pilus. Acquiring this is predicted to be a difficult task since the delivery of ssDNA occurs in a short time. Capturing the DNA transfer event is therefore difficult.

HORIZONTAL RETROTRANSFER OF EUKARYOTIC GENES INTO *AGROBACTERIUM*

Using the analogy of the mosquito, which feeds on blood but also fortuitously takes up as well as inoculates pathogenic microorganisms or viruses into the host, the machinery used to transfer T-DNA into the host may also take up by chance novel DNA and proteins from the host. Thus, although the transmission of the T-DNA from *A. tumefaciens* to plants is unidirectional, the same DNA delivery machinery might also pick up ancillary DNA from the host, i.e., the transfer machinery originally thought to be delegated for transferring the T-DNA into the recipient host might also pick up recipient DNA and transfer it back into the donor *Agrobacterium* cell. Thus, eukaryotic genes might have been acquired by a retrotransfer mechanism. Evidence for the presence of eukaryotic genes in *A. tumefaciens* and in *A. rhizogenes* is provided as follows.

Southern hybridization analysis of plant genomes revealed the presence of DNA sequences that are quite similar to genes present in the T_L-DNA of Ri plasmids (Spanò et al., 1982; White et al., 1982; Furner et al., 1986; Aoki et al., 1994, Meyer et al., 1995). In Ri plasmids, the T-DNA is split into two regions, T_L- and T_R-DNA, of which both are integrated into the genome of the host cells. Both regions contribute to the hairy root phenotype. The genes of the T_R-DNA code for the biosynthesis of auxin and cytokinin, whereas the genes of the T_L-DNA are needed for hairy root production and apparently play a negative regulatory role during pathogenesis (Lemcke and Schmulling, 1998). T-DNA homologues were found in *Nicotiana glauca*, *Nicotiana tabacum*, *Petunia* and carrot (Spanò et al., 1982; Furner et al., 1986). In *N. glauca* and in *N. tabacum*, the homologues *NgrolB*, *NgrolC*, *Ngorf13* and *Ngorf14*, and *trolB*, *trolC* and *torf13* were found by cloning and sequencing genomic DNA (Furner et al., 1986; Aoki et al., 1994; Meyer et al., 1995; Fründt et al., 1998; Aoki and Syno, 1999). In their respective *Nicotiana* species, these genes are present in the same arrangement as that in the core region of the T_L-DNA of the Ri plasmid (Fründt et al., 1998a,b) (Figure 5.3 (see color plates)). The predicted amino acid sequences of *rolB*, *rolC* and *orf13* homologues show identities ranging from 55–89% (Fründt et al., 1998a,b), indicating high degrees of sequence conservation at the DNA level for all three genes studied. As seen in Figure 5.3, *orf13*, *orf14* and *rolC* are translocated and inverted into the left side of the core region, but the gene order remains preserved. Through an evolutionary time span, the Ri plasmid could have sequestered a cluster of *rol* genes during interactions between *A. rhizogenes* and its plant host. The *rol* genes possess eukaryotic promoters recognized by the plant transcription machinery (Slightom et al., 1986; Yokoyama et al., 1994; Elmayan and Tepfer, 1995; Nagata et al., 1995; Hansen et al., 1997; Maeda et al., 1999). Also, an intron exists in *rolA* that is transcribed both in plants (Magrelli et al., 1994) and in bacteria (Pandolfini et al., 2000). These findings suggests *rol* genes could have been acquired via retrotransfer from the plant host. On the other hand, plant homologues of *rol* genes have only been found so far in tobacco, carrot and petunia species. The host range of *A. rhizogenes* is relatively broad, especially infecting woody crop plants; therefore, *rol* homologues could be more widely distributed than currently believed.

Besides the presence of *rol* homologues in plants, a regulatory gene called *ros* is present in the chromosomes of *Agrobacterium*, *Rhizobium* and *Sphingomonas* species (Bouhouche et al., 2000). Ros codes a 15 kDa protein that downregulates the expression of seven virulence genes and the T-DNA oncogene, *ipt*, in *A. tumefaciens*. Ros has several interesting features. It contains a C2H2 zinc finger that is essential for Ros to bind to an operator in the promoters of *vir* and *ipt* genes (Chou et al., 1998). The *ipt* gene is downregulated by Ros which binds to its apparent regulatory sequence recognized by the plant transcriptional machinery. Ros contains the unusual C2H2 zinc finger that is commonly found in animals and has the amino acid sequence motif of the finger with homologues in animals rather than in plants (Bouhouche et al., 2000) (Figure 5.4 (see color plates)). Analysis of the C2H2 zinc finger motifs of different species also suggested closer ties of the Ros gene to animals than plants (Figure 5.5 (see color plates)). The Japanese puffer fish, *Fugu rubripes*, is the closest non-bacterial species that contains the Ros homologue, suggesting that *A. tumefaciens* might have acquired the *ros* gene from a marine organism rather than a terrestrial source. It should prove very interesting to find the *ros* gene in marine organisms harboring marine species of *Agrobacterium*.

RESULTS AND DISCUSSION

That the *rol* genes originally found in the T_L-DNA of the Ri plasmid also occur in certain plants provides an argument in support of horizontal gene transfer. The direction of the *rol* gene flow is unknown, but it can be argued that since the *rol* genes codes root stimulating activity, the *rol* genes originated from the plant. On the other hand, since the *rol* genes are only found in certain plant species and not in others that are also hosts of *A. rhizogenes*, the *rol* genes must have come from *A. rhizogenes* or an earlier relative of this organism.

The origin of the *ros* gene would argue against the possibility of a plant origin for the *rol* gene. The C2H2 zinc finger, and its amino acid sequence motif of the finger closely akin to the zinc finger motifs in animals rather than in plants, suggests that the ancestor of *ros* is from an animal source. The distinct difference between the potential origins of *rol* and *ros* is reflected by their location within *Agrobacterium* species. The *rol* gene is a Ti plasmid gene whereas *ros* has a chromosomal location. With its plasmid conjugative transfer genes in the *tra* operons intact, the Ti plasmid can readily move about among different species. Hence, the *rol* gene might have been captured from a long-term inter-relationship with certain plant species such as wild-tobacco (*Nicotiana glauca*). In contrast, the *ros* gene codes a product that regulates Ti plasmid genes. Such a molecular relationship would suggest that *ros* was originally on the Ti plasmid, but long-term association of the Ti plasmid with *Agrobacterium* resulted in the transfer of *ros* to the chromosome. Molecular sequence and phylogenetic tree analysis of the C2H2 zinc finger motif commonly found in regulatory proteins in animals indicate Ros might have originated from a marine organism (Bouhouche et al., 2000). The *ros* gene is found in several plant root-associated bacteria and therefore may have radiated the *ros* gene into them following acquisition from the marine source.

That the Ti plasmid is equipped to deliver part of its DNA into foreign cells reflects the molecular positioning of the T-DNA to encode materials (opines) essential for the survival of the bacterium to ensure survival of the mother Ti plasmid. The DNA transmission machinery originally made to deliver DNA unidirectionally may have acquired DNA from the recipient. The bidirectional exchange of DNA would explain the presence of eukaryotic genes located on the Ti plasmid and on the *Agrobacterium* chromosome.

ACKNOWLEDGMENTS

I thank Roy Curtiss III for critical reading of the manuscript.

REFERENCES

Anthony, K.G., Klimke, W.A., Manchak, J. and Frost, L.S. (1999) Comparison of proteins involved in pilus synthesis and mating pair stabilization from the related plasmids F and R100-1: insights into the mechanism of conjugation. *J. Bacteriol.* **181**: 5149–5159.

Aoki, S., Kawaoka, A., Sekine, M. et al. (1994) Sequence of the cellular T-DNA in the untransformed genome of *Nicotiana gauca* that is homologous to ORFs 13 and 14 of the Ri plasmid and analysis of its expression in genetic tumours of *N. glauca* X *N. langsdorfii*. *Mol. Gen. Genet.* **243**: 706–710.

Aoki, S. and Syno, K. (1999) Horizontal gene transfer and mutation: ngrol genes in the genome of *Nicotiana glauca*. *Proc. Natl Acad. Sci. USA* **96**: 13229–13234.

Bouhouche, N., Syvanen, M. and Kado, C.I. (2000) The origin of prokaryotic C2H2 zinc finger regulators. *Trends Microbiol.* **8**: 77–81.

Chou, A.Y., Archedacon, J. and Kado, C.I. (1998) *Agrobacterium* transcriptional regulator Ros is a prokaryotic zinc finger protein that regulates the plant oncogene *ipt*. *Proc. Natl Acad. Sci. USA* **95**: 5293–5298.

Christie, P.J. and Vogel, J.P. (2000) Bacterial type IV secretion: conjugation systems adapted to deliver effector molecules to host cells. *Trends Microbiol.* **8**: 354–360.

Curtiss, III R. (1969) Bacterial conjugation. *Annu. Rev. Microbiol.* **23**: 69–136.

Eisenbrandt, R., Kalkum, M., Lai, E.-M. et al. (1999) Conjugative pili of IncP plasmids, and the Ti plasmid T pilus are composed of cyclic subunits. *J. Biol. Chem.* **274**: 22548–22555.

Elmayan, T. and Tepfer, M. (1995) Evaluation in tobacco of the organ specificity and strength of the *rolD* promoter, domain A of the 35S promoter and the 35S2 promoter. *Transgenic Res.* **4**: 388–396.

Fründt, C., Meyer, A.D., Ichikawa, T. and Meins, F. Jr. (1998a) Evidence for the ancient transfer of Ri plasmid T-DNA genes between bacteria and plants. In *Horizontal Gene Transfer* (eds, M. Syvanen and C.I. Kado), pp. 94–117, Chapman & Hall, London.

Fründt, C., Meyer, A.D., Ichikawa, T. and Meins, F. Jr. (1998b) A tobacco homologue of the Ri-plasmid orf13 gene casues cell proliferation in carrot root discs. *Mol. Gen. Genet.* **259**: 559–568.

Furner, I.J., Huffman, G.A., Amasino, R.M. et al. (1986) An *Agrobacterium* transformation in the evolution of the genus *Nicotiana*. *Nature* **319**: 422–427.

Hansen, G., Vaubert, D., Clerot, D. and Brevet, J. (1997) Wound-inducible and organ-specific expression of ORF13 from *Agrobacterium rhizogenes* 8196 T-DNA in transgenic tobacco plants. *Mol. Gen. Genet.* **254**: 337–343.

Kado, C.I. (1998) Origin and evolution of plasmids. *Antonie van. Leeuwenhoek* **73**: 117–126.

Kado, C.I. (2000) The role of the T-pilus in horizontal gene transfer and tumorigenesis. *Curr. Opin. Microbiol.* **3**: 643–648.

Kim, H. and Farrand, S.K. (1998) Opine catabolic loci from *Agrobacterium* plasmids confer chemotaxis to their cognate substrates. *Mol. Plant Microbe Interact.* **11**: 131–143.

Lai, E.-M. and Kado, C.I. (1998) Processed VirB2 is the major subunit of the promiscuous pilus of *Agrobacterium tumefaciens*. *J. Bacteriol.* **180**: 2711–2717.

Lai, E.-M. and Kado, C.I. (2000) The T-pilus of *Agrobacterium tumefaciens*. *Trends Microbiol.* **8**: 361–369.

Lemcke, K. and Schmulling, T. (1998) Gain of function assays identify non-*rol* genes from *Agrobacterium rhizogenes* TL-DNA that alter plant morphogenesis or hormone sensitivity. *Plant J.* **15**: 423–433.

Lessl, M., Balzer, D., Weyrauch, K. and Lanka, E. (1993) The mating pair formation system of plasmid RP4 defined by RSF1010 mobilization and donor-specific phage propagation. *J. Bacteriol.* **175**: 6415–6425.

Maeda, Y, Moriguchi, K., Kataoka, M. et al. (1999) Genome structure of Ri plasmid (2). Sequencing analysis of T-DNA and its flanking regions of pRi1724 in Japanese *Agrobacterium* strains. *Nucleic Acids Symp. Ser.* **42**: 67–68.

Magrelli, A., Langenkemper, K., Dehio, C. et al. (1994) Splicing of the *rolA* transcript of *Agrobacterium rhizogenes* in *Arabidopsis. Science* **266**: 1986–1988.

Meyer, A.D., Ichikawa, T. and Meins, F. Jr. (1995) Horizontal gene transfer: regulated expression of a tobacco homologue of the *Agrobacterium rhizogenes rolC* gene. *Mol. Gen. Genet.* **249**: 265–273.

Nagata, N., Kosono, S., Sekine, M. et al. (1995) The regulatory functions of the *rolB* and *rolC* genes of *Agrobacterium rhizogenes* are conserved in the homologous genes (*Ngrol*) of *Nicotiana glauca* in tobacco genetic tumors. *Plant Cell Physiol.* **36**: 1003–1012.

Pandolfini, T., Storlazzi, A., Calabria, E. et al.(2000) The spliceosomal intron of the *rolA* gene of *Agrobacterium rhizogenes* is a prokaryotic promoter. *Mol. Microbiol.* **35**: 1326–1334.

Slightom, J.L., Durand-Tardif, M., Jouanin, L. and Tepfer, D. (1986) Nucleotide sequence analysis of TL-DNA of *Agrobacterium rhizogenes* agropine type plasmid. Identification of open reading frames. *J. Biol. Chem.* **261**: 108–121.

Spano, L., Pomponi, M. and Costantino, P. et al. (1982) Identification of T-DNA in the root-inducing plasmid of the agropine type Agrobacterium rhizogenes 1855. *Plant Mol. Biol.* **1**: 291–300.

White, F.F., Ghidossi, G., Gordon, M.P. and Nester, E.W. (1982) Tumor induction by *Agrobacterium rhizogenes* involves transfer of plasmid DNA to the plant genome. *Proc. Natl Acad. Sci. USA* **79**: 3193–3197.

Wilkins, B. and Lanka, E. (1993) DNA processing and replication during plasmid transfer between Gram-negative bacteria. In *Bacterial Conjugation* (ed., D.B. Clewell), pp. 105–136, Plenum Press, New York.

Willetts, N. and Skurray, R. (1987) Structure and function of the F factor and mechanism of conjugation. In *Escherichia coli and* Salmonella typhimurium, *cellular and molecular biology* (eds, F.C. Neidhardt, J.L. Ingraham, K.B. Low et al.), pp. 1110–1133, American Society for Microbiology, Washington, DC.

Yokoyama, R. Hirose, T., Fujii, N. et al. (1994) The *rolC* promoter of *Agrobacterium rhizogenes* Ri plasmid is activated by sucrose in transgenic tobacco plants. *Mol. Gen. Genet.* **244**: 15–22.

CHAPTER 6

Horizontal Transfer of Proteins Between Species: Part of the Big Picture or Just a Genetic Vignette?

Richard J. Weld and Jack A. Heinemann

Horizontal gene transfer (HGT) normally conjures an image of DNA molecules exchanged between individuals not necessarily related to a common parent. Although this image is true for some gene transfer events, the process is complex, probably involving many different kinds of molecules. Macro-molecules besides nucleic acids can influence inheritance, thus, HGT must be considered as a mechanism for the dissemination of other macro-molecules besides DNA. In this chapter we discuss the role of proteins in DNA transfer systems and discuss the potential for multi-material genes to participate in HGT.

The concept that proteins have genetic qualities was resurrected by the discovery of prions. Prions have initiated a broader debate on whether proteins could indeed be an alternative fabric of heritable information in a way that is meaningfully analogous to DNA (Heinemann and Roughan, 2000). Indeed, prions have challenged the simple model that all genes are DNA (Campbell, 1998; Prusiner, 1998). The genetic molecules known as epigenes, which include proteins, are thus gaining a new credibility in current efforts to understand the molecular fabric of all genes (Klar, 1998).

In this chapter, we focus on the impact of proteins transferred together with nucleic acids. Sometimes transferred proteins can be assigned a specific role in DNA transmission to recipients, at other times protein transfer has the same, or nearly the same, implications as the transfer of DNA genes. We will not discuss the relationship between protein secretion systems and conjugation. The reader is directed to Chapter 1 by Ferguson and Heinemann for a coverage of that topic.

PROTEINS THAT PARTICIPATE IN DNA TRANSMISSION

Protein transfer between prokaryotes

Inter-specific and intra-specific DNA transfer can occur by plasmid-mediated conjugation (Heinemann, 1991). Conjugation is a process of plasmid DNA transfer between bacterial cells and the cells of other organisms that requires contact between donor and recipient (Heinemann, 1992; Zechner et al., 2000). Plasmid DNA transfer is initiated by strand-and-site-specific cleavage followed by strand displacement and replacement strand synthesis from the origin of transfer. The displaced plasmid DNA strand is then transferred to the recipient. It is now well established that during the process of conjugation, specific proteins are transferred from the donor to recipient (Rees and Wilkins, 1989, 1990; Heinemann, 1999).

Copyright © 2002 by Academic Press.
All rights of reproduction in any form reserved.

Early molecular experiments were unable to detect general transfer of proteins from donors to recipients during conjugation (Silver and Ozeki, 1962; Rosner et al., 1967). From such experiments researchers concluded that conjugation probably resulted in DNA transfer exclusively (Rosner et al., 1967). However, other molecular and genetic studies have demonstrated that specific proteins are transferred during conjugation (Fisher, 1962; Chatfield et al., 1982; Merryweather, 1986a; Rees and Wilkins, 1989; Heinemann, 1999). Two of these proteins (Pri and Sog primases) are directly involved in conjugation and are active in recipients (Chatfield et al., 1982). Transmission of RecA and the λ immunity protein, for which no clear role in conjugation has been defined, has also been demonstrated (Fisher, 1962; Heinemann, 1999).

Sog primase

A 210 kDa Sog primase is encoded by the *sog* gene on the IncI plasmid pColIb. Sog primase is required during transfer of pColIb for efficient DNA synthesis on the transferred strand within the recipient (Chatfield et al., 1982). Transfer of Sog primase from donor to recipient during conjugation has been inferred from measurements of chromosomal DNA synthesis in recipients (Chatfield and Wilkins, 1984). When recipient cells, lacking RNA primase activity and thus unable to synthesize chromosomal DNA, were mated with donor cells carrying an IncI plasmid encoding Sog primase, chromosomal DNA synthesis in recipient cells was observed. In contrast, little conjugation-dependent DNA synthesis in recipients was observed when they were mated with donors containing a *sog⁻* IncI plasmid. These results suggest that the observed DNA synthesis depended on the transfer of Sog protein during conjugation. However, this experimental system could not exclude the possibility that the observed chromosomal DNA synthesis occurred from oligoribonucleotides that were synthesized in the donor cells and then transferred to the recipient cells during conjugation (Chatfield and Wilkins, 1984).

Transfer of radiolabeled Sog primase from donor to recipient cells during conjugation has since been demonstrated (Merryweather et al.,

1986b; Rees and Wilkins, 1989). Proteins within donors were labeled with [^{35}S]-methionine prior to conjugation. After conjugation, proteins were extracted specifically from recipients, size fractionated and [^{35}S]-methionine-labeled proteins were visualized by autoradiography. A band on the autoradiograph corresponded to the known size of the Sog primase protein. In addition, approximately 20 other bands which were not representative of donor proteins (and therefore not merely contaminants from lysed donors) were detected (Merryweather et al., 1986b). The autoradiograph results suggest that a specific group of proteins, including the Sog primase, were transferred during conjugation. When the donors contained a plasmid encoding Sog, 0.9% of the radioactivity incorporated in donors was detected in recipients. In a control mating with plasmid-free donors, only 0.1% of the incorporated radioactivity was detected in recipients, suggesting that the transfer of proteins depended on conjugation. By comparing the radioactivity of labeled Sog protein in recipients with the radioactivity of Sog from an equivalent number of donors, it was estimated that at least 20% of the large Sog polypeptide in the donors (~250 molecules per bacterium) was transferred to the recipients (Rees and Wilkins, 1989). Sog primase was also detected in recipients by Western blotting using antiserum raised against Sog primase. Conversely, when Sog protein expressed in recipients was labeled, it was not detected by autoradiography in the donors following conjugation, suggesting that Sog protein transfer is unidirectional (Rees and Wilkins, 1989).

Recent work revealed that transfer of Sog primase was independent of DNA transfer but depended on authentic conjugation (Wilkins and Thomas, 2000). In an experimental system similar to that used by Chatfield and Wilkins (1984), primase activity was detected in recipients after matings with donors carrying a transfer-defective (Δnic) ColIb plasmid (Wilkins and Thomas, 2000). The transferred primase activity detected in recipients was approximately half that detected during DNA transfer-proficient conjugation. Two genetic experiments indicated that DNA-independent transfer of primase depended on conjugation. First, primase transfer was blocked by the presence, in recipients, of the CoIIb exclusion gene (*eex*). The mechanism by which *eex* blocks conjugation is unknown (Wilkins and Thomas, 2000). Secondly, the thin I1 conjugation pilus, which is required for mating contact between cells in liquid medium, was also required for DNA-independent transfer of primase activity in liquid medium.

Pri primase

Rees and Wilkins (1990) examined protein transfer during transmission of IncP and IncF plasmids. Proteins were labeled with [^{35}S]-methionine in donors prior to mating. After mating, proteins extracted from recipients were size fractionated and autoradiography was used to detect labeled proteins. More than 20 labeled proteins were transferred from the donors to the recipients during conjugation. However, a 120 kDa protein was the only detectable donor protein that was transferred to recipients specifically as a result of DNA transfer. When DNA transfer was blocked by phage T6, multiple other labeled proteins, but not the 120 kDa protein, were transferred to recipients. By mutation of the transferred plasmid, Rees and Wilkins (1990) demonstrated that the transferred protein was the product of the IncP plasmid-encoded *pri* gene. *pri*, like *sog*, is a transfer gene that encodes a protein with primase activity that is involved in the establishment of transferred plasmids in recipient cells (Lanka and Barth, 1981; Merryweather et al., 1986a). The transferred Pri protein was found in the cytoplasm of recipients, as determined by cell fractionation and an osmotic shock assay. In contrast to the results obtained with IncP plasmids, Rees and Wilkins (1990) did not observe protein transfer to the cytoplasm of recipients during transfer of an IncF plasmid. As observed after IncP plasmid transfer, conjugation resulted in the transfer of multiple proteins from IncF plasmid donors to the membrane fraction of recipients.

Protein transfer between eukaryotes

Virus transmission, whether through the extracellular infection cycle or other transmission routes, is the most direct means for proteins to transfer between eukaryotes. Other means include direct cell uptake or cell fusion. Tobacco mosaic virus (TMV) transmission through plasmodesmata demonstrate the role of protein in transmitting viral nucleic acids between eukaryotic cells. Yeast prions that can be transmitted during the yeast mating process and subsequently distributed to meiotic offspring exemplify the genetic capacity of proteins transmitted between eukaryotes.

Protein escorts

Protein chaperones may be involved in the intercellular transport of nucleic acids within plants. The transfer of TMV from infected cells to contiguous cells requires a virus-encoded movement protein (MP) (Deom et al., 1987; Meshi et al., 1987). One model for TMV intercellular transport is that MP binds to, and unfolds viral RNA, transports the RNA to and through, the plasmodesmata (Citovsky and Zambryski, 1993).

This protein chaperone model of TMV transport has some experimental support. First, MP-mutant TMV are specifically deficient for intercellular movement (Deom et al., 1987; Meshi et al., 1987). Secondly, MP binding to ssDNA and RNA has been demonstrated by gel-mobility-shift assays (Citovsky et al., 1990). High affinity MP binding is sequence non-specific and cooperative (Citovsky et al., 1990). Thirdly, expression of the MP gene within plant cells increases the size exclusion limit of the plasmodesmata from less than 3900 daltons to over 9400 daltons (Wolf et al., 1989). Finally, MP is localized to the periphery of protoplasts (Mas and Beachy, 1998) and accumulates in the plasmodesmata of leaf tissue (Tomenius et al., 1987).

A protein chaperone model has also been suggested for the intercellular transport of post-transcriptional gene silencing (PTGS) signals within plant tissue (Vionnet et al., 1998). Gene silencing is transmissible from cell to cell and may involve the intercellular movement of small nucleic acids (Vionnet et al., 1998). PTGS is associated with the presence of 25-nucleotide RNA molecules antisense to the silenced gene

(Hamilton and Baulcombe, 1999). The transmitted silencing signal may be a nucleic acid–protein complex incorporating the 25-nucleotide antisense RNA and a protein chaperone that moves from cell to cell via plasmodesmata (Hamilton and Baulcombe, 1999).

Protein transfer from prokaryotes to eukaryotes

T-DNA transport

A. tumefaciens is a plant pathogen that can transfer DNA to plant and fungal cells (Koukolikova-Nicola et al., 1985; Bundock et al., 1995) (Figure 6.1 (see color plates)). T-DNA transfer from bacteria to fungi and plants is probably functionally analogous to conjugation between bacteria (Stachel and Zambryski, 1986; Heinemann, 1991; Beijersbergen et al., 1992). The transferred DNA (T-DNA) is defined by 25 bp direct repeats (left and right border sequences) and is located on a large, extra-chromosomal Ti plasmid. The Ti plasmid also contains the virulence (*vir*) genes involved in T-DNA transfer (Stachel and Nester, 1986). The *vir* region contains at least 24 *vir* genes (Sonti et al., 1995; Gelvin, 2000). Proteins encoded by the *virD*, *virE* and *virF* genes are transferred to the recipient plant.

The T-DNA is thought to be transferred as single-stranded (ss) DNA since only ssT-DNA molecules have been detected in plant cells co-cultivated with *A. tumefaciens* (Tinland et al., 1994; Yusibov et al., 1994). VirD2 is bound to the 5' end of ssDNA of the T-DNA generated in *A. tumefaciens* and *E. coli* (Herrera-Estrella et al., 1988; Ward and Barnes, 1988; Young and Nester, 1988; Yusibov et al., 1994). Based on the *in vitro* ssDNA-binding properties of VirE2, it was initially thought that VirE2 protein may also bind to the T-DNA either in the recipient cell or prior to transfer (Citovsky et al., 1997). Recent studies have suggested that VirE2 is exported independently of the T-DNA and therefore VirE2 most likely coats the ssT-DNA in the plant cell rather than in the bacterium (Lee et al., 1999). Indirect evidence suggests that nuclear localization sequences (NLS) on VirE2 and VirD2 help transport the ssT-DNA to the nucleus after co-cultivation (Citovsky et al., 1992,

1994; Rossi et al., 1993, 1996; Ziemienowicz et al., 1999). The importance of NLS for T-DNA delivery to the nucleus is unclear because protein-free DNA introduced into eukaryotic cells will be efficiently transported to the nucleus. However, the putative escort protein may ensure the integrity of the 5' end of the T-DNA and may specifically catalyse T-DNA integration into the plant nuclear DNA.

There is indirect evidence supporting the suggestion that VirD2 and VirE2 are transported to plant cells and that they play an active role in T-DNA transport. VirE2 and VirD2 proteins expressed in plant cells as translational fusions to β-glucuronidase accumulate in the plant cell nucleus (Citovsky et al., 1992; Citovsky et al., 1994). Also, protein fusions between β-glucuronidase and the VirE2 NLSs expressed in tobacco cells are transported to the plant nucleus (Citovsky et al., 1992). Further, protein fusions between Cre recombinase and VirE2 or VirF are transferred to plant cells by the VirB/D4 transport system (Vergunst et al., 2000).

VirE2 protein expressed in tobacco cells complemented a *virE* mutation in *A. tumefaciens*, completely restoring tumorigenicity (Citovsky et al., 1992). Assuming that VirE2 was not transported to the *virE* mutant bacteria during co-cultivation, this result suggests that the function of VirE2 can be carried out in the plant cell and is consistent with transfer of VirE2 to the plant cell during T-DNA transfer. Given that the VirE2 protein expressed in plant cells was only detected in the nucleus (Citovsky et al., 1992), it might seem surprising, though not impossible, that VirE2 expressed in plant cells can completely complement a bacterial *virE* mutant if the relevant role of VirE2 is T-DNA transport through the cytoplasm. *In planta* VirE2 complementation provides further evidence that VirE2 coats the T-DNA in the plant cell rather than the bacterial cell.

Corroborative evidence of an *in planta* role for VirE2 comes from observations that virulence can be restored to *virE2* mutant *A. tumefaciens* strains when T-DNA transfer occurs in the presence of an *A. tumefaciens* strain lacking T-DNA but encoding *trans*-acting virulence functions (Otten et al., 1984; Christie et al., 1988; Lee et al., 1999). This extracellular complementation probably involves direct transfer of VirE2 to the plant cell rather than to the T-DNA donor (Binns et al., 1995). Direct VirE2 transport to the plant cell is supported by the observation that extracellular complementation only occurs if the "VirE2 donor" cell has chromosomal loci (chvA, chvB and exoC) required to initiate bacterium–plant cell attachment (Christie et al., 1988). Extracellular complementation of *virE2* requires *virB* genes in the VirE2 donor suggesting that VirE2 is exported via the T-DNA transport machinery (Binns et al., 1995). Otten et al. (1984) attempted to remove the possibility that these results could be explained by the transfer of the Ti plasmid from the T-DNA donor to the VirE donor by using a Ti plasmid that did not transfer between bacteria during parallel experiments (Otten et al., 1984). However, the possibility that the Ti plasmid was transferred to the VirE donor by *vir* functions after *vir* gene induction was not examined.

Further evidence of an *in planta* role for *vir* proteins was obtained by a study of the transfer of Cre recombinase-VirE2 and -VirF fusion proteins from *A. tumefaciens* to plant cells (Vergunst et al., 2000). Cre-mediated, *in planta*, recombination was used to monitor transfer of the fusion proteins to plant cells. Transfer of the fusion proteins was dependent on expression of the *virA*, *virG*, *virB* and *virD* operons and independent of T-DNA transfer (Vergunst et al., 2000). These results confirm the conclusions of previous studies that VirE2 is transferred from *A. tumefaciens* to plant cells independently of the T-DNA via the *virB*-encoded transport apparatus (Otten et al., 1984; Christie et al., 1988; Binns et al., 1995; Lee et al., 1999).

Nevertheless, the precise mechanism of Vir protein and T-DNA export is not yet resolved (Lai and Kado, 2000). Chen et al. (2000) detected VirE2, VirF and VirD2 proteins in the periplasm and supernatant of growing *A. tumefaciens* cells. Export of the Vir proteins did not require VirB proteins, suggesting a separate export pathway. However, as only 1% of the total cellular VirD2 and VirE2 proteins were detected in the supernatant, the biological significance of these results is uncertain (Christie and Vogel, 2000).

T-DNA integrity

It has been suggested that the principal function of VirE2 *in planta* is to protect the T-DNA, rather than to ensure T-DNA nuclear localization (Rossi et al., 1996). A deletion of the VirD2 nuclear localization sequence decreased T-DNA transfer (measured by frequency of transient expression) to about 5% of that of a strain containing the wild-type *virD2* gene regardless of the presence or absence of the *virE2* gene (Rossi et al., 1996). Rossi et al. also reported that in the absence of VirE2 protein, the T-DNA was more likely to be truncated. Fifteen of 20 T-DNA molecules integrated into plant genomic DNA after transfer from an *A. tumefaciens* strain with a *virE2* deletion had 3' truncations of up to 1 kb (Rossi et al., 1996). This result is consistent with a role for VirE2 in maintaining the integrity of the T-DNA, although that function does not necessarily occur within the plant cell. The effect of *in planta* expression of *virE2* on T-DNA integrity has not been explored.

T-DNA integration

It has been argued that T-DNA integration into nuclear DNA may be, in part, mediated by VirD2 (Tinland et al., 1995; Narasimhulu et al., 1996; Mysore et al., 1998). Specific mutations to the "omega region" of the VirD2 protein reduce stable T-DNA integration, possibly without a corresponding reduction in the efficiency of right border cleavage or nuclear localization of the T-DNA (Narasimhulu et al., 1996; Mysore et al., 1998). Narasimhulu et al. (1996) have shown that a specific mutation in the *virD2* omega sequence decreased the ratio of stable T-DNA expression to transient T-DNA expression in tobacco suspension culture cells when compared with expression following co-cultivation with an *A. tumefaciens* strain containing a functional *virD2* gene (Narasimhulu et al., 1996). On the assumption that the difference between the frequency of transient expression and stable expression was the frequency of T-DNA integration, Narasimhulu et al. (1996) argued a role for VirD2 in T-DNA integration.

This work was extended by Mysore et al. (1998) using RT-PCR amplification to study T-DNA expression in tobacco cell cultures. Mysore et al. could not detect transient expression from a promoterless *uid*A gene after T-DNA transfer from an *A. tumefaciens* strain containing a *virD2* omega mutation (Mysore et al., 1998). They argued that VirD2 was involved in integration because the promoterless *uid*A gene required integration for expression. However, this question could be better addressed by showing that there was a difference in the number of promoterless *uid*A expression events observed after transfer from the *A. tumefaciens* mutant compared with the number that would be expected if VirD2 was not involved in integration. Forty-two days after co-cultivation of tobacco suspension cells with *A. tumefaciens* (*virD2* omega mutation), no T-DNA was detected in tobacco genomic DNA (Mysore et al., 1998). However, this result is hard to interpret without a comparison to a co-cultivation system with similar transfer frequencies and wild-type integration frequencies.

Jarchow et al. (1991) have suggested that other virulence proteins (such as VirF) are transferred to plant cells during T-DNA transfer and are responsible for determining the host range of *A. tumefaciens* pathogenesis. Lack of *virF* in *A. tumefaciens* can be complemented by expressing *virF* in recipient plant cells. Nopaline *A. tumefaciens* strains (which lack the *virF* gene) are avirulent to *N. glauca* plants (Regensburg-Tuink and Hooykaas, 1993). Expression of *virF* in transgenic *N. glauca* plants converts this non-host to a host for nopaline strains of *A. tumefaciens* (Regensburg-Tuink and Hooykaas, 1993). As with similar studies with VirE2, this observation suggests that VirF has a function within the plant cell and is consistent with VirF transfer to plant cells during T-DNA transfer. Lack of *virF* in *A. tumefaciens* can also be complemented by co-inoculation of *virF* mutants with bacteria encoding *vir* genes but lacking T-DNA (Otten et al., 1985; Melchers et al., 1990). Assuming that the T-DNA was not transferred to the complementing bacteria, extracellular complementation also suggests that VirF is exported from *A. tumefaciens* during co-cultivation with plant material (Melchers et al., 1990).

PROTEINS WITH GENE-LIKE QUALITIES

The second part of this chapter reviews proteins that arguably could be described as components of horizontally mobile genetic elements.

Self-templated genes

Prions are infectious particles that transmit between individuals by, for example, cell fusion and protein consumption, thus demonstrating a genetic capacity. They are an example of protein-templated inheritance. The heritable "prion" quality is mapped onto the three-dimensional structure of already polymerized amino acids rather than transmitted through the linear order of nucleotides (Keyes, 1997). Therefore, the relationship between the prion (as a gene) and polypeptides is analogous to the template relationship between DNA and free nucleotides. In contrast, the relationship between prions (as proteins) and DNA is the same as the relationship between DNA and DNA polymerase. In the latter case, each is the product of a biochemical assembly process (Figure 6.2).

Yeast prions are excellent examples of the inheritable qualities possessed by some proteins. The first example of a non-mammalian prion was [URE3] of *Saccharomyces cerevisiae* (Wickner, 1994). Yeast lacking aspartate transcarbamylase are able to grow on ureidosuccinate if its uptake is not inhibited by ammonium. The *ure2*[+] phenotype was demonstrated by yeast mutants that could grow on ureidosuccinate in the presence of ammonium. Some recessive mutants segregated the phenotype as expected of a normal Mendelian marker (2:2), designated *URE2*. In contrast, other mutants transmitted a dominant *ure2*[+] phenotype through mitosis and meiosis (4:0). The second gene was designated [URE3] because it was dominant and transmitted as a non-chromosomal element.

[URE3] is an alternative conformation of Ure2p, the *URE2* gene product. What makes [URE3] dominant is its capacity to convert Ure2p into [URE3] presumably through protein–protein interaction. Thus, [URE3] is a non-DNA and template-type gene. Once acquired, [URE3] is

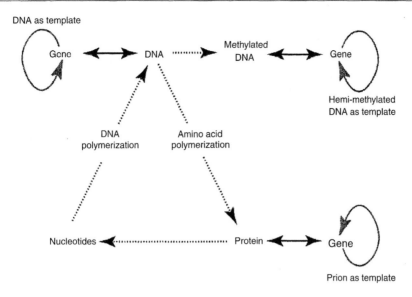

FIGURE 6.2 Information storage vs. informational macromolecule synthesis. DNA is an informational macromolecule because it serves as a template for its accurate duplication (solid lines). It can also serve as part of a template for the propagation of methylation patterns (solid lines). DNA duplication is dependent upon proteins in the same way that prion synthesis is dependent upon DNA and methylation patterns must be created by methylases (broken lines). Prions and DNA are genes when their structural information is transmitted to descendants (solid lines).

stably inherited. Hence, introduction of the protein through cell fusion or other means, such as during interkingdom conjugation, would be sufficient to propagate the element.

Non-templated factors in inheritance

Proteins can influence inheritance even without being templates for their own replication, just as DNA can sometimes influence inheritance without being a gene template. Some "genes" are inherited as a composite of interacting molecules, such as inheritable transcriptional loops (e.g. Novick and Weiner, 1957; Ptashne et al., 1982; Pillus and Rine, 1989; Heinemann, 1999; Heinemann and Roughan, 2000), rather than as self-templating molecules. The molecular basis for inherited phenotype (i.e. the gene) can be something other than merely DNA sequence. In these cases, even if all the traditional DNA "genes" were the same in two DNA identical cousins, they each could display a different phenotype in the same environment. This is because the sibling parents had activated different

self-perpetuating transcriptional loops that remain active in their respective offspring. Excellent examples of heritable transcriptional loops are provided by bacteriophage λ of *Escherichia coli* (Figure 6.3).

Protein components

Damaged DNA induces a change in the *E. coli* RecA protein that results in its ability to destabilize the tetrameric cI repressor of phage λ. The cI protein promotes its own expression as a transcriptional inducer and represses expression of the λ cro protein (Ptashne et al., 1982).

cI expression causes λ to remain stably integrated in the host chromosome as a prophage. The prophage reproduces in synchrony with the bacterium as a lysogen. λ reproducing lytically express genes under the control of the Cro protein which is repressed by, and acts as a transcriptional repressor of, *cI*. Lytic growth results in the production of progeny phage subsequently released through cell lysis, with ultimate transmission to other susceptible hosts. cI

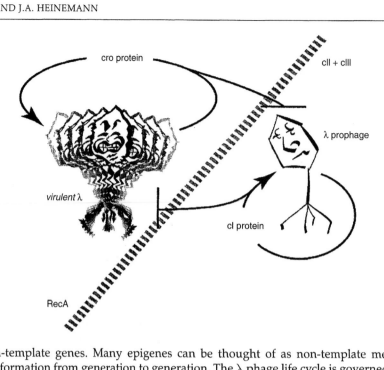

FIGURE 6.3 Non-template genes. Many epigenes can be thought of as non-template mechanisms for the transmission of information from generation to generation. The λ phage life cycle is governed by two mutually exclusive feedback loops that can perpetuate themselves across phage or cellular generations. The "cro loop" maintains the phage in a lytic cycle. Upon entry into a new host, cro protein is expressed constitutively causing the activation of lytic cycle genes and repressing *cI*. Under certain conditions, cII and cIII proteins can switch the phage gene expression pattern in favor of the "cI loop". The cI protein represses *cro* gene expression and induces its own expression. Under certain conditions, the RecA protein can switch the phage gene expression pattern in favor of the "cro loop".

and cro are the central elements of two self-perpetuating, mutually exclusive, transcriptional loops. Prophage reproduce synchronously with the host cell since each daughter cell inherits copies of the cI protein during cell division; lytic phage reproduce by infection and cell lysis since each phage upon entry into a new host can constitutively express cro protein in the absence of cI. Thus, we can think of these loops as multi-material genes since each phage is DNA identical but will bequeath a different phenotype to its offspring depending upon whether it was lytic or a prophage in the previous generation.

RecA transfer between bacteria can inheritably switch λ from one transcriptional loop to the other. Heinemann (1999) used this property of λ to demonstrate that the *E. coli* RecA protein was transferred from donors to recipients during conjugative transfer of an IncP plasmid (Heinemann, 1999). In that

experiment, λ lysogens with a deletion of the chromosomal *recA* gene were used as recipients. Transfer of RecA was thus measured by the frequency of λ induction in recipients. The frequency of λ induction was nearly 1000 times higher in matings involving *recA*+ donors, compared with *recA* donors, of plasmids. Transfer of RecA was dependent on the presence of the transferable plasmid in donor cells. Moreover, the high frequency of λ induction was not due to stable transmission of the *recA* gene. In control experiments, no recipient cells resistant to mitomycin-C were detected (RecA confers resistance to mitomycin-C). However, it is not certain whether RecA transfer depended on DNA transfer or some other conjugation process. It is possible that rather than transferring as a complex bound to the DNA, membrane-bound RecA was transferred through membrane–membrane contact (Heinemann, 1999).

DNA components

The conjugative transfer of damaged DNA to a λ lysogen is also sufficient to destabilize cI and activate the prophage (Borek and Ryan, 1958). DNA damaged by exposing the donor bacteria to UV radiation and transferred before being repaired was sufficient to activate RecA in the recipient and reset the balance between the cI-cro loops. In this example, the informational content of the transferred DNA was not synonymous with its ability to serve as a template for new DNA synthesis but as a component in the establishment of a self-perpetuating transcriptional loop gene.

The transfer of molecules between organisms can therefore have consequences both initially (e.g., death to the cell following prophage induction) or in future generations (e.g., dissemination of the phage following prophage induction). Molecules, such as DNA and prions, can serve as templates for their own replication. In other instances, molecules, such as DNA and proteins, are components of more complex but heritable systems. Thus, the study of gene exchange between organisms cannot be restricted to the inheritance of nucleic acids (Heinemann and Roughan, 2000).

DISCUSSION

Conjugation between bacteria probably does not involve general, unrestricted exchange of protein (Silver and Ozeki, 1962; Rosner et al., 1967; Merryweather et al., 1986b; Rees and Wilkins, 1989, 1990). The vast majority of proteins that have been observed to transfer from donors to recipients have been localized to the membrane fraction of recipients (Rees and Wilkins, 1990). Therefore it is likely that during conjugation there is an exchange of membrane-bound proteins. DNA transfer from bacteria to eukaryotes also appears to involve at least the transfer of specific cytoplasmic proteins (Stroun, 1971; Herrera-Estrella et al., 1988; Ward and Barnes, 1988; Yusibov et al., 1994; Citovsky et al., 1997).

Virulence proteins are probably transferred from *A. tumefaciens* to plant cells during T-DNA transfer. Complementation experiments suggest that VirE2 and VirF could function in plant cells (Otten et al., 1984; Christie et al., 1988; Citovski et al., 1992; Regensburg-Tuink and Hooykaas, 1993; Binns et al., 1995; Lee et al., 1999). The main roles of VirE2 and VirD2 are possibly to bind and protect the ssT-DNA and assist in nuclear localization (Citovsky et al., 1992, 1994; Rossi et al., 1996). However, a role for VirD2, or any other bacterial protein in T-DNA integration, is controversial (Bravo-Angel et al., 1998) and remains to be unequivocally demonstrated (Heinemann, 1999). Whether or not bacterial proteins other than virulence proteins are transferred to plant cells during T-DNA transfer appears to be an unexplored possibility.

Proteins, such as VirD2, VirE2 and DNA primases, may move between organisms during DNA transfer and enhance its expression and transmission. While these proteins are important catalysts in the process of genetic transformation, they do not have genetic qualities nor do they alter the expression of host genes. In contrast, some transferred proteins (e.g. RecA, [URE3]) can influence the inheritable phenotype of the recipient cell without a concomitant alteration to the organism's DNA composition. Some of these proteins (e.g. [URE3] and other prions) are arguably horizontally mobile genetic elements and may, in a limited way, be subject to selection and evolutionary pressures distinct from those acting on their hosts (Levin and Bergstrom, 2000). How common such elements are and how frequently they transfer between organisms is unknown and could be difficult to estimate as they are not amenable to traditional genetic (i.e. DNA) analysis.

ACKNOWLEDGMENTS

We thank A. Harker for critical reading of the manuscript. JAH wishes to express a special thanks to M. Stroun and P. Anker for their generous support both financially and intellectually and to the editors for their tolerance. This work was supported in part by the Marsden Fund (M1042 to JAH), the University of Canterbury Postdoctoral Grant (that supported RJW) and a New Zealand Crop and Food Research grant (to JAH).

REFERENCES

Beijersbergen, A., Dulk-Ras, A.D., Schilperoort, R.A. and Hooykaas, P.J.J. (1992) Conjugative transfer by the virulence system of *Agrobacterium tumefaciens*. *Science* **256**: 1324–1327.

Binns, A.N., Beaupre, C.E. and Dale, E.M. (1995) Inhibition of VirB-mediated transfer of diverse substrates from *Agrobacterium tumefaciens* by the IncQ plasmid RSF1010. *J. Bacteriol.* **177**: 4890–4899.

Borek, E. and Ryan, A. (1958) The transfer of irradiation-elicited induction of a lysogenic organism. *Proc. Natl Acad. Sci. USA* **44**: 374–377.

Bravo-Angel, A.M., Hohn, B. and Tinland, B. (1998) The omega sequence of VirD2 is important but not essential for efficient transfer of T-DNA by *Agrobacterium tumefaciens*. *Mol. Plant-Microbe Interact.* **11**: 57–63.

Bundock, P., Dulk-Ras, A. and Hooykaas, P.J.J. (1995) Trans-kingdom T-DNA transfer from *Agrobacterium tumefaciens* to *Saccharomyces cerevisiae*. *EMBO J.* **14**: 3206–3214.

Campbell, A.M. (1998) Prions as examples of epigenetic inheritance. *ASM News* **64**: 314–315.

Chatfield, L.K. and Wilkins, B.M. (1984) DNA primase of plasmid CoIIb is involved in conjugal DNA synthesis in donor and recipient bacteria. *Mol. Gen. Genet.* **197**: 461–466.

Chatfield, L.K., Orr, E., Boulnois, G.J. and Wilkins, B.M. (1982) Conjugative transfer of IncI1 plasmid DNA primase. *J. Bacteriol.* **152**: 1188–1195.

Chen, L., Li, C.M. and Nester, E.W. (2000) Transferred DNA (T-DNA)-associated proteins of *Agrobacterium tumefaciens* are exported independently of *virB*. *Proc. Natl Acad. Sci. USA.* **97**(13): 7545–7550.

Christie, P.J. and Vogel, J.P. (2000) Bacterial type IV secretion: conjugative systems adapted to deliver effector molecules to host cells. *Trends Microbiol.* **8**(8): 354–360.

Christie, P.J., Ward, J.E., Winans, S.C. and Nester, E.W. (1988) The *Agrobacterium tumefaciens* virE2 gene product is a single-stranded-DNA-binding protein that associates with T-DNA. *J. Bacteriol.* **170**: 2659–2667.

Citovsky, V. and Zambryski, P. (1993) Transport of nucleic acids through membrane channels: snaking through small holes. *Annu. Rev. Microbiology* **47**: 167–197.

Citovsky, V., Knorr, D., Schuster, G. and Zambryski, P. (1990) The P30 movement protein of tobacco mosaic virus is a single-strand nucleic acid binding protein. *Cell* **60**: 637–647.

Citovsky, V., Zupan, J., Warnick, D. and Zambryski, P. (1992) Nuclear localization of *Agrobacterium* VirE2 protein in plant cells. *Science* **256**: 1802–1805.

Citovsky, V., Warnick, D. and Zambryski, P. (1994) Nuclear import of *Agrobacterium* VirD2 and VirE2 proteins in maize and tobacco. *Proc. Natl Acad. Sci. USA* **91**: 3210–3214.

Citovsky, V., Guralnick, B., Simon, M.N. and Wall, J. (1997) The molecular structure of *Agrobacterium* VirE2-single stranded DNA complexes involved in nuclear import. *J. Mol. Biol.* **271**: 718–727.

Deom, C.M., Oliver, M.J. and Beachy, R.N. (1987) The 30-kilodalton gene product of tobacco mosaic virus potentiates virus movement. *Science* **237**: 389–394.

Fisher, K.W. (1962) Conjugal transfer of immunity to phage λ multiplication in *Escherichia coli* K-12. *J. Gen. Microbiol.* **28**: 711–719.

Gelvin, S.B. (2000) *Agrobacterium* and plant genes involved in T-DNA transfer and integration. *Annu. Rev. Plant Phys.* **51**: 223–256.

Hamilton, A. and Baulcombe, D. (1999) A species of small antisense RNA in posttranscriptional gene silencing in plants. *Science* **286:** 950–952.

Heinemann, J.A. (1991) Genetics of gene transfer between species. *Trends Genet.* **7**, 181–185.

Heinemann, J.A. (1992) Conjugation genetics. In *Encyclopedia of Microbiology* (ed. J. Lederberg), 1st edn, pp. 547–558, Academic Press, San Diego.

Heinemann, J.A. (1999) Genetic evidence of protein transfer during bacterial conjugation. *Plasmid* **41**: 240–247.

Heinemann, J.A. and Roughan, P.D. (2000) New hypotheses on the material nature of horizontally mobile genes. *Ann. New York Acad. Sci.* **906**: 169–186.

Herrera-Estrella, A., Chen, Z., Van Montagu, M. and Wang, K. (1988) VirD proteins of *Agrobacterium tumefaciens* are required for the formation of a covalent DNA-protein complex at the 5′ terminus of T strand molecules. *EMBO J.* **7**: 4055–4062.

Jarchow, E., Grimsley, N.H. and Hohn, B. (1991) *virF*, the host-range-determining virulence gene of *Agrobacterium tumefaciens*, affects T-DNA transfer to *Zea mays*. *Proc. Natl Acad. Sci. USA* **88**: 10426–10430.

Kado, C.I. (2000) The role of the T-pilus in horizontal gene transfer and tumorigenesis. *Curr. Opin. Microbiol.* **3**(6): 643–648.

Keyes, M.E. (1997) The prion challenge to the 'central dogma' of molecular biology, 1965–1991. *Stud. Hist. Phil. Biol. Biomed. Sci.* **30**: 1–19.

Klar, A. (1998) Propagating epigenetic states through meiosis: where Mendel's gene is more than a DNA moiety. *Trends Genet.* **14**: 299–301.

Koukolikova-Nicola, Z., Shillito, R.D. and Hohn, B. (1985) Involvement of circular intermediates in the transfer of T-DNA from *Agrobacterium tumefaciens* to plant cells. *Nature* **313**: 191–196.

Lai, E-M. and Kado, C.I. (2000) The T-pilus of *Agrobacterium tumefaciens*. *Trends Microbiol.* **8**(8): 361–369.

Lanka, E. and Barth, P.T. (1981) Plasmid RP4 specifies a deoxyribonucleic acid primase involved in its conjugal transfer and maintenance. *J. Bacteriol.* **148**: 769–781.

Lee, L.-Y., Gelvin, S. and Kado, C.I. (1999) pSa causes oncogenic suppression of *Agrobacterium* by inhibiting VirE2 protein export. *J. Bacteriol.* **181**: 186–196.

Levin, B.R. and Bergstrom, C.T. (2000) Bacteria are different: observations, interpretations, speculations, and opinions about the mechanisms of adaptive evolution in procaryotes. *Proc. Natl Acad. Sci. USA* **97**: 6981–6985.

Mas, P. and Beachy, R.N. (1998) Distribution of TMV movement protein in single living protoplasts immobilized in agarose. *Plant J.* **15**: 835–842.

Melchers, L.S., Maroney, M.J., den Dulk-Ras, A. et al. (1990) Octopine and nopaline strains of *Agrobacterium tumefaciens* differ in virulence; molecular characterisation of the Vir locus. *Plant Mol. Biol.* **14**: 249–259.

Merryweather, A., Barth, P.T. and Wilkins, B.M. (1986a) Role and specificity of plasmid RP4-encoded DNA primase in bacterial conjugation. *J. Bacteriol.* **167**: 12–17.

Merryweather, A., Rees, C.E.D., Smith, N.M. and Wilkins, B.M. (1986b) Role of sog polypeptides specified by plasmid CoIIb-P9 and their transfer between conjugating bacteria. *EMBO J.* **5**: 3007–3012.

Meshi, T., Watanabe, Y., Saito, T. et al. (1987) Function of the 30kd protein of tobacco mosaic virus: involvement in cell-to-cell movement and dispensability for replication. *EMBO J.* **6**: 2557–2563.

Mysore, K.S., Bassuner, B., Deng, X. et al. (1998) Role of the *Agrobacterium tumefaciens* VirD2 protein in T-DNA transfer and integration. *Mol. Plant–Microbe Interact.* **11**: 668–683.

Narasimhulu, S.B., Deng, X., Sarria, R. and Gelvin, S.B. (1996) Early transcription of *Agrobacterium* T-DNA genes in tobacco and maize. *Plant Cell* **8**: 873–886.

Novick, L. and Weiner, M. (1957) Enzyme induction as an all-or-none phenomenon. *Proc. Natl Acad. Sci. USA* **43**: 553–566.

Otten, L., DeGreve, H., Leemans, J. et al. (1984) Restoration of virulence of *vir* region mutants of *Agrobacterium tumefaciens* strain B653 by coinfection with normal and mutant *Agrobacterium* strains. *Mol. Gen. Genet.* **175**: 159–163.

Otten, L., Piotrowiak, G., Hooykaas, P.J.J. et al. (1985) Identification of an *Agrobacterium tumefaciens* pTiB6S3 *vir* region fragment that enhances the virulence of *pTiC58*. *Mol. Gen. Genet.* **199**: 189–193.

Pillus, L. and Rine, J. (1989) Epigenetic inheritance of transcriptional states in *S. cerevisiae*. *Cell* **59**: 637–647.

Prusiner, S.B. (1998) Prions. *Proc. Natl Acad. Sci. USA* **95**: 13363–13383.

Ptashne, M., Johnson, A. and Pabo, C. (1982) A genetic switch in a bacterial virus. *Sci. Am.* **247**: 128–140.

Rees, C.E.D. and Wilkins, B.M. (1989) Transfer of *tra* proteins into the recipient cell during bacterial conjugation mediated by plasmid CoIIb-P9. *J. Bacteriol.* **171**: 3152–3157.

Rees, C.E.D. and Wilkins, B.M. (1990) Protein transfer into the recipient cell during bacterial conjugation: studies with F and RP4. *Mol. Microbiol.* **4**: 1199–1205.

Regensburg-Tuink, A.J.G. and Hooykaas, P.J.J. (1993) Transgenic *N. gluaca* plants expressing bacterial virulence gene *vir*F are converted into hosts for nopaline strains of *A. tumefaciens*. *Nature* **363**: 69–71.

Rosner, J.L., Adelberg, E.A. and Yarmolinsky, M.B. (1967) An upper limit on β-galactosidase transfer in bacterial conjugation. *J. Bacteriol.* **94**: 1623–1628.

Rossi, L., Hohn, B. and Tinland, B. (1993) The VirD2 protein of *Agrobacterium tumefaciens* carries nuclear localization signals important for transfer of T-DNA to plants. *Mol. Gen. Genet.* **239**: 345–353.

Rossi, L., Hohn, B. and Tinland, B. (1996) Integration of complete transferred DNA units is dependent on the activity of virulence E2 protein of *Agrobacterium tumefaciens*. *Proc. Natl Acad. Sci. USA* **93**: 126–130.

Silver, S. and Ozeki, H. (1962) Transfer of deoxyribonucleic acid accompanying the transmission of colicinogenic properties by cell mating. *Nature* **195**: 873–874.

Sonti, R.V., Chiurazzi, M., Wong, D. et al. (1995) *Arabidopsis* mutants deficient in T-DNA integration. *Proc. Natl Acad. Sci. USA* **92**: 11786–11790.

Stachel, S.E. and Nester, E.W. (1986) The genetic and transcriptional organisation of the *vir* region of the A6 plasmid of *Agrobacterium tumefaciens*. *EMBO J.* **5**: 1445–1454.

Stachel, S.E. and Zambryski, P. (1986) *Agrobacterium tumefaciens* and the susceptible plant cell: a novel adaptation of extracellular recognition and DNA conjugation. *Cell* **47**: 155–157.

Stroun, M. (1971) *Biochem. Biophys. Res. Commun.* **44**: 571–578.

Tinland, B., Hohn, B. and Puchta, H. (1994) *Agrobacterium tumefaciens* transfers single-stranded transferred DNA (T-DNA) into the plant cell nucleus. *Proc. Natl Acad. Sci. USA* **91**: 8000–8004.

Tinland, B., Schoumacher, F., Gloeckler, V. et al. (1995) The *Agrobacterium tumefaciens* virulence D2 protein is responsible for precise integration of T-DNA into the plant genome. *EMBO J.* **14**: 3585–3595.

Tomenius, K., Clapham, D. and Meshi, T. (1987) Localization by immunogold cytochemistry of the virus-coded 30k protein in plasmodesmata of leaves infected with tobacco mosaic virus. *Virology* **160**: 363–371.

Vergunst, A.C., Schrammeijer, B., den Dulk-Ras, A. et al. (2000) VirB/D4-dependent protein translocation from Agrobacterium into plant cells. *Science* **290**: 979–982.

Vionnet, O., Vain, P., Angell, S. and Baulcombe, D. (1998) Systemic spread of sequence-specific transgene RNA degradation in plants is initiated by localised introduction of ectopic promoterless DNA. *Cell* **95**: 177–187.

Ward, E.R. and Barnes, W.M. (1988) VirD2 protein of *Agrobacterium tumefaciens* very tightly linked to the 5′ end of T-strand DNA. *Science* **242**: 927–930.

Wickner, R.B. (1994) URE3 as an altered URE2 protein: evidence for a prion analog in *Saccharomyces cerevisiae*. *Science* **264**: 566–569.

Wilkins, B.M. and Thomas, A.T. (2000) DNA-independent transport of plasmid primase protein between bacteria by the I1 conjugative system. *Mol. Microbiol.* **38(3)**: 650–657.

Wolf, S., Deom, C., Beachy, R. and Lucas, W. (1989) Movement protein of tobacco mosaic virus modifies plasmodesmatal size exclusion limit. *Science* **246**: 377–379.

Young, C. and Nester, E.W. (1988) Association of the VirD2 protein with the 5′ end of T strands in *Agrobacterium tumefaciens*. *J. Bacteriol.* **170(8)**: 3367–3374.

Yusibov, V.M., Steck, T.R., Gupta, V. and Gelvin, S.B. (1994) Association of single-stranded transferred DNA from *Agrobacterium tumefaciens* with tobacco cells. *Proc. Natl Acad. Sci. USA* **91**: 2994–2998.

Zechner, EL., de la Cruz, F., Eisenbtandt, R. et al. (2000) Conjugative-DNA transfer processes. In *The Horizontal Gene Pool: Bacterial Plasmids and Gene Spread* (ed., C.M. Thomas), pp. 87–174, Harwood Academic Publishers, Amsterdam.

Ziemienowicz, A., Gorlich, D., Lanka, E. et al. (1999) Import of DNA into mammalian nuclei by proteins originating from a plant pathogenic bacterium. *Proc. Natl Acad. Sci. USA* **96**: 3729–3733.

CHAPTER 7

Transformation in Aquatic Environments

Martin Day

Transformation, one of three mechanisms that promote gene exchange in bacterial populations, is performed by many Gram-positive and Gram-negative species. It is an active, highly evolved and dedicated process, governed by chromosomal genes. "Free" DNA, released from either a living or a lysed "dead" cell, is taken up only by a competent recipient cell. The environment is "awash" with eukaryotic and prokaryotic DNA, thus mechanisms to identify the DNA as "self" or "non-self" have evolved. In some species this "free" DNA has an internal sequence that allows it to be selectively identified and taken up. The period of competence is governed in many species by the physiological status of the individual that may, in turn, be regulated through quorum sensing by that of their population. This state is also regulated (induced or repressed) by a range of environmental parameters, such as nutrient status, ionic composition, temperature, pH, etc. The few studies performed in microcosms and *in situ* have resolved some of the practical difficulties and demonstrated gene exchange by transformation to be a measurable event. Thus, like other transfer systems the potential for contributing to evolution is clear.

INTRODUCTION

Phenotypic changes occur to bacteria due to the impact of the environment, on the expression of genes within the genome. Novel phenotypes may arise through mutation or through the acquisition, mediated by some form of parasexual process, of homologous DNA. In the former case some alteration, via an insertion, duplication, deletion or transposition event, is made to the DNA sequence, resulting in a change to the cellular phenotype. In the latter case the DNA sequence is also changed but the change is confined to the sequence transferred and exchanged by recombination. For the sequence to become integrated it requires homology with the recipient genome and thus will insert by homologous recombination mediated by the RecA protein. If the sequence has a replication origin, i.e. if it is a plasmid, and no homology with other gene sequences within the cell, then to become established it must be able to recombine internally. This can be achieved in one of two ways. If the sequence has an internal direct repeat this provides the homology needed for recombination to allow for its reconstruction as a replicon, but the recombinational process deletes some of the genetic information carried so the molecule becomes smaller. Alternatively two copies of the plasmid molecule, transferred coincidentally and linearized at different sites, may then undergo recombination to yield the original plasmid. Thus gene transfer processes are potentially mutagenic, as the genome may be altered from the wild-type as a result of the recombination process. Reflecting briefly on the haploid life cycle and capacity for asexual reproduction in bacteria makes it possible to see why they have a variety of transfer and mutational mechanisms: these produce the genetic diversity, equivalent to that formed by sexual reproductive processes.

Horizontal Gene Transfer
ISBN: 0-12-680126-6

Copyright © 2002 by Academic Press.
All rights of reproduction in any form reserved.

Griffith (1928) first described transformation and in 1944 its use became transiently important to the determination of the role of DNA in genetics (Avery et al., 1944). Although transformation plays a role in horizontal gene transfer in many bacteria of medical, industrial and environmental interest (Wolfgang et al., 1999), it remains open to debate whether this is its main function. Besides acquisition of new traits competence development may act to enable the repair of damaged and mutated genes. Redfield (1993) tested the hypothesis that the primary function of transformation is DNA repair and found competence is not regulated by DNA damage in *Bacillus subtilis* and *Haemophilus influenzae*. Thus if competence did not evolve for recombinational repair, then did it evolve as a system to acquire nucleotides from DNA taken up? Redfield (1993) observed that several of the competence regulatory genes in *H. influenzae* are triggered by cAMP and that in *B. subtilis* these genes regulate starvation induced genes. Thus did the evolution of competence occur in response to nutritional needs when resources were scarce? This evolutionary strategy would save energy in comparison to *de novo* synthesis (MacFadyen et al., 1996; Solomon and Grossman, 1996). Redfield (1993) also considered that even though the amount of DNA taken up by competent cells is small, transformation would reduce the mutational load on a recipient cell. Thus it remains an open question: did transformation arise as a by-product of other essential cellular mechanisms or as a gene transfer/repair mechanism?

GENE TRANSFER PROCESSES

To acquire an understanding of gene exchange in microbial systems it is important to view DNA transfer as potentially a combination of processes. Briefly there are three mechanisms involved. The first, conjugation requires the donor to have a particular type of plasmid or a transposon to mobilize the donor DNA. These provide a mechanism that recognizes a recipient, establishes close physical contact and transfers DNA into it. To achieve success both cells must be physiologically active (Ippen-Ihler, 1989).

The second mechanism is transduction and it requires the donor cell to release phage carrying host genomic DNA. It requires the donor to be active during the synthesis of the phage (Kokjohn, 1989) and the recipient to be closely related, but they can be both temporally and spatially separated from each other. In both cases for recombination to occur the recipient and transferred DNA have to be homologous. Phage particles, during their passage from the producer-cell to the recipient, effectively carry DNA in hibernation. They generally show some higher level of resistance, in comparison to free-DNA, to adverse environmental conditions (Stotzsky, 1989).

Transformation is the third. The first step in gene transfer by transformation is the release of DNA from a living or a lysed "dead" cell. The persistence and integrity of intact DNA, in the environment, and its biological availability will determine to what degree it participates and retains a role in gene exchange (Leff et al., 1992). As a general rule a bacterium must be in a physiologically active state in order to be competent. This property is dependent on a set of chromosomally encoded genes (Hahn et al., 1987; Dubnau, 1991) which are coordinately expressed and physiologically regulated. Thus the transfer process responds, like an enzyme system, to induction and repression signals that enables a cell actively to regulate its participation. Bacteria appear to be the only group of organisms capable of undergoing transformation naturally (Lorenz and Wackernagel, 1994) and due to the extracellular DNA step, transduction may be discriminated from conjugation and transduction because of its sensitivity to DNAase (Albritton et al., 1982). Although the process is distinct from conjugation and transduction the genetic consequences are identical, the genome may be altered by the recombinational integration of a novel sequence.

TRANSFORMATION

Transformation is an active and dedicated process of gene exchange, governed by chromosomal genes that allows the uptake of exogenous "free" DNA by a competent cell. Competence is often a transient and tightly regulated stage through which a proportion of a population passes due to physiological changes occurring within those individual cells. It is a mechanism of gene exchange widespread among individual strains of many species, but not universally among all strains of a species (Wolfgang et al., 1999). Well over 43 species from 23 genera are known to acquire competence (Lorenz and Wackernagel, 1994) and it is likely that as the breadth of testing and the sensitivity of the assay rises more strains and species will be shown to participate (Stone and Kwaik, 1999; Newberry, 2001). Due to both a lack of sensitivity in the assay procedure and of research into indigenous strains the process of transformation has not been generally envisaged as having a significant contribution to the genetics of microbial populations. It must be the only sexual process not seen as sexy! A comparison of transformable species is shown in Table 7.1 with their taxonomic positions. There is an absence of transformation-proficient species in the Crenarchaeota (Euryarchaeota) and of several groups in the Division Bacteria (Thermogales, Fusobacterium, Fibrobacter, Spirochaetes and Planctomyces/Chlamydia) (Lorenz and Wackernagel, 1994). Within other groups and sub-groupings the occurrence of transformable species is patchy. This reflects, in part, a lack of work in this area to establish the genetic capacities of many of these more exotic bacterial types. There are several stages to the transformation process.

Competence

Competence is regulated externally by competence factors (pheromones) and internally by physiological changes promoted by nutritional status (Stewart, 1992). The great disparity in conditions required for competence development reflects regulation strategies evolved by organisms isolated from different ecological habitats (Stotzky, 1989). Transformation in *B. subtilis* has been well studied and approximately 40 genes govern early, middle and late stages of competence (Hahn et al., 1996; Solomon and Grossman, 1996). Some of these operons have multiple

purposes, involved with sporulation, motility and degradative enzyme production (Msadek et al., 1991; Ogura and Tanaka, 1996; Liu and Zuber, 1998). Their expression may be mutually exclusive, for example one stimulus may express competence in a high cell density environment, but may express motility if cell density is low, allowing migration to an alternative environment (Liu and Zuber, 1998). Thus transformation appears to be intimately related to other survival strategies. Early and mid competence genes in *B. subtilis* are involved in the regulation of the other competence specific proteins, whilst the late competence genes are thought to be involved in the structural apparatus required for binding, uptake and transport of DNA (Dubnau, 1991; Hahn et al., 1996; Chung et al., 1998). In some species, e.g. *B. subtilis* and *Streptococcus pneumoniae*, the pheromone induces competence when a certain population level is reached, possibly acting by "quorum sensing" (Turgay et al., 1997). Competence development in *B. subtilis* requires two extracellular peptides, competence stimulating peptide (CSP), and ComX pheromone (Solomon and Grossman, 1996). The production and expression of CSP and the competence genes is stimulated by conditions such as nutritional status (Albano et al., 1987; Alloing et al., 1996; Cheng et al., 1997), magnesium ions (Mg^{2+}) and temperature (Buitenwerf and Venema, 1977). Exogenous provision of competence factor to a noncompetent culture of *S. pneumoniae* will induce a competent state (Yother et al., 1986) and the production of a transporter molecule (Alloing et al., 1996). The CSP is strain specific with individual phenotypes recognized by a signaling domain (ComD), thus stimulating competence development only in those strains recognizing that particular pheromone (Håvarstein et al., 1997). Streptococci also regulate the termination of competence with release of a protein (Chen and Morrison, 1987).

Other bacteria (*Azotobacter vinelandii*, *Acinetobacter calcoaceticus* and *H. influenzae*) all regulate competence internally as a result of changing physiology. *H. influenzae* may produce a competence factor (Stewart, 1992), but regulation is due in part to catabolite repression (cAMP), with maximum competence achieved by a nutrient downshift or transient anaerobic

TABLE 7.1 Naturally transformable prokaryotic species

Species isolated from terrestrial or aquatic habitats	Transformation frequency[a]	Woess grouping[b]
Protolithotrophic		
Agmenellum quadruplicatum	4.3×10^{-4}	
Anacystis nidulans	8.0×10^{-4}	CFB group
Chlorobium limicola	1.0×10^{-5}	cyanobacteria
Nostoc muscorum	1.2×10^{-3}	cyanobacteria
Synechocystis sp. strain 6803	5.0×10^{-4}	cyanobacteria
Synechocystis sp. strain OL50	2.0×10^{-4}	
Chemolithotrophic		
Thiobacillus thioparus	$10^{-3} - 10^{-2}$	; purple
Thiobacillus sp. strain Y	1.7×10^{-3}	; purple
Heterotrophic		
Achromobacter spp.	+	
Acinetobacter calcoaceticus	7.0×10^{-3}	; purple
Azotobacter vinelandii	9.5×10^{-2}	; purple
Bacillus subtilis	3.5×10^{-2}	LGC
Bacillus licheniformis	1.2×10^{-2}	LGC
Lactobacillus lactis	2.3×10^{-5}	LGC
Mycobacterium smegmatis	$10^{-7} - 10^{-6}$	HGC
Pseudomonas stutzeri	7.0×10^{-5}	; purple
Rhizobium meliloti	7.0×10^{-4}	; purple
Streptomyces	+	HGC
Thermoactinomyces vulgaris	2.7×10^{-3}	
Deinococcus radiodurans	2.1×10^{-2}	Thermus
Thermus thermophilus	1.0×10^{-2}	Thermus
Thermus flavus	8.8×10^{-3}	Thermus
Thermus caldophilus	2.7×10^{-3}	Thermus
Thermus aquaticus	6.4×10^{-4}	Thermus
Vibrio sp. strain D19	2.0×10^{-7}	; purple
Vibrio sp. strain WJT-1C[d]	2.5×10^{-4}	; purple
Vibrio parahaemolyticus	1.9×10^{-9}	; purple
Methylotrophic		
Methylobacterium organophilum	5.3×10^{-3}	; purple
Archaebacteria		
Methanobacterium thermoautotrophicum	+	Euryarchaeota
Methanococcus voltae	8.0×10^{-6}	Euryarchaeota
Clinical isolates of pathogenic species		
Campylobacter jejuni	2.0×10^{-4}	; purple
Campylobacter coli	1.2×10^{-3}	; purple
Haemophilus influenzae	7.0×10^{-3}	; purple
Haemophilus parainfluenzae	8.6×10^{-3}	; purple
Helicobacter pylori	5.0×10^{-4}	& ; purple
Moraxella spp	+	; purple
Neisseria gonorrhoeae	1.0×10^{-4}	; purple
Neisseria meningitidis	1.1×10^{-2}	; purple
Staphylococcus aureus	5.5×10^{-6}	
Streptococcus pneumoniae	2.9×10^{-2}	HGC
Streptococcus sanguis	2.0×10^{-2}	HGC
Streptococcus mutans	7.0×10^{-4}	HGC

Modified from Lorenz and Wackernagel (1994).

[a]Transformation frequency (of a chromosomal marker given as transformants/viable cell).

[b]CBF = Cytophaga/Flexibacteria/Bacteroides group; HGC = high G + C Gram-positive; LGC = low G + C Gram-positive. Although these groupings have been superseded now, the terms provide an idea of the taxonomic breadth of the organisms able to promote exchange by transformation.

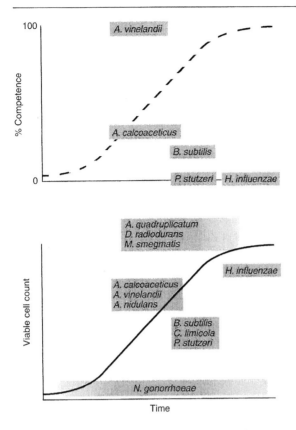

FIGURE 7.1 A comparison between transformable species of the percentage of competent cells in a population with the stage in their growth cycle when they attain competence.

growth (MacFadyen et al., 1996). Gwinn et al. (1996) found a number of *H. influenzae* genes were involved in maintaining the levels of cAMP needed for competence development but these were not directly involved in regulation of competence.

There is no evidence of a competence factor for *A. calcoaceticus* (Palmen et al., 1992), but competence is maximal after a nutrient spike following a cessation of growth (Palmen et al., 1994a) and is dependent on divalent ions; Ca^{2+}, Mg^{2+} or Mn^{2+} (Lorenz et al., 1992).

In most transformable bacteria the state of competence is transient (Figure 7.1), only in *Neisseria gonorrhoeae* is it clearly constitutive (Sparling, 1966). In *H. influenzae* competence develops when the cells cease to divide (Smith et al., 1981). In *A. calcoaceticus* (Palmen et al., 1992), *A. vinelandii* (Page and Sadoff, 1976) and *Anacystis*

nidulans (Chauvat et al., 1983) it occurs early to late exponential phase, and in the transition from exponential to stationary phase in *B. subtilis* (Smith et al., 1981), *Chlorobium limicola* (Ormerod, 1988) and *Pseudomonas stutzeri* (Carlson et al., 1983). In *Agmenellum quadruplicatum*, *Deinococcus radiodurans* and *Mycobacterium smegmatis* (Norgard and Imaeda, 1978; Page and Grant, 1987) competence occurs throughout the exponential phase and declines thereafter.

The proportion of competent cells within a culture varies between species (Figure 7.1). For example it has been estimated that the competent fraction of *A. vinelandii* (Doran and Page, 1983) is 100%, while within *B. subtilis* (Smith et al., 1981) and *A. calcoaceticus* (Palmen et al., 1993) it is 10–25%. In *H. influenzae* (Redfield, 1991) and *P. stutzeri* it is well below 1% (Lorenz and Wackernagel, 1992).

The timing and level of competence achieved by species is regulated and responds to a variety of relevant environmental factors. For example, a critical cell density is required to induce competence in *S. pneumonia* and *B. subtilis* (Smith et al., 1981). A *Vibrio* sp. (Frischer et al., 1993) shows a capacity to retain competence for over 10 days at 29°C in artificial seawater, and the transfer frequencies fell only 200-fold over a 7 day period at 7°C. It was possible that dying cells were providing a short pulse of nutrients that could "kick" the competence period into action periodically (Lorenz et al., 1992). In the photosynthetic bacteria *A. nidulans* and *A. quadruplicatum* exclusion from light results in a loss of competence (Chauvat et al., 1983; Essich et al., 1990). The frequency of transformation by *A. nidulans* is enhanced in medium containing iron (Golden and Sherman, 1984), and in *A. quadruplicatum* competency decreases when cells are deprived of a source of nitrogen (Essich et al., 1990). Thiobacilli also show reduced transformation with deprivation of nitrogen (Yankofsky et al., 1983). *Methanobacterium thermoautrophicum*, an archaebacterium, does not develop competence in liquid medium but transformation occurs on a solid Gellan gum surface and was specific for homologous DNA (Worrell et al., 1988). These examples show competence is a physiologically regulated process, thus an understanding of the environmental factors moderating its "window of activity" should provide for an understanding of its role and efficacy in nature. Thus competence appears a component part of a global response to the microbe sensing a change in its immediate environment. The term "quorum sensing" has recently been introduced and covers an organism's ability to sense changes in its environment and to respond to these changes. Transformation is definitely a cellular mechanism, and one which needs to assay the external environment to be efficient (Håvarstein and Morrison, 1999).

Availability of exogenous DNA

Many transformable bacteria actively excrete or release DNA during growth (Hirsch, 1999). The spontaneous release of DNA from bacterial cells in the aquatic environment is also normal (Paul et al., 1989; Paul and David, 1989) and is largely in the form required for transformation, i.e. it is linear and double stranded (Paul and Carlson, 1984; DeFlaun et al., 1987; Karl and Bailiff, 1989; Paul and David, 1989). Release of DNA has been shown to occur with *Brucella* spp., *Flavobacterium* spp., *Alcaligenes* spp., *Micrococcus* spp., *Pseudomonas* spp. (Catlin, 1956, 1960; Takahashi and Gibbons, 1957), *B. subtilis* (Sinha and Iyer, 1971; Lorenz et al., 1991), *Burkoldaria cepacia* and *Bradyrhizobium japonicum* (Paul and David, 1989) cells during normal growth, and these organisms are probably not unique. Although for *A. calcoaceticus* (Palmen and Hellingwerf, 1995) only small amounts of DNA were liberated over the entire exponential phase. Thus the presence of free DNA in the environment may arise from several processes: cell lysis, predation and phage replication (Börsheim, 1993), as well as normal physiological activity. The significance of an active release of DNA may represent a mechanism for targeting delivery to competence periods of recipients (Lorenz et al., 1991). In an evolutionary sense the development of gene transfer mechanisms, where the release and uptake of DNA in the environment is coordinated, would make transformation a biologically significant event (Stewart and Carlson, 1986). The active excretion of DNA from living cells may balance the costs incurred by taking up damaged DNA released into the environment through a process of cell death and lysis, as this may be subject to more degradative pressures and possibly contain a higher proportion of mutations. In a stable environment mortality and growth occur at much the same rate (Servais et al., 1983), thus free DNA is normally present.

The second important factor is that DNA has to be of an appropriate size. For example DNA between 10–60 kb gave equal transformation frequencies with *P. stutzeri*, but decreased over tenfold, when the size was reduced between 1–10 kb (Carlson et al., 1983). *A. calcoaceticus* was slightly different and gave a reduction in frequency directly proportional to size between 40–10 kb (Lonsdorf et al., cited in Lorenz and Wackernagel, 1994).

DNA binding and uptake

At the cell surface double-stranded DNA associates quickly with the external surface, especially in the presence of the divalent ions, Mg^{2+} and Ca^{2+}, but monovalent ions also have a role (Romanowski et al., 1993). In both *B. subtilis* and *S. pneumoniae* the DNA is held in a state that is resistant to gentle washing but still sensitive to DNAase. There are about 50 DNA uptake sites in *B. subtilis* and 30–80 in *S. pneumoniae* (Smith et al., 1981; Dubnau, 1991). Blebs or transformasomes are externally positioned membrane derived vesicles, which in many species contain plasmid and chromosomal genes (Dorward and Garon, 1990). In *H. influenzae* these blebs appear during the onset of competence and are shed as the cell leaves this state (Kahn et al., 1983). These blebs take up double-stranded DNA but only from the same or closely related species, and specificity is determined by a DNA receptor that recognizes an 11-bp sequence distributed at 600 sites throughout the genome (Sisco and Smith, 1979; Danner et al., 1980). There is evidence that AT-rich flanking regions increase transformation frequency 48-fold compared with GC flanking sequences (Danner et al., 1982). In *H. influenzae* there is selection for chromosomal sequences because superhelical and nicked plasmids are not taken up as efficiently (Gromkova and Goodgal, 1979). Albritton and colleagues (1982) have shown that cell-to-cell transformation occurs in *H. influenzae* and it is relatively insensitive to DNAase. DNA uptake by *N. gonorrhoeae* is similar (Goodman and Scocca, 1988).

In all cases, DNA enters into the cell in the 3'-to-5' direction and degradation of the opposite strand occurs concurrently at about 100 nt s^{-1} (Mejean and Claverys, 1993).

Frequency of transformation

A DNA sequence, transformed into a cell, needs to be integrated into a self-replicating element, by the recombination system controlled by the *rec*A gene (Mahajan, 1988), to stabilize it. To visualize transformants in the laboratory recombination must also result in a phenotypic change. Recombination is essential for the formation of transformants and is eliminated in *rec*A mutants of *A. calcoaceticus* and *P. stutzeri* (Vosman and Hellingwerf, 1991; Palmen et al., 1992). Upon entry into the cytoplasm the single-stranded sequence displaces the corresponding homologous strand in the duplex and recombination occurs. In *Escherichia coli* (King and Richardson, 1986) 20–40 base pairs was found to be the minimum length of sequence homology, compared with over 70 in *B. subtilis* (Khasanov et al., 1992), to allow a reasonable rate of *in vivo* recombination. The level of RecA in competent cells of *B. subtilis* is tenfold higher than normal (Lovett et al., 1989). The size of sequence displaced is 8.5 kb (Smith et al., 1981; Dubnau, 1991) which is about 70% of that transferred (Dubnau, 1991), signifying the integration of 8–10 genes. Observations show the frequency of transformation is lower for plasmid and phage molecules than for the single-hit response of chromosomal or plasmid dimers (Saunders and Guild, 1981). This indicates the establishment of plasmid and phage genomes is achieved by the same mechanism, but requires a two-hit or has second order dependency. This effectively means at least two copies of the replicon, linearized at different points, must be present within the cell so they can hybridize and recombine to form a duplex and circular molecule.

There are also marker-specific transformation frequencies, which vary over 20-fold, in *S. pneumoniae* (Gasc et al., 1987). A comparison of transformation frequencies of a streptomycin-resistance mutation from *H. influenzae* to nine other *Haemophilus* spp. showed a 25 000-fold variation (Albritton et al., 1984). *B. subtilis* was transformed (10^6-fold range) with DNA from a range of other *Bacillus* species (Harford and Mergeay, 1973). *A. vinelandii* was transformed by DNA from related species at 0.002–21% (Bishop et al., 1977; Doran and Page, 1983; Page, 1985), but from other strains not at all (DeLey, 1992). Why should this be? Some of the reduction in transformation frequency can potentially be explained by poor homology. There are other contributing explanations. Or example such a frequency bias was observed in a homologous recombination transduction experiment (Masters and Broda, 1971). In the host genome there are more origin sequences than terminal sequences in moderately to fast growing cells. This is due to further rounds of chromosome replication being initiated before the initial replication round has terminated. As phage only replicate in growing cells, this replication produces bias in relative gene concentrations, and it means that there are more transducing particles with sequences around the origin than ones at the terminus. Consequently this results in a four- to sixfold differential in transduction frequencies between genes around the origin and the terminus. It is doubtful if this bias is completely responsible for the transformation frequency differences. Clearly factors, other than taxonomic distance, are relevant in determining transfer frequencies. This brief description of transformation highlights similarities in the various component parts of the process. Other species show the process of transformation is superficially similar but it differs in detail. For example *P. stutzeri* exhibits selectivity in DNA binding and uptake, but not to the level of *H. influenzae* and *N. gonorrhoeae* (Mathis and Scocca, 1982) and *A. calcoaceticus* shows no specificity in either uptake or binding (Lorenz et al., 1992; Börsheim, 1993).

Barriers to genetic exchange

Genetic exchange between microorganisms does not take place indiscriminately; there are barriers that exist which tend to maintain species integrity (Saunders et al., 1990), with most

TABLE 7.2 High molecular weight DNA is present in aquatic systems (Lorenz and Wackernagel, 1994)

Habitat	Molecular size (kb)	DNA concentration (μg l^{-1})	Half-life (h)
Fresh water	ND	0.5–25.6	4–5.5
Estuarine	0.15–35.2	10–19	3.4–5.5
Offshore/ocean	0.24–14.3	0.2–1.9	4.5–83
Freshwater sediment	1.0–23.0	1.0	–
Marine sediment	–	–	140–235

ND = not detected; – = unknown

transformable bacteria integrating DNA only from closely related species (Stewart and Carlson, 1986). Thus gene transfer between distantly related bacteria may fail internally due to homology or externally through surface recognition requirements. The contribution that restriction and modification makes to limit the frequency of transfer remains to be determined and is a matter of debate (Nielsen, 1998). Restriction and modification (RM) is an enzymologically controlled process that permits the host cell to distinguish self from non-self duplex DNA and degrade the latter (Meselson and Yuan, 1968). DNA that is enzymologically modified in a cell will survive the transfer to another similar cell. Potential hosts carrying different RM systems will thus discriminate between incoming DNA and degrade those sequences not carrying the appropriate modification signals. The activity of an RM system can reduce the transfer frequency by 10^5-fold (Roberts, 1985). It is important in influencing the amount of DNA available for recombination when transfer is achieved by plasmid and phage. However transforming DNA would appear to be insensitive to restriction enzymes (Prozorov, 1999) since it is single-stranded and thus unless it self-anneals to form duplexes or some restriction systems recognized single-stranded DNA, it will be resistant to restriction. Lorenz and Wackernagel (1994) discuss the effects of "restriction" on gene exchange. But it may be that single-stranded DNA does not need to be identified and restricted since if it is not stabilized by recombination it will be degraded within the cell. Thus the distinction between foreign and self is determined by homology which, particularly in respect to chromosomal sequences,

tends to be more conservative (Saunders et al., 1990). Restriction, however, is recognized as an important influence on gene transfer (Fry and Day, 1990a).

This aspect of homology is important for other reasons. The incoming sequence needs to be recognized initially by homology but secondly needs to have recognizable promoters, ribosome binding sites and permissive codon usage patterns to be expressed (Hirsch, 1990). In addition the repair systems that recognize extensive mismatching when recombination occurs during heteroduplex formation may represent a substantial barrier to heterologous genetic exchange (Maynard-Smith et al., 1991).

DNA and the aquatic environment

DNA in the environment is both a nutrient and a source of genes for those strains able to take it up. Table 7.2 shows DNA is degraded/hydrolyzed at a rapid rate (hours) in wastewater (Phillips et al., 1989; Fibi et al., 1991), seawater (Maeda and Taga, 1973, 1974; Bazelyan and Ayzatullin, 1979; Paul et al., 1987; Turk et al., 1992) and fresh water (Paul et al., 1989). In marine sediments (Maeda and Taga, 1973, 1974; Novitzky, 1986) it is degraded at much lower rates. Although DNA may be protected from hydrolysis it can also become biologically inaccessible by binding to clays and other minerals (Lorenz and Wackernagel, 1994). In addition DNA-degrading bacteria may form a substantial fraction (10^5 ml^{-1}) in seawater populations (Maeda and Taga, 1974) and over 90% in other aquatic environments (Greaves and Wilson, 1970; Maeda and Taga, 1974). A longer half-life of "free DNA," in aqueous solution, is supported by the observation that most DNA degrading activity is associated with particles (Maeda and Taga, 1973; Lovett et al., 1989). However, pBR322 can be degraded in as little as 20 min (Phillips et al., 1989) in an aquatic microcosm, indicating that the immediate environment and microbial population structure can be critical. Thus the presence of DNA is dynamic in the environment; it is subject to decay and replacement and is present in the environment at sizes acceptable for efficient transformation to occur.

Cell proximity

Although DNA binds to particulate matter it is probable that more transformation occurs in habitats where active cells are close together. Total bacterial counts average 3×10^6 ml^{-1} in water and 8×10^9 g^{-1} in sediment (Van Es and Meyer-Riel, 1982). In the epilithon, the slimy microbial community on the surface of submerged surfaces, about 5×10^9 bacteria ml^{-1} may be present in a 200-μm thick film (Fry and Day, 1990b). In a biofilm *Streptococcus mutans* cells were transformed up to 600-fold higher than planktonic cells, indicating the peptide pheromone system promoting competence development works best when the cells are crowded together (Li et al., 2001).

Life in competence

In the natural environment bacteria live under nutrient limitation, with occasional bursts of excess! As a consequence individual cells experience "short" periods of growth followed/preceded by periods of starvation (Roszak and Colwell, 1987; Kaprelyants et al., 1993). To enter the state of competence requires *B. subtilis* (Smith et al., 1981) and *A. vinelandii* (Page and Sadoff, 1976) to be in an active metabolic state, but this should not be equated with a constant nutrient rich environment (Dubnau, 1991). A variety of studies have shown competence depends on the imposition of a range of physiological stresses. For example, iron availability is one factor governing competence in *A. vinelandii*

(Page, 1985) and *A. nidulans* (Lorenz and Wackernagel, 1992). When grown aerobically or anaerobically and starved of iron they become competent (Page and von Tigerstrom, 1978). Some environments, for example the rhizosphere, are nutrient rich and relatively iron-free (O'Sullivan and O'Gara, 1992). When starved of iron and molybdenum, competence levels were higher in *A. vinelandii*, than when just starved of iron (Page, 1985). Divalent ions such as Ca^{2+} also have a role in competence development in *A. vinelandii* (Page and Doran, 1981). Logically regulation/induction of competence would not involve the build up of a competence protein as the potential for dilution by diffusion, before a critical concentration was reached, would be too high.

MICROCOSM STUDIES

There have been very few experiments performed. Paul and colleagues (1991) have isolated a high frequency transformation mutant WJT-1C from the *Vibrio* strain D19 (Frischer et al., 1990). This strain was transformed with pure DNA from a Tn5 mutant of the bhr plasmid RSF1010 (Paul et al., 1991). In a sterile sediment and water microcosm, transformation only occurred at a low frequency (2×10^{-7}) in non-sterile water. Nutrient amendment raised the transformation frequency in water (3×10^{-5}), but had no effect on the sediment frequency ($>1 \times 10^{-10}$; Paul et al., 1991).

Transformation occurred, at tenfold higher than the reversion rate, in *P. stutzeri* contained in a sterile marine water/sediment microcosm, amended with chromosomal DNA (carrying a rifampicin-resistance gene (Stewart and Sinigalliano, 1990). Transformation was also observed in a variety of freshwater and marine sediments and prevented by DNAase. Interestingly, under the same conditions the marine *Vibrio* spp. D19, a soil *P. stutzeri* and *V. parahaemolyticus* all performed poorly (Stewart and Sinigalliano, 1990).

B. subtilis was hardly transformable in a freshwater aquifer in the absence of Mg^{2+} (Romanowski et al., 1993), but *Vibrio* sp. (Frischer et al., 1993) did so in marine and *A.*

calcoaceticus (Fry et al., 1992) did so in fresh water. Thus this infers that as species become genetically adapted to their habitats/ecological sites, their transformation activities have evolved in concert.

THE *IN SITU* ASSAY

DNA may be provided in various ways: pure, as a crude lysed cell lysate (Juni, 1972), pasteurized cells or as whole viable cells (Williams, 1993). In Cardiff we have routinely adopted the following approaches. In laboratory transformation matings, early stationary phase cells were mixed and deposited on filters. These filters were placed on PCA and incubated, to allow growth through an appropriate phase for transformation to occur. The cells were suspended and enumerated for donors, recipients and transformants. Initially transformation occurring on selective media (post experiment!) proved a problem, particularly in whole-cell and *in situ* experiments. The addition of DNAase1 (0.5 g ml^{-1}) inhibited transformation with lysates and between whole cells in liquid, thus confirming that the process was transformation. However it did not always prevent transformation occurring on selective media in

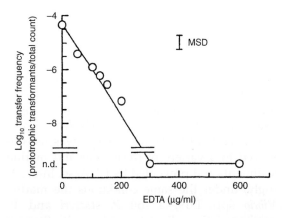

FIGURE 7.2 The effect of EDTA on cell-to-cell transformation frequencies to prototrophy. HGW1501 and HGW1521 (pQM17) were mixed and plated on B22 agar plus EDTA (300 μg ml^{-1}) for 24 h at 20°C. MSD between log$_{10}$ geometric means = 0.508 log units; n.d., no transfer detected.

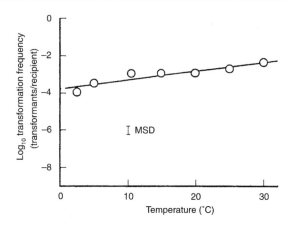

FIGURE 7.3 The effect of temperature on the frequency of transformation frequency of HGW1521 (pQM17) to prototrophy by cell lysate. Each point is the mean of three replicates. MSD ($P = 0.05$) was 0.256 (log units).

control matings with pasteurized donor cells or between whole cells (Williams, 1993).

Thus for *in situ* experiments an unambiguous protocol for monitoring *in situ* transfer events was required. Increasing EDTA concentrations (Figure 7.2) linearly inhibited transformation. EDTA (300 μg ml^{-1}) inhibits plate transformation, but also inhibits/delays cell growth (>100 cell ml^{-1}; Figure 7.3) on minimal media. At higher cell densities fewer effects were seen. When culture filtrates (0.5% w/v) were added to the media, or two plates were inoculated (one selective and the other nutrient based) and taped together so they shared the same atmosphere, transformants grew at the rate of the controls. On this medium transformants were not detected in control experiments when donor and recipient were plated out together.

Laboratory experiments showed both temperature and pH to have importance. In *A. calcoaceticus* transformation was detected at temperatures as low as 2°C in the laboratory, at a frequency of 10^{-4} per recipient (Williams, 1993) and this increased linearly with temperature such that at 30°C it was <10^{-2} (Figure 7.3). *In situ* experiments (Figure 7.4) compare frequencies of HGW1521 transformed to prototrophy with lysates and whole cells. The laboratory mutation rate to prototrophy for this gene is 5×10^{-9} on PCA (Table 7.3). Incubation *in situ* gives a

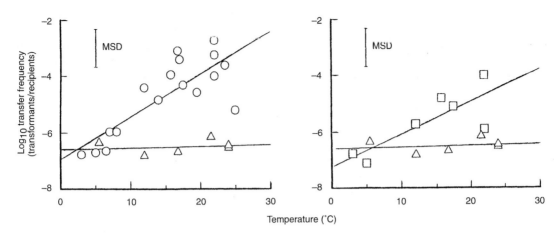

FIGURE 7.4 Effect of temperature on the frequency of transformation *in situ* using cell lysates (O), whole cells (□) and the indigenous microbial population (△) as the source of DNA for transformation. The experiments below 20°C were performed at Cardiff and those above at Hillsborough river, USA. Each point represents the mean of three replicates. The MSD = 1.36 log units × *y*.

TABLE 7.3 A comparison of transformation frequencies obtained from laboratory, microcosm and *in situ* matings

Phenotype	Source of DNA	Laboratory[a]	Microcosm sterile[a]	Microcosm non-sterile[a]	*In situ*[a]
his⁻	Lysate	1×10^{-2} (M)	1×10^{-4} (M)	1×10^{-5} (M)	1×10^{-2} (A) 1×10^{-4} (E)
Hgʳ	Lysate	1×10^{-6} (M)			1×10^{-7} (A) ND (E)
his⁻	Viable cells	1×10^{-4} (M)			$< 1 \times 10^{-4}$ (A) ND (E)
Hgʳ	Viable cells	5×10^{-7} (M)			1×10^{-7} (A) ND (E)
his⁻	Pasteurized cells	7×10^{-4} (M)			
Hgʳ	Pasteurized cells	1×10^{-6} (M)			
his⁻	No cells	5×10^{-9} (M)			2×10^{-7} (E)
Hgʳ	No cells	2×10^{-9} (M)			1×10^{-7} (E)

[a]Representative transformation frequencies for a histidine auxotrophic gene (chromosome) and mercury resistance (plasmid pQM17). A = aquatic, experiments in water; E = epilithon, experiments in a biofilm; M = experiments on a membrane. ND = not detected, i.e. the frequency of transfer are at or below the expected reversion/mutation frequency.

"reversion" frequency of 2×10^{-7} (Table 7.3, Figure 7.4), thus either the *in situ* mutation rate is higher or there is DNA present in a suitable form available for the formation of prototrophic transformants. They could both contribute to the lower line (Figure 7.4), which is the summation of "reversion to prototrophy" and "transformation by indigenous DNA" naturally present in the epilithon. The transformation frequency in the presence of exogenous DNA and "donor" cells is significantly higher and clearly influenced by temperature. In the laboratory *A. vinelandii* transforms optimally between 26 and 37°C (Page and Sadoff, 1976) and *P. stutzeri* increases exponentially between 12 and 20°C, with an optimum between 20 and 37°C (Lorenz and Wackernagel, 1992). In a *Vibrio* spp. the transformation frequencies were high between 15 and 33°C, but fell 100-fold as the temperature rose a further 4°C (Frischer et al., 1993).

Both *A. vinelandii* and *P. stutzeri*, cultured in media with specific pH values, have a narrow peak of high transformation efficiency around pH 7.0 (Page and Sadoff, 1976; Page, 1982; Lorenz and Wackernagel, 1992). The transfer frequency in *A. calcoaceticus* fell three orders of magnitude (Palmen et al., 1993) as the pH changed from 6.7 to 5.4. However pH changes over this range with the same organism, in a different experimental system, had no effect (Rochelle et al., 1988). As the pH of the culture medium of *S. pnemoniae* becomes more alkaline (pH 7.3–8.0) the competence period moves earlier in the growth phase and also at lower cell densities (Chen and Morrison, 1987) inferring the competence factor has more physiological activity in environments with higher pH values. Thus the relationship between the development of competence and environmental stimuli is not straightforward, and results with the same organism using different protocols are likely to be different.

Finally *B. subtilis* was hardly transformable in a freshwater aquifer in the absence of Mg^{2+} (Romanowski et al., 1993), but *Vibrio* spp. (Frischer et al., 1990) did so in a marine environment and *A. calcoaceticus* (Williams, 1993) did so in fresh water. This is a simplistic analysis, but as the latter were both tested in a relevant environment and the former was not, this suggests that as species have become genetically adapted to their habitats/ecological sites, their transformation activities have evolved in concert.

PLANNING *IN SITU* STUDIES

The way an *in situ* experiment is carried out is governed to a large extent by the environment chosen. The numbers of replicates needed may be more than can be physically handled, the design of selective media can confound and then there are the vagaries of weather, its effects on water flow rate, temperature, ionic concentrations, etc. and the "small boy effect." The latter occurs when somebody covertly watches you set up your *in situ* incubation; in Wales it is usually a small boy. When you leave the site, the "small boy" proceeds to wade in and destructively examine the experiment! Thus there is a need to establish as simple and secure a

procedure as is practically feasible and yet one that retains scientific quality.

Performing *in situ* experiments, as described, requires the use of robust strains of *A. calcocaceticus* with good selectable phenotypes. The source of transforming DNA (cells or lysate) may be deposited onto a membrane, as may the recipient. The membranes are placed face-to-face and may be incubated either on agar or, if in a microcosm or *in situ*, on the surface of a stone (Bale et al., 1988). When used *in situ* the stone needs to be marked and anchored.

The controls needed are simple. First, set up parallel experiments, one without donors and the other without recipients. The strains used must have secondary unselected phenotypes, which distinguish them from indigenous bacteria that might grow on the selective media. Molecular procedures for further identification are excellent, if they are economic in time, but generally they are best kept to confirm a proportion of putative transformants.

The choice of strains is critical. A combination of strains that provides the highest transfer rates is optimal. Utilize natural strains from the habitat if possible, as they have ecological relevance. An examination of the literature, especially in conjugation, shows a great deal of conservatism in this aspect. Choice has favored the utilization of well known and established laboratory strains (e.g. *E. coli*) and genetic systems (e.g. RP4) in *in situ* experiments (e.g. Van Elsas et al., 1988). In many cases this choice is apparently made without real consideration of the relevance of environment(s) from where the organisms and transfer systems were derived. The justification (largely unwritten) for this approach appears initially to place reliance on the information known about laboratory models. This means it is quicker to establish experiments as there is no need to identify and characterize a novel transfer system. The second (pessimistic) justification is there is little faith in being able to identify a transfer system for the environment under investigation.

The use of as natural a system as possible is eminently sensible; it means the work has direct and immediate ecological relevance. This provides a potential, in later experiments, to manipulate the strains and conditions in various ways.

In situ transfer experiments require a transfer system that provides a good and "normal" unmodified transfer rate. Gene transfer frequencies obtained from experiments carried out *in situ*, in a natural microbial community, are widely reported to be "depressed" by several orders of magnitude compared with laboratory controls. This means that the higher the laboratory transfer frequency ($\geq10^6–10^9$), the higher level of sensitivity available in *in situ*.

FREQUENCY

There are three sources of DNA for transformation available in different ways. First, there is the gradient in concentration of "old and new" DNA, derived directly from the lysis of a range of organisms, both prokaryotes and eukaryotes. The "DNA population" is historical, it has an age and a diversity which may not be temporarily consistent. In addition there is DNA, released directly from within the community of organisms. This DNA source is related to the immediate community structure, its members and their individual population sizes. Thus these sources of DNA will, it seems, contribute unequally, depending on the "physiological status" (health) of the community. In addition the evidence infers that the term "transformable strains" may be a better descriptive term than "transformable species," as not all members of a species are proficient in taking up DNA. Superimpose on these communities daily (and seasonal) rhythms, which contribute to periods of growth, starvation and death. Add the various regulatory constraints of the transformational process shown by different species and it is clearly a complicated situation. It is confusing from our standpoint and it might be just the same at the molecular level!

Is it rational at present to attempt to estimate *in situ* transformation frequencies/rates? It is difficult to see how a measurement designed for use in controlled conditions in the laboratory can be extrapolated directly and validly to a complex environment. Viable counts taken at "0" time, and used to calculate transfer frequencies do not give the same frequencies when counts taken at 6, 8 or 24 h are used. This is due to a combination of factors, such as mating time, cell population growth, death, ratio of cell types, repeated matings and transfers-on, changes in physiological status, temperature, selection (intended or not), etc. It is probably sensible routinely to enumerate populations before and after, and use the after counts to calculate frequencies. At least the number that survived is known.

I continue to be extremely uncertain what the significance of a transfer frequency calculated in an *in situ* experiment means. To establish a frequency that has ecological significance requires that an ecological perturbation (some effect) has to be demonstrated. A simple analogy would be the determination of the critical dose relationship between the numbers of pathogenic viruses, or bacteria, required to produce disease in an organism. Thus we are at the point where we wish to use/extrapolate *in situ* frequencies of gene transfer, but have little idea of their environmental relevance.

CONCLUSIONS

A general feature of gene transfer experiments in the laboratory is the maintenance of continuous environmental conditions. *In situ* experiments are characterized by an inherently uncontrollable discontinuous environment. The discontinuities in the environment at the microbial level are likely to lead to differing physiologies in adjacent cells, potentially influencing the efficiency and efficacy of genetic communication between them. Thus can those data generated in the laboratory be expected to provide anything but an indication of transfer that will be observed *in situ*?

Thus logically the success of gene transfer by transformation will depend on

(1) the donor and recipient and their immediate physiological states,
(2) the past, current and impending environmental conditions,
(3) the current density of the gene(s) in question,
(4) selection pressure and
(5) serendipity.

What are the chances of one transformant cell, in a biofilm population of millions, surviving to have selection pressure for the gene imposed upon it? The growth/amplification of the transformant within the population will depend on the benefits derived from the expression of the novel gene in response to the selective pressures imposed on the recombinant. It will be the amplification of the transformant that will provide evidence for successful *in situ* exchange. The challenge is now to extend relevant model laboratory systems into their environments and achieve a description of the ecological role of transformation.

ACKNOWLEDGMENTS

I would like to express my sincere thanks to my colleagues, Professors J. Fry and R.V. Miller, together with Drs K. Ashelford and H.G. Williams, for their "scientific enthusiasm" throughout our collaborations. Parts of the work reported here were supported by a Natural Environmental Research Council studentship, an EC contract (BIOT-CT91–0284) and a MAFF contract (CSA 3346).

REFERENCES

Albano, M., Hahn, J. and Dubnau, D. (1987) Expression of competence genes in *Bacillus subtilis*. *J. Bacteriol.* **169**: 3110–3117.

Albritton, W.L., Setlow, J.K. and Slaney, L. (1982) Transfer of *Haemophilus influenzae* chromosomal genes by cell-to-cell contact. *J. Bacteriol.* **152**: 1066–1070.

Albritton, W.L., Setlow, J.K. Thomas, M. et al. (1984) Heterospecific transformation in the genus *Haemophilus*. *Mol. Gen. Genet.* **193**: 358–363.

Alloing, G., Granadel, C., Morrison, D.A. and Claverys, J.-P. (1996) Competence pheromone, oligopeptide permease, and induction of competence in *Streptococcus pneumoniae*. *Mol. Microbiol.* **21**(3): 471–478.

Avery, O.T., MacLoed, C.M. and McCarthy, M. (1944) Studies on the chemical nature of the substance inducing transformation in pneumococcal types. *J. Exp. Med.* **79**: 137–159

Bale, M.J., Day, M.J. and Fry, J.C. (1988) Transfer and occurrence of large mercury resistance plasmids in epilithon. *Appl. Environ. Microbiol.* **54**: 972–978.

Bazelyan, V.L. and Ayzatullin, T.A. (1979) Kinetics of enzymatic hydrolysis of DNA in sea water. *Oceanology* **19**: 30–33.

Bishop, P.E., Dazzo, F.B., Appelbaum, E.R. et al. (1977) Intergeneric transfer of genes involved in the *Ri\obium*-legume symbiosis. *Science* **198**: 938–940.

Börsheim, K.Y. (1993) Native marine bacteriophages. *FEMS Microbiol. Ecol.* **102**: 141–159.

Buitenwerf, J. and Venema, G. (1977) Transformation in *Bacillus subtilis*: Biological and physical evidence for a novel DNA-intermediate in synchronously transforming cells. *Mol. Gen. Genet.* **156**: 145–155.

Carlson, C.A., Pierson, L.S., Rosen, J.I. and Ingraham, J.L. (1983) *Pseudomonas stutzeri* and related species undergo natural transformation. *J. Bacteriol.* **153**: 93–99.

Catlin, B.W. (1956) Extracellular deoxyribonucleic acid of bacteria and a deoxyribonuclease inhibitor. *Science* **124**: 441–442.

Catlin, B.W. (1960) Transformation of *Neisseria meningitidis* by deoxyribonucleates from cells and from culture slime. *J. Bacteriol.* **79**: 579–590.

Chauvat, F., Astier, C., Vedel,. F. and Joset-Espardellier, F. (1983) Transformation in the cyanobacterium *Synechococcus* R2: improvement of efficiency; role of pUH24 plasmid. *Mol. Gen. Genet.* **191**: 39–45.

Chen, J.D. and Morrison, D.A. (1987) Modulation of competence for genetic transformation in *Streptococcus pneumonia*. *J. Gen. Microbiol.* **133**: 1959–1967.

Cheng, G., Campbell, E.A., Naughton, A.M. et al. (1997) The *com* locus controls genetic transformation in *Streptococcus pneumoniae*. *Mol. Microbiol.* **29**: 905–913.

Chung, Y.S., Breidt, F. and Dubnau, D. (1998) Cell surface localisation and processing of the ComG proteins, required for DNA binding during the transformation of *Bacillus subtilis*. *Mol. Microbiol.* **29**(3): 905–913.

Danner, D.B., Deich, R.A., Sisco, K.L. and Smith, H.O. (1980) An eleven-base pari sequence determines the specificity of DNA uptake in *Haemophilus* transformation. *Gene* **11**: 311–318.

Danner, D.B., Smith, H.O. and Narang, S.A. (1982) Construction of DNA recognition sites active in *Haemophilus* transformation. *Proc. Natl Acad. Sci. USA* **79**: 2393–2397.

DeFlaun, M.F., Paul, J.H. and Jeffrey, W.H. (1987) Distribution and molecular weight of dissolved DNA in subtropical estuarine and oceanic environments. *Mar. Ecol. Prog. Ser.* **38**: 65–73.

DeLey, J. (1992) The proteobacteria: ribosomal RNA cistron similarities and bacterial taxonomy. In *The Prokaryotes* (eds, A. Balows, H.G. Trüper, M. Dworkin et al.), pp. 2111–2140, Springer-Verlag, New York.

Doran, J.L. and Page, W.J. (1983) Heat sensitivity of *Azotobacter* vinelandii genetic transformation. *J. Bacteriol.* **155**: 159–168.

Dorward, D.W. and Garon, C.F. (1990) DNA is packaged within membrane-derived vesicles of gram-negative but not gram-positive bacteria. *Appl. Environ. Microbiol.* **56**: 1960–1962.

Dubnau, D. (1991) Genetic competence in *Bacillus subtilis*. *Microbiol. Rev.* **55**: 395–424.

Essich, E., Sevens, E. Jr. and Porter, R.D. (1990) Chromosomal transformation in the cyanobacterium *Agmenellum quadruphicarum*. *J. Bacteriol.* **172**: 1916–1922.

Fibi, M.R., Bröker, M., Schulz, R. et al. (1991) Inactivation of recombinant plasmid DNA from human erythropoietin-producing mouse cell line grown on a large scale. *Appl. Microbiol. Biotech.* **35**: 622–630.

Frischer, M.E., Thurmond, J.M. and Paul, J.H. (1990) Natural plasmid transformation in a high-frequency-of-transformation marine *Vibrio* strain. *Appl. Environ. Microbiol.* **56**: 3439–3444.

Frischer, M.E., Thurmond, J.M. and Paul, J.H. (1993) Factors affecting competence in a high frequency of transformation marine *Vibrio. J. Gen. Microbiol.* **139**: 753–761.

Fry J.C., Day, M.J. and Williams, H.G. (1992) Plasmid and chromosomal gene transfer by transformation in the aquatic environment. In *DNA Transfer and Gene Expression in Microorganisms* (eds, E. Balla, G. Berencsi and A. Szentirmai), pp.111–121, Intercept.

Fry, J.C. and Day, M.J. (1990a) Microbial ecology genetics and risk assessment. In *Release of Genetically Engineered and other Microorganisms* (eds, J.C. Fry and M.J. Day) ch. 12, pp. 160–167, Cambridge University Press, Organisms.

Fry, J.C. and Day, M.J. (1990b) Plasmid transfer in the epilithon. In *Bacterial genetics in natural environments* (eds, J.C. Fry and M.J. Day), pp. 55–80, Chapman and Hall, London.

Gasc, A.M., Garcia, P., Baty, D. and Sicard, A.M. (1987) Mismatch repair during pneumococcal transformation of small deletions produced by site-directed mutagenesis. *Mol. Gen. Genet.* **210**: 369–372.

Golden, S.S. and Sherman, L.A. (1984) Optimal conditions for genetic transformation of the cyanobacterium *Anacystis nidulans* R2. *Bacteriol.* **158**: 36–42.

Goodman, S.D. and Scocca, J.J. (1988) Identification and arrangement of the DNA sequence recognized in specific transformation of *Neisseria gonorrhoeae. Proc. Natl Acad. Sci. USA* **85**: 6982–6986.

Greaves, M.P. and Wilson, M.J. (1970) The degradation of nucleic acids and montmorillonite-nucleic-acid complexes by soil microorganisms. *Soil Biol. Biochem.* **2**: 257–268.

Griffith, F. (1928) The significance of pneumococcal types. *J. Hyg.* **27**: 113–159.

Gromkova, R. and Goodgal, S. (1979) Transformation by plasmid and chromosomal DNAs in *Haemophilus parainfluenzae. Biochem. Biophys. Res. Commun.* **88**: 1428–1434.

Gromkova, R. and Goodgal, S. (1979) Uptake of plasmid deoxyribonucleic acid by *Haemophilus. Bacteriol.* **146**: 79–84.

Gwinn, M.L., Yi, D., Smith, H.O. and Tomb, J.-F. (1996) Role of the two-component signal transduction and the phosphoenolpyruvate:carbohydrate phosphotransferase systems in competence development of *Haemophilus influenzae* Rd. *J. Bacteriol.* **178**: 6366–6368.

Hahn, J., Albans, M. and Dubnau, D. (1987) Isolation and characterization of Tn*917 lac*-generated competence mutants of *Bacillus subtilis. Bacteriol.* **169**: 3104–3109.

Hahn, J., Luttinger, A. and Dubnau, D. (1996) Regulatory inputs for the synthesis of ComK, the competence transcription factor of *Bacillus subtilis. Mol. Microbiol.* **21**(4): 763–775.

Harford, N. and Mergeay, M. (1973) Interspecific transformation of rifampicin resistance in the genus *Bacillus. Mol. Gen. Genet.* **120**: 151–155.

Håvarstein, L.S. and Morrison, D.A. (1999) Quorum sensing and peptide pheromones in streptococcal competence for genetic transformation. In *Cell-cell Signalling in Bacteria* (eds, G.M. Dunny and S.C. Winans), pp.9–26, ASM Press, Washington, DC.

Håvarstein,L.S., Hakenbeck, R. and Gaustad, P. (1997) Natural competence in the genus *Streptococcus*: evidence that streptococci can change phenotype by interspecies recombinational exchange. *J. Bacteriol.* **179**: 6589–6594.

Hirsch, P.R. (1990) Factors limiting gene transfer in bacteria. In *Bacterial Genetics in Natural Environments* (eds, J.C. Fry and M.J. Day), 1st edn, vol. 1, pp. 31–40, Chapman and Hall, London.

Haverstein 1977

Ippen-Ihler, K. (1989) Bacterial conjugation. In *Gene Transfer in the Environment* (eds, S.B. Levy and R.V. Miller), pp. 33–72, McGraw-Hill Book Co., New York.

Juni, E. (1972) Interspecies transformation of *Acinetobacter*. Genetic evidence for a ubiquitous genus. *J. Bacteriol.* **112**: 917–931

Kahn, M.E., Barany, F. and Smith, H.O. (1983) Transformasomes: specialized membraneous structures which protect DNA during *Haemophilus* transformation. *Proc. Natl Acad. Sci. USA* **80**: 6927–6931.

Kaprelyants, A.S., Gottschal, J.C. and Kell, D.B. (1993) Dormancy in nonsporulating bacteria. *FEMS Microbiol. Rev.* **104**: 271–286.

Karl, D.M. and Bailiff, M.D. (1989) The measurement and distribution of dissolved nucleic acids in aquatic environments. *Limnol. Oceanogr.* **34**: 543–558.

Khasanov, F.K., Zvingila, D.J., Zainullin., A.A. et al. (1992) Homologous recombination between plasmid and chromosomal DNA in *Bacillus subtilis* requires approximately 70 bp of homology. *Mol. Gen. Genet.* **234**: 494–497.

King, S.R. and Richardson, J.P. (1986) Role of homology and pathway specificity for recombination between plasmids and bacteriophage. *Mol. Gen. Genet.* **204**: 141–147.

Kokjohn, T.A. (1989) Transduction: mechanism and potential for gene transfer in the environment. In *Gene Transfer in the Environment* (eds, S.B. Levy and R.V. Miler), pp. 73–97, McGraw-Hill Book Co., New York.

Leff, L.G., McArthur, J.V. and Shimkets, L.J. (1992) Information spiraling: movement of bacteria and their genes in streams. *Microb. Ecol.* **24**: 11–24.

Li, Y-H., Lau, P.C.Y., Lee. J.H. et al. (2001) Natural genetic transformation of *Streptococcus mutans* growing in biofilms. *J. Bacteriol.* **183**: 897–908.

Liu, J. and Zuber, P. (1998) A molecular switch controlling competence and motility: Competence regulatory factors ComS, MecA, and ComK control [D]-dependent gene expression in *Bacillus subtilis. J. Bacteriol.* **180**: 4243–4251.

Lorenz, M.G., Gerjets, D. and Wackernagel, W. (1991) Release of transforming plasmid and chromosomal DNA from two cultured soil bacteria. *Arch. Microbiol.* **156**: 319–326.

Lorenz, M.G., Reipschläger, K. and Wackernagel, W. (1992) Plasmid transformation of naturally competent *Acinetobacter calcoaceticus* in non-sterile soil extract and groundwater. *Arch. Microbiol.* **157**: 355–360.

Lorenz, M.G. and Wackernagel, W. (1992) DNA binding to various clay mineral and retarded enzymatic degradation of DNA in a sand/clay microcosm. In *Gene Transfer and Environment* (ed., M.J. Gauthier), pp. 103–113, Springer-Verlag KG, Berlin.

Lorenz, M.G. and Wackernagel, W. (1993) Transformation as a mechanism for bacterial gene transfer in soil and sediment-studies with a sand/clay microcosm and the cyanobacterium *Synechocystis* OL50. In *Trends in Microbial Ecology* (eds, R. Guerrero and C. Pedros-Alio), pp. 325–330, Spanish Society for Microbiology, Barcelona.

Lorenz, M.G. and Wakernagel, W. (1994) Bacterial gene transfer by natural genetic transformation in the environment. *Microbiol. Rev.* **58**: 563–602.

Lovett, C.M., Love, P.E. and Yasbin, R.E. (1989) Competence-specific induction of the *Bacillus subtilis* RecA protein analog: evidence for dual regulation of a recombination protein. *J. Bacteriol.* **171**: 2318–2322.

MacFadyen, L.P., Dorocicz, R., Reizer, J. et al. (1996) Regulation of competence development and sugar utilisation in *Haemophilus influenzae* Rd by a phospho-enolpyruvate:fructose phosphotransferase system. *Mol. Microbiol.* **21**: 941–952.

Maeda, M. and Taga, N. (1973) Deoxyribonuclease activity in seawater and sediment. *Mar. Biol.* **20**: 58–63.

Maeda, M. and Taga, N. (1974) Occurrence and distribution of deozyribonucleic acid hydrolyzing bacteria in seawater. *J. Exp. Mar. Biol. Ecol.* **14**: 157–169.

Mahajan, S.K.I. (1988) Pathway of homologous recombination in *Escherichia coli*. In. *Genetic recombination* (eds, G.R. Smith and R. Kucherlapati), pp. 88–140, American Society for Microbiology, Washington, DC.

Masters, M. and Broda, P. (1971) Evidence for the bidirectional replication of the *Escherichia coli* chromosome. *Nature* **232**: 137–140.

Mathis, L.S. and Scocca, J.J. (1982) *Haemophilus influenzae* and *Neisseria gonorrhoeae* recognize different specificity determinants in the DNA uptake step of genetic transformation. *J. Gen. Microbiol.* **128**: 1159–1161.

Maynard Smith, J., Dowson, C.G. and Spratt, B.G. (1991) Localised sex in bacteria. *Nature* **349**: 29–31.

Mejean, V. and Claverys, J.-P. (1993) DNA processing during entry in transformation of *Streptococcus pneumoniae*. *J. Biol. Chem.* **268**: 5594–5599.

Meselson, M. and Yuan, R. (1968) DNA restriction enzyme from *E. coli*. *Nature* **217**: 1110–1114.

Msadek, T., Kunst, F., Klier, A. and Rapoport, G. (1991) DegS-DegU and ComP-ComA modulator-effector pairs control expression of the *Bacillus subtilis* pleiotropic regulatory gene *degQ*. *J. Bacteriol.* **173**: 2366–2377.

Newberry (2001) PhD thesis, University of Wales, Cardiff, Wales.

Nielsen, K.M. (1998) Barriers to horizontal gene exchange by natural transformation in soil bacteria. *APMIS.* (suppl. 106) **84**: 77–84.

Norgard, M.V. and Imaeda, T. (1978) Physiological factors involved in the transformation of *Mycobacterium smegmatis. J. Bacteriol.* **133**: 1254–1262.

Novitsky, J.A. (1986) Degradation of dead microbial biomass in a marine sediment. *Appl. Environ. Microbiol.* **52**: 504–509.

Ogura, M. and Tanaka, T. (1996) *Bacillus subtilis* DegU acts as a positive regulator for *comK* expression. *FEBS Letters* **397**: 173–176.

Ormerod, J.G. (1988) Natural genetic transformation in *Chlorobium*. In Green photosynthetic bacteria (eds, J.M. Olson, J.G. Ormerod, J. Amesz, et al.), pp. 315–319, Plenum Press, New York.

O'Sullivan, D.J. and O'Gara, F. (1992) Traits of fluorescent *Pseudomonas* spp. involved in suppression of plant root pathogens. *Microbiol. Rev.* **56**: 662–676.

Page, W.J. (1982) Optimal conditions for competence development in nitrogen-fixing *Azotobacter vinelandii. Can. J. Microbiol.* **28**: 389–397.

Page, W.J. (1985) Genetic transformation of molybdenum-starved *Azotobacter vinelandii*: increased transformation frequency and recipient range. *Can. J. Microbiol.* **31**: 659–662.

Page, W.J. and Doran, J.L. (1981) Recovery of competence in calcium-limited *Azotobacter vinelandii. J. Bacteriol.* **146**: 33–40.

Page, W.J. and Grant, G.A. (1987) Effect of mineral iron on the development of transformation competence in *Azotobacter vinelandii. FEMS Microbiol. Lett.* **41**: 257–261.

Page, W.J. and Sadoff, H.L. (1976) Physiological factors affecting transformation of *Azotobacter vinelandii. J. Bacteriol.* **125**: 1080–1087.

Page, W.J. and von Tigerstrom, M. (1978) Induction of transformation competence in *Azotobacter vinelandii* iron-limited cultures. *Can. J. Microbiol.* **24**: 1590–1594.

Palmen, R. and Hellingwerf, K.J. (1995) *Acinetobacter calcoaceticus* liberates chromosomal DNA during induction of competence by cell lysis. *Current Microbiol.* **30**: 7–10.

Palmen, R., Vosman, B., Kok, R. et al. (1992) Characterization of transformation-deficient mutants of *Acinetobacter calcoaceticus. Mol. Microbiol.* **6**: 1747–1754.

Palmen, R., Vosman, B., Buijsman, B. et al. (1993) Physiological characterization of natural transformation in *Acinetobacter calcoaceticus. J. Gen. Microbiol.* **139**: 295–305.

Palmen, R., Buijsman, B. and Hellingwerf, K.J. (1994) Physiological regulation of competence induction for natural transformation in *Acinetobacter calcoaceticus. Arch. Microbiol.* **162**: 344–351.

Paul, J.H. and Carlson, D.J. (1984) Genetic material in the marine environment: implication for bacterial DNA. *Limnol. Oceanogr.* **29**: 1091–1097.

Paul, J.H., Jeffrey, W.H. and DeFlaun, M.F. (1987) Dynamics of extracellular DNA in the marine environment. *Appl. Environ. Microbiol.* **53**: 170–179.

Paul, J.H. and David, A.W. (1989) Production of extracellular nucleic acids by genetically altered bacteria in aquatic-environment microcosms. *Appl. Environ. Microbiol.* **55**: 1865–1869.

Paul, J.H., Frischer, M.E. and Thurmond, J.M. (1991) Gene transfer in marine water column and sediment microcosms by natural plasmid transformation. *Appl. Environ. Microbiol.* **57**: 1509–1515.

Paul, J.H., Jeffrey, W.H., David, A.W. et al. (1989) Turnover of extracellular DNA in eutrophic and oligotrophic freshwater environments of south west Florida. *Appl. Environ. Microbiol.* **55**: 1823–1828.

Paul, J.H., Jeffrey, W.M. and DeFlaun, M.F. (1987) Dynamics of extracellular DNA in the marine environment. *Appl. Environ. Microbiol.* **53**: 70–179.

Phillips, S.J., Dalgarn, D.S. and Young, S.K. (1989) Recombinant DNA in wastewater: pBR322 degradation kinetics. *J. Water Pollut. Control Fed.* **62**: 1588–1595.

Proserov, A.A. (1999) Horizontal gene transfer in bacteria: laboratory modelling, natural populations and data from genome analysis. *Microbiology* **68**: 551–564.

Redfield, R.J. (1988) Evolution of bacterial transformation: Is sex with dead cells ever better than no sex at all? *Genetics* **119**: 213–221.

Redfield, R.J (1993) Evolution of natural transformation – testing the DNA-repair hypothesis in *Bacillus subtilis* and *Haemophilus influenzae. Genetics* **133**: 755–761.

Roberts, R.J. (1985) Restriction and modification enzymes and their recognition sequences. *Nucleic Acid Res.* **13**: 165–200.

Rochelle, P.A., Day, M.J. and Fry, J.C. (1988) Occurrence, transfer and mobilisation in epilithic strains of *Acinetobacter* of mercury-resistance plasmids capable of transformation. *J. Gen. Microbiol.* **134**: 2933–2941.

Romanowski, G., Lorenz, M.G. and Wackernagel, W. (1993) Plasmid DNA in a groundwater aquifer microcosm – adsorption, DNAse resistance and natural genetic transformation of *Bacillus subtilis. Mol. Ecol.* **2**: 171–181.

Roszak, D.B. and Colwell, R.R. (1987) Survival strategies of bacteria in the natural environment. *Microbiol. Rev.* **51**: 365–379.

Saunders, C.W. and Guild, W.R. (1981) Pathway of plasmid transformation in pneumococcus: open circular and linear molecules are active. *J. Bacteriol.* **146**: 517–526.

Saunders, J.G., Morgan, J.A.W., Winstanley, C. et al. (1990) 1. Genetic approaches to the study of gene transfer in microbial communities. In *Bacterial Genetics in Natural Environments* (eds, J.C. Fry and M.J. Day), 1st edn, pp. 3–21, Chapman and Hall, London.

Servais, P., Billen, G. and Vives-Rigo, J. (1983) Rate of bacterial mortality in aquatic environments *Appl. Environ. Microbiol.* **49**: 1448–1454

Sinha, R.P. and Iyer, V.N. (1971) Competence for genetic transformation and the release of DNA from *Bacillus subtilis.* Biochim. *Biophys. Acta* **232**: 61–71.

Sisco, K.L. and Smith, H.O. (1979) Sequence-specific DNA uptake in *Haemophilus* transformation. *Proc. Natl Acad. Sci. USA* **76**: 972–976.

Smith, H.O., Danner, D.B. and Deich, R.A. (1981) Genetic transformation. *Annu. Rev. Biochem.* **50**: 41–68.

Solomon, J.M. and Grossman, A.D. (1996) Who's competent and when: regulation of natural genetic competence in bacteria. *Trends Genet.* **12**(4, April): 150–155.

Sparling, P.F. (1966) Genetic transformation of *Neisseria gonorrhoeae* to streptomycin resistance. *J. Bacteriol.* **92**: 1364–1371.

Stewart, G.J. (1989) The mechanism of natural selection. In Releases of Genetically Engineered and other Microorganisms, Plant and Microbial Biotechnology Research Series 2 (eds, M.J. Day and J.C. Fry), pp. 82–93, Cambridge University Press, Cambridge.

Stewart, G.J. (1992) Gene transfer in the environment: transformation. In *Release of Genetically Engineered and Other Micro-organisms. Plant and Microbial Bio-technology Research Series 2* (eds, J. Lynch, M.J. Day and J.C. Fry), pp. 82–93, Cambridge University Press.

Stewart, G.J. and Carlson, C.A. (1986) The biology of natural transformation. *Annu. Rev. Microbiol.* **40**: 211–235.

Stewart, G.J. and Sinigalliano, C.D. (1989) Detection and characterisation of natural transformation in the marine bacterium *Pseudomonas stutzeri* strain Zobell. *Arch. Microbiol.* **152**: 520–526.

Stewart, G.J. and Sinigalliano, C.D. (1990) Detection of horizontal gene transfer by natural transformation in native and introduced species of bacteria in marine and synthetic sediments. *Appl. Environ. Microbiol.* **56**: 1818–1824.

Stone, B.J. and Kwaik, Y.A. (1999) Natural competence for DNA transformation *by Legionella pneumophila* and its association with expression of type IV pili. *J. Bact.* **181**: 1395–1402.

Stotzky, G. (1989) Gene transfer among bacteria in soil. In *Gene Transfer in the Environment* (eds, S.B. Levy and R.V. Miller), pp. 165–222, McGraw-Hill Book Co., New York.

Takahashi, I. and Gibbons, N.E. (1957) Effect of salt concentration on the extracellular nucleic acids of *Micrococcus halodenitrificans. Can. J. Microbiol.* **3**: 687–694.

Turgay, K., Hamoen, L., Venema, G. and Dubnau, D. (1997) Biochemical characterisation of a molecular switch involving the heat shock protein ClpC, which controls the activity of ComK, the competence transcription factor of *Bacillus subtilis. Genes Dev.* **11**: 119–128.

Turk, V., Rehnstam, A.S., Lundberg, E. and Hagström, A. (1992) Release of bacterial DNA by marine nano-flagellates, an intermediate step in phosphorus regeneration. *Appl. Environ. Microbiol.* **58**: 3744–3750.

Van Elsas, J.D., Trevors, J.T., Starodub, M.E. and Van Overbeek, L.S. (1988) Transfer of plasmid RP4 between pseudomonads after introduction into soil: Influence of spatial and temporal aspects of inoculation. *FEMS Microb. Ecol.* **73**: 1–12.

Van Es, R.B. and Meyer-Riel, L.A. (1982) Biomass and metabolic activity of heterotrophic marine bacteria. *Adv. Microb. Ecol.* **6**: 111–170.

Vosman, B. and Hellingwerf, K.J. (1991) Molecular cloning and functional characterization of a *recA* analog from *Pseudomonas stutzeri* and construction of a *P. stutzeri recA* mutant. *Antonie van Leeuwenhoek. J. Microbiol.* **59**: 115–123.

Williams, H.G. (1993) Plasmid and chromosomal gene transfer by transformation in river epilithon. PhD thesis, University of Wales, Cardiff, Wales.

Wolfgang, M., van Putten, J.P.M., Hayes, S.F. and Koomey, M. (1999) The *com*P locus of *Neisseria gonorrhoeae* encodes a type IV prepilin that is dispensable for pilus biogenesis but essential for natural transformation. *Mol. Microbiol.* **31**(5): 1345–1357.

Worrell, V.E., Nagle, D.P., McCarthy, D. and Eisenbraun, A. (1988) Genetic transformation system in the archaebacterium *Methanobacterium thermoautotrophicum* Marburg. *J. Bacteriol.* **170**: 635–656.

Yankofsky, S.A., Gurevich, R., Grimland, N. and Stark, A.A. (1983) Genetic transformation of obligately chemolithotrophic thiobacilli. *J. Bacteriol.* **153**: 652–657.

Yother, J., McDaniel, L.S. and Briles, D.E. (1986) Transformation of encapsulated *Streptococcus pneumoniae. J. Bacteriol.* **168**: 1463–1465.

8

Pseudolysogeny: A Bacteriophage Strategy for Increasing Longevity *in situ*

Robert V. Miller and Steven A. Ripp

Bacteriophages occur in high numbers in natural environments and are significant mediators of microbial survival, horizontal gene transfer and evolution. These findings have stimulated renewed interest in identifying natural bacterial–bacteriophage interactions. Bacteriophage particles remain infective for only short periods of time in natural milieus. Burst sizes of phages produced in lytic infections decrease as cells starve and latency periods increase. However, hosts starved for extended periods of time (up to 5 years) remain susceptible to phage infection. These characteristics suggest that bacteriophages have unique strategies for maintaining their high concentrations in nature that have not been appreciated from laboratory studies. Perhaps the most significant of these is increases in the effective half-life of their genomes in nature by establishing pseudolysogenic relationships between bacteriophages and their starved hosts. Pseudolysogeny is defined here as a class of phage–host interactions where the viral nucleic acid remains viable but not as a prophage. Neither is a lytic infection elicited. Instead, an unstable relationship exists where the phage nucleic acid simply resides in limbo within the cell in a non-active form. Due to the cell's starved state, there appears to be insufficient energy available for the phage to initiate either a temperate or lytic life cycle, pseudolysogeny substantially increases phage half-lives, leading to maintenance of viruses that otherwise would not exist in the environment. Pseudolysogeny explains, at least in part, the large bacterial virus populations observed in natural environments.

INTRODUCTION

Bacterial viruses (bacteriophages) have been long recognized as important in the ecology of microorganisms (D'Herelle, 1949). Virus-mediated horizontal transfer of genetic material (transduction) is now recognized as a significant gene-exchange mechanism among bacteria in many natural habitats. Transduction has been observed in fresh water (Morrison et al., 1978; Saye et al., 1987, 1990; Ripp and Miller, 1995; Miller, 1998) environments, marine habitats (Jiang and Paul, 1998), in wastewater treatment facilities (Osmond and Gealt, 1988), on the surfaces of plants (Kidambi et al., 1994), within shellfish (Baross et al., 1978), and in the kidneys of mice (Jarolmen et al., 1965; Novic and Morse, 1967). Both plasmid and chromosomal DNA are transferred among bacteria by transduction, suggesting that this mechanism of gene transfer may substantially impact bacterial evolution (Miller, 1998). Continuous culture model systems support this hypothesis by demonstrating the ability of transduction to maintain novel phenotypes in the bacterial gene pool from which they would otherwise be eliminated (Replicon et al., 1995).

Since its discovery (Zinder and Lederberg, 1952), transduction was considered unlikely to be important to bacterial ecology and evolution

Horizontal Gene Transfer
ISBN: 0-12-680126-6

Copyright © 2002 by Academic Press.
All rights of reproduction in any form reserved.

because it was believed that the environmental frequencies of bacteriophages would not support phage–host interactions of any significance (Miller and Sayler, 1992). However, recent studies have demonstrated that the numbers of bacterial viruses in environmental ecosystems are much higher than suspected, often reaching levels of 10^8 particles per milliliter or higher in natural marine and aquatic environments (Bergh et al., 1989; Proctor and Fuhrman, 1990; Paul et al., 1991; Miller and Sayler, 1992; Miller et al., 1992). Such high levels of viruses (often greater than the concentrations of bacterial hosts encountered in these environments (Miller et al., 1992) would support substantial levels of phage–host interaction. These observations highlight the potential importance of bacteriophages in the natural ecology and evolution of microbial communities (Miller and Sayler, 1992).

Discovery of the high levels of bacteriophage particles in the environment led to many new questions about the interactions between bacteria and these pathogens. Because laboratory obtained data could not explain the maintenance of such high concentrations of viruses *in situ*, a number of studies have been undertaken to explore phage–host interactions in natural habitats. From these studies, the importance of the phenomenon of pseudolysogeny has been recognized in the natural ecology of bacteriophages. Pseudolysogeny describes a phage–host interaction in which the phage, upon infecting its host, elicits an unstable, non-productive state (Baess, 1971). Neither lysogeny nor lytic growth is initiated. We believe that this response occurs because the host cell is starved and cannot provide the phage with the necessary energy and substrates required for alternative responses. Cellular starvation in environmental ecosystems is commonplace. When a nutrient source is provided, however, the phage acquires essential energy to initiate virion formation or establish stable lysogeny. The overall result of pseudolysogeny is to extend the viral genome's life time in the ecosystem, explaining, in part at least, the large environmental phage populations.

In this chapter, we explore several aspects of environmentally important interactions between bacteria and their viral parasites in an attempt to gain insight into the origins of the high environmental titers of bacteriophages. Pseudolysogeny is emphasized here and its importance in providing a reservoir for viral genomes in many unfriendly environments is discussed.

SURVIVAL OF BACTERIOPHAGE VIRIONS IN THE ENVIRONMENT

In many of the initial studies in which high levels of virus-like particles were observed in the environment, the level of infectivity of these particles was not explored. This led to renewed interest in determining the effective lives of bacterial virions in natural environments. Upon investigation, the half-lives of bacterial virus particles were found to be very short in natural habitats. Saye et al. (1987) showed that the infective half-life of virions of the *Pseudomonas aeruginosa* phage F116 was substantially reduced in lake water. F116 virions that had a half-life of greater than 15 days in Luria Broth, only remained infective for less than one day in water obtained from Fort Loudon Lake near Knoxville, TN. Ogunseitan et al. (1990) observed that *P. aeruginosa* phage UT1 isolated directly from Fort Loudon Lake had an infective half-life of 29 h or less in sterilized Fort Loudon Lake water. The survival rate of virions was further reduced to 18 h when the natural bacterial community of Fort Loudon Lake was included in the incubation compared with incubation in sterilized lake water. Suttle and Chen (1992) determined bacteriophage virions in marine waters lost 50% of their viability in approximately 30 h. These and other studies have made it clear that the presence of high levels of virus particles in natural aquatic environments cannot be explained simply by an extremely long infective life of their virions.

EFFECT OF STARVATION ON BACTERIOPHAGE–HOST INTERACTIONS

Kokjohn et al. (1991) and Schrader et al. (1997) explored the abilities of several *P. aeruginosa*- and *Escherichia coli*-specific bacteriophages to infect and multiply on long-term, severely-starved hosts. These studies revealed that the latency periods of these infections were significantly elongated in starved hosts. *P. aeruginosa* phage F116 had a latency period of 100–110 min in well-fed laboratory hosts but at least 240 min were observed between infection and lysis of starved hosts. The latency period of *P. aeruginosa* phage UT1 increased from 70–80 min in exponential hosts to greater than 110 min in starved hosts (Kokjohn et al., 1991). These results correlate well with the extended generation times of the starved host cells. Similar results were observed with other *P. aeruginosa* and *E. coli* phages (Schrader et al., 1997). Here again latent periods of phage infection were increased by as much as fourfold in severely-starved hosts. *P. aeruginosa* cells starved for up to 5 years were found to support bacteriophage growth (Schrader et al., 1997), suggesting that starvation does not offer bacteria a refuge from bacteriophage infection in the environment.

In addition to elongated latency periods, burst sizes of progeny phage particles are severely impacted in starved hosts. Bacteriophage ACQ has a burst size of 10^3 in exponential phase *P. aeruginosa* hosts, but only an average of 10^2 particles are produced in starved hosts (Schrader et al., 1997). *P. aeruginosa* phage F116 burst size decreases from 27 phage particles per infected cell in fed hosts to an average of four particles in starved hosts, while phage UT1 bursts drop to six particles in starved hosts compared with an average of 65 virions produced from each infected exponentially-growing host (Kokjohn et al., 1991).

FEAST AND FAMINE IN THE MICROBIAL WORLD

In the environment, bacterial cells are most likely to be confronted with chronic starvation broken at random intervals by short periods of growth due to transient availability of nutrients. We conducted a study (Miller and Ripp, 1998) in which we added yeast extract to a starved culture containing two *P. aeruginosa* strains. The first strain was lysogenic for bacteriophage F116 and the second served as a recipient cell

sensitive to infection by F116. The culture medium consisted of *Pseudomonas* Minimal Medium (Miller and Ku, 1978) lacking sodium citrate (PMM-c). A very low level of nutrient (yeast extract at $10^{-5}\%$) was added initially allowing the cells to starve quickly. Each week, a spike of nutrient consisting of yeast extract at the very low concentration of $10^{-5}\%$ was added to the culture. These conditions were designed to simulate the feast–famine conditions, typically encountered by microorganisms in the natural environment. Within 1–2 days of this spike, an increase in the phage-to-bacterium ratio (PBR) and in the percentage of cells exhibiting lysogenic characteristics was seen (Figure 8.1). Nutrient spikes continued on a weekly basis up to 110 days, with PBR and lysogeny (measured by determining the number of clones releasing phage virions) increases occurring after each weekly nutrient spike. No increases were observed in sister cultures that were not spiked with nutrient. Starved *E. coli* cells show a similar response when infected with bacteriophage λ (Kourilsky, 1973). These data illustrate that nutrient availability dramatically affects phage–host interactions and that, under starvation conditions, the establishment of true lysogeny and the production of virions are minimal at best. They suggest that bacterial viruses must have developed reservoirs that allow their continued existence during times of starvation of their hosts.

EVIDENCE FOR A PSEUDOLYSOGENIC STATE BETWEEN BACTERIOPHAGES AND THEIR HOSTS IN NATURE

Environmental studies have clearly revealed that neither rapid, highly-productive lytic infections, extremely long-lived virions, nor high levels of induction of lysogens to lytic growth can solely account for the high concentrations of viruses observed in nature. Thus we began a search for natural reservoirs of environmental bacteriophages. Seventy per cent of *P. aeruginosa* cells directly isolated from Fort Loudon Lake were found to contain phage-specific homologous DNA sequences

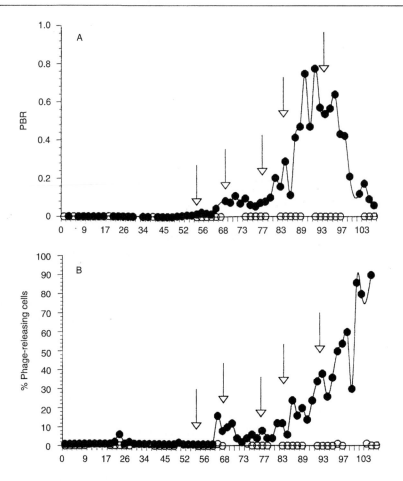

FIGURE 8.1 (A) Phage-to-bacterium ratio (PBR) and (B) percentage of cells actively releasing F116 phage in microcosms containing PMM-c. One-half of the microcosms were spiked with nutrient at times indicated by arrows. The other half was not spiked with nutrient. Bacteriophage counts were determined from 0.45 μm filtered samples that remove bacterial cells. Cells actively releasing phages were identified by replica-plating colonies onto a lawn of indicator bacteria. O, no nutrient added; ●, nutrient added. (Reprinted with permission from Miller and Ripp, 1998.)

when probed with nucleic acid from Fort Loudon Lake phages (Ogunseitan et al., 1992). However, of these, only 1–7% actually exhibited the characteristics associated with lysogeny. The condition of the remainder suggested they had a pseudolysogenic relationship with the phage genomes they contained.

Pseudolysogeny was first described by Twort (1915) but was not fully defined until Romig and Brodetsky (1961) used the term to describe the relationship between various soil bacilli and their viruses. Its true importance to environmental phage–host relationships and its

ecological consequences have only recently been appreciated.

Pseudolysogeny defines a condition in which the starved host cell coexists with a viral genome in an unstable relationship for extended periods (Baess, 1971; Ripp and Miller, 1997, 1998). In such cases, the phage genome has the potential of enduring for extended periods of starvation of a host, that, through evolutionary modifications, has adapted itself for survival under harsh environmental conditions (Roszak and Colwell, 1987). In so doing, the pseudoprophage finds a safe haven in which to survive. As nutrient supplies become more favorable, the unstable pseudolysogens are converted to stable lysogens or the lytic response is activated, resulting in the release of progeny virions.

Pseudolysogeny has been observed in many species of bacteria. In addition to the observations made by Romig and Brodestsky (1961) that soil isolates of *Bacillus subtilis* demonstrate pseudolysogeny, Bramucci et al. (1977) identified a pseudolysogenic phage from a soil isolate of *Bacillus pumilus*. Pseudolysogeny has been observed in several species of mycobacteria as well (Baess, 1971; Grange, 1975; Grange and Bird, 1975). Drozhevkina et al. (1984) inspected over 1000 isolates of *Vibrio cholerae* strains from ponds, sewage, and fecal samples. They found that approximately 1% of the strains were pseudolysogens. Pseudolysogeny of *Azotobacter vinlandii* by phage A21 was studied by Thompson et al. (1980a,b). They observed conversion of the pseudolysogenic state to a stable one of true lysogeny. Unstable pseudolysogens have also been identified in *Acholeplasma laidlawii* (Roger, 1983). Wall et al. (1975) identified "phage carries" which they believed to be pseudolysogens among a collection of strains of the photosynthetic bacterium *Rhodopseudomonas capsulata*. A similar phage-carrier state was observed among *Bacteroides frageilis* strains using phages isolated from sewage (Booth et al., 1979). Pseudolysogens of group A streptococci were found to be associated with toxigenic conversion (Nida and Ferretti, 1982). Phage isolated from soil and animal droppings have been shown to exist in a state of pseudolysogeny in *Myxococcus virescens* and *Myxococcus fulvus* (Brown et al., 1976). Pseudolysogenic states have also been identified in soil *Streptomyces* spp. (Marsh and Wellington, 1992). Thus, it is becoming clear that pseudolysogeny is spread widely among divergent environments and a wide variety of bacterial hosts.

PHYSIOLOGICAL STUDIES ON THE PSEUDOLYSOGENIC STATE IN TEMPERATE BACTERIOPHAGE

We carried out a number of studies using *P. aeruginosa* and its temperate bacteriophage F116,

investigating the physiology of pseudolysogeny. The first experiments were designed to determine if long-term pseudolysogenic relationships could be resolved into active virion production or stable, truly lysogenic states. Even after 43 days of existence in a pseudolysogenic state in a starved host, nutrient spikes still led to production of virions and establishment of lysogeny (Miller and Ripp, 1998). The second set of experiments were carried out in continuous culture (Replicon et al., 1995) and allowed study of pseudolysogenic populations for extended periods of time (Miller and Ripp, 1998). Chemostats were run at a generation time (turnover time) of 14 h. Samples were removed and assayed for the number of cells actively releasing phage (i.e. true lysogens) and, through colony hybridization, the total number of cells containing phage F116 DNA. The difference between these two values represents the number of pseudolysogens in the sample. A large but variable proportion of the population of cells that contained F116 DNA appeared to exist in a pseudolysogenic state (18–83%).

Individual cells from those colonies exhibiting pseudolysogeny were tested for hybridization to F116 DNA. Only a small number of cells from the original pseudolysogenic colonies were found actually to contain phage F116 DNA. Thus, it appears that the pseudoprophage is not a stable entity in pseudolysogens. We hypothesize that due to the starved state of the host cell, there is not enough energy available for the infecting phage genome to become stably established in the host cell. Therefore, this pseudoprophage may not be replicated in synchrony with the host genome. This phenomenon is similar to abortive transduction, where transduced DNA is not replicated with the cell's genome (Arber, 1994). Only one of the two daughter cells from each division cycle acquires the exogenote DNA

In another set of experiments, similar results were obtained when microcosms were incubated *in situ* in a small, semi-oligotrophic freshwater lake (Miller and Ripp, 1998). Only microcosms incubated on the lake's surface and spiked with yeast extract exhibited large concentrations of cells infected with phages. Of these surface microcosms, a high percentage contained phage DNA but did not actively release virions (pseudolysogens). The microcosms incubated on the lake bottom produced virtually no phage-releasing cells. However, approximately 20% of this starved population also exhibited pseudolysogenic characteristics by the end of the experiment. The lack of phage-producing cells in these bottom-incubated microcosms was most likely due to the lack of exposure to solar UV radiation, which has previously been shown to induce F116 prophages to lytic growth (Kokjohn and Miller, 1985; Kidambi et al., 1996).

PSEUDOLYSOGENY WITH VIRULENT BACTERIOPHAGES

A vast literature clearly demonstrates the widespread occurrence of pseudolysogeny in bacteria infected with temperate bacteriophages. Pseudolysogeny is likely to serve as an environmental reservoir for these bacteriophages. We wished to determine if pseudolysogeny could also be established by virulent phages in starved hosts. Previously, experiments in which UT1, a virulent freshwater *P. aeruginosa* phage, was mixed with indigenous bacterial

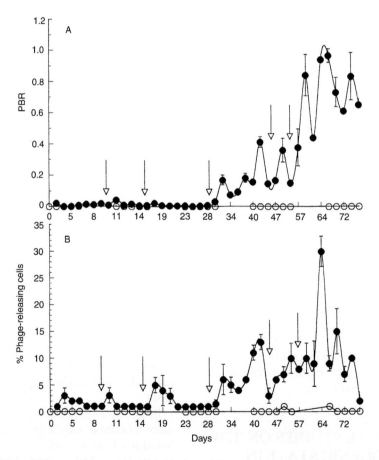

FIGURE 8.2 (A) PBR and (B) percentage of cells releasing UT1 phage in a microcosm containing PMM-c. One-half of the microcosms were spiked with nutrient at times indicated by arrows. The other half was not spiked with nutrient. Bacteriophage counts were determined from 0.45 μm filtered samples that remove bacterial cells. Cells actively releasing phages were identified by replica-plating colonies onto a lawn of indicator bacteria. ○, no nutrient added; ●, nutrient added. (Reprinted with permission from Ripp and Miller, 1997.)

hosts in lake-water microcosms sustained over a 45-day period suggested that such long-term maintenance of virulent phages was possible (Ogunseitan et al., 1990). Bacterial densities in these microcosms initially decreased but then stabilized. Colony hybridizations of host cells using phage UT1 DNA probes indicated that 45% of the total recoverable colonies contained UT1 genetic material. This suggests that in nature an equilibrium between starved hosts and phages can be implemented such that virulent phages do not eradicate their hosts, but rather coexists with them

To determine if pseudolysogeny could be established between a lytic phage and a starved host, we undertook a series of experiments similar to those we conducted with the temperate bacteriophage F116 but this time using the virulent phage UT1 (Ripp and Miller, 1997, 1998). First, the effects of long-term starvation on virion production were explored. Phage UT1, which always exhibits a lytic phenotype on well-fed host cells (Ogunseitan et al., 1990), was incubated with starved hosts for an 80-day period. Some cultures were periodically spiked with nutrient (yeast extract at 10^{-5}%) in order to simulate the feast or famine conditions typically encountered by bacterial populations in freshwater environments (Figure 8.2). Control cultures, receiving no additional nutrient supplements, were also established, and after 80 days of incubation, these microcosms showed no substantial increases in the frequency of phages or phage-releasing cells within the bacterial population. However, microcosms receiving nutrient spikes produced pronounced increases in phage populations. In addition, a number of host cells actively releasing phage virions were observed, usually within 12 h of nutrient addition. Thus, initiation of active UT1 infections was directly related to nutrient availability.

To verify this relationship further, another set of microcosms containing only *P. aeruginosa* hosts was created. After a 25-day-starvation period, phage UT1 was added followed by an immediate nutrient spike or a spike after 24 h. Concurrent with the hypothesis that pseudolysogenic phages would be activated by nutrient addition only when they were given the opportunity first to infect and then maintain

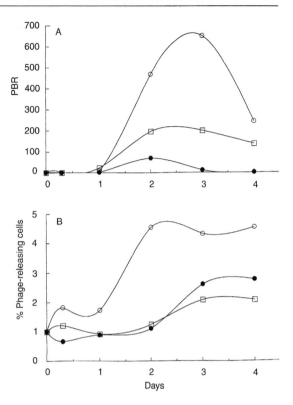

FIGURE 8.3 A submicrocosm was removed from a main batch culture inoculated with *P. aeruginosa* hosts and incubated for 25 days (day 0 represents day 25 of the main microcosm. (A) PBR and (B) percentage of cells actively releasing UT1 phages. (All values divided by t_0.) ○, phage UT1 was added 24 h after the submicrocosm was removed, allowed to infect for 6 h and then removed by centrifugation. Nutrient was added 24 h later. ●, phage UT1 was added and removed 6 h later as explained above; there was no nutrient addition. □, phage UT1 was added and removed 6 h later; nutrient was added at the same time as phage addition. (Reprinted with permission from Ripp and Miller, 1997.)

themselves within the starved host cell, we discovered that the greatest virion production and frequency of phage-releasing cells occurred in microcosm where nutrient spiking took place 24 h post-UT1 addition (Figure 8.3). In those microcosms receiving nutrient spikes coincident with phage addition, low virion concentrations resulted due to UT1 only establishing itself lytically in these microcosms.

The microcosm experiments provided initial evidence that phage UT1 could enter into a

TABLE 8.1 Comparison of the percentage of cells actively releasing phage virions (percentage activated) to the percentage of cells shown to contain a UT1 phage genome but not actively releasing phage virions (percentage not activated). These cells are pseudolysogenic. The fraction of cells containing phage genomes represents the number of cells positively identified through colony hybridization to contain phage DNA per total number of cells tested

Day	Fraction of cells containing phage genome	Percentage activated	Percentage not activated
0	0/200	0	0
3	1/49	0	100
9	32/164	17	83
16	31/120	32	68
24	25/122	35	65

Reproduced with permission from Ripp and Miller (1997).

pseudolysogenic state in starved cells that could later be resolved by addition of nutrients. Further confirmation was achieved through the use of continuous culture experiments that better simulated host-cell starvation conditions within an environmental system. A *P. aeruginosa* host and phage UT1 were incubated for one month in a minimal nutrient medium at a generation time of 10 h. At various time points during this period, samples were removed and tested for the number of cells spontaneously releasing phage as well as the number of cells containing phage genomes (via colony hybridization with a UT1 DNA probe). At all sampling points, more colonies were shown to contain a UT1 genome than were actually releasing phage particles (Table 8.1). Further, when these pseudolysogenic colonies were streaked onto rich media, only a small fraction of the resulting individual colonies were shown to contain UT1 DNA when colony hybridized. Thus, as was observed in studies with the temperate phage F116, the UT1 pseudoprophage seemed to undergo a more or less random transfer during cell division, indicating that the phage genome in the pseudolysogenic host cell is in a form that does not allow fidelity of scheduled DNA replication coordinated with host DNA replication and cell division.

In another series of continuous culture microcosm experiments, nutrient concentrations were varied to confirm that starvation conditions were necessary for the formation of a pseudolysogenic state (Table 8.2). In chemostats influxed with higher concentrations of nutrient (10^{-2} to 10^{-3}%

yeast extract), few pseudolysogenic cells were observed. However, as nutrient concentrations decreased from 10^{-4} to 10^{-6}% yeast extract, the numbers of pseudolysogenic cells increased.

As a final step in assessing the pseudolysogenic relationship between starved hosts and phage UT1, microcosm chambers were established *in situ* in a freshwater lake. Lake water was filter sterilized and used to prepare four microcosms inoculated with *P. aeruginosa* and UT1. Two microcosms were suspended on the lake surface and two others at a depth of 2 m. One microcosm from each set was periodically spiked with nutrient while the other remained untouched. Samples were removed periodically over a 43-day incubation period and assayed for cells spontaneously releasing phage and for cells containing phage genomes but not releasing phage (pseudolysogens) (Table 8.3). High concentrations of cells containing phage UT1 DNA were observed, while a percentage of these did not actively release phage virions. Nutrient spiking in chambers 2 and 4 did not produce significantly larger numbers of pseudolysogens, probably due to a sufficient influx of nutrients from the *in situ* lake-water environment into the chambers such that supplemental nutrient addition had little to no added effect.

CONCLUSIONS

Pseudolysogenic relationships between both temperate and virulent bacteriophages and

TABLE 8.2 Comparison of the percentage of UT1-phage releasing cells (% activated) to pseudolysogenic cells (% not activated) in chemostat microcosms influxed with varying concentrations of nutrient

Day | Yeast extract (%)

Day	10^{-2}			10^{-3}			10^{-4}			10^{-5}			10^{-6}		
	Fraction	% act.	% not act.	Fraction	% act.	% not act.	Fraction	% act.	% not act.	Fraction	% act.	% not act.	Fraction	% act.	% not act.
10	83/400	91	9	125/420	95	5	54/460	90	10	30/280	78	22	33/360	92	8
15	92/770	92	8	137/880	92	8	282/830	85	15	75/1100	80	20	412/2600	88	12
22	20/820	100	<1	360/1800	98	2	62/280	84	16	78/1300	76	24	342/1100	72	28
28	60/860	100	<1	294/1200	95	5	42/400	84	16	45/190	80	20	330/660	66	34
35	87/430	100	<1	52/120	94	6	100/230	85	15	82/130	75	25	40/100	54	46

Reproduced with permission from Ripp and Miller (1998).

TABLE 8.3 Comparison of the percentage of UT1-releasing cells (% activated) to pseudolysogenic cells (% not activated) in *in situ* incubated lake water chambers

	Surface						Bottom					
	Chamber 1			Chamber 2			Chamber 3			Chamber 4		
Day	Fraction of cells containing phage genome	% act.	% not act.	Fraction of cells containing phage genome	% act.	% not act.	Fraction of cells containing phage genome	% act.	% not act.	Fraction of cells containing phage genome	% act.	% not act.
0	0/170	0	0	0/108	0	0	0/310	0	0	0/198	0	0
9	6/310	0	100	0/140	0	0	0/220	0	0	6/430	0	100
15	247/570	51	49	187/740	92	8	65/140	74	26	30/720	25	75
21	350/570	92	8	360/420	74	26	280/520	94	6	40/370	100	0
27	300/480	100	0	15/109	64	36	244/280	100	0	223/740	80	20
29	490/650	93	7	310/440	93	7	224/610	81	19	153/300	78	22
43	64/86	84	16	112/112	82	18	236/400	92	8	68/83	93	7

Reproduced with permission from Ripp and Miller (1997).

their starved host appear to be widespread among bacterial species and can occur with both temperate and virulent phages. As a direct result, bacteriophages are able to survive within their hosts for extended periods. This prolonged lifespan of phage genomes in the environment may provide an expanded natural reservoir of viruses leading to large phage populations being observed in natural environments (Bergh et al., 1989; Proctor and Fuhrman, 1990; Paul et al., 1991; Miller et al., 1992; Bratbak and Heldal, 1993; Hennes and Suttle, 1995).

REFERENCES

Arber, W. (1994) Bacteriophage transduction. In *Encyclopedia of Virology* (eds, R.G. Webster and A. Granoff), vol. 1, pp. 107–113, Academic Press, London.

Baess, I. (1971) Report on a pseudolysogenic mycobacterium and a review of the literature concerning pseudolysogeny. *Acta Path. Microbiol. Scand.* **79**: 428–434.

Baross, J.A., Liston, J. and Morita, R.Y. (1978) Incidence of *Vibrio parahaemolyticus* bacteriophages and other *Vibrio* bacteriophages in marine samples. *Appl. Environ. Microbiol.* **36**: 492–499.

Bergh, Ø., Børsheim, K.Y., Bratback, G. and Heldal, M. (1989) High abundance of viruses found in aquatic environments. *Nature* **340**: 467–468.

Booth, S.J., van Tassel, R.L., Johnson, J. and Wilkins, T.D. (1979) Bacteriophages of *Bacteroides*. *Rev. Infect. Dis.* **1**: 325–336.

Bramucci, M.G., Keggins, K.M. and Lovett, P.S. (1977) Bacteriophage conversion of spore negative mutants to spore-positive in *Bacillus pumilus. J. Virol.* **22**: 194–202.

Bratbak, G. and Heldal, M. (1993) Total count of viruses in aquatic environments. In *Handbook of Methods in Aquatic Microbial Ecology* (eds, P. Kemp, B.F. Sherr, E.B. Sherr and J.J. Cole), pp. 153–158, Lewis, Boca Raton, FL.

Brown, N.L., Burchard, R.P., Morris, D.W. et al. (1976) Phage and defective phage of strains of *Myxococcus. Arch. Microbiol.* **108**: 271–279.

D'Herelle, F. (1949) The bacteriophage. *Sci. News* **14**: 44–59.

Drozhevkina, M.S., Kharitonova, T.I., Voronezhskaia, L.G. and Kirdeev, V.K. (1984) Lysogeny studies of *Vibrio cholerae* NAG. *Zh. Mikrobiol. Epidemiol. Immunobiol.* **12**: 50–54.

Grange, J.M. (1975) Pseudolysogeny in *Mycobacterium diernhoferi* ATCC19341. *J. Gen. Microbiol.* **89**: 387–391.

Grange, J.M. and Bird, R.G. (1975) The nature and incidence of lysogeny in *Micobacterium fortuitum. J. Med. Microbiol.* **8**: 215–223.

Hennes, K.P. and Suttle, C.A. (1995) Direct counts of viruses in natural waters and laboratory cultures by epifluorescence microscopy. *Limnol. Ocenogr.* **40**: 1050–1055.

Jarolmen, H., Bonke, A. and Crowell, R.L. (1965) Transduction *of Staphylococcus aureus* to tetracycline resistance *in vivo. J. Bacteriol.* **89**: 1286–1290.

Jiang, S.C. and Paul, J.C. (1998) Gene transfer by transduction in the marine environment. *Appl. Environ. Microbiol.* **64**: 2780–2787

Kidambi, S.P., Ripp, S. and Miller, R.V. (1994) Evidence for phage-mediated gene transfer among *Pseudomonas aeruginosa* strains on the phylloplane. *Appl. Environ. Microbiol.* **60**: 496–500.

Kidambi, S.P., Booth, M.G., Kokjohn, T.A. and Miller, R.V. (1996) recA-dependence of the response *of Pseudomonas aeruginosa* to UVA and UVB irradiation. *Microbiol.* **142**: 1033–1040.

Kokjohn,T.A. and Miller, R.V. (1985) Molecular cloning and characterization of the recA gene of *Pseudomonas aeruginosa. J. Bacteriol.* **163**: 568–572.

Kokjohn, T.A., Sayler, G.S., and Miller, R.V. (1991) Attachment and replication of *Pseudomonas aeruginosa* bacteriophages under conditions simulating aquatic environments. *J. Gen. Microbiol.* **137**: 661–666.

Kourilsky, P. (1973) Lysogenization by bacteriophage lambda, I. Multiple infection and the lysogenic response. *Mol. Gen. Genet.* **122**: 183–195.

Marsh, P. and Wellington, E.M.H. (1992) Interactions between actinophage and their streptomycete hosts in soil and the fate of phage borne genes. In *Gene Transfer and Environment* (ed., M.J. Gauthier), pp. 135–142, Springer-Verlag, Berlin.

Miller, R.V. (1998) Bacterial gene swapping in nature. *Sci. Am.* **278**: 46–51.

Miller, R.V. and Ku, C.-M.C. (1978) Characterization of *Pseudomonas aeruginosa* mutants deficient in the establishment of lysogeny. *J. Bacteriol.* **134**: 875–883.

Miller, R.V. and Ripp, S. (1998) The importance of pseudolysogeny to *in situ* bacteriophage–host interactions. In *Horizontal Gene Transfer* (eds, M. Syvanen and C.I. Kadopp), pp. 179–191, Chapman and Hall, London.

Miller, R.V., Ripp, S., Replicon, J. et al. (1992) Virus-mediated gene transfer in freshwater environments. In *Gene Transfer and Environment* (ed., M.J. Gauthier), pp. 51–62, Springer-Verlag, Berlin.

Miller, R.V. and Sayler, G.S. (1992) Bacteriophage–host interactions in aquatic systems. In *Genetic Interactions Among Microorganisms in the Natural Environment* (eds, E.M. Wellington and J.D. van Elsas), pp. 176–193, Pergamon Press, Oxford.

Morrison, W.D., Miller, R.V. and Sayler, G.S. (1978) Frequency of F116 mediated transduction of *Pseudomonas aeruginosa* in a freshwater environment. *Appl. Environ. Microbiol.* **36**: 724–730.

Nida, S.K. and Ferretti, J.J. (1982) Phage influence on the synthesis of extracellular toxins in group A streptococci. *Infect. Immun.* **36**: 745–750.

Novick, R.P. and Morse, S.i. (1967) *In vivo* transmission of drug resistance factors between strains of *Staphylococcus aureus. J. Exp. Med.* **125**: 45–59

Ogunseitan, O.A., Sayler, G.S., and Miller, R.V. (1990) Dynamic interactions of *Pseudomonas aeruginosa* and bacteriophages in lakewater. *Microb. Ecol.* **19**: 171–185.

Ogunseitan, O.A., Sayler, G.S., and Miller, R.V. (1992) Application of DNA probes to analysis of bacteriophage distribution patterns in the environment. *Appl. Environ.*

Microbiol. **58**: 2046–2052.

Osmond, M.A and Gealt, M.A. (1988) Wastewater bacteriophages transduce genes from the chromosome and a recombinant plasmid. In *Abst. Ann. Mt. Am. Soc. Microbiol.* 1988, p. 254, American Society for Microbiology, Washington, DC.

Paul, J.H., Jiang, S.C. and Rose, J.B. (1991) Concentration of viruses and dissolved DNA from aquatic environments by vortex flow filtration. *Appl. Environ. Microbiol.* **57**: 2197–2204.

Proctor, L.M. and Fuhrman, J.A. (1990) Viral mortality of marine bacteria and cyanobacteria. *Nature* **343**: 60–62.

Replicon, J., Frankfater, A. and Miller, R.V. (1995) A continuous culture model to examine factors that affect transduction among *Pseudomonas aeruginosa* strains in freshwater environments. *Appl. Environ. Microbiol.* **36**: 724–730.

Ripp, S. and Miller, R.V. (1995) Effects of suspended particulates on the frequency of transduction among *Pseudomonas aeruginosa* in a freshwater environment. *Appl. Environ. Microbiol.* **61**: 1214–1219.

Ripp, S. and Miller, R.V. (1997) The role of pseudolysogeny in bacteria–host interactions in a natural freshwater environment. *Microbiology* **143**: 2065–2070.

Ripp, S. and Miller, R.V. (1998) Dynamics of pseudo-lysogenic response in slowly growing cells of *Pseudomonas aeruginosa*. *Microbiology* **144**: 2225–2232.

Roger, A. (1983) Instability of host-virus association in *Acholeplasma laidlawiii* infected by a mycoplasma virus of the Gourlay group L1. *Zentralbl. Bakteriol. Mikrobiol. Hyg. (A)* **254**: 139–145.

Romig, W.R. and Brodestsky, A.M. (1961) Isolation and preliminary characterization of bacteriophages of *Bacillus subtilis*. *J. Bacteriol.* **82**: 135–141.

Roszak, D.B. and Colwell, R.R. (1987) Survival strategies of bacteria in the natural environment. *Microbiol. Rev.* **51**: 365 379.

Saye, D.J., Ogunseitan, O., Sayler, G.S. and Miller, R.V. (1987) Potential for transduction of plasmids in a natural freshwater environment: effect of donor concentration and a natural microbial community on transduction *in Pseudomonas aeruginosa*. *Appl. Environ. Microbiol.* **53**: 987–995.

Saye, D.J., Ogunseitan, O.A. Sayler, G.S. and Miller, R.V. (1990) Transduction of linked chromosomal genes between *Pseudomonas aeruginosa* during incubation *in situ* in a freshwater habitat. *Appl. Environ. Microbiol.* **56**: 140–145.

Schrader, J.S., Schrader, J.O., Walker, J.J. et al. (1997) Bacteriophage infection and multiplication occur *in Pseudomonas aeruginosa* starved for 5 years. *Can. J. Microbiol.* **43**: 1157–1163.

Suttle, C.A. and Chen, F. (1992) Mechanisms and rates of decay of marine viruses in seawater. *Appl. Environ. Microbiol.* **58**: 3721–3729.

Thompson, B.J., Domingo, E. and Warner, R.C. (1980a) Pseudolysogeny of *Azotobacter* phages. *Virology* **102**: 267–277.

Thompson, B.J., Wagner, M.S., Domingo, E. and Warner, R.C. (1980b) Pseudolysogenic conversion of *Azotobacter vinelandii* by phage A21 and the formation of a stably converted form. *Virology* **102**: 278–285.

Twort, F.W. (1915) An investigation on the nature of ultramicroscopic viruses. *Lancet* **11**: 1241–1243.

Wall, J.D., Weaver, P.F. and Gest, H. (1975) Gene transfer agents, bacteriophages, and bacteriocins of *Rhodopseudomonas capsulata*. *Arch. Microbiol.* **105**: 217–224.

Zinder, N.D. and Lederberg, J. (1952) Genetic exchange in *Salmonella*. *J. Bacteriol.* **64**: 679–699.

Mosaic Genes and Chromosomes

One of the more conspicuous manifestations of horizontal gene transfer is the existence of mosaic genes and chromosomes. Mosaic patterns of chromosome evolution were first inferred over 30 years ago from comparative electron microscope studies of heteroduplex DNA from different bacterial viruses (Campbell and Botstein, 1982). At that time, phage evolution was viewed in terms of interchangeable functional modules. Reviews of more recent comparisons based on completely sequenced phage chromosomes are covered in Chapter 12 by Hendrix et al. and Chapter 13 by Mosig and Calendar. It is now clear that the mosaic pattern extends to bacterial chromosomes. The existence of horizontally transferred genes in bacteria was not easily detected and has only become apparent in the last decade, as sufficient sequence information has become available. Some of the earliest reports, that bacterial genomes contained mosaic gene patterns, came from studies on pathogenesis factors (Spratt et al, 1992; Li et al., 1994). These are genes that are not generally needed for bacterial growth and survival except in the highly specialized environment of its eukaryotic host. Hollingshead covers this area in Chapter 11 and describes the mosaic nature of cell surface proteins in the *Streptococci*. Solnick and Young, in Chapter 10, review pathogenicity islands found in bacteria that infect the gastrointestinal tract. Pathogenicity islands appear to be a subset of a larger class of genomic regions that are rich in insertion sequences, prophages and accessory gene regions, as described by Lawrence in Chapter 9.

The chapters in this section describe many genes in bacteria that are important to survival in specialized ecological niches and appear to be derived from remotely related bacteria. The question of whether or not this constitutes evolution in the sense that progressive or novel traits have been contributed to a population is not clear. For example, the sudden appearance of new genes (along with their associated prophages and insertion sequences) in a bacterial lineage does not necessarily mean that these elements are completely novel, but rather may simply represent a cycle of gain and loss experienced by these genetic elements (or "accessory" genes as they have been called).

Allen Campbell raised the question of whether many inferred changes in bacterial populations should be considered simply cyclic processes or whether these changes should be considered to be what we normally think of as progressive evolution. As he says (Campbell, 1982): "It may be profitable for bacterial geneticists to consider carefully where they want to draw the line between evolutionary change and variation that is part of the ongoing population biology of the species." He goes on to say that "perhaps, on the other hand, it is impracticable to make the distinction between evolutionary change and nonevolutionary change and we should simply chronicle variation as such. Even in that event, I would prefer to see that decision made consciously and explicitly rather than, as it seems at the moment, implicitly and by default." This question is still relevant to the discussion of our mosaic gene patterns.

At first glance, the distinction between cyclic processes and progressive evolution may seem overly academic, not leading to any important insights other than opening a sterile debate about selection and whether it acts at the level of the species or the gene. However, if these groups of accessory genes and pathogenicity islands that are described in the chapters in this section are part of cyclic processes, then it does raise some interesting possibilities. One of the criteria used to identify if a gene may have been involved in a transfer event is whether or not the GC content of that gene deviates significantly from that of the genome in which it resides. Many of the regions and pathogenicity islands described in Chapter 9 and 10 in enteric genomes were identified in this manner. The fact that these regions have highly diverged GC contents suggests that the donor species would have to be distantly related to the hosts. For example, in the cases involving *Escherishchia coli* the donor would presumably be unrelated to Gram-negative enterics which all have similar GC contents. This fact could be used to argue that these regions are highly foreign. This is not necessarily correct. What if these genes are accessory genes that are circulated among, for example, the enterics? If so, long-term evolutionary survival of these genes could select for deviant GC contents. The majority of the accessory regions described here have unusually high AT contents. Is it possible that high AT content could be selected *per se* at the level of the recombination reactions that are needed to integrate these regions during their natural cycling? Two explanations have been advanced to support this possibility. The first is that less energy is required to denature high AT content duplex DNA, and the second is that the probability of a base-pair mismatch is significantly decreased between two high AT-content heterologous DNA partners (Syvanen, 1994).

Chapter 14 by Matic et al. is included here because they review an interesting property of *mutS* null mutants. These mutants not only have high mutation rates but they also promote heterologous recombination and possibly promote the recombination events that give rise to the mosaic gene patterns. The *mutS* gene is found in the pathogenicity islands, even though *mutS* is not an accessory gene but is required for robust growth. The intriguing finding that *mutS* mutants are frequently encountered in natural populations of *E. coli*. These mutants possibly arise due to the relative instability of such regions. Matic et al. propose that *mutS* evolved to be easily lost so that populations of *E. coli* will frequently give rise to these *mutS* mutants. These mutants then provide a service for the larger population as a source of new variants that arise both from the mutator phenotype and from the promotion of heterologous recombination with the accessory gene reservoirs. The *mutS* mutant itself never displaces the population since it has a growth disadvantage, but it continuously arises because it produces useful variants. This controversial proposal seems to be in conflict with traditional evolutionary genetic theory. However, models of this type have frequently been proposed by molecular biologists when questions as to the evolutionary significance of structures like plasmids, insertion sequences, introns and viruses are posed. Indeed, all of the mechanisms that promote horizontal gene transfer could have evolved because of naturally selected variants that arise from these mechanisms.

REFERENCES

Campbell, A. (1982) Some general questions about movable elements and their implications. In *Evolution Now: A Century after Darwin* (ed., J. Maynard Smith), pp. 42–57, Freeman, San Francisco.

Campbell, A. and Botstein, D. (1982) Evolution of the lambdoid phages. In *Lambda II* (eds R.W. Hendrix et al.), Cold Spring Harbor; ibid., *Evolution Now: A Century after Darwin* (ed., J. Maynard Smith), pp. 365–380, Freeman, San Francisco.

Li, J., Nelson, K., McWhorter, A.C., Whittam, T.S. and Selander R.K. (1994) Recombinational basis of serovar diversity in *Salmonella enterica*. Proc. Natl Acad. Sci. USA **91**: 2552–2556.

Spratt, B.G., Bowler, L.D., Zhang, Q.Y., Zhou, J. and Smith, J.M. (1992) Role of interspecies transfer of chromosomal genes in the evolution of penicillin resistance in pathogenic and commensal *Neisseria* species. J. Mol. Evol. **34**: 115–125.

Syvanen, M. (1994) Horizontal gene transfer: evidence and possible consequences. *Annu. Rev. Genetics* **28**: 237–261.

The Dynamics of Bacterial Genomes

Jeffrey G. Lawrence

The availability of complete genome sequences for large numbers of microorganisms has catalyzed a paradigm shift on how evolutionary biologists view the bacterial chromosome. No longer a mere collection of genes, each to be studied independently, a bacterial chromosome can be considered a complex document that both establishes its unique set of biological functions and reflects the long series of evolutionary events that shaped its current composition. Examination of genes *en masse*, and across multiple species, reveal evolutionary processes not evident in studies of individual genes. Two sets of processes affect the character of microbial genomes, mutation and recombination, and their roles in microbial evolution are quite different.

Mutation

Mutational processes alter the sequences of existing genes. Ultimately, mutations produce all of the variation seen among extant organisms. However, as predicted by the neutral theory of molecular evolution (Kimura, 1983) – and verified experimentally for vast numbers of genes – most of the variation found in natural populations has no quantifiable effect on organismal fitness. That is, most nucleotide changes make little or no measurable, functional difference in their encoded products. Even when examining multiple species, only a small number of differences between any two forms of a gene may affect that gene's function or expression, if at all. Therefore, while mutational processes are

ubiquitous, their impact in cellular evolution is necessarily smaller than their numbers suggest.

Gene exchange

Recombinational processes serve to reassort genes among genomes. Intraspecific recombination acts to distribute variant alleles – including, but not restricted to, those arising by mutational processes – among members of a bacterial species (Milkman and Stoltzfus, 1988; Milkman and Bridges, 1990, 1993; Dykhuizen and Green, 1991; Guttman and Dykhuizen, 1994; Selander et al., 1996; Milkman, 1997). The frequency of intraspecific recombination varies tremendously among taxa, but can be quite high (Smith et al., 1993; Feil et al., 2000). While this reassortment affects allele distribution within a microbial species, it does not serve to introduce novel information into the species' gene pool. Discussion regarding the delineation of bacterial species can be found elsewhere (Dykhuizen, 1998; Cohan, 2000; Lawrence, 2001).

In contrast, lateral (or horizontal) genetic transfer entails the mobilization of genes across species boundaries (Syvanen, 1990, 1994; Syvanen and Kado, 1998). While recombination processes would seem to be of lesser importance than mutation processes – after all, no new information is really created; existing information is merely reassorted – its impact on the evolution of particular genomes can be profound. Once viewed as a rarity, with little impact on microbial evolution, bacterial genome sequences have allowed for preliminary quantification of

Copyright © 2002 by Academic Press.
All rights of reproduction in any form reserved.

the scope of lateral gene transfer (Koonin and Galperin, 1997; Aravind et al., 1998; Huynen and Bork, 1998; Doolittle, 1999a,b, 2000; Fitz-Gibbon and House, 1999; Snel et al., 1999; Wolf et al., 1999; Ochman et al., 2000), and has fueled the reevaluation of its effect on microbial genome evolution.

Although transfer of genes among lineages may be uncommon, it can occur across vast phylogenetic distances (Buchanan-Wollaston et al., 1987; Heinemann and Sprague, 1989; Figge et al., 1999) and can introduce genes or gene clusters whose products mediate complex metabolic feats, like cytochrome maturation (Kranz and Goldman, 1998), phosphonate degradation (Metcalf and Wanner, 1993), or the adoption of a pathogenic lifestyle (Groisman and Ochman, 1994, 1997; Barinaga, 1996); therefore, its impact can be great (Lawrence, 1997, 1999b). Hence, lateral gene transfer can dramatically alter the metabolic character of a species in a manner not available to mutational processes (Lawrence, 1997, 1999a; Lawrence and Roth, 1998, 1999; Ochman et al., 2000). Below I present a brief overview of the mechanics of the horizontal gene transfer, since this topic has been reviewed in depth elsewhere (Doolittle, 1999a,b, 2000; Ochman et al., 2000), but elaborate more on the effects rampant gene transfer may have on the evolution of microbial genes and genomes.

DETECTION OF HORIZONTALLY TRANSFERRED DNA

To assess the impact of horizontally transferred DNA, foreign genes must first be identified and quantified. Two disparate approaches have been employed to detect horizontally transferred DNA. Each approach has its merits, but each has its drawbacks as well.

Identification of phylogenetic incongruency

Lateral genetic transfer across large phylogenetic distances will result in an usually high level of similarity between genes found in otherwise unrelated taxa. Hence, one may infer that the converse is true, that genes showing such unusual similarity in phylogenetic studies may have been subject to horizontal transfer. While this conclusion is often warranted, interpretation of sequence similarity – especially when weak – can be difficult, and testing this hypothesis (especially in systematic surveys of genome sequences) can be computationally intensive. Moreover, the success of this approach depends entirely upon the depth and breadth of the nucleotide sequence database.

Yet these data, when available, are quite convincing in demonstrating lateral gene exchange. For example, proteins encoded by the 18-gene *Escherichia coli phn* operon, specifying a phosphonate utilization system, are 50–90% related to proteins encoded by *Rhizobium* genome. A lateral transfer event likely resulted in the high degree of similarity observed. An additional strength of the phylogenetic approach is that it may readily identify genes which have been introduced into the genome sufficiently long ago that any additional evidence of its unusual pedigree has been abolished. For example, the *gapA* gene of enteric bacteria was clearly obtained from the eukaryotic domain several hundred million years ago (Doolittle et al., 1990), yet appears to be an otherwise unremarkable member of these genomes. Phylogenetic analyses have identified large numbers of genes in the genomes of *Aquifex* (Aravind et al., 1998) and *Thermotoga* (Nelson et al., 1999) inferred to have been derived from the Archaea. Yet the vagaries of phylogenetic analyses lead to other interpretations of these data (Logsdon and Fuguy, 1999).

Identification of atypical genes

In addition to phylogenetic tests, many recent lateral transfer events can be detected by virtue of their most obvious result: they introduce foreign DNA which has evolved in another genomic context. This property of horizontally transferred DNA can be detected independent of the nucleotide sequence database, and does not rely upon interpretations of phylogenetic comparisons.

How are recently acquired genes identified? Each organism experiences a particular set of directional mutation pressures, which shapes the evolution of their genomic sequences (Sueoka, 1961, 1962, 1988, 1992, 1993, 1995, 1999; Sueoka et al., 1959). That is, the array of nucleotide substitutions that arise within any one genome results from the mutational proclivities of the native DNA polymerase, the relative abundances of dNTPs, the relative efficiencies of mismatch repair systems, and other factors. Genes native to – or long-term residents of – a microbial genome will share all the compositional peculiarities that result from that organism's mutational quirks. Recently introduced foreign genes can be detected as atypical in this context, as they will show a different set of properties that reflect their donor organism's mutational biases. Several methods have been employed to detect such atypical genes, including the following:

1. *Nucleotide composition.* Microbial genomes vary widely in nucleotide composition as a result of directional mutation pressures. Composition is typically measured as %G + C, but individual nucleotide compositions can also be examined, especially when correcting for strand asymmetries. Therefore, the atypical nucleotide composition of some genes may reflect long-term evolution under a different set of directional mutation pressures, that is, in another genome. This method has been employed by Lawrence and Ochman (Ochman and Lawrence, 1996; Lawrence and Ochman, 1997; Lawrence and Roth, 1998), among others.

2. *Di- and tri-nucleotide compositions.* Somewhat more sophisticated algorithms examine di-nucleotide and tri-nucleotide frequencies, especially within protein coding regions (Karlin and Burge, 1995; Karlin et al., 1998). Here, sufficient information can be gleaned that a genomic "fingerprint" can identify native genes, and genes not conforming to these patterns are suspect. This method is intrinsically more powerful than merely examining nucleotide composition, since it may detect atypical genes bearing typical nucleotide compositions that would be otherwise overlooked.

3. *Codon usage bias.* In concert with mutational biases, differences in presence and abundance of different tRNA species can contribute to differences in codon usage preferences among organisms (Ikemura, 1980, 1981, 1982, 1985). This bias is strongest among highly expressed genes (Sharp and Li, 1987; Karlin and Mrazek, 2000). Therefore, if a gene exhibits codon usage bias – but not the bias inherent in the majority of chromosomal genes – then this gene may have been recently acquired. This method has been employed by Lawrence and Ochman (Ochman and Lawrence, 1996; Lawrence and Ochman, 1997; Lawrence and Roth, 1998), among others (Medigue et al., 1991; Whittam and Ake, 1992; Moszer et al., 1999).

4. *Markov Models.* Borodovsky (Hayes and Borodovsky, 1998) has extended the sensitivity of these analyses by using fifth-order Markov Models to examine chromosomal genes and sort them into several classes based on complex compositional patterns. One of these classes contains genes known to be of foreign origin (e.g. prophage genes, or those encoding transposases), and may represent a "catch-all" class of genes that do not belong to classes of highly expressed or weakly expressed chromosomal genes.

Regardless of the method used for detection, one may propose that genes appear as atypical because they have evolved in a separate genomic context and were only recently introduced into their current genome. However, the confidence we have with this conclusion is correlated to our ability to reject any other mechanism by which atypical patterns could arise. Therefore, caution must be used in making conclusions regarding lateral transfer based on only one criterion. Correspondence between methods detecting atypical sequences and those detecting phylogenetic incongruencies establish the most robust assessment of horizontal exchange. For example, the *E. coli* phn operon described above was first suspected of being horizontally acquired due to its atypical nucleotide composition and codon usage bias.

Using nucleotide composition, codon usage bias and dinucleotide frequency approaches, Lawrence and Ochman concluded that nearly 18% of the *Escherichia coli* genome – some 755 of 4288 genes – were recently introduced in at least 234 separate events (Lawrence and Ochman, 1998); many of these genes also showed phylogenetic incongruencies consistent with lateral gene exchange. Some of these events predated the divergence of *E. coli* with its sister species, *Salmonella enterica*, but many of them occurred more recently, and contribute to the physiological distinctiveness of these two organisms. Similar methods were employed to examine the genomes of 18 other Bacteria and Archaea (Ochman et al., 2000).

Not surprisingly, the amount of atypical DNA uncovered depends upon which genome is examined. Some organisms, like *Rickettsia prowazekii*, are almost devoid of atypical DNA, suggesting that few genes have been introduced recently; this dearth of recently acquired foreign DNA may reflect the relatively sheltered, obligately intracellular lifestyle exploited by this organism, although there is evidence for more ancient transfers (Wolf et al., 1999). Other genomes, like those of *Aquifex aeolicus*, *Thermotoga maritima* and *Synechocystis*, also showed high proportions of foreign DNA in their genomes, suggested a greater rate of horizontal transfer. Additional analyses also support the conclusion that large amounts of horizontally transferred DNA are present in the genomes of *Aquifex* (Aravind et al., 1998) and *Thermotoga* (Nelson et al., 1999; Worning et al., 2000).

Agents of gene transfer

How did it happen? Successful lateral gene transfer encompasses three distinct stages that can be facilitated by specific prokaryotic genetic elements (Ochman et al., 2000). First, DNA must be introduced into an organism's cytoplasm; three general mechanisms of DNA introduction are recognized: (a) *conjugation*, whereby resident plasmids transfer DNA directly from one organism to another by direct, cell–cell contact; plasmids that have integrated into the bacterial chromosome can mediate transfer of chromosomal genes from one strain to another; (b) *transduction*, whereby bacteriophages negotiate indirect exchange via transducing particles, which are formed when resident phages erroneously package chromosomal DNA into capsids which mediate their subsequent introduction into another host's cytoplasm; and (c) *transformation*, whereby a recipient cell takes up foreign DNA directly from the environment.

Secondly, DNA must be integrated into the microbial genome to allow long-term propagation; this process is distinct from the introduction of DNA into the cytoplasm. Bacteriophage genomes may be integrated directly into the host chromosome by *int*-mediated site-specific recombination. Aside from introducing bacteriophage genes into the host chromosome – including genes expressed from the prophage that may benefit the host cell (Hendrix et al., 2000) – bacteriophages that have imprecisely excised from their previous genome may carry with them chromosomal genes of their former host. Alternatively, integration of introgressed DNA may be facilitated by transposable elements, which can capture extrachromosomal elements by replicative transposition. The association of foreign DNA with native transposons has been observed in *E. coli* (Lawrence and Ochman, 1998) and in other genomes (Ochman et al., 2000). Barring these mechanisms, illegitimate recombination events may capture foreign DNA at sites of DNA damage; this mechanism has been observed for integration of mitochondrial DNA sequences into the yeast genome (Ricchetti et al., 1999).

Lastly, the foreign genes must be expressed in a manner that benefits the recipient organism. That is, natural selection acts as the final arbiter of lateral gene transfer, allowing the loss of genes that fail to provide a useful function to their new host. Only genes which improve the fitness of the recipient will be maintained and constitute a "successful" horizontal exchange. This criterion likely restricts gene flow by horizontal transfer more than any other, since few organisms may benefit from the acquisition of many genes. Moreover, most gene products must work with other proteins to provide a useful function. For example, the transfer of only one of the genes encoding the histidine biosynthetic apparatus would not benefit a His⁻ cell, and this transfer would not be successful.

This last hurdle is reduced dramatically by the organization of prokaryotic genes into operons, groups of transcribed genes whose products contribute to a single selectable function or phenotype (like the *E. coli hisGDCBHAFIE* operon). The operon represents a portable, promiscuous package of DNA that allows for the dissemination of genes whose products cooperate to provide for specific functions (Lawrence, 1997, 1999b, 2000; Lawrence and Roth, 1996b). In addition, it has been proposed that the mere process of horizontal gene exchange mediates the clustering of genes into operons (Lawrence and Roth, 1996b), implicating horizontal gene transfer as a powerful influence on the organization of bacterial genes.

RATE OF HORIZONTAL GENE TRANSFER

For lateral gene transfer to have a significant impact on microbial evolution, it must successfully introduce beneficial genes at an appreciable rate. As discussed above, an acquired gene must provide a useful function for it to persist. The lack of a beneficial function does not prevent a DNA fragment from integrating into the genome. Therefore, the mere presence of gene in a bacterial chromosome does not mean that it provides any selectable phenotype. For example, transposons are introduced into bacterial genomes at high rate (Lawrence et al., 1992; Naas et al., 1994), without providing beneficial cellular functions. Therefore, the demonstration that acquired genes have been important in shaping microbial evolution is tantamount to demonstrating that they have persisted in large numbers following introduction.

Inferences from compositional data

While phylogenetic studies can identify cases of likely transfer, they do not provide a straightforward means for quantifying the rate of transfer, especially in taxa lacking groups of well-characterized sister species. In contrast, examination of atypical gene sequences provides a direct mechanism for elucidating the rate of recent DNA acquisition. Using nucleotide composition data, Lawrence and Ochman (1997, 1998) developed methods for assessing how long introduced sequences have dwelled in their new host genomes following lateral transfer. As detailed above, directional mutation pressures result in variation in nucleotide composition across genomes; Muto and Osawa (1987) first recognized that this variation followed quantifiable patterns. That is, the nucleotide composition of different classes of sites – each of which experiences a different degree of selection – will have characteristic nucleotide compositions. As shown in Figure 9.1a, as genomic $\%G + C$ increases, third codon positions (which containing predominantly synonymous sites) vary widely while second codon positions (which cannot be altered without altering the identity of the encoded amino acid) vary little. These patterns provide a predictable signature to genes in directional mutational equilibrium, regardless of the overall nucleotide composition of the genome. (The deviations of individual observations from the overall relationships seen in Figure 9.1a include errors introduced by small sample sizes and the presence of horizontally transferred DNA. A refinement of this analysis using complete genome sequences recapitulates these relationships (Lawrence, unpublished results), where residual deviations are a function of genetic headroom (Lawrence, 2001, and see below).

Immediately after horizontal transfer, genes will reflect the compositional patterns of their donor genomes. Yet over time, atypical sequences experience the directional mutational pressures of their new host genome and evolve over time to resemble native genes. During this period of amelioration, foreign sequences are in directional mutational disequilibrium, and resemble neither their donor genome nor their recipient genome. More importantly, their compositions do not follow the predictable relationships of nucleotide compositions to which genomes in mutational equilibrium conform.

These deviations from the Muto and Osawa relationships can be measured and used to predict the residence time of foreign genes within bacterial genomes by determining how much time had elapsed since sequences fit these relationships (Figure 9.1b). For any class of sites (e.g. first, second or third codon positions), the deviation in nucleotide compositions will decrease over time as

$$\Delta GC^{HT} =$$
$$[(IV + 1/2)/(IV + 1)]*R*(GC^{Native} - GC^{HT}) \quad (1)$$

where IV is the transition/transversion ratio, R is the rate of nucleotide substitution, and GC^{HT} and GC^{Native} are the compositions of acquired and native sequences, respectively.

Inferring rates of transfer

Using this approach, Lawrence and Ochman (1997, 1998) calculated the rate of horizontal transfer in *Escherichia coli* to be ~16 kb/Myr. This rate was calculated by fitting an exponential decay curve to the distribution of horizontally transferred genes found in the *E. coli* genome sorted by time of introduction (Figure 9.2). Since this rate is normalized to a divergence time of 100 Myr for the *E. coli* and *Salmonella* lineages (Ochman and Wilson, 1987, 1988; Moran et al., 1993), it predicts that ~1.6 Mb of DNA (about one-third the size of its current genome) have been introduced into the *E. coli* genome – and at least temporarily persisted – since its

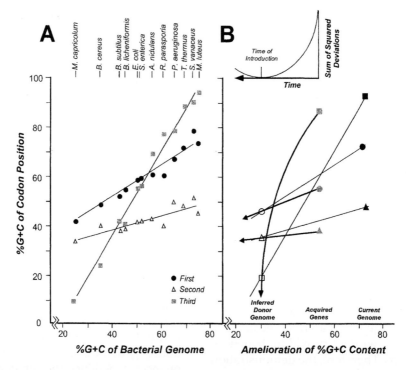

FIGURE 9.1 (A) The relationships between the nucleotide composition of a bacterial genome and those of the three codon positions. Data are after Muto and Osawa (1987) except for *E. coli* and *S. enterica*, which were calculated from 100% and 25% of the genome sequences, respectively, after known horizontally-transferred sequences were removed (Lawrence and Ochman, 1997). (B) An overview of gene amelioration (Lawrence and Ochman, 1997, 1998). Acquired genes (gray points) are atypical for the genome (dark points) in which they are found. The codon-position-specific nucleotide compositions of acquired genes are back-ameliorated until the minimum deviation (by least-squares analysis) from the Muto and Osawa relationships are obtained (open points). The heavy lines denote codon-position-specific nucleotide compositions during back-amelioration. The inset graph shows the deviation of those curves from the Muto and Osawa relationships as a function of time. (Figure after Lawrence and Roth, 1999.)

divergence from *Salmonella*. While the bulk of acquired sequences have been subsequently lost by deletion, many still persist in the genome, indicating that they have conferred long-term advantageous phenotypes on their new host.

Comparatively, then, lateral transfer and mutational processes have introduced similar numbers of variant bases into the *E. coli* genome. Although both mutational processes and lateral transfer processes introduce a large amount of variant DNA into the bacterial genome, the quality of information supplied by these two processes is strikingly different. The bulk of substitutions that arise by mutation are effectively neutral, while information introduced by lateral transfer can include fully functional genes and operons that may allow exploitation of novel resources and environments (Lawrence, 1997, 1999a; Lawrence and Roth, 1998, 1999). As a result, one may infer that horizontal gene transfer may play a significant role in driving the phenotypic evolution and ecological differentiation of this bacterial lineage. Can this result be generalized to other lineages?

Using similar methods, the amount of atypical DNA (which we infer as likely to have been

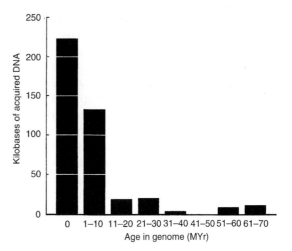

FIGURE 9.2 The amount of foreign DNA found in the *Escherichia coli* genome as a function of time of introduction. (Figure after Lawrence and Ochman, 1998.)

introduced by lateral transfer) can be quantified among other completely sequenced microbial genomes (Figure 9.3). Surveys of 19 other bacterial genomes reveal a range of rates of lateral gene transfer among completely sequenced genomes (Ochman et al., 2000). Some species appear to

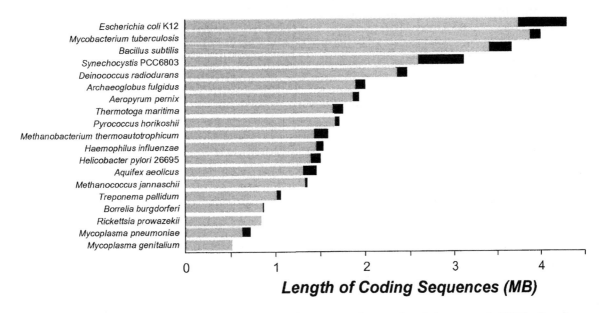

FIGURE 9.3 The amount of foreign DNA in bacterial genomes; figure after Ochman et al. (2000). Gray bars denote sequences ancestral to, or long term residents of, microbial genomes; black bars represent atypical sequences likely introduced by lateral gene transfer.

have low rates of recent gene acquisition (e.g. *Rickettsia*, *Borrelia*), while others have experienced high rates of transfer (e.g. *E. coli*, *Synechocystis*); some organisms with large genome size (e.g. *Mycobacterium tuberculosis*) have comparative lower amounts of foreign DNA, while some organisms with smaller genomes (e.g. *Aquifex aeolicus*) have larger amounts. The widespread evidence for horizontal transfer among many bacterial genomes demonstrates that this process has had a strong impact in overall microbial evolution, and is not restricted to genomes of large size or to organisms adapted to a particular environment or lifestyle.

GENOME EXPANSION AND CONTRACTION

Genomes do not grow in size

The constant acquisition of DNA by lateral transfer implies that bacterial genomes should be increasing in size. In addition, genomes can increase in size due to a number of other processes, including tandem duplications, asymmetric exchange over during intraspecific recombination, insertion of prophages, accumulation of transposons, and integration of plasmids. Although these processes can lead to an overall increase in genome size, this increase in size is not evident. Surveys among closely related taxa demonstrate that their genomes are notably similar in size (Bergthorsson and Ochman, 1995, 1998; Ochman and Bergthorsson, 1998). For example, although genomes of characterized bacteria vary from less than 500 kb to more than 10 000 kb in size, the genomes size of natural variants of *E. coli* vary far less, measuring 4968 ± 253 kb (Bergthorsson and Ochman, 1998), despite the influx of 16 kb/Myr over at least the past 100 Myr. The genome of the intracellular parasite *Buchnera* varies even less (630–643 kb), possibly reflecting the reduced exposure of this organism to foreign DNA (Wernegreen et al., 2000).

Among enteric bacteria predictably uncover cases where genes have been lost from particular lineages. This is to be expected as organisms invade different niches, since genes that confer selectable functions in one environment may fail to provide a benefit in another ecological context; these genes would be subject to loss by mutation and genetic drift. The *phoA* gene, encoding alkaline phosphatase, was lost from the *Salmonella* lineage while being maintained in the genomes of virtually all other enteric bacteria (DuBose and Hartl, 1990), and the genes for glycerol dehydratase appear to have been lost from the ancestor of *E. coli* and *Salmonella* (Lawrence and Roth, 1996a). Alternatively, genes may be lost if their functions actively interfere with the adoption of a novel ecological role. For example, the OmpT surface protease was likely lost from the pathogenic bacteria *Shigella* because its function interferes with virulence (Nakata et al., 1993). *Shigella* also lost the *cadA* gene, likely because its product – lysine decarboxylase – also diminishes virulence (Maurelli et al., 1998).

Therefore, as expected, the constant influx of DNA is offset by persistent gene loss, resulting in a steady-state genome size. Consistent with this model, tandem duplications have been measured to resolve at moderate frequencies (Galitski and Roth, 1997), surveys of transposable elements show that they are deleted at a rapid rate (Lawrence et al., 1992), and amelioration studies demonstrate a constant loss of horizontally acquired sequences (Lawrence and Ochman, 1998). In addition, very few pseudogenes are found in most genomes (*Rickettsia* and other exceptions are discussed below), demonstrating that genes lost to mutational processes are rapidly removed by deletion. This issue is discussed in a later section.

Constraints on genome size

So why don't bacterial genomes increase steadily in size, and why do steady state genome sizes vary among organisms? That is – aside from actively problematic genes like *cadA* in *Shigella* – why is information discarded? The population genetic constraints on genome size are quite clear: there is a maximum amount of information that can be maintained in a bacterial genome at any one time. Stated simply, in a finite population, mutations at a finite number of sites can be counterselected; that is, an organism cannot maintain an infinite amount of information free from mutation. Given a finite population size, information will be lost by mutational processes; genetic drift will ultimately fix variant alleles at some loci while selection prevents their fixation at other loci. This loss may be offset by high rates of intraspecific recombination or by larger effective population sizes, both of which slow the stochastic loss of information by genetic drift (Lawrence, 2001; Lawrence and Roth, 1999). Therefore, the maximum amount of genomic information (G) that can be maintained can be described as functions of the effective population size (N_e), mutation rate (μ), and rate of intraspecific recombination (r):

$$G \alpha [f(N_e)^* g(r)]/h(\mu) \qquad (2)$$

Genes which fail to make a sufficiently large contribution to average organismal fitness will be lost through mutation and genetic drift, as their maintenance in the population is insufficiently advantageous to prevent their inevitable loss.

Genetic headroom: genomic information and rate of horizontal transfer

The breadth of genomic information describes classes of sites, not merely a catalog of genes. For example, each protein-coding gene encompasses a rich set of information – including transcription and translation start signals, codons corresponding to critical structural or catalytic residues, codon usage bias etc. – which can bear varying selection coefficients (e.g. mutations that disrupt a catalytic site are far more detrimental than mutations which alter the codon usage bias of that gene). Yet the information maintained in the codon usage bias of a highly expressed gene may be more valuable than the entire function provided by a weakly expressed gene whose product is rarely used; it is for this reason that some genes may be lost entirely from a genome, while other genes are not only maintained, but exhibit high levels of codon usage bias. Since genomes are limited in the amount of information they can carry, the additional information provided by laterally transferred genes must inevitably be offset by information loss for the new genes to persist. A genome cannot maintain both its complement

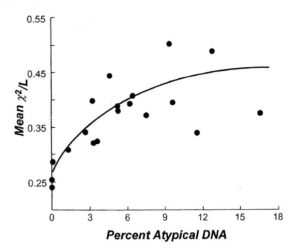

FIGURE 9.4 The correlation between genetic headroom and the rate of horizontal transfer; figure after Lawrence (2001). Data represent complete-genome analyses for the organisms listed in Figure 9.3. Average codon usage bias was calculated from length-normalized χ^2 values for each gene using codon-position-specific nucleotide compositions as expected values. Atypical DNA was calculated as described (Lawrence and Ochman, 1997, 1998).

of ancestral information as well as the information required by newly acquired genes to confer beneficial functions. The information which maximally benefits the organism – that is, sites with the highest selection coefficients – will be maintained, at the expense of mutations accumulating at other sites.

Genetic headroom has been defined as information bearing very low selection coefficients that can be removed from purifying selection without altering the metabolic capabilities of the organism (Lawrence, 2001). This information is manifested as codon usage bias, codon context bias, and other sites that do not contribute directly to metabolic capabilities (Lawrence, 2001). When genetic headroom is crudely measured as the overall level of codon usage bias (e.g. the average normalized χ^2 of codon usage), it is a good predictor of the amount of horizontally transferred DNA in a microbial genome (Figure 9.4). As expected, this correlation between genetic headroom and percentage atypical DNA is independent of genome size. Why is this correlation seen?

In organisms with large genetic headroom, the additional information required by newly acquired genes is offset by the accumulation of mutations that affect codon usage bias and codon context bias (but not primary amino acid sequences), since this class of information bears the lowest selection coefficients. In organisms with little genetic headroom, the maintenance of additional information is offset by loss of protein-coding sequences, or other classes of information bearing high selection coefficients. In effect, the maintenance of laterally transferred genes is more expensive for genomes with little genetic headroom, since valuable information must be discarded for the new information to persist. Very few functions will be sufficiently valuable to replace native functions, resulting in a lower rate of successful lateral transfer in organisms with low genetic headroom. In contrast, experimentation with laterally transferred genes is relatively inexpensive for organisms with large genetic headroom, since the information being offset carries low selective coefficients. Many genes may provide a sufficient selective benefit to guarantee persistence, and rates of lateral transfer will be correspondingly higher (Figure 9.3).

As a result, organisms with large genetic headroom can explore novel ecologies with impunity, since the information that is transiently sacrificed does not handicap the organism with respect to its metabolic capabilities. If the niche is successful, ancestral genes and their information may be discarded (e.g. the *Shigella cadA* and *ompT* genes) as the organism adapts to a new ecological role. Alternatively, if the new niche is not successful, the acquired genes will be discarded. In either case, ancestral physiology will be maintained during this exploratory phase, and codon usage bias can be re-established in those genes where it was transiently abandoned. Without genetic headroom, an organism may compromise ancestral physiology to pursue novel ecological routes. If the lineage is not successful, it cannot readily return to its ancestral state. In this way, the magnitude of genetic headroom can promote microbial diversification by allowing purifying selection to be reapportioned between counterselecting synonymous substitutions in native genes and counterselecting non-synonymous substitutions in acquired genes.

Gene loss versus gene deletion

The population genetic constraints outlined above explain why infinite amounts of information cannot be maintained in any one genome. When information is lost, a gene may become non-functional by one of two processes: (a) point mutations may accumulate, eliminating gene function, or (b) deletions may eliminate the gene sequence altogether. As detailed above, deletions have been documented to eliminate many genes from enteric bacterial genomes, and serve to offset genes acquired by lateral transfer. Yet relaxation of selection on a gene's information content does not necessitate the physical removal of its DNA; therefore, it is not clear why mechanisms supporting high rates of gene deletion are maintained in microbial genomes.

Deletions occur by recombination between directly repeated sequences in the genome (Roth et al., 1996); rates of deletion are controlled by the nature of the sequence identity required to support successful strand invasion. The complexity of the recombination apparatus varies among species, with some organisms (e.g. *Escherichia coli*, *Salmonella enterica*) harboring multiple, independent pathways for recombination (Galitski and Roth, 1997), while others have few of these genes (Lawrence, unpublished data). One may predict that rates of deletion will vary among species, depending on the makeup of the recombination apparatus. Lower rates of deletion would be manifested in the accumulation of pseudogenes, non-functional sequences wherein mutations are accumulating. The deletion rates in some genomes – e.g. *Buchnera*, *Mycobacterium leprae* or *Rickettsia prowazekii* – are sufficiently low to allow for the gradual accumulation of large numbers of pseudogenes (Andersson and Andersson, 1999a, 1999b; Cole, 1998). The relative absence of pseudogenes from the *E. coli* genome shows that non-functional DNA is rapidly eliminated, consistent with the observation of rapid transposon deletion (Lawrence et al., 1992).

Harboring a recombination apparatus that can delete chromosomal DNA is inherently a dangerous tactic. Recombination between long, naturally repeated sequences (rRNA, tRNA, etc.) can be lethal since essential genes found between these repeats would be lost; deletion between long regions of identity can occur readily (Galitski and Roth, 1997). Mechanisms that allow for deletion between smaller regions of identity are also detrimental, potentially disrupting virtually any gene. Therefore, even infrequent deletion of random DNA must incur a selective cost to the organism, as important or essential genes will inevitably be affected, resulting in a non-viable cell. What is the benefit that must offset this cost?

DNA loss promotes parasite removal

For some time, the benefit of DNA loss in bacteria has been attributed to genome "streamlining," whereby smaller chromosomes allow for faster replication and require a smaller energetic investment to propagate. However, it is now clear that the rate of DNA replication does not limit the rate cell division; cells merely initiate multiple replication forks. Moreover, the energetic cost in the replication of small amounts of additional DNA is miniscule compared with the cost of physiological maintenance, and is dwarfed by the costs of mRNA and protein turnover alone.

Yet DNA deletion does provide a potentially very powerful benefit to the cell in the removal of dangerous genetic parasites, including transposons, bacteriophages, and addiction modules. As described above, these elements are introduced by lateral gene exchange. Yet they represent a sort of genetic time bomb that will destroy a cell if not inactivated by mutation; the activation of a prophage results in almost certain cell lysis, and the continued transposition of mobile genetic elements will inevitable affect an essential gene. Therefore, some portion of the progeny of lysogens and transposon-bearing cells will be killed by these elements, reducing the fitness of this lineage. DNA deletion mechanisms can provide a selective benefit, since they allow for the removal of prophages (and transposons) before activation of their lytic genes (or transposases) kills the cell. Deletion processes may be particularly effective at removing prophages and transposons since these elements are associated with repeated sequence at their boundaries – direct *attL* and *attR* sites in bacteriophages and inverted repeats in transposons – that may mediate their removal (by homologous recombination between *att* sites or cruciform removal between inverted repeats).

How does horizontal transfer impact the rate of DNA loss? Horizontal transfer begins with the introduction of DNA into the cytoplasm and then into the genome (see above); these processes are mediated by transposons and bacteriophages, which are also introduced by incoming DNA. If the accumulation of transposons and prophages selects for an accelerated rate of deletion, then exposure to the agents of horizontal exchange may be correlated to a more rapid rate of DNA loss. As predicted, the failure to detect pseudogenes in most genomes is correlated with their higher rates of horizontal transfer. Large numbers of pseudogenes are found in the genomes of in *Rickettsia prowazekii* (Andersson and Andersson, 1999a, 1999b), *Buchnera*, and some *Mycobacteria* (Cole, 1998); these species also bear comparatively smaller amounts of horizontally transferred DNA (*Rickettsia* and *M. tuberculosis* are shown in Figure 9.2). These species may experience lower rates of exposure to bacteriophages, transposons and other foreign DNA as a result of their intracellular lifestyles, resulting in lower rates of gene acquisition. Without a load of genetic parasites, high rates of DNA deletion would only incur their detrimental effects, since this penalty would not being offset by the benefits of parasite removal. As a result, one would expect their rates of DNA deletion to decrease, and for pseudogenes to accumulate in their genomes.

SPECIATION AND THE EVOLUTION OF FUNCTIONAL NOVELTY

As detailed above, rampant horizontal gene transfer results in a far more dynamic genome than previously inferred, primarily from the comparison of genetic maps. For example, the linkage maps of *Escherichia coli* and *Salmonella enterica* are remarkably congruent (Berlyn et al., 1996; Sanderson et al., 1996), despite approximately 120 Myr of separation. These similarities suggested a static, evolutionarily stable genome, whereby differences arose primarily by mutational processes that did not disrupt the underlying genetic map. Indeed the correspondence of their maps does result from each species' retention of ancestral genes which contribute to important or essential processes. Yet interspersed with these genes are numerous lineage-specific genes, which were introduced by horizontal genetic transfer into one lineage, or were deleted from the other. It is likely that these sequences – unique combinations of genes not found in closely related taxa – have mediated the divergence of these organisms.

Impact on microbial speciation

Although mutation and lateral transfer both modify the content and composition of the genome, the kinds of information these processes furnish are quite different. While mutations can alter the function of existing genes, it can do so only incrementally. Often, many steps would be required to alter the role of even one gene, let alone a suite of genes whose products contribute to a complex function or phenotype. In contrast, lateral transfer can introduce – in single steps – entire clusters of genes which can mediate complex metabolic tasks. The encapsulation of metabolic functionality into operons represents an enormous potential for organismal diversification when these clusters are acquired by naive lineages. For example, the *Salmonella* lineage acquired via horizontal transfer a 20-gene operon for the biosynthesis of coenzyme B_{12}, and linked operon for the B_{12}-dependent degradation of propanediol (Lawrence and Roth, 1995, 1996a). Similarly, the acquisition of pathogenicity islands has allowed the rapid assembly of functions required to exploit a pathogenic lifestyle (Barinaga, 1996; Groisman and Ochman, 1994, 1996, 1997; Ochman, 1997; Ochman et al., 2000). Given the enormous complexity of B_{12} biosynthesis, *Salmonella* could not reasonably evolve this capability by mutational

alteration of existing genetic material in such a short period of time. Given this remarkable potential, horizontal transfer is quite likely to mediate microbial speciation, since the competitive exploitation of new ecological niches is less likely to be facilitated by the slow, stepwise accrual of mutations than by the acquisition of fully functional genes by horizontal transfer (Lawrence, 1997, 1999a, 2001; Lawrence and Roth, 1998, 1999). For example, all metabolic traits known to distinguish *Escherichia coli* and *Salmonella enterica* correspond to genes either acquired by horizontal transfer or lost by deletion; none appear to result by the lineage-specific modification of ancestral genetic material.

Surveys of extant genomes support the hypothesis that organisms have historically acquired new functions by horizontal transfer rather than by mutational alteration of existing information. Extant protein sequences clearly cluster into families, each with similar functions. For example, NAD^+-binding dehydrogenases show clear segregation based on substrate specificity (e.g. malate dehydrogenase, lactate dehydrogenase), and these enzymes are distinct from FAD-binding dehydrogenases (Rossman et al., 1974). So, while exceptions can be noted (Wu et al., 1999), lactate dehydrogenases have not arisen multiple times from malate dehydrogenases (or vice versa), even though single mutation can enable this transition (Golding and Dean, 1998; Wilks et al., 1988). The sporadic distribution of orthologous proteins among organisms can be explained by horizontal transfer of genes among organisms, thereby avoiding the cumbersome proposal of an ancestral genome containing all of these genes, most of which were lost from most genomes (requiring this organism to bear an enormously unwieldy genome).

Duplication and divergence

Yet lateral gene transfer serves only to reassort genes among organisms, not to create novel information directly. For a gene to evolve a new function – for example, the very first divergence of malate and lactate dehydrogenases – the novel variation must have arisen by mutation. Historically, this process has been described as gene duplication and divergence, whereby an intragenomic duplication creates a copy of a gene that is free from selective constraints (since the other copy provides the function), thereby allowing substitutions to incur and new functions to be generated. Duplication and divergence thus allows a gene to evolve new functions while previous functions are continually provided by a duplicate copy. It would seem, then, that mutational processes must dictate the creation of functional novelty.

Yet there are implicit problems with the classic duplication and divergence model. Principally, there is no selection to maintain two duplicate copies of a gene in the same genome if one copy serves no function (if the duplicate copies are required to increase gene dosage, e.g. rDNA loci, then they are not free to evolve new functions). In addition to the accrual of deleterious mutations which abolish gene function, deletions efficiently remove tandemly duplicated sequences from bacterial genomes. This model requires transiently duplicated genes to obtain the tremendously rare beneficial mutation that allows adoption of a novel function prior to loss by mutation or deletion. Moreover, this mutation would have to confer a sufficiently strong advantage for it to be maintained in the genome sufficiently long to obtain the necessary additional mutations that allow effective and efficient performance of the alternate function. This process is analogous to a sympatric speciation model, whereby species must achieve ecological distinctiveness in the face of reproductive mixing.

Cell division requires gene duplication

The duplication and divergence model originated when it was recognized that related proteins performed different functions in the same organism. Given the widespread process of gene exchange, it is not necessary to require that genes duplicate and adopt new functions while resident in the same cytoplasm. Every gene is duplicated when the chromosome is replicated and cells divide, and every gene diverges as genes accumulate mutations in separate cytoplasmic contexts. Therefore, genes may diverge and perform alternate functions – the ability to

perform their ancestral functions if these are essential – before gene transfer unites the paralogues in the same genome. This transfer may be horizontal, where integration proceeds by non-homologous recombination) or intraspecific, where an asymmetric homologous exchange between repeated sequences allows both copies to be maintained. This process is analogous to an allopatric speciation model, whereby species are allowed to achieve ecological distinctiveness in separate locations prior to any potential mixing in the same geographic location (i.e. genome). In this way, gene transfer may facilitate the evolution of novel gene functions by moving genes into genomic contexts where preadapted alleles may be further refined without the danger of recombination or mutational loss.

CONCLUSIONS

Lateral genetic transfer can be simply described as the mobilization of DNA across species boundaries. Yet as described above, it has the potential for profound alterations in the organization of bacterial genes and genomes, including the organization of genes into operons (the selfish operon model), rapid intragenomic gene turnover due to the limitations of maximal information content and the limitations of genetic headroom, DNA loss by constant exposure to genetic parasites (e.g. transposons and bacteriophages), mediation of gene duplication and divergence, and the inevitable speciation (ecological differentiation) of microbial lineages due to turnover of the gene content.

REFERENCES

Andersson, J.O. and Andersson, S.G.E. (1999a) Genome degradation is an ongoing process in *Rickettsia*. *Mol. Biol. Evol.* **16**: 1178–1191.

Andersson, J.O. and Andersson, S.G.E. (1999b) Insights into the evolutionary process of genome degradation. *Curr. Opin. Genet. Dev.* **9**: 664–671.

Aravind, L., Tatusov, R.L., Wolf, Y.I, et al. (1998) Evidence for massive gene exchange between archaeal and bacterial hyperthermophiles. *Trends Genet.* **14**: 442–444.

Barinaga, M. (1996) A shared strategy for virulence. *Science* **272**: 1261–1263.

Bergthorsson, U. and Ochman, H. (1995) Heterogeneity of genome size among natural isolates of *Escherichia coli. J. Bacteriol.* **177**: 5784–5789.

Bergthorsson, U. and Ochman, H. (1998) Distribution of chromosome length variation in natural isolates of *Escherichia coli. Mol. Biol. Evol.* **15**: 6–16.

Berlyn, M.K.B., Low, K.B. and Rudd, K.E. (1996) Linkage map of *Escherichia coli* K-12, edition 9. In *Escherichia coli and Salmonella: Cellular and Molecular Biology* (eds, F.C. Neidhardt, R. Curtiss III, J.L. Ingraham et al.), 2nd edn, pp. 1715–1902, ASM Press, Washington, DC.

Buchanan-Wollaston, V., Passiatore, J.E. and Canon, F. (1987) The *mob* and *oriT* mobilization functions of a bacterial plasmid promote its transfer to plants. *Nature* **328**: 170–175.

Cohan, F.M. (2001) Bacterial species and speciation. *Sys. Biol.* **50**: 513–524.

Cole, S.T. (1998) Comparative mycobacterial genomics. *Curr. Opin. Microbiol.* **1**: 567–571.

Doolittle, W.F. (1999a) Lateral genomics. *Trends Cell. Biol.* **9**: M5–8.

Doolittle, W.F. (1999b) Phylogenetic classification and the universal tree. *Science* **284**: 2124–2129.

Doolittle, W.F. (2000) The nature of the universal ancestor and the evolution of the proteome. *Curr. Opin. Struct. Biol.* **10**: 355–358.

Doolittle, R.F., Feng, D.F., Anderson, K.L. and Alberro, M.R. (1990) A naturally occurring horizontal gene transfer from a eukaryote to a prokaryote. *J. Mol. Evol.* **31**: 383–388.

DuBose, R.F. and Hartl, D.L. (1990) The molecular evolution of alkaline phosphatase: correlating variation among enteric bacteria to experimental manipulations of the protein. *Mol. Biol. Evol.* **7**: 547–577.

Dykhuizen, D.E. (1998) Santa Rosalia revisited: why are there so many species of bacteria? *Antonie van Leeuwenhoek* **73**: 25–33.

Dykhuizen, D.E. and Green, L. (1991) Recombination in *Escherichia coli* and the definition of biological species. *J. Bacteriol.* **173**: 7257–7268.

Feil, E.J., Smith, J.M., Enright, M.C. and Spratt, B.G. (2000) Estimating recombinational parameters in streptococcus pneumoniae from multilocus sequence typing data. *Genetics* **154**: 1439–1450.

Figge, R.M., Schubert, M., Brinkmann, H. and Cerff, R. (1999) Glyceraldehyde-3-phosphate dehydrogenase genes in eubacteria and eukaryotes: evidence for intra- and inter-kingdom gene transfer. *Mol. Biol. Evol.* **16**: 429–440.

Fitz-Gibbon, S.T. and House, C.H. (1999) Whole genome-based phylogenetic analysis of free-living microorganisms. *Nucl. Acids Res.* **27**: 4218–4222.

Galitski, T. and Roth, J.R. (1997) Pathways for homologous recombination between chromosomal direct repeats in *Samonella typhimurium. Genetics* **146**: 751–767.

Golding, G.B. and Dean, A.M. (1998) The structural basis of molecular adaptation. *Mol. Biol. Evol.* **15**: 355–369.

Groisman, E.A. and Ochman, H. (1994) How to become a pathogen. *Trends Microbiol* **2**: 289–294.

Groisman, E.A. and Ochman, H. (1996) Pathogenicity islands: bacterial evolution in quantum leaps. *Cell* **87**: 791–794.

Groisman, E.A. and Ochman, H. (1997) How *Salmonella* became a pathogen. *Trends Microbiol.* **5**: 343–349.

Guttman, D.S. and Dykhuizen, D.E. (1994) Clonal divergence in *Escherichia coli* as a result of recombination, not mutation. *Science* **266**: 1380–1383.

Hayes, W.S. and Borodovsky, M. (1998) How to interpret an anonymous bacterial genome: machine learning approach to gene identification. *Genome Res.* **8**: 1154–1171.

Heinemann, J.A. and Sprague, G.F.J. (1989) Bacterial conjugative plasmids mobilize DNA transfer between bacteria and yeast. *Nature* **340**: 205–209.

Hendrix, R.W., Lawrence, J.G., Hatfull, G.F. and Casjens, S. (2000) The origins and ongoing evolution of viruses. *Trends Microbiol.* **8**: 504–508

Huynen, M.A. and Bork, P. (1998) Measuring genome evolution. *Proc. Natl Acad. Sci. USA* **95**: 442–444.

Ikemura, T. (1980) The frequency of codon usage in *E. coli* genes: correlation with abundance of cognate tRNA. In *Genetics and Evolution of RNA Polymerase, tRNA and Ribosomes* (eds, S. Osawa, H. Ozeki, H. Uchida and T. Yura), pp. 519–523, University of Tokyo Press, Tokyo.

Ikemura, T. (1981) Correlation between the abundance of *Escherichia coli* transfer RNAs and the occurence of the respective codons in its protein genes. *J. Mol. Biol.* **146**: 1–21.

Ikemura, T. (1982) Correlation between the abundance of yeast transfer RNAs and the occurrence of the respective codons in protein genes. Differences in synonymous codon choice patterns of yeast and *Escherichia coli* with reference to the abundance of isoaccepting transfer RNAs. *J. Mol. Biol.* **158**: 573–597.

Ikemura, T. (1985) Codon usage and tRNA content in unicellular and multicellular organisms. *Mol. Biol. Evol.* **2**: 13–34.

Karlin, S. and Burge, C. (1995) Dinucleotide relative abundance extremes: a genomic signature. *Trends Genet.* **11**: 283–290.

Karlin, S. and Mrazek, J. (2000) Predicted highly expressed genes of diverse prokaryotic genomes. *J. Bacteriol.* **182**: 5238–5250.

Karlin, S., Mrazek, J. and Campbell, A.M. (1998) Codon usages in different gene classes of the *Escherichia coli* genome. *Mol. Microbiol.* **29**: 1341–1355.

Kimura, M. (1983) *The Neutral Theory of Molecular Evolution.* Cambridge University Press, Cambridge.

Koonin, E.V. and Galperin, M.Y. (1997) Prokaryotic genomes: the emerging paradigm of genome-based microbiology. *Curr. Opin. Genet. Dev.* **7**: 757–763.

Kranz, R.G. and Goldman, B.S. (1998) Evolution and horizontal transfer of an entire biosynthetic pathway for cytochrome *c* biogenesis: *Helicobacter, Deinococcus,* Archae and more. *Mol. Microbiol.* **27**: 871–874.

Lawrence, J.G. (1997) Selfish operons and speciation by gene transfer. *Trends Microbiol.* **5**: 355–359.

Lawrence, J.G. (1999a) Gene transfer, speciation, and the evolution of bacterial genomes. *Curr. Opin. Microbiol.* **2**: 519–523.

Lawrence, J.G. (1999b) Selfish operons: the evolutionary impact of gene clustering in the prokaryotes and eukaryotes. *Curr. Opin. Genet. Dev.* **9**: 642–648.

Lawrence, J.G. (2000) Clustering of antibiotic resistance genes: Beyond the selfish operon. *ASM News* **66**: 281–286.

Lawrence, J.G. (2001) Catalyzing bacterial speciation: correlating lateral transfer with genetic headroom. *Syst. Biol.* **50**: 479–496.

Lawrence, J.G. and Ochman, H. (1997) Amelioration of bacterial genomes: rates of change and exchange. *J. Mol. Evol.* **44**: 383–397.

Lawrence, J.G. and Ochman, H. (1998) Molecular archaeology of the *Escherichia coli* genome. *Proc. Natl Acad. Sci. USA* **95**: 9413–9417.

Lawrence, J.G. and Roth, J.R. (1995) The cobalamin (coenzyme B_{12}) biosynthetic genes of *Escherichia coli*. *J. Bacteriol.* **177**: 6371–6380.

Lawrence, J.G. and Roth, J.R. (1996a) Evolution of coenzyme B_{12} synthesis among enteric bacteria: evidence for loss and reacquisition of a multigene complex. *Genetics* **142**: 11–24.

Lawrence, J.G. and Roth, J.R. (1996b) Selfish operons: Horizontal transfer may drive the evolution of gene clusters. *Genetics* **143**: 1843–1860.

Lawrence, J.G. and Roth, J.R. (1998) Roles of horizontal transfer in bacterial evolution. In *Horizontal Transfer* (eds, M. Syvanen, and C.I. Kado), pp. 208–225, Chapman and Hall, London.

Lawrence, J.G. and Roth, J.R. (1999) Genomic flux: genome evolution by gene loss and acquisition. In *Organization of the Prokaryotic Genome* (ed., R.L. Charlebois), pp. 263–289, ASM Press, Washington, DC.

Lawrence, J.G., Ochman, H. and Hartl, D.L. (1992) The evolution of insertion sequences within enteric bacteria. *Genetics* **131**: 9–20.

Logsdon, J.M. and Fuguy, D.M. (1999) *Thermotoga* heats up lateral gene transfer. *Curr. Biol.* **9**: R747-R751.

Maurelli, A.T., Fernández, R.E., Bloch, C.A. et al. (1998) "Black holes" and bacterial pathogenicity: a large genomic deletion that enhances the virulence of *Shigella* spp. and enteroinvasive *Escherichia coli*. *Proc. Natl Acad. Sci. USA* **95**: 3943–3948.

Medigue, C., Rouxel, T., Vigier, P. et al. (1991) Evidence for horizontal gene transfer in *Escherichia coli* speciation. *J. Mol. Biol.* **222**: 851–856.

Metcalf, W.W. and Wanner, B.L. (1993) Evidence for a fourteen-gene, *phnC* to *phnP* locus for phosphonate metabolism in *Escherichia coli*. *Gene* **129**: 27–32.

Milkman, R. (1997) Recombination and population structure in *Escherichia coli*. *Genetics* **146**: 745–750.

Milkman, R. and Bridges, M.M. (1990) Molecular evolution of the *E. coli* chromosome. III. Clonal frames. *Genetics* **126**: 505–517.

Milkman, R. and Bridges, M.M. (1993) Molecular evolution of the *E. coli* chromosome. IV. Sequence comparisons. *Genetics* **133**: 455–468.

Milkman, R. and Stoltzfus, A. (1988) Molecular evolution of the *Escherichia coli* chromosome. II. Clonal segments. *Genetics* **120**: 359–366.

Moran, N.A., Munson, M.A., Baumann, P. and Ishikawa, H. (1993) A molecular clock in endosymbiotic bacteria is calibrated using insect hosts. *Proc. Royal Soc. Lond. B.* **253**: 167–171.

Moszer, I., Rocha, E.P. and Danchin, A. (1999) Codon usage and lateral gene transfer in *Bacillus subtilis*. *Curr. Opin. Microbiol.* **2**: 524–528.

Muto, A. and Osawa, S. (1987) The guanine and cytosine content of genomic DNA and bacterial evolution. *Proc. Natl Acad. Sci. USA* **84**: 166–169.

Naas, T., Blot, M., Fitch, W.M. and Arber, W. (1994) Insertion sequence-related genetic variation in resting *Escherichia coli* K-12. *Genetics* **136**: 721–730.

Nakata, N., Tobe, T., Fukuda, I. et al. (1993) The absence of a surface protease, OmpT, determines the intercellular spreading ability of *Shigella*: the relationship between the *ompT* and *kcpA* loci. *Mol. Microbiol.* **9**: 459–468.

Nelson, K.E., Clayton, R.A., Gill, S.R. et al. (1999) Evidence for lateral gene transfer between Archaea and bacteria from genome sequence of *Thermotoga maritima*. *Nature* **399**: 323–329.

Ochman, H. (1997) Miles of isles. *Trends Microbiol.* **5**: 222.

Ochman, H. and Bergthorsson, U. (1998) Rates and patterns of chromosome evolution in enteric bacteria. *Curr. Opin. Microbiol.* **1**: 580–583.

Ochman, H. and Lawrence, J.G. (1996) Phylogenetics and the amelioration of bacterial genomes, in *Escherichia coli and Salmonella typhimurium: Cellular and molecular biology* (eds, F.C. Neidhardt, R. Curtiss III, J.L. Ingraham et al.), 2nd edn, pp. 2627–2637, American Society for Microbiology, Washington, DC.

Ochman, H. and Wilson, A.C. (1987) Evolutionary history of enteric bacteria, in *Escherichia coli and Salmonella typhimurium: cellular and molecular biology*, (eds, F.C. Neidhardt, J.L. Ingraham, K.B. Low et al.), pp. 1649–1654, American Society for Microbiology, Washington, DC.

Ochman, H. and Wilson, A.C. (1988) Evolution in bacteria: evidence for a universal substitution rate in cellular genomes. *J. Mol. Evol.* **26**: 74–86.

Ochman, H., Lawrence, J.G. and Groisman, E. (2000) Lateral gene transfer and the nature of bacterial innovation. *Nature* **405**: 299–304.

Ricchetti, M., Fairhead, C. and Dujon, B. (1999) Mitochondrial DNA repairs double-strand breaks in yeast chromosomes. *Nature* **402**: 96–100.

Rossman, M.G., Moras, D. and Olsen, K.W. (1974) Chemical and biological evolution of a nucleotide-binding protein. *Nature* **250**: 194–199.

Roth, J., Benson, N., Galitski, T. et al. (1996) Rearrangements of the bacterial chromosome – Formation and applications. In *Escherichia Coli and Salmonella: Cellular and Molecular Biology* (eds, F.C. Neidhardt, R. Curtiss. III, J.L. Ingraham et al.), 2nd edn, pp. 2256–2276, ASM Press, Washington, DC.

Sanderson, K.E., Hessel, A., Liu, S.-L. and Rudd, K.E. (1996) The genetic map of *Salmonella typhimurium*, edition VIII. In *Escherichia Coli and Salmonella Typhimurium: Cellular and Molecular Biology* (eds, F.C. Neidhardt, R. Curtiss III, J.L. Ingraham et al.), 2nd edn, pp. 1903–1999, American Society for Microbiology, Washington, DC.

Selander, R.K., Li, J. and Nelson, K. (1996) Evolutionary genetics of *Salmonella enterica*. In *Escherichia Coli and Salmonella Typhimurium: Cellular and Molecular Biology* (eds, F.C. Neidhardt, R. Curtiss III, J.L. Ingraham et al.), 2nd edn, pp. 2691–2707, American Society for Microbiology, Washington, DC.

Sharp, P.M. and Li, W.-H. (1987) The codon adaptation index – a measure of directional synonymous codon usage bias, and its potential applications. *Nucleic Acids Res.* **15**: 1281–1295.

Smith, J.M., Smith, N.H., O'Rourke, M. and Spratt, B.G. (1993) How clonal are bacteria? *Proc. Natl Acad. Sci. USA* **90**: 4384–4388.

Snel, B., Bork, P. and Huynen, M. (1999) Genome phylogeny based on gene content. *Nature Genet.* **21**: 108–110.

Sueoka, N. (1961) Variation and heterogeneity of base composition of deoxy-ribonucleic acids: a compilation of old and new data. *J. Mol. Biol.* **3**: 31–40.

Sueoka, N. (1962) On the genetic basis of variation and heterogeneity in base composition. *Proc. Natl Acad. Sci. USA* **48**: 582–592.

Sueoka, N. (1988) Directional mutation pressure and neutral molecular evolution. *Proc. Natl Acad. Sci. USA* **85**: 2653–2657.

Sueoka, N. (1992) Directional mutation pressure, selective constraints, and genetic equilibria. *J. Mol. Evol.* **34**: 95–114.

Sueoka, N. (1993) Directional mutation pressure, mutator mutations, and dynamics of molecular evolution. *J. Mol. Evol.* **37**: 137–153.

Sueoka, N. (1995) Intrastrand parity rules of DNA base composition and usage biases of synonymous codons. *J. Mol. Evol.* **40**: 318–325.

Sueoka, N. (1999) Two aspects of DNA base composition: G + C content and translation-coupled deviation from intra-strand rule of A = T and G = C. *J. Mol. Evol.* **49**: 49–62.

Sueoka, N., Marmur, J. and Doty, P. (1959) Heterogeneity in deoxyribonucleic acids. II. Dependence of the density of deoxyribonucleic acids on guanine-cytosine. *Nature (London)* **183**: 1429–1431.

Syvanen, M. (1990) Migrant DNA in the bacterial world. *Cell* **60**: 7–8.

Syvanen, M. (1994) Horizontal gene flow: evidence and possible consequences. *Annu. Rev. Genet.* **28**: 237–261.

Syvanen, M. and Kado, C.I. (eds) (1998) *Horizontal Gene Transfer*, Chapman and Hall, London.

Wernegreen, J.J., Ochman, H., Jones, I.B. and Moran, N.A. (2000) Decoupling of genome size and sequence divergence in a symbiotic bacterium. *J. Bacteriol.* **182**: 3867–3869.

Whittam, T.S. and Ake, S. (1992) Genetic polymorphisms and recombination in natural populations of *Escherichia coli*. In *Mechanisms of Molecular Evolution* (eds, N. Takahata, and A.G. Clark), pp. 223–246, Japan Scientific Society Press, Tokyo.

Wilks, H.M., Hart, K.W., Feeney, R. et al. (1988) A specific, highly active malate dehydrogenase by redesign of a lactate dehydrogenase framework. *Science* **242**: 1541–1544.

Wolf, Y.I., Aravind, L. and Koonin, E.V. (1999) Rickettsiae and Chlamydiae: evidence of horizontal gene transfer and gene exchange. *Trends Genet.* **15**: 173–175.

Worning, P., Jensen, L.J., Nelson, K.E. et al. (2000) Structural analysis of DNA sequence: evidence for lateral gene transfer in *Thermotoga maritima*. *Nucl. Acids Res.* **28**: 706–709.

Wu, G., Fiser, A., ter Kuile, B. et al. (1999) Convergent evolution of *Trichomonas vaginalis* lactate dehydrogenase from malate dehydrogenase. *Proc. Natl Acad. Sci. USA* **96**: 6285–6290.

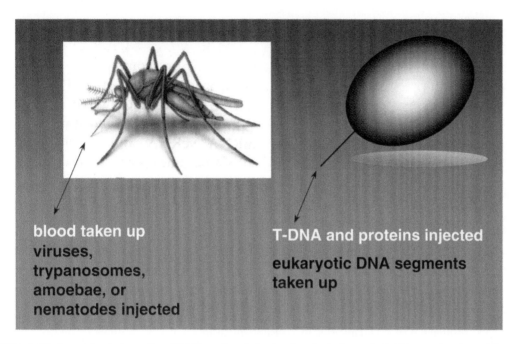

FIGURE 5.1 Horizontal retrotransfer of DNA and proteins by *Agrobacterium* mimics the mechanism of fortuitous organismal and viral transmission during feeding by the mosquito, whereby uptake of genetic and proteinaceous material can be bidirectional.

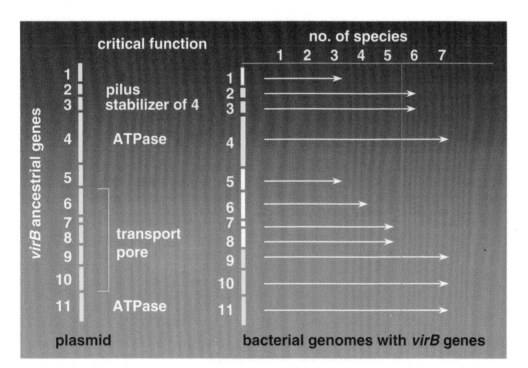

FIGURE 5.2 Degree of conservation of *virB* genes encoding critical functions needed for the transmission

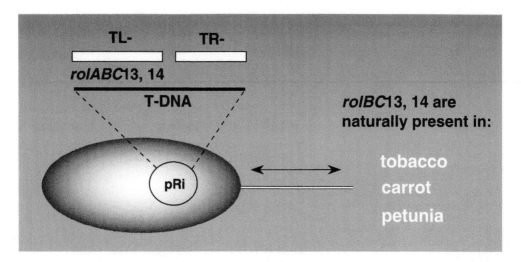

FIGURE 5.3 Conserved *rol* genes A, B and C, and open reading frames 13 and 14 in the core region of the T_L-DNA of *Agrobacterium* rhizogenes and in the genome of *Nicotiana glauca* (wild tobacco), *Daucus carota* var. *sativa* (carrot), and *Petunia hybrida* (garden petunia). These *rol* genes are colinear to those genes are occurring naturally in the plant genomes. *rol* genes are often repeated and inverted in *N. glauca*, suggesting that gene shuffling might have occurred.

```
Halobacterium salinarium        EETEEE.APED  MVQCRVCGEY  YQAI..TEPH  LQTH.DMTIQ  E
Haloferax mediterranei          QFLDDE.TPED  MVQCLVCGEY  YQAI..TEPH  LQTH.DMTIK  K
Methanococcus jannaschii 1      EKLEIQ.KEGF  FYKCPYCNYT  NADVKAIKKH  IKSKHYDIIA  K
Methanococcus jannaschii 2      KLELQK.NDIG  FYKCPFCDYT  NADAKVVRKH  VKSKHLEEIE  K
Agrobacterium tumefaciens       VSVRKS.VQDD  HIVCLECGGS  FKSL...KRH  LTTHHSMTPE  E
Agrobacterium radiobacter       VSVRKS.VQDD  HIVCLECGGS  FKSL...KRH  LTTHHSMTPE  E
Rhizobium meliloti              VSVRKS.VQDD  QITCLECGGT  FKSL...KRH  LMTHHNLSPE  E
Rhizobium sp.NRG234             VSIRKS.VQDD  QITCLECGGA  FKSL...KRH  LMTHHNLSPE  D
Rhizobium etli                  VSVRKS.VQDE  QITCLECGGN  FKSL...KRH  LMTHHSLSPE  E
Sphingomonas aromaticivorans    VSVRAS.VKPD  AVTCLDCGAK  MKML...KRH  LGTDHGMTPA  A
Ascobolus immersus              PSVKET.GNPK  ETLCVPCGKK  FRDF...KAH  MLTHQDERPE  K
Fugu rubripes                   HELKHEKGQEN  ..VCVECGLD  FPTLAQLKRH  LTTHRGPTLY  R
Mus musculus                    KKPSKPVKKNH  ..ACEMCGKA  FRDVYHLNRH  KLSHSDEKPF  E
Homo sapiens 1                  KKPSKPVKKNH  ..ACEMCGKA  FRDVYHLNRH  KLSHSDEKPF  E
Xenopus laevis 1                HSRTHTGY..N  PYVCTECGKR  FSSNSGLRRH  MRTHTGVKPY  A
Xenopus laevis 2                HLRIHTGE..T  PFVCPECGKG  FRDASFLKSH  LSIHTGEKPF  V
Homo sapiens 2                  HEIIHTGE.KL  .YDCKECGKT  FFSLKRIRRH  IITHSGYTPY  K
Drosophila melanogaster         KKSAYSLAPNR  .VSCPYCQRM  FPWSSSLRRH  ILTHTGQKPF  K
```

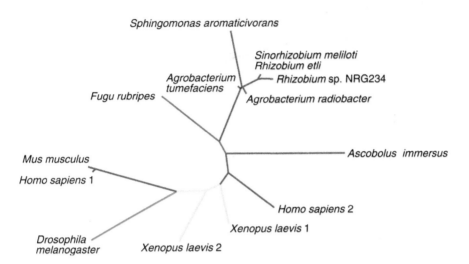

IGURE 5.5 Distance tree derived from a corrected-distance matrix. Multi-sequence alignment was determined with the Pileup program and was edited manually. The programs used were the Kimura Distance, the Growtree nd Treeview programs as described previously (Bouhouche et al., 2000).

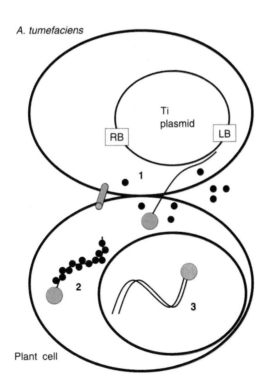

IGURE 6.1 Protein and T-DNA transfer from *A. tumefaciens* to plant cells: ●, VirD2 protein; ●,VirE2 protein; ▬, T-pilus "injectosome" (Kado 2000; Lai and Kado 2000); LB, Left T-DNA border; RB, Right T-DNA border; **1,** 'irD2 is bound to the T-DNA 5′ end within *A. tumefaciens* and is probably transported into the plant cell as part of protein/DNA complex. Other *vir* proteins such as VirE2 and VirF may be transported separately into the plant ell. **2,** VirD2 and VirE2 may be bound to the T-DNA and possibly have a role in guiding the T-DNA specifically to

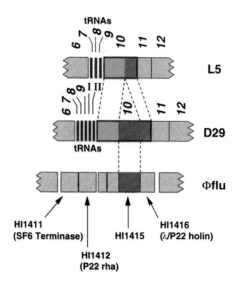

FIGURE 12.2 Relationship between mycobacteriophages L5 and D29 and *H. influenzae* gene HI1415. Segments of the genomes of the closely related mycobacteriophages L5 and D29 are shown with genes *6–12* represented as boxes (L5 genes *7–9* and D29 genes *7–9*, I and II encode tRNAs as indicated). Below is shown a segment of the *H. influenzae* genome from within the putative cryptic prophage φflu (shown in full in Figure 12.3). Gene *10* of D29 is larger than L5 gene *10* due to an additional 600 bp within the coding region which results in a gene product (gp10) that is 200 amino acids larger than L5 gp10. The upstream parts of gp10 (green) are significantly more similar than the downstream segments (blue) having 81% and 50% amino acid identity respectively. The "insert" in D29 gene *10* (red) encodes a protein sequence that has 34% identity with the product of *H. influenzae* HI1415.

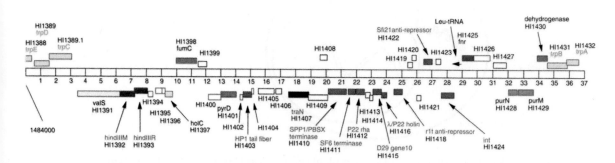

FIGURE 12.3 Organization of φflu, a cryptic prophage of *H. influenzae*. A 37 kb segment of the *H. influenzae* genome is shown corresponding to coordinates 1484000 to 1521000 (Fleischmann et al., 1995). Genes and their putative functions are shown according to the previously published annotation (Fleischmann et al., 1995), and are shown either above (rightwards) or below (leftwards) depending on the direction of transcription. Genes

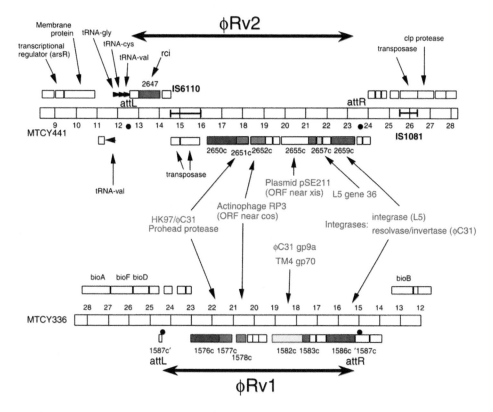

FIGURE 12.4 Cryptic prophages of *M. tuberculosis* H37Rv. Portions of two segments of the *M. tuberculosis* H37Rv genome present in cosmids MYCY441 (φRv2) and MTCY336 (φRv1) are shown (accession numbers Z80225 and Z95586 respectively). Horizontal bars represent the genomes with markers at 1 kb intervals; numbers correspond to the cosmid coordinates. Genes are shown either above or below the bar depending on the direction of transcription, with genes below transcribed leftwards. Related genes are shown in similar colors. Putative attachment junctions, *attL* and *attR*, delineate the boundaries of the cryptic prophages φRv2 and φRv1. The location of genes and their putative functions are from annotations by the Sanger Center.

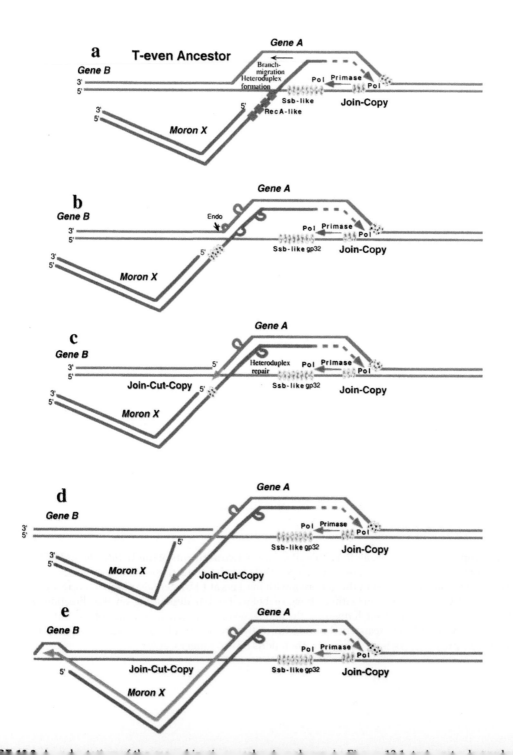

FIGURE 13.2 A real[...]tion of the way of joining genes by transduction as in Figure 13.1 A. A systed may be [...]

```
T4     MAHFNECAHLIEGVDKAQNEYWDILGDEKDPLQVMLDMQRFLQIRLANVR-
RB15   MAYFNECSQLIEGADKAQNEYWDILGDEKDPLQVMLDMQKSLQVRLAKDKP
T2     MAHFNECAHLIEGVDKATRAYAENIMHNIDPLQVMLDMQRHLQIRLANDKP
LZ5    MAHFNECAHLIEGVDKANRAYAENIMHNIDPLQVMLDMQRHLQIRLANDKP
T6     MAHFNECAHLIEGVDKATRAYAENIMHNIDPLQVMLDMQRSLQIRLANDKP

T4     EYCYHPDKLETAGDVVSWMREQKDCIDDEFRELLTSLGEMSRGEKEASAV
RB15   EYNRHPDDLATAGEVVDWLRNQKDYIDDEFRELLTSLGGMSNGEKEASAV
T2     ETNRHPDSLETAGEVLAWLRNQDDYIADETRELYTSLGGMSNGEKEASAV
LZ5    ETNRHPDSLETAGEVLAWLRNQDDYIADETRELYTSLGGMSNGEKEASAV
T6     ETNRHPDSLETAGEVLAWLRNQDDYIADETRELYTSLGGMSNGEKEASAV

T4     WKKWKARYIEAQEKRIDEMSPEDQLEIKFELVDIFHFVLNMFVGLGMNAEEI
RB15   WKPWKSQHGERRETLITDLSPQDQLEIKFEMIDILHFVLNMFQGLGLTAEEI
T2     WKPWKKRYSEMQSKKIQDLSPEDQLEIKFELIDQFHFFMNKFIALGMSAEEI
LZ5    WKPWKKRYSEMQSKKIQDLSPEDQLEIKFELIDQFHFFMNKFIALGMSAEEI
T6     WKPWKKRYSEMQSKKIQDLSPEDQLEIKFELIDQFHFFMNKFIALGMSAEEI

T4     FKLYYLKNKHNFERQDNGYX
RB15   FKLYYLKNAENFARQDRGYX
T2     FKLYYLKNAENFARQDRGYX
LZ5    FKLYYLKNAENFARQDRGYX
T6     FKLYYLKNAENFARQDRGYX
```

FIGURE 13.3 Alignments of the amino acid sequences of dCTPases of different T-even phages. Identical amino acids are printed in black. Differences are indicated by different colors.

```
              CAACAAGAAAA--GGTTCAAGACCGTT      2018
              CAGCCAGAGTATCGG--CGATATCCTG      90

2019   TG-AC-CCTGAATTTGGAT-----GTGATTTATCAGA   2048
  91   CGCACACCGG---TCGGCTCACGGGTGAT---------  116

2049   CCAGCTTTTTGAAAATATGACTCCTC-TTACTG-CTG   2083
 117   ---GCGTCGTG-----ATTACGGCTCGTTGCTGGCTG   145

2084   ACACGGTTGAAC----GCA-ATATCGAAAGCGC---A   2112
 146   CAATGATTGACCAGCCGCAGACCCCG---GCGCTTGA   179

2113   GTAAGAAACTATGAGCCACGTATTGATAAATTAGCAG   2149
 180   GTTGCAGATTA-AAGTCGCCTGTT----ACATGGCAG   211

2150   TT----AATGTGATACCCG
 212   TGCTGAAATGGGAACCCCG
```

FIGURE 13.4 Alignment of the base sequences of phage T4 gene *25* (upper line) and P2 gene *W* (lower line) in the gene segment corresponding to the protein similarity (base no. 65 to 231 of P2 gene *W*). Identities are indicated in

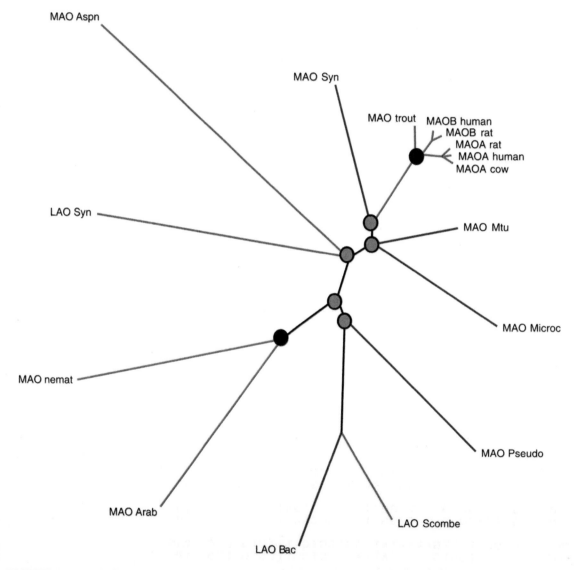

FIGURE 25.2 A phylogenetic tree of monoamine oxidases illustrating possible horizontal gene exchange between vertebrates and bacteria. The rootless least-square tree (Fitch and Margoliash, 1967) was constructed using the FITCH program of the PHYLIP package (Felsenstein, 1996), with 1000 bootstrap replications analyzed using the SEQBOOT, PROTDIST and CONSENSE of PHYLIP. Multiple alignments were constructed using the CLUSTALW program (Thompson et al., 1994). The black circles show the nodes with at least >70% bootstrap support and the green circles show the nodes with 50–70% support (1000 replications). Red branches, eukaryotic proteins, blue branches, bacterial proteins. MAO, monoamine oxidase, LAO, L-amino acid oxidase.

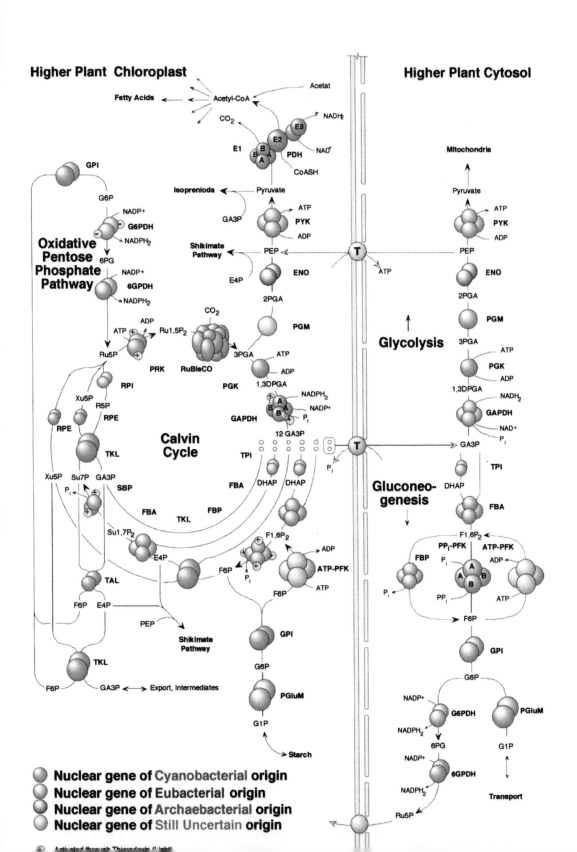

Higher Plant Chloroplast

Higher Plant Cytosol

Nuclear gene of Cyanobacterial origin
Nuclear gene of Eubacterial origin
Nuclear gene of Archaebacterial origin
Nuclear gene of Still Uncertain origin

FIGURE 27.2 (On previous page.) Summary of gene phylogenies for enzymes of compartmentalized sugar phosphate metabolism in plants. Suggested evolutionary origins for the nuclear genes are color-coded. Enzymes regulated through the thioredoxin system are indicated. Many enzymes in the figure are allosterically regulated, but no allosteric regulation is indicated here. Enyzme abbreviations are: FBA fructose-1,6-bisphosphate aldolase; FBP fructose-1,6-bisphosphatase; GAPDH glyceraldehyde-3-phosphate dehydrogenase; PGK 3-phospho-glycerate kinase; PRI ribose-5-phosphate isomerase; PRK phosphoribulokinase; Rubisco ribulose-1,5-bisphosphate carboxylase/oxygenase; RPE ribulose-5-phosphate 3-epimerase; SBP sedoheptulose-1,7-bisphosphatase; TKL transketolase; TPI triosephosphate isomerase; TAL transaldolase; GPI glucose-6-phosphate isomerase; G6PDH glucose-6-phosphate dehydrogenase; 6GPDH 6-phosphogluconate dehydrogenase; PGluM phosphogluco-mutase; PGM phosphoglyceromutase; PFK phosphofructokinase (pyrophosphate and ATP-dependent); ENO enolase; PYK pyruvate kinase; PDC pyruvate dehydrogenase complex (E1, E2, E3 components), T translocator. PDC is a multienzyme complex, but only one set of components is drawn here. Note that chloroplast isoenzymes of PGM, ENO, and PYK have not been demonstrated in spinach leaves, but for convenience we have included those enzymes in this figure, since they have been well characterized in the plastids of other higher plants (Plaxton, 1996). For further details see Martin and Schnarrenberger (1997), Martin and Herrmann (1998), Nowitzki et al. (1998) and Krepinsky et al. (2001).

FIGURE 27.3 (Opposite.) A tree of genomes. Each prokaryotic genome is represented as a single colored line, different colors symbolize different groups of prokaryotes. A working hypothesis for the origin of eukaryotic genes as outlined primarily in Martin and Müller (1998) and taking lateral gene transfer into account (modified from Martin, 1999a). The figure extends symbiotic associations (merging of genomes into the same cellular confines is indicated by the merging of colored lines) to include schematic indication of several independent secondary symbioses for the acquisition of plastids during eukaryotic history (Martin et al., 1998; Zauner et al., 2000). Importantly, among eukaryotes, colored lines indicate merely that prokaryotic genomes existed at one time within the cellular confines of a given eukaryotic lineage, not that they have persisted to the present as an independently compartmented genome (because most genes from organelles are transferred to the nucleus). For example some eukaryotes with secondary symbionts are schematically indicated with six lines, but only have four distinctly compartmented genomes. Similarly, eukaryotes that lack mitochondria apparently possessed such organelles in the past but only have one genome – that in the nucleus (for example *Giardia*, *Trichomonas*, and *Entamoeba*, see text). Furthermore it is important to note that mitochondrion-lacking eukaryotes have arisen through loss of the organelle in many independent lineages (Embley and Martin, 1998; Hashimoto et al., 1998; Roger, 1999). In the enlargement, lateral gene transfer *between eubacteria* prior to – and implicitly, but not shown, subsequent to – the origin of mitochondria (blue lines) and plastids (green lines) is schematically represented. Genomes are represented as heavy lines, individual gene transfer events (regardless of possible numbers of genes involved) as thin lines. The phylogeny of eukaryotes indicated is mostly schematic, containing various elements from van der Peer et al. (1996), Keeling and Palmer (2000), Martin et al. (1998) and other sources, and is not intended to be close to correct. The various schematic plastid organizational types are shown merely to indicate that dramatic differences in organelle structure can arise from a single ancestral symbiotic state.

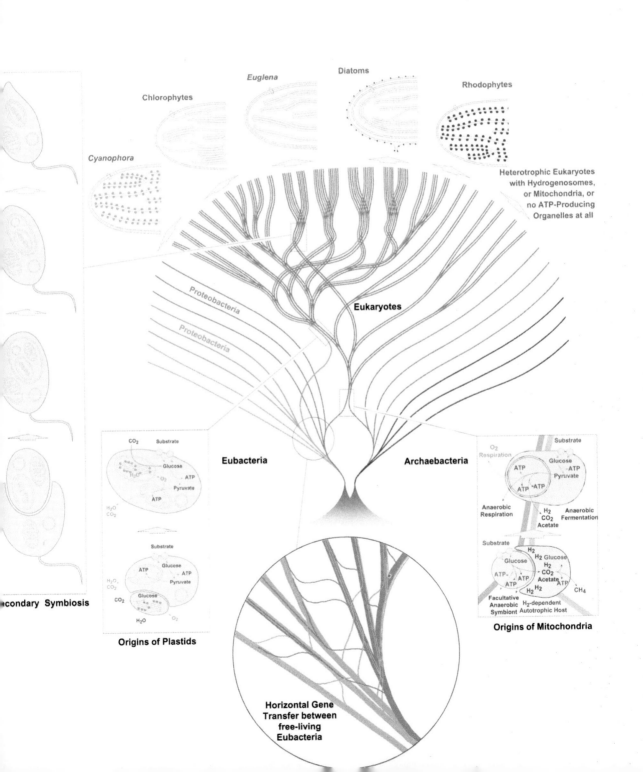

Chlorophytes

Euglena

Diatoms

Rhodophytes

Cyanophora

Heterotrophic Eukaryotes
with Hydrogenosomes,
or Mitochondria, or
no ATP-Producing
Organelles at all

Proteobacteria

Proteobacteria

Eukaryotes

Eubacteria

Archaebacteria

CO$_2$ Substrate

Glucose

H$_2$O   O$_2$ ATP

Pyruvate

ATP

H$_2$O
CO$_2$

Substrate

ATP   Glucose

H$_2$O   ATP
CO$_2$ Pyruvate

Glucose

CO$_2$

H$_2$O   O$_2$

Secondary Symbiosis

Origins of Plastids

O$_2$ Substrate
Respiration

ATP   Glucose
ATP
ATP   ATP   Pyruvate

Anaerobic
Respiration   H$_2$ Anaerobic
CO$_2$ Fermentation
Acetate

Substrate   H$_2$
H$_2$ Glucose
Glucose   H$_2$
ATP   H$_2$ CO$_2$
ATP   ATP   Acetate   ATP
H$_2$ H$_2$ CH$_4$

Facultative   H$_2$-dependent
Anaerobic   Autotrophic Host
Symbiont

Origins of Mitochondria

**Horizontal Gene
Transfer between
free-living
Eubacteria**

CHAPTER 10

Bacterial Pathogenicity Islands and Infectious Diseases

Jay V. Solnick and Glenn M. Young

The significance of lateral gene transfer for infectious diseases was first appreciated in the 1960s, when it was realized that bacterial genes coding for virulence factors (Smith and Halls, 1968) and for resistance to antimicrobial agents (Ochiai et al., 1959; Akiba et al., 1960) could be transmitted on plasmids. Subsequent studies of uropathogenic *Escherichia coli* suggested that large chromosomal fragments containing blocks of virulence genes might also be horizontally transferred (Hacker et al., 1983; Low et al., 1984). These DNA sequences were called "pathogenicity islands" (PAIs) by Jörg Hacker and colleagues at the University of Würzburg (Hacker et al., 1990). From an evolutionary view, PAIs clearly represent a subgroup of what have been called "genomic islands" (Morschhäuser et al., 2000), which are large clusters of chromosomal genes that are exchanged among members of a species and confer some selective advantage, such as antimicrobial resistance (Ito et al., 1999; Katayama et al., 2000) or symbiosis with a host (Sullivan and Ronson, 1998). Nevertheless, it is useful to consider the special case of genomic islands that confer pathogenicity because it has broad application to the understanding and perhaps eventually control of a variety of bacterial pathogens and the diseases they cause. The purpose of this chapter is to discuss selected bacterial PAIs and their significance for infectious diseases. Readers interested in a more detailed discussion may consult a recent review (Kaper and Hacker, 1999).

IDENTIFICATION OF PATHOGENICITY ISLANDS

PAIs carry multiple virulence factors on a relatively large fragment of chromosomal DNA, typically at least 10–20 kb and sometimes up to 200 kb (Hacker et al., 1997). Smaller fragments containing fewer pathogenicity loci can also be found, and have sometimes been called "islets" (Moncrief et al., 1998). PAIs are usually present in the genome of pathogenic bacteria but not in the genome of the same or closely related non-pathogenic species, although the presence of a PAI does not necessarily confer the ability to cause disease. Known PAIs carry genes that encode adhesins, invasins, iron uptake systems, toxins, and complex protein secretion machinery. With the proliferation of complete bacterial genome sequences, PAIs are increasingly recognized by sequence analyses that reveal homologues with these classes of known virulence genes. Sometimes multiple PAIs with specialized functions may be present in a single species, such as SPI 1–5 found in *Salmonella enterica* (Wallis and Galyov, 2000).

Several features of PAIs suggest that they have been acquired by horizontal gene transfer. First, the G + C content and sometimes the codon usage of a PAI differ from that of other portions of the genome. Secondly, the insertion site for a PAI is often the 3' end of tRNA loci, which are known to serve as attachment sites for integration of phage and

Copyright © 2002 by Academic Press.
All rights of reproduction in any form reserved.

conjugative transposons (Cheetham and Katz, 1995). A striking example of this is tRNA^{selC}, which serves as the integration site for PAI-1 of uropathogenic *E. coli* (Blum et al., 1994), the LEE island of enteropathogenic *E. coli* (McDaniel and Kaper, 1997), the SPI-3 island of *S. enterica* (Blanc-Potard and Groisman, 1997), and the SHI-2 island of *Shigella flexneri* (Moss et al., 1999; Vokes et al., 1999). Thirdly, PAIs often carry genes encoding mobility factors, such as integrase, transposase, or insertion sequences, which presumably indicate their origin in bacteriophage or conjugative plasmids. Finally, PAIs are often flanked by small directly repeated DNA sequences that may have been generated during recombination into the host genome.

PATHOGENICITY ISLANDS AND INFECTIOUS DISEASES

Uropathogenic *E. coli*

E. coli commonly inhabits the intestinal tract of humans and other mammals. It was often considered to be a commensal organism that was associated with disease only under unusual circumstances that result in a breach of normal immune defenses. However, this view has changed with the identification and characterization of virulence factors that contribute to specific aspects of pathogenicity, which indicate that specific strains of *E. coli* have evolved to be clever pathogens. Uropathogenic *E. coli* (UPEC) has been a model organism for the identification of genes responsible for the evolution of bacterial pathogenicity. UPEC causes urinary tract infections in humans and is the most common cause of cystitis in women. This group of pathogens can also cause pyelonephritis, a more serious infection of the kidneys. The first formal description of PAIs was for UPEC strains 536 and J96. Other PAIs of UPEC have subsequently been identified in strain CFT073.

The four known PAIs of strain 536 range in size from 25 kb to 190 kb. They are composed of genes that encode virulence factors including two α-hemolysins, three different fimbrial adhesins, and a siderophore-type iron uptake system. PAI I_{536} and II_{536} carry the hemolysin (*hly*) and the P-related fimbrial genes (*prf*) (Hacker et al., 1983, 1990; Knapp et al., 1986). PAI III_{536} carries the S fimbrial adhesive gene cluster (Blum et al., 1994). PAI IV_{536} contains genes necessary for the synthesis and uptake of an iron scavenging system. It is evolutionarily related to a pathogenicity island carrying the ferric yersiniabactin uptake system that is present in highly pathogenic species of *Yersinia* (Fetherston et al., 1992; Carniel et al., 1996). It has also been associated with some diarrheal strains of enteroaggregative *E. coli* (EAEC) and isolates of *E. coli* associated with septicemia (Schubert et al., 1998). The target site of chromosomal insertion of the four identified PAIs of UPEC 536 has been determined. Each PAI maps to a different position on the chromosome, but shares the common feature of being located adjacent to a tRNA gene (Table 10.1).

For UPEC strains J96 and CT073, PAIs containing similar clusters of virulence genes have been described. Sequences similar to PAI IV_{536} have been detected in strain J96 by hybridization analysis. In addition, PAIs that encode α-hemolysins, fimbriae, and cytotoxic necrotizing factor 1 (CNF1) are present (Blum et al., 1995). Similarly, strain CT073 has sequences encoding α-hemolysin and fimbriae that are contained within a PAI (Guyer et al., 1998). Interestingly, the chromosomal location of these PAIs and the repertoire of virulence genes within the PAIs of these strains exhibit differences. This indicates that UPEC strains, which display similar pathogenicity phenotypes, are not necessarily a clonal population of organisms. Nevertheless, PAI-associated traits are implicated in or have been shown to contribute to virulence. Fimbriae most likely influence the localization of the bacteria in the host by affecting adherence to specific types of cellular receptors. UPEC strains frequently contain P fimbriae genes (*pap*) that encode an adhesin that influences the development of pyelonephritis in the cynomolgus monkey model (Roberts et al., 1994). P fimbriae appear to influence the development of an infection by binding to specific isoreceptors containing the urothelial α-gal-(1–4) β-gal disaccharide (Svenson et al., 1983; Roberts et al., 1997). Likewise, EPEC strains

TABLE 10.1 Characteristics of selected pathogenicity islands (PAI)

Species	Designation	Size (kb)	Function or relevant characteristics	Locations/adjacent genes
Uropathogenic *E. coli* 536	PAI I$_{536}$	70	Hemolysin	*selC*
	PAI II$_{536}$	190	Hemolysin, P-fimbriae	*leuX*
	PAI III$_{536}$	25	S-fimbriae	*thrW*
	PAI IV$_{536}$	45	Iron acquisition system related to *Yersinia* HPI	*asnT*
Enteropathogenic *E. coli*	LEE	35–43	Type III secretion system, attaching and effacing histopathology	*selC* and *pheU*
Salmonella enterica	SPI-1	40	Type III secretion system, host cell invasion, macrophage apoptosis	Between *flhA* and *mutS*
	SPI-2	40	Type III secretion system, systemic infection, intracellular survival (macrophage)	*valV*
	SPI-3	17	Intracellular survival (macrophage)	*selC*
	SPI-4	25	Intracellular survival (macrophage)	Putative tRNA gene
	SPI-5	7	Inflammation, host cell invasion	*serT*
Shigella flexneri	SHI-I	51	Enterotoxin, IgA protease	IS elements
	SHI-II	ND	Aerobactin-dependent iron acquisition system	ND
	Invasion gene cluster	37	Type III secretion system, host cell invasion, cell-cell spread	Plasmid
Yersinia spp.	HPI	43–102	Yersiniabactin-dependent iron acquisition system	
	pYV (Yop gene cluster)		Type III secretion system, cytotoxin, macropahage apoptosis, inhibition of macropahge uptake and cytokine release	Plasmid
Helicobacter pylori	Cag	40	Type IV secretion system	*glr*
Listeria monocytogenes	*hly* element	9.6	Hemolysin, actin nucleation, phospholipase C	Between *prs* and *orfB*
Staphylococcus aureus RN4282	*tst* element	15.2	Toxic shock toxin	17 bp direct repeats
Staphylococcus aureus Methicillin resistant	Mec DNA	52	Antibiotic resistance	*orfX*/15 bp direct repeats
Clostridium difficile	*tcdA/B* element	19.6	Enterotoxin, cytotoxin	115 bp/cdu2/2′ and cdd2–3

ND = not determined.

typically produce α-hemolysin, which in addition to its ability to lyse erythrocytes, also induces internal calcium oscillations in renal epithelial cells and stimulates production of the cytokines interleukin (IL)-6 and IL-8 (Uhlen et al., 2000). This indicates that hemolysin may have a role in inflammation.

Representative Gram-negative enteric pathogens

The identification of PAIs among bacteria known to cause enteric diseases has occurred at a rapid pace over the past 10 years. More than 10 PAIs have been identified in strains of E. coli, Salmonella enterica, Shigella spp. and Yersinia spp. The repertoire of PAIs present in each pathogen and the function of the genes contained in individual PAI often exhibit considerable differences. This, in part, is the reason for individual characteristics of the diseases caused by the bacteria. However, detailed studies of PAIs are beginning to show some common features. Some PAIs are dedicated to maintaining the nutritional status of the bacterium in the host environment. An example of this type of PAI has been described for highly pathogenic strains of Yersinia, including the plague pathogen Y. pestis, which causes lethal infections in mice (Fetherston et al., 1992; Carniel et al., 1996). The association of this PAI with lethal disease has led to it being called the High-Pathogenicity Island (HPI) (Carniel et al., 1996). Included within the Yersinia HPI are genes required for the acquisition of iron that is tightly sequestered in the host environment. The SHI II island of Sh. flexneri also contains genes encoding an iron uptake system (Moss et al., 1999; Vokes et al., 1999). Other PAIs encode dedicated protein secretion systems that export polypeptides from the bacterium to the extracellular environment or direct protein secretion into the cytoplasm of eukaryotic cells.

One type of protein secretion system that has been identified in several enteric pathogens is the type III secretion system (for reviews see Hueck, 1998; Nguyen et al., 2000). PAI-containing gene clusters encoding these secretion systems have also been identified in bacteria causing a wide range of other diseases in animals and plants. The virulence factors that are secreted by these systems exhibit a range of virulence properties that are tailored to the specific needs of each pathogen. Their study has enriched our understanding of how PAIs of different pathogens can be similar, but has also revealed how PAIs are fine-tuned in each bacterium. The proteins secreted by this mechanism by different bacteria, often called effectors, can be quite divergent. They include cytotoxins, protein kinases, and protein phosphorylases. For enteropathogenic and enterohemorrhagic E. coli (EPEC and EHEC) the type III secretion system genes are contained within the locus of enterocyte effacement, or LEE PAI, which is responsible for the characteristic intestinal histopathology known as "attaching and effacing" (McDaniel et al., 1995). The attaching and effacing histopathological changes elicited by EPEC and EHEC are the effacement of the intestinal epithelial cell microvilli and intimate adherence of the bacterium to the epithelial cell membrane. Additionally, there are distinct changes in the cytoskeleton of the epithelial cells that result in the bacterium residing on a structure resembling a pedestal. In part, the type III secretion system is necessary for targeted secretion of the Tir protein into epithelial cells (Kenny et al., 1997). Tir subsequently serves as a receptor for intimin, a bacterial surface protein encoded by the eae locus, which is also contained on the LEE PAI.

In Salmonella enterica there have been five Salmonella pathogenicity islands (SPI-1 to 5) identified (Mills and Lee, 1995; Ochman et al., 1996; Shea et al., 1996; Blanc-Potard and Groisman, 1997; Wong et al., 1998; Wood et al., 1998). SPI-1 and SPI-2 are known to contain genes that encode a type III secretion system. The SPI-1 system is also required for invasion of nonphagocytic cells and for induction of apoptosis in macrophages (Callazo and Galan, 1997). This secretion system also exports several effector proteins encoded by genes within SPI-1, and is required for the secretion of SopB (SigD), a protein encoded within SPI-5 that promotes fluid accumulation and inflammation in infected ileal mucosa (Galyov et al., 1997; Hong and Miller, 1998). The type III secretion system genes contained in SPI-2 are required for Salmonella survival in macrophages and are necessary to cause systemic disease (Ochman et al., 1996; Shea et al., 1999). Although these genes are not readily expressed under standard laboratory conditions, they are induced in infected macrophages (Cirillo et al., 1998). An interesting consequence of this regulation is that this PAI-encoded secretion system operates in the intracellular environment of the eukaryotic cells, rather than from an extracellular location as is the case for most other type III secretion systems.

In other enteric pathogens, PAIs are not restricted to the bacterial chromosome. For enteroinvasive *E. coli* (EIEC), *Shigella* spp. and *Yersinia* spp., clusters of pathogenicity genes that encode type III secretion systems are located on large virulence plasmids. Although plasmid-encoded gene clusters do not conform to the strict definition of a PAI, they are important to consider in any discussion focused on the evolution of bacterial pathogens. Similar to chromosomally located PAIs, these plasmid PAIs are a stable part of the bacterial genome and contain genes dedicated to virulence. Current evidence suggests they were acquired by a horizontal mechanism of DNA transfer and, due to their stability, there does not appear to be a strong selection for integration into the chromosome. Similar to SPI-1 of *Salmonella*, the virulence plasmid-encoded type III secretion systems of *Shigella* and enteroinvasive *E. coli* (Sansonetti and Egile, 1998) are necessary for invasion of epithelial cells (Sansonetti et al., 1981, 1982; Harris et al., 1982) and for the induction of apoptosis in infected macrophages (Zychlinski et al., 1992). Other genes clustered in the same region of the plasmid are required for intracellular movement of the bacteria within infected cells. Essential to this process is *icsA*, which encodes an outer membrane protein that acts as a catalyst for actin polymerization (Goldberg and Theriot, 1995; Kocks et al., 1995). Bacterial-induced polymerization of host cell actin at the base of bacterial attachment then leads to spread from cell to cell.

Exhibiting some contrasting pathogenic characteristics, the genes contained on the *Yersinia* virulence plasmid (pYV) encode a type III secretion system that appears to be dedicated to maintaining the characteristic extracellular life-style of the bacterium in infected hosts. It is responsible for the targeted export of at least six effectors that appear to limit activation of the host immune response. The functions of some of the effectors have been delineated. The YopE and YopH proteins appear to disarm eukaryotic cells such as macrophages by interfering with processes involved in phagocytosis (Rosqvist et al., 1988, 1991; Black and Bliska, 1997; Persson et al., 1997). Secretion of the YopJ (also called YopP) effector by the type III secretion system has been shown to cause apoptosis of infected macrophages (Mills et al., 1997; Monack et al., 1997). YopJ (YopP) has also been shown to block TNF-α and IL-8 release (Schulte et al., 1996; Boland and Cornelis, 1998; Palmer et al., 1998; Schesser et al., 1998).

The study of proteins secreted by PAI-encoded type III secretion systems has revealed some of the reasons for the differences in pathogenic traits among enteric pathogens. However, the similarities between the proteins that form the subunits of the secretion apparatus have revealed possible evolutionary links to their origin. Depending on the bacterial species, 11 to 21 different subunits compose the supramolecular structure or apparatus responsible for polypeptide transport (Hueck, 1998). Despite this range, all known type III secretion systems contain a core set of 10 proteins that are homologous. Furthermore, these proteins are believed to form the base part of the apparatus and are similar to proteins that form the basal body of the bacterial flagellum, indicating an evolutionary relationship (Hueck, 1998; Anderson et al., 1999). This premise is supported by electron microscopic studies of the *Salmonella* SPI-1 and *Shigella* virulence plasmid-encoded apparatus, which revealed they are morphologically similar to the flagellar basal body (Kubori et al., 1998; Blocker et al., 1999; Tamano et al., 2000). The basal body is a portion of the bacterial flagellum known to function as a protein export apparatus during morphogenesis (Macnab, 1996, 1999; Fan et al., 1997; Minamino and Macnab, 1999). It transports flagellar subunits from the bacterial cytoplasm to the tip of the growing organelle. There is also functional evidence that strikingly demonstrates the flagellum has the capacity to secrete extracellular virulence factors (Young et al., 1999). Secretion of the virulence factor YplA by *Y. enterocolitica* is dependent on the flagellar system (Young et al., 1999). YplA, which may be best described as an effector, is a phospholipase that does not contribute to bacterial motility (Young et al., 1999), but enhances bacterial survival and promotes inflammation during the early stages of infection (Schmiel et al., 1998). Since evolution of the flagellum is considered to predate other type III secretion systems, it seems that the ancestral genes of PAIs encoding this category of protein secretion system originated from flagellar genes (Macnab, 1999).

Helicobacter pylori

H. pylori commonly infects human gastric mucosa, and in 10% to 15% of infected persons is associated with the development of peptic ulcer disease or gastric adenocarcinoma (Dunn et al., 1997). About 60% of US strains have the *cagA* gene, which is associated with but not required for vacuolating cytotoxin activity. Sequencing out from *cagA* revealed that it is part of a 40 kb PAI that is more often found in strains isolated from patients with clinical disease (Censini et al., 1996). Six of the 31 genes on the Cag PAI have homology to components of the type IV secretion systems found in *E. coli*, *Agrobacterium tumefaciens*, *Bordetella pertussis* and others, which specialize in the transport of DNA and other macromolecules across the bacterial membrane or into other cells (Covacci et al., 1999; Christie and Vogel, 2000; Lai and Kado, 2000). Although the effector molecule is unknown, these and other genes on the Cag PAI are required for *H. pylori* to induce gastric epithelial cells to produce interleukin-8 (IL-8), which causes inflammation by recruitment of polymorphonuclear leukocytes (Segal et al., 1997; Crabtree et al., 1999; Li et al., 1999). The CagA protein itself is also exported by the type IV apparatus and injected into the host cell, where it is tyrosine phosphorylated and likely interrupts eukaryotic signal transduction pathways (Segal et al., 1999; Odenbreit et al., 2000; Stein et al., 2000). In this respect, the interaction between *H. pylori* and host cells resembles that found in enteropathogenic *E. coli*, which secretes Tir into the host membrane where it serves as a receptor for intimin binding and alters cell signaling (Kenny et al., 1997).

Gram-positive bacteria

Gram-positive bacteria may also harbor genomic islands, which in some cases satisfy the criteria for a PAI. Probably the best-studied of these is the 9.6 kb pathogenicity locus found in *Listeria monocytogenes*, a human pathogen that causes bacteremia and meningoencephalitis in neonates, pregnant women, the elderly, and others with impaired cell-mediated immunity. *L. monocytogenes* is readily adapted to experimental study that has yielded a remarkably clear picture of its intracellular lifestyle. After initial cell entry, which is mediated by genes not on the PAI (*inlAB* and *iap*), *L. monocytogenes* enters a phagocytic cell vacuole. Escape from this vacuole into the cell cytoplasm requires the gene (*hly*) for the pore-forming protein, lysteriolysin O (LLO), which is related to streptolysin O and other cytolysins (Gedde et al., 2000). *L. monocytogenes* multiplies and moves through the cytoplasm by initiation of actin filament polymerization at its cell surface. Actin polymerization is dependent upon the bacterial ActA surface protein, which interacts at its NH_2-terminal domain with eukaryotic actin related proteins (Arp2/3 complex) that are important for membrane protrusion (Welch et al., 1998). In effect, *L. monocytogenes* hijacks the cellular machinery for membrane protrusion in order to move through the host cell. Neighboring cells then ingest pseudopod-like projections in the host cell membrane. Phospholipases of the C type (PlcA and PlcB) have also been implicated in escape from a vacuole and cell-to-cell spread (Smith et al., 1995). The genes for each of these proteins, together with a pleiotropic activator (PrfA), are present on the PAI of *L. monocytogenes*. Genes related to those on the PAI *of L. monocytogenes* are also found on the PAI of *L. ivanovii*, which causes disease in cows and sheep and is the only other pathogenic member of the *Listeria* genus. Interestingly, the nonpathogenic *L. seeligeri* contains a similar PAI, which emphasizes the point that simply the presence of a PAI does not necessarily confer pathogenicity (Gouin et al., 1994). Lack of pathogenicity in *L. seeligeri* appears to be due to low level or absent expression of hemolysin and phospholipase C (Geoffroy et al., 1989; Notermans et al., 1991).

Infection with strains of *Staphylococcus aureus* that produce toxic shock syndrome toxin-1 (TSST-1) causes a syndrome of high fever, severe hypotension, watery diarrhea, and erythroderma. *Tst*, the gene for TSST-1, is a chromosomal element that is part of a 15.2 kb PAI that is absent in TSST-1 negative strains of *S. aureus* (Lindsay et al., 1998). The *tst* PAI in *S. aureus* RN4282 is flanked by 17 nucleotide direct repeats and inserts with specific orientation into a unique chromosomal location near *tyrB*, but not at a tRNA site as is often the case in Gram-negative bacteria. The RN4282 element, designated SaPI-1, also contains an enterotoxin gene with homology to other superantigens, and an open reading frame with homology to *vapE*, which is part of a putative PAI in the animal pathogen, *Dichelobacter nososus* (Katz et al., 1991). Resistance of *S. aureus* and coagulase negative staphylococci to methicillin is also caused by horizontal gene transfer. These strains contain an altered penicillin binding protein, (PBP2a) that has decreased affinity for β-lactam antibiotics. The genetic determinant for PBP2a is *mecA*, which is part of a 52 kb DNA cassette that also codes for clusters of other antibiotic resistance genes (Ito et al., 1999). Like SaPI-1, this cassette is inserted into a unique chromosomal location, but in an open reading frame of unknown function, designated *orfX*. Recent evidence suggests that the evolutionary precursor to *mecA* may be a closely related gene in an animal species, *Staphylococcus sciuri* (Wu et al., 1998).

Antibiotic-associated diarrhea is often caused by colonic overgrowth with *Clostridium difficile*, a Gram-positive spore-forming rod which causes a spectrum of disease ranging from relatively asymptomatic carriage to toxic megacolon, perforation, and death (Bartlett, 1992). *C. difficile* isolated from symptomatic patients contains a 19.6 kb PAI that is inserted in the same location and orientation in all toxigenic strains (Hammond and Johnson, 1995; Braun et al., 1996), and is absent in non-pathogenic strains. The locus carries genes for toxins A and B (*tcdA* and *tcdB*), which serve as an enterotoxin and cytotoxin, respectively. Small open reading frames flanking the toxin genes (*tcdD* and *tcdC*) are thought to be important in regulation of toxin production, while the function of a third open reading frame (*tcdE*) located between the toxin genes is unknown.

IMPLICATIONS FOR EMERGING INFECTIOUS DISEASES

In 1992 the Institute of Medicine issued its now-famous report that called attention to new, re-emerging, and drug-resistant infections that posed an increasing threat to human health (Lederberg et al., 1992). The causes of emerging infectious diseases are complex and often are related to changes in human behavior that permit a microorganism to cross a species barrier and spread among humans. Human immunodeficiency virus (HIV) and Lyme disease are two well-known examples. But other emerging infections have arisen because of horizontal gene transfer that has altered the pathogenicity of a microorganism. Acquisition of SPI-1 some 100 million years ago appears to define the divergence of pathogenic *Salmonella* from the commensal *Escherichia* (Morschhäuser et al., 2000). The emergence of enterohemorrhagic *E. coli* (EHEC) offers a more recent example. First recognized as a human pathogen in 1982, *E. coli* O157:H7 has become a common cause of diarrhea that in some cases is associated with hemolytic uremia and death. *E. coli* O157:H7 is essentially EPEC that has been lysogenized with a bacteriophage carrying shiga like toxin (SLT). These examples also serve to illustrate the point that pathogenic organisms frequently evolve in quantum leaps when a gene or cluster of genes is acquired by horizontal gene transfer (Groisman and Ochman, 1996; Finlay and Falkow, 1997; Ochman et al., 2000).

SUMMARY AND CONCLUSIONS

Pathogenic bacteria frequently contain large chromosomal elements that contain clusters of horizontally transferred genes called PAIs, which are important in their ability to cause disease. Although the strategies differ, many of these genes encode adhesins, toxins, and type III or type IV protein secretion systems. Novel bacterial pathogens with similar systems will likely be identified. Understanding this phenomenon may permit development of novel therapies and vaccines that are directed against targets present in pathogenic but absent in non-pathogenic members of a species.

REFERENCES

Akiba, T., Koyama, K. and Ishiki, Y. (1960) On the mechanism of the development of multiple drug-resistant clones of *Shigella*. *Jap. J. Microbiol.* **4**: 219–227.

Anderson, D.M., Fouts, D.E., Collmer, A. and Schneewind, O. (1999) Reciprocal secretion of proteins by the bacterial type III machines of plant and animal pathogens suggests a universal recognition of mRNA targeting signals. *Proc. Natl Acad. Sci. USA* **96**: 12829–12843.

Bartlett, J.G. (1992) Antibiotic-associated diarrhea. *Clin. Infect. Dis.* **15**: 573–581.

Black, D.S. and Bliska, J.B. (1997) Identification of p130Cas as a substrate of *Yersinia* YopH (Yop51), a bacterial protein tyrosine phosphatase that translocates into mammalian cells and targets focal adhesions. *EMBO J.* **16**: 2730–2744.

Blanc-Potard, A.B. and Groisman, E.A. (1997) The *Salmonella* selC locus contains a pathogenicity island mediating intramacrophage survival. *EMBO J.* **16**: 5376–5385.

Blocker, A., Gounon, P., Larquet, E. et al. (1999) The tripartite type III secreton of *Shigella flexneri* inserts IpaB and IpaC into host membranes. *J. Cell. Biol.* **147**: 683–693.

Blum, G., Ott, M., Lischewski, A. et al. (1994) Excision of large DNA regions termed pathogenicity islands from tRNA-specific loci in the chromosome of an *Escherichia coli* wild-type pathogen. *Infect. Immun.* **62**: 606–614.

Blum, G., Falbo, V., Caprioli, A. and Hacker, J. (1995) Gene clusters encoding cytotoxic ecotizing factor type I, Prs-fimbriae and α-homolysin form pathogenicity island II of the uropathogenic *E. coli* strain J96. *FEMS Microbiol. Lett.* **126**: 189–196.

Boland, A. and Cornelis, G.R. (1998) Role of YopP in suppression of tumor necrosis factor alpha release by macropahes during *Yersinia* infection. *Infect. Immun.* **66**: 1878–1884.

Braun, V., Hundsberger, T., Leukel, P. et al. (1996) Definition of the single integration site of the pathogenicity locus in *Clostridium difficile*. *Gene* **181**: 29–38.

Callazo, C.M. and Galan, J.E. (1997) The invasion-associated type-III protein secretion system in *Salmonella* – a review. *Gene* **192**: 51–59.

Carniel, E., Guilvout, I. and Prentice, H. (1996) Characterization of a large chomosomal "high-pathogenicity island" in Biotype 1B *Yersinia enterocolitica*. *J. Bacteriol.* **178**: 6743–6751.

Censini, S., Lange, C., Xiang, Z. et al. (1996) cag, a pathogenicity island of *Helicobacter pylori*, encodes type I-specific and disease-associated virulence factors. *Proc. Natl Acad. Sci. USA* **93**: 14648–14653.

Cheetham, B.F. and Katz, M.E. (1995) A role for bacteriophages in the evolution and transfer of bacterial virulence. *Mol. Microbiol.* **18**: 201–208.

Christie, P.J. and Vogel, J.P. (2000) Bacterial type IV secretion: conjugation systems adapted to deliver effector molecules to host cells. *Trends Microbiol.* **8**: 354–360.

Cirillo, D.M., Valdivia, R.H., Monack, D.M. and Falkow, S. (1998) Macrophage-dependent induction of the *Salmonella* pathogenicity island 2 type III secretion system and its role in intracellular survival. *Mol. Microbiol.* **30**: 175–188.

Covacci, A., Telford, J.L., Del Giudice, G. et al. (1999) *Helicobacter pylori* virulence and genetic geography. *Science* **284**: 1328–1333.

Crabtree, J.E., Kersulyte, D., Li, S.D. et al. (1999) Modulation of *Helicobacter pylori* induced interleukin-8 synthesis in gastric epithelial cells mediated by cag PAI encoded VirD4 homologue. *J. Clin. Pathol.* **52**: 653–657.

Dunn, B.E., Cohen, H. and Blaser, M.J. (1997) *Helicobacter pylori*. *Clin. Microbiol. Rev.* **10**: 720–741.

Fan, F., Ohnishi, K., Francis, N.R. and Macnab, R.M. (1997) The FliP and FliR proteins of *Salmonella typhimurium*, putative components of the type III flagellar export apparatus, are located in the flagellar basal body. *Mol. Microbiol.* **26**: 1035–1046.

Fetherston, J.D., Schuetze, P. and Perry, R.D. (1992) Loss of pigmentation phenotype in *Yersinia pestis* is due to spntaneous deletion of 102 kb of chromosomal DNA which is flanked by a repetitive element. *Mol. Microbiol.* **6**: 2693–2704.

Finlay, B.B. and Falkow, S. (1997) Common themes in microbial pathogenicity revisited. *Microbiol. Mol. Biol. Rev.* **61**: 136–169.

Galyov, E.E., Wood, M.W., Rosqvist, R. et al. (1997) A secreted effector protein of *Salmonella dublin* is translocated into eukaryotic cells and mediates inflammation and fluid secretion in infected ileal mucosa. *Mol. Microbiol.* **25**: 903–912.

Gedde, M.M., Higgins, D.E., Tilney, L.G. and Portnoy, D.A. (2000) Role of listeriolysin O in cell-to-cell spread of *Listeria monocytogenes*. *Infect. Immun.* **68**: 999–1003.

Geoffroy, C., Gaillard, J.L., Alouf, J.E. and Berche, P. (1989) Production of thiol-dependent haemolysins by *Listeria monocytogenes* and related species. *J. Gen. Microbiol.* **135**: 481–487.

Goldberg, M.B. and Theriot, A.J. (1995) *Shigella flexneri* surface protein IscA is sufficient to direct actin-based motility. *Proc. Natl Acad. Sci. USA* **92**: 6572–6576.

Gouin, E., Mengaud, J. and Cossart, P. (1994) The virulence gene cluster of *Listeria monocytogenes* is also present in *Listeria ivanovii*, an animal pathogen and *Listeria seeligeri*, a nonpathogenic species. *Infect. Immun.* **62**: 3550–3553.

Groisman, E.A. and Ochman, H. (1996) Pathogenicity islands: bacterial evolution in quantum leaps. *Cell* **87**: 791–794.

Guyer, D.M., Kao, J.-S. and Mobley, H.L. T. (1998) Genomic analysis of a pathogenicity island in uropathogenic *Escherichia coli* CFT073: distribution of homologous sequences among isolates from patients with pyelonephritis, cystitis, and catheter-associated bacteriuria and from fecal samples. *Infect. Immun.* **66**: 4411–4417.

Hacker, J., Knapp, S. and Goebel, W. (1983) Spontaneous deletions and flanking regions of the chromosomally inherited hemolysin determinant of an *Escherichia coli* O6 strain. *J. Bacteriol.* **154**: 1145–1152.

Hacker, J., Bender, L., Ott, M. et al. (1990) Deletions of chromosomal regions coding for fimbriae and hemolysins occur *in vitro* and *in vivo* in various extra-intestinal *Escherichia coli* isolates. *Microb. Pathog.* **8**: 213–225.

Hacker, J., Blum-Oehler, G., Mühldorfer, I. and Tschäpe, H. (1997) Pathogenicity islands of virulent bacteria: structure, function and impact on microbial evolution. *Mol. Microbiol.* **23**: 1089–1097.

Hammond, G.A. and Johnson, J.L. (1995) The toxigenic element of *Clostridium difficile* strain VPI 10463. *Microb. Pathog.* **19**: 203–213.

Harris, J.R., Wachsmuth, I.K., Davis, B.R. and Cohen, M.L. (1982) High-molecular-weight plasmid correlates with *Escherichia coli* enteroinvasiveness. *Infect. Immun.* **37**: 1295–1298.

Hong, K.H. and Miller, V.L. (1998) Identification of a novel *Salmonella* invasion locus homologous to *Shigella ipgDE*. *J. Bacteriol.* **180**: 1793–1802.

Hueck, C.J. (1998) Type III protein secretion systems in bacterial pathogens of animals and plants. *Microbiol. Mol. Biol. Rev.* **62**: 379–433.

Ito, T., Katayama, Y. and Hiramatsu, K. (1999) Cloning and nucleotide sequence determination of the entire mec DNA of pre-methicillin-resistant *Staphylococcus aureus* N315. *Antimicrob. Agents Chemother.* **43**: 1449–1458.

Kaper, J.B. and Hacker, J. (eds) (1999) *Pathogenicity Islands and Other Mobile Genetic Elements*, pp. 1–346, ASM Press, Washington, DC.

Katayama, Y., Ito, T. and Hiramatsu, K. (2000) A new class of genetic element, staphylococcus cassette chromosome mec, encodes methicillin resistance in *Staphylococcus aureus*. *Antimicrob. Agents Chemother.* **44**: 1549–1555.

Katz, M.E., Howarth, P.M., Yong, W.K. et al. (1991) Identification of three gene regions associated with virulence in *Dichelobacter nodosus*, the causative agent of ovine footrot. *J. Gen. Microbiol.* **137**: 2117–2124.

Kenny, B., DeVinney, R., Stein, M., Reinscheid, D.J., Frey, E.A. and Finlay, B.B. (1997) Enteropathogenic *E. coli* (EPEC) transfers its receptor for intimate adherence into mammalian cells. *Cell* **14**: 511–520.

Knapp, S., Hacker, J., Jarchau, T. and Goebel, W. (1986) Large unstable inserts in the chromosome affect virulence properties of uropathogenic *scherichia coli* strain 536. *J. Bacteriol.* **168**: 22–30.

Kocks, C., Marchand, J.B., Gouin, E. et al. (1995) The unrelated surface proteins ActA of *Listeria monocytogenes* and IcsA of *Shigella flexneri* are sufficient to confer actin-based motility on *Listeria innocua* and *Escherichia coli* respectively. *Mol. Microbiol.* **18**: 413–423.

Kubori, T., Matsushima, Y., Nakamura, D. et al. (1998) Supramolecular structure of the *Salmonella typhimurium* type III protein secretion system. *Science* **280**: 602–605.

Lai, E.M. and Kado, C.I. (2000) The T-pilus of *Agrobacterium tumefaciens*. *Trends Microbiol.* **8**: 361–369.

Lederberg, J., Shope, R.E. and Oaks Jr, S.C. (eds) (1992) *Emerging Infections: Microbial Threats to Health in the United States*, pp. 1–294, National Academy Press, Washington, DC.

Li, S.D., Kersulyte, D., Lindley, I.J. et al. (1999) Multiple genes in the left half of the cag pathogenicity island of *Helicobacter pylori* are required for tyrosine kinase-dependent transcription of interleukin-8 in gastric epithelial cells. *Infect. Immun.* **67**: 3893–3899.

Lindsay, J.A., Ruzin, A., Ross, H.F. et al. (1998) The gene for toxic shock toxin is carried by a family of mobile pathogenicity islands in *Staphylococcus aureus*. *Mol. Microbiol.* **29**: 527–543.

Low, D., David, V., Lark, D. et al. (1984) Gene clusters governing the production of hemolysin and mannose-resistant hemagglutination are closely linked in *Escherichia coli* serotype O4 and O6 isolates from urinary tract infections. *Infect. Immun.* **43**: 353–358.

Macnab, R.M. (1996) Flagella and motility. In *Escherichia Coli and Salmonella Typhimurium; Cellular and Molecular Biology* (ed., F.C. Neidhardt), 2nd edn, vol. 1, pp. 123–145, ASM Press, Washington DC.

Macnab, R.M. (1999) The bacterial flagellum: reversible rotary propellor and type III export apparatus. *J. Bacteriol.* **181**: 7149–7153.

McDaniel, T.K., Jarvis, K.G., Donnenberg, M.S. and Kaper, J.B. (1995) A genetic locus of enterocyte effacement conserved among diverse enterobacterial pathogens. *Proc. Natl Acad. Sci. USA* **28**: 1664–1668.

McDaniel, T.K. and Kaper, J.B. (1997) A cloned pathogenicity island from enteropathogenic *Escherichia coli* confers the attaching and effacing phenotype on *E. coli* K-12. *Mol. Microbiol.* **23**: 399–407.

Mills, D.M. and Lee, C.A. (1995) A 40 kilobase chromosomal fragment encoding *Salmonella* invasion genes is absent from the corresponding region of the *Escherichia coli* K12 chromosome. *Mol. Microbiol.* **15**: 749–759.

Mills, S.D., Boland, A., Sory, M.P. et al. (1997) *Yersinia enterocolitica* induces apoptosis in macrophages by a process requiring functional type III secretion and translocation mechanisms and involving YopP, presumably acting as an effector protein. *Proc. Natl Acad. Sci. USA* **94**: 12638–12643.

Minamino, T. and Macnab, R.M. (1999) Components of the *Salmonella* flagellar export apparatus and classification of export substrates. *J. Bacteriol.* **181**: 1388–1394.

Monack, D., Mecsas, J., Ghori, N. and Falkow, S. (1997) *Yersinia* signals macrophage to undergoe apoptosis and YopJ is necessary for this cell death. *Proc. Natl Acad. Sci. USA* **94**: 12638–12643.

Moncrief, J.S., Duncan, A.J., Wright, R.L. et al. (1998) Molecular characterization of the fragilysin pathogenicity islet of enterotoxigenic *Bacteroides fragilis*. *Infect. Immun.* **66**: 1735–1739.

Morschhäuser, J., Köhler, G., Ziebuhr, W. et al. (2000) Evolution of microbial pathogens. *Phil. Trans. R. Soc. Lond. B.* **355**: 694–704.

Moss, J.E., Cardozo, T.J., Zychlinsky, A. and Groisman, E.A. (1999) The selC-associated SHI-2 pathogenicity island of *Shigella flexneri*. *Mol. Microbiol.* **33**: 74–83.

Nguyen, L., Paulsen, I.T., Tchieu, J. et al. (2000) Phylogenetic analyses of the constituents of Type III protein secretion systems. *J. Mol. Microbiol. Biotechnol.* **2**: 125–144.

Notermans, S.H., Dufrenne, J., Leimeister-Wächter, M. et al. (1991) Phosphatidylinositol-specific phospholipase C activity as a marker to distinguish between pathogenic and nonpathogenic *Listeria* species. *Appl. Environ. Microbiol.* **57**: 2666–2670.

Ochiai, K., Yamanaka, T., Kiumra, K. and Sawada, O. (1959) Inheritance of drug reistance (and its transfer) between *Shigella* and *E. coli* strains. *Nihon Iji Shimpo* **1861**: 34.

Ochman, H., Soncini, F.C., Solomon, F. and Groisman, E.A. (1996) Identification of pathogenicity island required for *Salmonella* survival in host cells. *Proc. Natl Acad. Sci. USA* **93**: 7800–7804.

Ochman, H., Lawrence, J.G. and Groisman, E.A. (2000) Lateral gene transfer and the nature of bacterial innovation. *Nature* **405**: 299–304.

Odenbreit, S., Püls, J., Sedlmaier, B. et al. (2000) Translocation of *Helicobacter pylori* CagA into gastric epithelial cells by type IV secretion. *Science* **287**: 1497–1500.

Palmer, L.E., Hobbie, S., Galan, J.E. and Bliska, J.B. (1998) YopJ of *Yersinia pseudotuberculosis* is required for the inhibition of macrophage TNF-alpha production and downregulation of the MAP kinases p38 and JNK. *Mol. Microbiol.* **27**: 953–965.

Persson, C., Carballeira, N., Wolf-Watz, H. and Fallman, M. (1997) The PTPase YopH inhibits uptake of *Yersinia*, tyrosine phosphorylation of p130Cas and FAK, and the associated accumulation of these proteins in peripheral focal adhesions. *EMBO J.* **16**: 2307–2318.

Roberts, J.A., Marklund, B.I., Ilver, D. et al. (1994) The Gal(alpha 1–4)Gal-specific tip adhesin of *Escherichia coli* P-fimbriae is needed for pyelonephritis to occur in the normal urinary tract. *Proc. Natl Acad. Sci. USA* **91**: 11889–118893.

Roberts, J.A., Kaack, M.B., Baskin, G. et al. (1997) Epitopes of the P-fimbrial adhesin of *E. coli* cause different urinary tract infections. *J. Urol.* **158**: 1610–1613.

Rosqvist, R., Bolin, I. and Wolf-Watz, H. (1988) Inhibition of phagocytosis in *Yersinia pseudotuberculosis*: a virulence plasmid-encoded ability involving the Yop2b protein. *Infect. Immun.* **56**: 2139–2143.

Rosqvist, R., Forsberg, A. and Wolf-Watz, H. (1991) Intracellular targeting of the *Yersinia* YopE cytotoxin in mamalian cells induces actin filament disruption. *Infect. Immun.* **59**: 4562–4569.

Sansonetti, P.J. and Egile, C. (1998) Molecular basis of epithelial cell invasion by *Shigella flexneri*. *Antonie van Leeuwenhoek* **74**: 191–197.

Sansonetti, P.J., Kopecko, D.J. and Formal, S.B. (1981) *Shigella sonnei* plasmids: evidence that a large plasmid is necessary for virulence. *Infect. Immun.* **34**: 75–83.

Sansonetti, P.J., Kopecko, D.J. and Formal, S.B. (1982) Involvement of a plasmid in the invasive abiltiy of *Shigella flexneri*. *Infect. Immun.* **35**: 852–860.

Schesser, K., Spiik, A.K., Dukuzumuremyi, J.M. et al. (1998) The yopJ locus is required for *Yersinia*-mediated inhibition of NF-kappaB activation and cytokine expression: YopJ contains a eukaryotic SH2-like domain that is essential for its repressive activity. *Mol. Microbiol.* **28**: 1067–1079.

Schmiel, D.H., Wagar, E., Karamanou, L. et al. (1998) Phospholipase A of *Yersinia enterocolitica* contributes to pathogenesis in a murine mouse model. *Infect. Immun.* **66**: 3941–3951.

Schubert, S., Rakin, H., Karch, H. et al. (1998) Prevalence of the "high-pathogenicity island" of *Yersinia* species among *Escherichia coli* strains that are pathogenic to humans. *Infect. Immun.* **66**: 480–485.

Schulte, R., Wattiau, P., Hartland, E.L. et al. (1996) Differential secretion of interleukin-8 by human epithelial cell lines upon entry of virulent or nonvirulent *Yersinia enterocolitica*. *Infect. Immun.* **64**: 2106–2113.

Segal, E.D., Lange, C., Covacci, A. et al. (1997) Induction of host signal transduction pathways by *Helicobacter pylori*. *Proc. Natl Acad. Sci. USA* **94**: 7595–7599.

Segal, E.D., Cha, J., Lo, J. et al. (1999) Altered states: involvement of phosphorylated CagA in the induction of host cellular growth changes by *Helicobacter pylori*. *Proc. Natl Acad. Sci. USA* **96**: 14559–14564.

Shea, J.E., Hensel, M., Gleeson, C. and Holden, D.W. (1996) Identification of a virulence locus encoding a second type III secretion system in *Salmonella typhimurium*. *Proc. Natl Acad. Sci. USA* **93**: 2593–2597.

Shea, J.E., Beuzon, C.R., Gleeson, C. et al. (1999) Influence of the *Salmonella typhimurium* pathogenicity island 2 type III secretion system on bacterial growth in the mouse. *Infect. Immun.* **67**: 213–219.

Smith, G.A., Marquis, H., Jones, S. et al. (1995) The two distinct phospholipases C of *Listeria monocytogenes* have overlapping roles in escape from a vacuole and cell-to-cell spread. *Infect. Immun.* **63**: 4231–4237.

Smith, H.W. and Halls, S. (1968) The transmissible nature of the genetic factor in *Escherichia coli* that controls haemolysin production. *J. Gen. Microbiol.* **48**: 319–334.

Stein, M., Rappuoli, R. and Covacci, A. (2000) Tyrosine phosphorylation of the *Helicobacter pylori* CagA antigen after cag-driven host cell translocation. *Proc. Natl Acad. Sci. USA* **97**: 1263–1268.

Sullivan, J.T. and Ronson, C.W. (1998) Evolution of rhizobia by acquisition of a 500-kb symbiosis island that integrates into a phe-tRNA gene. *Proc. Natl Acad. Sci. USA* **95**: 5145–5149.

Svenson, S.B., Hultberg, H., Kallenius, G. et al. (1983) P-fimbriae of pyelonephritogenic *Escherichia coli*: identification and chemical characterization of receptors. *Infection* **11**: 61–67.

Tamano, K., Aizawa, S.I.K.E., Nonaka, T. et al. (2000) Supramolecular structure of the *Shigella* type III secretion machinery: the needle part is changeable in length and essential for delivery of effectors. *EMBO J.* **19**: 3876–3887.

Uhlen, P., Laestadius, A., Jahnukainen, T. et al. (2000) Alpha-haemolysin of uropathogenic *E. coli* induces Ca^{2+} oscillations in renal epithelial cells. *Nature* **405**: 694–697.

Vokes, S.A., Reeves, S.A., Torres, A.G. and Payne, S.M. (1999) The aerobactin iron transport system genes in *Shigella flexneri* are present within a pathogenicity island. *Mol. Microbiol.* **33**: 63–73.

Wallis, T.S. and Galyov, E.E. (2000) Molecular basis of *Salmonella*-induced enteritis. *Mol. Microbiol.* **36**: 997–1005.

Welch, M.D., Rosenblatt, J., Skoble, J. et al. (1998) Interaction of human Arp2/3 complex and the *Listeria monocytogenes* ActA protein in actin filament nucleation. *Science* **281**: 105–108.

Wong, K.K., McClelland, M., Stillwell, L.C., et al. (1998) Identification and sequence analysis of a 27-kilobase chromosomal fragment containing a *Salmonella* pathogenicity island located at 92 minutes on the chromosome map of *Salmonella enterica* serovar *typhimurium* LT2. *Infect. Immun.* **66**: 3365–3371.

Wood, M.W., Jones, M.A., Watson, P.R. et al. (1998) Identification of a pathogenicity island required for *Salmonella* enteropathogenicity. *Mol. Microbiol.* **29**: 8883–8891.

Wu, S., de Lencastre, H. and Tomasz, A. (1998) Genetic organization of the *mecA* region in methicillin-susceptible and methicillin-resistant strains of *Staphylococcus sciuri*. *J. Bacteriol.* **180**: 236–242.

Young, G.M., Schmiel, D.H. and Miller, V.L. (1999) A new pathway for the secretion of virulence factors: the flagellar export apparatus functions as a type III secretion system. *Proc. Natl Acad. Sci. USA* **96**: 6456–6461.

Zychlinski, A., Prevost, M.C. and Sansonetti, P.J. (1992) *Shigella flexneri* induces apoptosis in infected macrophages. *Nature* **358**: 167–168.

CHAPTER 11

Mosaic Proteins, Not Reinventing the Wheel

Susan Hollingshead

Horizontal transfer is a rapid process that allows the acquisition of novel phenotypes. During its relatively short lifetime, a single bacterium can acquire a particular determinant and pass on its newfound ability to all its descendents. As long as the acquired determinant originated outside the bacterium itself the transfer is horizontal. Often this process involves the acquisition of complete genes or of operons (note several other chapters in this book). Even more frequently, though, the process involves acquisition of only segments of genes. In both cases, mosaic patterns are created in the recipient bacterium, the former process creating mosaic operons and chromosomes and the latter, mosaic genes and proteins. This review will focus on the latter process as it occurs in bacterial genomes with a focus on the evolution of surface protein genes that function in the interface between host and pathogen.

The advantage to the recipient bacterium in acquiring a determinant through the mosaic process is the equivalent of not having to reinvent the wheel. The creation of a new mosaic pattern within a gene has an immediate effect on the amino acid sequence of the encoded protein. Horizontal transfer thus efficiently enables a protein to add a new function or alter a previous function. Functional modules added together in this way are consistent with both the modular theory of evolution (first noted by Haldane (1932)) and the concept of protein family evolution through gene duplication (Ohno, 1970). In eukaryotes, the domains pieced together by the mosaic process may often be separate exons. By contrast, in prokaryotes, the domains creating mosaic proteins are within the boundaries of a single coding segment. The new phenotype, even one that might have required several mutational steps when it was first formed in the donor, becomes immediately available to the recipient. The wheel is received intact, without the need to retrace the smaller steps of evolution.

FACTORS IN FORMING MOSAIC GENES AND PROTEINS

Success in adding a mosaic segment vitally depends upon the origin of the DNA to be added or exchanged. If the DNA originates from a sibling member of the same bacterial species it will face fewer barriers to acquisition than if it originates from another species in the same genus, or from another bacterial genus altogether (Figure 11.1). Same-species exchange is more likely to succeed in part because the donor and recipient organisms are in the same environment. There is little to no geographic barrier. For cross-species exchange, however, organisms are disadvantaged because they do not share the same environmental niche. There are many additional non-environmental barriers to cross-species genetic exchange. Organisms may differ in their capacity to receive DNA by gene transfer processes such as transformation, transduction or conjugation. These sexual isolation barriers are shown in Figure 11.1 as walls with

Copyright © 2002 by Academic Press.
All rights of reproduction in any form reserved.

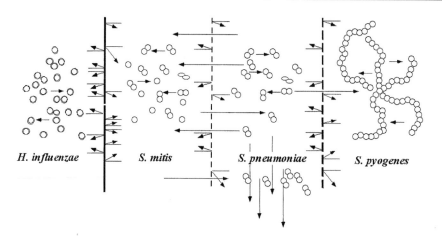

FIGURE 11.1 Barriers to mosaic gene formation, horizontal gene transfer, and the gene pool concept. The origin of new domains that form mosaic blocks is limited by environmental niche, genetic barriers to gene transfer, and mechanics of DNA uptake and recombination. Available domains circulate in a gene pool that operates primarily with other bacteria of the same species. The wider gene pool might include closely related species for which DNA can contribute to the gene pool with a certain amount of filtering. The horizontal arrows indicate horizontal transmission and the vertical arrows indicate the vertical transmission of mosaic determinants.

differing porosities. The chinks that allow horizontal transfer are smaller in the cross-species walls than in the same-species walls and become progressively smaller for more distantly related organisms.

After DNA has been received through horizontal transfer, the incoming DNA can replace a resident DNA segment by homologous recombination or may sometimes be added to a new location through homology-directed illegitimate recombination (Mortier-Barriere et al., 1997). There are potential barriers to acquisition of DNA segments that will create mosaic genes and proteins. Several barriers for DNA transformation in *Streptococcus pneumoniae* were recently reviewed (Claverys et al., 2000; Majewski et al., 2000). Specific barriers and their regulatory controls are likely to differ from species to species, but some general characteristics are common to most bacterial species. The genetic distance between the incoming DNA segment and the resident DNA segment is inversely related to the likelihood of acquisition by the recombination process. DNA may be restricted, repaired or mutated by enzymes such as those involved in mismatch repair before it is assimilated by recombination. The efficiency of repair or the stringency of the system may be modulated or regulated when the bacterium is under stress, allowing greater diversification in difficult survival times. Considering the number of barriers to horizontal transfer, when homologous mosaic segments are found to be present in distantly related bacteria, they are often likely to represent very rare events.

The transfer of DNA horizontally within the same species accounts for the vast majority of recombination noted in studying DNA sequences. Some instances of recombinational exchange within a species go undetected because the sequence acquired is so similar to its predecessor. The ratio of change in DNA sequence due to horizontal transfer and recombination over change due to mutational processes is a useful parameter of the evolutionary dynamic of bacterial populations (Maynard Smith et al., 1993). Factors that control this ratio differ from species to species and throughout the bacterial kingdom. If this ratio is high, for example, it is difficult to follow the evolutionary history of the species because the clonal population structure is obscured through the repeated recombinational exchanges. A recent method for measuring this ratio in several bacterial groups was devised (Feil et al., 1999, 2000). Authors highlight the finding that recombinational exchanges appear to be more frequent for all bacteria than was previously thought, but their efforts also suggest that *N. meningitidis*, *S. pneumoniae*, *S. pyogenes* and *S. aureus* all appear to experience a higher ratio of recombinational exchange over mutational change than that of *Escherichia coli* and *Haemophilus influenzae* (Feil et al., 2001). The formation of mosaic genes and proteins is favored by this higher ratio.

THE GENE POOL CONCEPT

The concept of a global gene pool has emerged for organisms in which the prevailing factors allow frequent gene transfer (Figure 11.1). Within this global gene pool are auxiliary genes that are present in some members but absent in others of the same species. The auxiliary genes may be passed like a baton from strain to strain through horizontal transfer. At times when their encoded traits are selectively advantageous, cells carrying the baton might spread geographically. Alternatively, encoded traits that are less beneficial might be lost in the wake of clones with better fitness. Moreover, the unit transferred need not be the whole gene, but might approximate a functional domain. If the domain is of selective value in a changing environment, then its horizontal spread will be favored at certain times. During other times, the domain may be lost from a single lineage. However, its continued presence in the gene pool allows it to be re-acquired on the proper occasion from another bacterium that is sampling from and contributing to the same pool. The sampling of functional domains acquired from the common gene pool also allows the evolutionary testing of new combinations of domains.

SELECTION, OR WHEN DO YOU NEED THE WHEEL?

Functional domains are passed around in the auxiliary gene pool because they encode traits that are advantageous at times but inconsequential at other times. There is little need for a wheel during a passage across the ocean. In fact,

a trait might be slightly burdensome when it is not needed. While it is the recombination process that creates mosaic proteins, it is natural selection that allows their continued existence in the population.

One example of the importance of selection is the case of high level (>2 μg/ml minimal inhibitory concentration) penicillin-resistance in *Streptococcus pneumoniae*. Penicillin resistance (Pen^R) is partially due to the existence of specific alleles of a family of penicillin binding proteins that bind penicillin and its derivatives with greatly reduced efficiency as compared with the native PBP alleles in Pen^S strains. The PBPs are transpeptidases that carry out the cross-linking of amino acids in the cell wall of bacteria. Resistant alleles of PBP1a, 2x and 2a were determined to be mosaic proteins that had been formed in part by multiple recombinational exchanges with the PBPs from *Streptococcus mitis* and *Streptococcus oralis*. It has recently become clear, however, that the acquisition of this high level penicillin resistance was not accomplished by the simple modification through evolution of the resistant PBP alleles or even by the acquisition of the right combination of alleles through horizontal transfer. Rather, the usefulness of acquiring the resistant allele was greatly tempered by the need to have an altered branch chain amino acid upon which the resistant PBP allele could work (Filipe et al., 2000; Weber et al., 2000). This combination of traits created a strain in which the resistant PBPs could now function, analogous to having a road upon which the acquired wheels could roll.

CHIMERIC PROTEINS – CENTAURS FROM MINOTAURS

The mosaic formation process sometimes allows the adaptation of a particularly useful domain such as one for a surface localization to be combined repeatedly with a number of other functional domains. One particularly telling example of combining different functional domains immediately brings to mind the mythic chimeras known as centaurs because of the head to tail nature of the domains joined. Among Gram-positive organisms, many surface proteins are recognized as such by the presence of a sequence motif LPxTG followed by a hydrophobic domain with potential to span a lipid bilayer, usually near the C-terminus of an amino acid sequence. This sequence motif is recognized by the enzyme known as sortase that can cut the protein between the threonine and the glycine in the sortase motif and attach it to an amino acid in the cell wall cross-bridge (Figure 11.2). The threonine is linked via a transpeptidation in an ester linkage to one of the cross-bridge amino acids, which differ from species to species (Mazmanian et al., 1999, 2000; Ton-That et al., 1999). All of the Gram-positive bacteria studied to date have a sortase gene and use this mechanism to attach at least some of their surface proteins.

Within the species *Streptococcus pneumoniae*, a cell wall polysaccharide known as C-polysaccharide (C-PS) that is restricted to this species has been studied. It consists of β-D-Glc*p*-

(1–3)-AAT-α-D-Gal*p*-(1–4)-α-D-GalNAc-(1–3)-β-D-GalNAc-(1–1)-D-Rib-ol-(5-P- with phosphocholine linked in the O-6 position to the GalNAc amino-sugars (Kamerling, 2000). A similar lipoteichoic acid (LTA) in a membrane-bound form is known as the F-antigen (F-Ag). The phosphocholine on C-PS and F-Ag is the site by which a family of surface proteins is bound to the pneumococcal surface (Yother and White, 1994; Garcia et al., 2000; Gosink et al., 2000). There are 16 choline-binding proteins and 19 sortase-type proteins in the genome sequence of a capsule serotype 4 strain that was recently sequenced (Tettelin et al., 2001). An additional 36 proteins with proposed lipid attachment sites are also found within the genome sequence. The choline binding means of attachment is limited to organisms that decorate the wall teichoic acids and LTA with phosphocholine that includes only some members of the streptococci and clostridia.

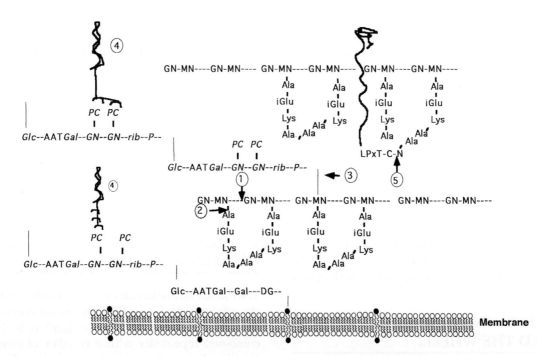

FIGURE 11.2 Cartoon of pneumococcal cell wall. Sites for murein hydrolase cleavage of the peptidoglycan backbone are shown at position 1 (lytA), 2 (lytB) and 3 (lytC). Shown at position 4 are sites of phosphocholine, where choline binding proteins are non-covalently associated with the cell wall polysaccharide or lipoteichoic acids. Shown at position 5 is the proposed site for sortase transpeptidation, linking cell surface proteins with a LPxTG motif to interpeptidoglycan amino acid crossbridges. Blue is LTA, purple is C-polysaccharide or wall TA, red is choline binding protein (CBP) and green is sortase-type surface protein.

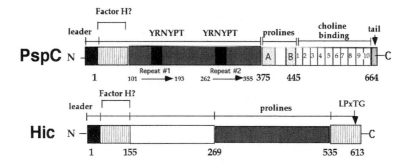

FIGURE 11.3 Chimeric protein HIC compared with representatives of the PspC and PspA choline binding proteins. YRNYPT represent a hexapeptide thought to be associated with binding of PspC to secretory component in fluid phase or to the polymeric Ig receptor on eukaryotic cell surfaces. Factor H binding domain is putative because of the identical sequence in molecules with this binding activity. The proline rich regions of each are repetitive but vary in motif. Bacterial cell surface attachment is by CBP repeats to phosphocholine for PspC but by the LPxTG motif using sortase for the Hic protein.

At least two members of the choline binding protein family (PspA and PspC) interact with host innate or adaptive immunity components, while other members of the larger CBP family are murein hydrolases (LytA, LytB and LytC). The protein PspA binds to human lactoferrin (Hammerschmidt et al., 1999) and inhibits complement deposition (Tu et al., 1999). The protein PspC (also known as SpsA, CbpA and PbcA) has been attributed with the ability to bind immobilized sialic acid or lacto-*N*-neotetraose on human cells (Rosenow et al., 1997), human secretory component (SC) (Hammerschmidt et al., 1997), human C3 protein (Cheng et al., 2000) and human Factor H (Dave et al., 2001).

Recently, a surface protein called Hic from *S. pneumoniae* strain A66 was found to bind the human complement regulatory protein, Factor H (Janulczyk et al., 2000). The DNA sequence of the Hic protein shows it to be an interesting chimeric protein equivalent to the centaur (Figure 11.3). Near the C-terminus, Hic has a LPxTG sortase motif and would be attached to the surface by the sortase mechanism described above, while the N-terminal portion of the Hic protein is nearly identical to PspC/CbpA/SpsA molecules, most of which also bind Factor H (Dave et al., in press). A hexapeptide motif YRNYPT was recently shown to be the binding site for SC to SpsA and by inference would also be the binding site for attachment to the polymeric Ig receptor which carries secretory component at its extended end (Hammerschmidt et al., 2000; Zhang et al., 2000). This hexapeptide is missing in the Hic variant. In this chimera, it looks as if functional domains for interaction with the innate immunity system of the host have been combined with other functional domains for cell surface placement. A close look at the mosaic structure of this gene locus reveals that the signal peptidase domain, which is the same for all PspC, SpsA, CbpA, PbcA and Hic alleles, is also very similar to the signal peptidase domain of the surface protein known as the β-antigen in the group B streptococci (Brooks-Walter et al., 1999). The β-antigen or Arp is an IgA Fc binding protein with a sortase motif for surface attachment (Lindahl et al., 1991).

SEWING DOMAINS TOGETHER LIKE BEADS ON A STRING – SURFACE PROTEINS

If the domains for cell surface attachment are analogous to the hooves and horse-like body of the centaur, then the N-terminal domains of the surface proteins found on bacteria have enough chimeric patterns to outdo even the three-headed Chimera of Greek legend fame. Bacterial surface proteins have been studied extensively because of their importance to the control of disease. In order to illustrate the concept it is useful to consider one of the largest families of mosaic surface proteins and what is known about binding or functional domains. Such a family is the M protein family of the group A streptococci (Fischetti, 1989; Robinson and Kehoe, 1992).

Within the M protein family, a large number of functional domains that bind components within human plasma and or serum have been identified. Included are domains that bind albumin, plasminogen, fibrinogen, the Fc region of IgG1, IgG2, IgG3, IgG4, IgA, complement regulatory proteins C4BP, MCP, Factor H and the list gets longer as more is understood (Lindahl et al., 2000). In some cases, there are multiple mosaic segments that have the same apparent binding activity, such as to the Fc region of IgG1, 2 and 4, but have different primary amino acid sequences involved in the binding domain. There have been over 150 different primary sequences of proteins in the M protein family sequenced to date (Facklam et al., 1999). The binding domains allowing interaction with the host proteins listed above are found in many different combinations in different M family members. One protein might bind IgA and fibrinogen and another Factor H and IgG. Then, in another strain, a unique M family member might bind IgA, IgG and Factor H, while another binds only fibrinogen. In this family, the mosaic evolution process appears to have experimented with different combinations in search of the perfect fit for the environment in which it existed.

The multidomain structure of surface proteins is not limited to protein-binding domains. Some domains are enzymatic. In *Staphylococcus aureus*, for example, one surface protein is a glyceraldehyde-3-phosphate dehydrogenase (GAPDH) and is also a transferrin receptor protein involved in the uptake of iron (Modun et al., 2000). In group A streptococci, the GAPDH enzymatic activity is in the same gene with a protein-binding domain that binds to plasmin (Pancholi and Fischetti, 1993). Altogether, five of the enzymatic activities of the glycolytic pathway can be detected on the surface of group A streptococci although their surface-associated activities for the bacterial cell are not fully understood (Pancholi and Fischetti, 1998). The enzymatic functional domains on the surface are related to their cytoplasmic counterparts, but in evolution, they have usually gained a surface-attachment domain such as the sortase-type domain by the mosaic process. The surface-association of these enzymes suggests that it was occasionally advantageous to have their activity outside of the cell, although this would imply a change in functional association from their common role in glycolysis.

RADIATION OF THE CBP FAMILY WITHIN *S. PNEUMONIAE*

An examination of the family of choline binding proteins within the pneumococcal genome gives the following information (Tettelin et al., 2001, unpublished). There are 16 choline binding proteins predicted from the type 4 genome sequence based on a search for the motifs WYY, WYYL, YYF, and YYL along with a requirement for more than one copy of such motifs to be present. The genes are somewhat dispersed about the chromosome although 9/16 are within 350 base pairs of the predicted origin of replication. The dispersed location of several CBPs into what now look like multigene functional operons suggests that the gene family has been present in this species for a long time period and that ectopic gene movements have now been assimilated into working operons with the nearby gene units. All but one of the 16 CBPs are directed in transcription leading away from the origin of replication as are 78% of all the genes in the genome in this bacterium. Two different pairs of choline binding proteins (CbpC with CbpJ and CbpF with CbpG) are arranged in tandem with each other, suggesting gene evolution by gene duplication. Although no longer found in tandem arrangement, sequence comparison of the choline binding domain suggests that PspA and PspC share one ancestry while LytA, LytB and LytC share a different ancestry within the larger CBP family. The CBP domain is usually near the C-terminus of each protein with the exception of LytB, LytC, and two others. LytB and LytC are both reported to require choline for an enzymatic process, rather than just for surface attachment. There is a single protein with an apparent CBP

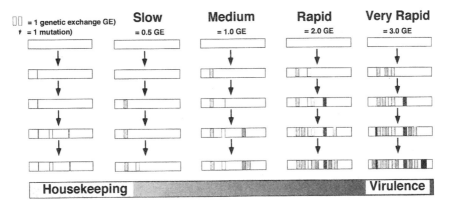

FIGURE 11.4 Mosaic process. The figure shows potential of the mosaic process as it is ongoing within a single gene locus. The lines indicate mutations while small boxes indicate genetic exchanges of small segments of the type that lead to mosaic proteins. A higher frequency of mosaic segment exchange leads to a more complex mosaic pattern. These more variable mosaic genes are associated with genes that interact with the environment, such as genes which interact with host proteins if the bacterium is a pathogen. Lesser mosaic genes are also seen in "housekeeping" genes where there might be more constraints for maintaining a particular function while accumulating change. The presence of mosaic segments in housekeeping genes is also useful as a chronometer to mark the relative rate of recombination and mutation. When genetic exchanges are infrequent or slow the clonal frame is for the most part conserved over time. When genetic exchanges are frequent and rapid, they might obscure clonal frame within a short time period.

region that also has a sortase motif also used in surface attachment.

MOSAIC PROTEINS AND VACCINES – IS THE MIX LIKE WATER AND OIL OR LIKE NITROCELLULOSE AND FIRE?

Large-scale genomic sequencing is providing massive datasets through which the evolution of protein families may be addressed. Comparative genome hybridization on DNA microarrays and comparative genome sequencing within different strains of the same bacterial species will add even further to the abundant information available. One protein family "trend" is that within any given species, genes that perform housekeeping functions are relatively stable and vary little in sequence composition from strain to strain while genes that perform roles in virulence are more variable (Figure 11.4).

This gene conservation spectrum from invariable to variable is accompanied by a similar scale representing the peppering of mosaic segments in variable genes that have evolved through the mosaic process. Those genes at the variable end of the scale have more complex mosaic patterns. This is a function of the selection that allows the mosaic segments to accumulate. New functional domains encoded in the mosaic segments often are maintained because of their positive selective value. The positive selection might be an auxiliary metabolic function or a capability that may allow survival in an alternative environmental niche

When the subset of variable genes within any particular species is examined, this group is often found to represent genes involved in interaction with the environment of the particular bacterial species or in interaction with the host if the species is a commensal or a pathogen. The variability in this case represents functional gene modifications that have allowed niche expansion or escape from periodic selection during upheavals in the micro-environment. Often, these variable genes and their encoded proteins are found to be those crucial to the host–pathogen interaction. For example, among humans, the MHC alleles and the immunoglobulin loci are highly variable (Murphy, 1993) In bacterial pathogens, surface proteins are often variable with those involved in a critical host interaction, often being among the most variable.

Although there is a tendency in selecting candidate proteins to use for vaccines towards identifying common surface proteins so that the vaccine will be effective against all strains, this selection is counterintuitive to the evolutionary process that created the variability and conservation in the first place. Many common surface antigens may have been able to maintain their conserved state because when the host has responded to them in the past, that response has not been particularly advantageous to the host. Thus, antibodies directed at common antigens may not be as effective as those directed at variable antigens. Moreover, if variable mosaic proteins are effective targets for the host immune system then the spectrum of variability within the mosaic protein family can be determined and used to design an effective vaccine. The variable mosaic segments can be added together in the vaccine to cover all of the strains that might have variable segments.

The mosaic process will continue to be an effective strategy by which bacteria will avoid the need to reinvent useful traits from scratch. New variable traits may be generated as the pathogenic bacterial population struggles to maintain a presence in the host in the face of host immunity elicited by vaccination. Intervention strategies both for new vaccines and new antibiotics will be more successful with awareness of this possibility. It is fortunate that, even with the ability to pick up new functional domains through horizontal transfer, bacterial evolution is a slower process than that of an RNA virus such as influenza or HIV. Because of this, our ability to identify changes and modify strategies should be more easily accomplished than for rapidly evolving RNA viruses.

REFERENCES

Brooks-Walter, A., Briles, D.E. and Hollingshead, S.K. (1999) The pspC gene of Streptococcus pneumoniae encodes a polymorphic protein, PspC, which elicits cross-reactive antibodies to PspA and provides immunity to pneumococcal bacteremia. Infect. Immun. 67: 6533–6542.

Cheng, Q., Finkel, D. and Hostetter, M.K. (2000) Novel purification scheme and functions for a C3-binding protein from Streptococcus pneumoniae. Biochemistry 39: 5450–5457.

Claverys, J.P., Prudhomme, M., Mortier-Barriere, I. and Martin, B. (2000) Adaptation to the environment: Streptococcus pneumoniae, a paradigm for recombination-mediated genetic plasticity? Mol. Microbiol. 35: 251–259.

Dave, S., Brooks-Walter, A., Pangburn, M.K. and McDaniel, L.S. (2001) PspC a pneumococcal surface protein binds human factor H. Infect. Immun. 69: 3435–3437.

Facklam, R., Beall, B., Efstratiou, A. et al. (1999) emm typing and validation of provisional M types for group A streptococci. Emerg. Infect. Dis. 5: 247–253.

Feil, E.J., Maiden, M.C., Achtman, M. and Spratt, B.G. (1999) The relative contributions of recombination and mutation to the divergence of clones of Neisseria meningitidis. Mol. Biol. Evol. 16: 1496–1502.

Feil, E.J., Smith, J.M., Enright, M.C. and Spratt, B.G. (2000) Estimating recombinational parameters in Streptococcus pneumoniae from multilocus sequence typing data. Genetics 154: 1439–1450.

Feil, E.J., Holmes, E.C., Bessen, D.E. et al. (2001) Recombination within natural populations of pathogenic bacteria: short-term empirical estimates and long-term phylogenetic consequences. Proc. Natl Acad. Sci. USA 98: 182–187.

Filipe, S.R., Pinho, M.G. and Tomasz, A. (2000) Characterization of the murMN operon involved in the synthesis of branched peptidoglycan peptides in Strepto- coccus pneumoniae. J. Biol. Chem. 275: 27768–27774.

Fischetti, V.A. (1989) Streptococcal M protein: molecular design and biological behavior. Clinical Microbiol. Rev. 2: 285–314.

Garcia, J.L., Sanchez-Beato, A.R., Medrano, F.J. and Lopez, R. (2000) Versatility of the choline-binding domain. In Streptococcus pneumoniae (ed., A. Tomasz), pp. 231–244, Mary Ann Liebert, Inc., New York.

Gosink, K.K., Mann, E.R., Guglielmo, C. et al. (2000) Role of novel choline binding proteins in virulence of Strepto-coccus pneumoniae. Infect. Immun. 68: 5690–5695.

Haldane, J.B.S. (1932) The Causes of Evolution. Longmans and Green, London.

Hammerschmidt, S., Talay, S.R., Brandtzaeg, P. and Chhatwal, G.S. (1997) SpsA, a novel pneumococcal surface protein with specific binding to secretory Immunoglobulin A and secretory component. Mol. Microbiol. 25: 1113–1124.

Hammerschmidt, S., Bethe, G., Remane, P.H. and Chhatwal, G.S. (1999) Identification of pneumococcal surface protein A as a lactoferrin-binding protein of Streptococcus pneumoniae. Infect. Immun. 67: 1683–1687.

Hammerschmidt, S., Tillig, M.P., Wolff, S. et al. (2000) Species-specific binding of human secretory component to SpsA protein of Streptococcus pneumoniae via a hexapeptide motif. Mol. Microbiol. 36: 726–736.

Janulczyk, R., Iannelli, F., Sjoholm, A.G. et al. (2000) Hic, a novel surface protein of Streptococcus pneumoniae that interferes with complement function. J. Biol. Chem. 275: 37257–37263.

Kamerling, J.P. (2000) Pneumococcal polysaccharides: a chemical view. In *Streptococcus pneumoniae* (ed., A. Tomasz), pp. 81–114, Mary Ann Liebert, New York.

Lindahl, G., Akerstrom, B., Stenberg, L. et al. (1991) Genetics and biochemistry of protein Arp, an IgA receptor from group A streptococci. In *Genetics and Molecular Biology of Streptococci, Lactococci, and Enterococci* (eds, G.M. Dunny, P.P. Cleary and L.L. McKay), pp. 155–159, American Society of Microbiology, Washington, DC.

Lindahl, G., Sjobring, U. and Johnsson, E. (2000) Human complement regulators: a major target for pathogenic microorganisms. *Curr. Opin. Immunol.* **12**: 44–51.

Majewski, J., Zawadzki, P., Pickerill, P. et al. (2000) Barriers to genetic exchange between bacterial species: *Streptococcus pneumoniae* transformation. *J. Bacteriol.* **182**: 1016–1023.

Maynard Smith, J., Smith, N.H., O'Rourke, M. and Spratt, B.G. (1993) How clonal are bacteria? *Proc. Natl Acad. Sci. USA* **90**: 4384–4388.

Mazmanian, S.K., Liu, G., Ton-That, H. and Schneewind, O. (1999) *Staphylococcus aureus* sortase, an enzyme that anchors surface proteins to the cell wall. *Science* **285**: 760–763.

Mazmanian, S.K., Liu, G., Jensen, E.R. et al. (2000) *Staphylococcus aureus* sortase mutants defective in the display of surface proteins and in the pathogenesis of animal infections. *Proc. Natl Acad. Sci. USA* **97**: 5510–5515.

Modun, B., Morrissey, J. and Williams, P. (2000) The staphylcoccal transferrin receptor: a glycolytic enzyme with novel functions. *Trends Microbiol.* **8**: 231–237.

Mortier-Barriere, I., Humbert, O., Martin, B., Prudhomme, M. and Claverys, J.P. (1997) Control of recombination rate during transformation of *Streptococcus pneumoniae*: an overview. *Microb. Drug Resist.* **3**: 233–242.

Murphy, P.M. (1993) Molecular mimicry and the generation of host defense protein diversity. *Cell* **72**: 823–826.

Ohno (1970) *Evolution by Gene Duplication*. Springer-Verlag, Berlin.

Pancholi, V. and Fischetti, V.A. (1993) Glyceraldehyde-3-phosphate dehydrogenase on the surface of group A streptococci is also an ADP-ribosylating enzyme. *Proc. Natl Acad. Sci. USA* **90**: 8154–8158.

Pancholi, V. and Fischetti, V.A. (1998) Alpha-enolase, a novel strong plasmin(ogen) binding protein on the surface of pathogenic streptococci. *J. Biol. Chem.* **273**: 14503–14515.

Robinson, J.H. and Kehoe, M.A. (1992) Group A streptococcal M proteins: virulence factors and protective antigens. *Immunol. Today*, **13**: 362–367.

Rosenow, C., Ryan, P., Weiser, J.N. et al. (1997) Contribution of novel choline-ginding proteins to adherence, colonization and immunogenicity of *Streptococcus pneumoniae*. *Mol. Microbiol.* **25**: 819–829.

Tettelin, H., Nelson, K., Paulsen, I.T. et al. (2001) Complete genome sequence of a virulent isolate of *Streptococcus pneumoniae*. *Science* **293**: 498–506.

Ton-That, H., Liu, G., Mazmanian, S.K. et al. (1999) Purification and characterization of sortase, the transpeptidase that cleaves surface proteins of *Staphylococcus aureus* at the LPXTG motif. *Proc. Natl Acad. Sci. USA* **96**: 12424–12429.

Tu, A.H., Fulgham, R.L., McCrory, M.A. et al. (1999) Pneumococcal surface protein A inhibits complement activation by *Streptococcus pneumoniae*. *Infect. Immun.* **67**: 4720–4724.

Weber, B., Ehlert, K., Diehl, A. et al. (2000) The *fib* locus in *Streptococcus pneumoniae* is required for peptidoglycan crosslinking and PBP-mediated beta-lactam resistance. *FEMS Microbiol. Lett.* **188**: 81–85.

Yother, J. and White, J.M. (1994) Novel surface attachment mechanism for the *Streptococcus pneumoniae* protein PspA. *J. Bact.* **176**: 2976–2985.

Zhang, J.R., Mostov, K.E., Lamm, M.E. et al. (2000) The polymeric immunoglobulin receptor translocates pneumococci across human nasopharyngeal epithelial cells. *Cell* **102**: 827–837.

C H A P T E R 12

Evolutionary Relationships Among Diverse Bacteriophages and Prophages: All The World's a Phage

Roger W. Hendrix, Margaret C.M. Smith, R. Neil Burns, Michael E. Ford and Graham F. Hatfull

We report DNA and predicted protein sequence similarities, implying homology, among genes of dsDNA bacteriophages and prophages spanning a broad phylogenetic range of host bacteria. The sequence matches reported here establish genetic connections, not always direct, among the lambdoid phages of *Escherichia coli*, phage ϕC31 of *Streptomyces*, phages of *Mycobacterium*, a previously unrecognized cryptic prophage, ϕFlu, in the *Haemophilus influenzae* genome, and two small prophage-like elements, ϕRv1 and ϕRv2, in the genome of *M. tuberculosis*. The results imply that these phage genes, and very possibly all of the dsDNA tailed phages, share common ancestry. We propose a model for the genetic structure and dynamics of the global phage population in which all dsDNA phage genomes are mosaics with access, by horizontal exchange, to a large common genetic pool, but in which access to the gene pool is not uniform for all phage.

The double-stranded DNA-containing (dsDNA) bacteriophages are very likely the most numerically abundant group of similar organisms in the biosphere, and nearly 4500 different dsDNA phages capable of infecting a large diversity of bacterial hosts have been described (Coetzee, 1987). However, these phages have proven difficult to classify, in part due to the breadth of their genetic variation. For example, phages with similar morphologies, modes of replication and overall genomic architectures may be completely unrelated at the nucleotide level. By comparing the genomes of several newly characterized phages and cryptic prophages it appears that the vast majority of dsDNA tailed-phages have common ancestry, and that they undergo profuse exchange of functional genetic elements drawn from a large shared pool.

The classification of bacteriophages by their host range, morphology or available life-cycles has led to conflicting conclusions regarding their origins and evolution (Coetzee, 1987). Phages can be found in virtually all places where their bacterial hosts exist, although only a small number have been investigated in detail (Bergh et al., 1989; Ackerman, 1996). As phages with near-identical genomes are rarely isolated from independent sources in nature, the term "species" is of limited use in describing relationships among phages (Casjens et al., 1992). Groups of phages related to each other by common gene organization and some degree of sequence similarity clearly do exist (for example, the "lambdoid" or λ-like group) and evidence for horizontal transfer among tail fiber genes has been reported (Haggard-Ljungquist et al.,

133

Copyright © 2002 by Academic Press.
All rights of reproduction in any form reserved.

1992; Sandmeier et al., 1992; Tetart et al., 1996; Monod et al., 1997). However, it has been unclear whether, and in what ways, different groups of phages – particularly phage groups with phylogenetically distant hosts – are related to each other.

The complete genomic sequences of several phages closely related to phage lambda have been determined, and shown to have similar genomic organizations (R. Juhala et al., in preparation). However, the genomes are clearly mosaic in nature, with regions of obvious sequence similarity interspersed with segments that are apparently unrelated (Simon et al., 1971; Campbell and Botstein, 1988; Highton et al., 1990; Campbell et al., 1994; Juhala, R. et al., in preparation). This argues for the existence of extensive horizontal genetic exchange among members of this group of phages. Other bacteriophages – such as mycobacteriophage L5 – have no clear sequence similarity with these phages but do possess a similar genomic architecture, raising the possibility that they share common ancestry with the λ-like phages (Hatfull and Sarkis, 1993; Hatfull and Jacobs, 1994).

Recently, we determined the genome sequences of mycobacteriophages D29 (a relative of L5; Ford et al., 1998a) and TM4 (Ford et al., 1998b), coliphages HK97 and HK022 (closely related members of the lambdoid group; R. Juhala et al., in preparation), and actinophage φC31 (host: *Streptomyces* spp.; Burns and Smith, in preparation). Although none of the mycobacteriophages has any recognizable sequence similarity to HK97 or HK022 (nor to other λ-like phage genomes), φC31 provides a clear bridge between these two apparently distant groups. On the one hand, the head protein genes of φC31 have a very similar organization to those of HK97 and HK022, and some of these are clearly homologous on the basis of sequence similarity. Thus, the putative terminase, portal, protease, and capsid genes from HK97 and φC31 are collinear and share 28, 29, 28, and 20% amino acid sequence identity, respectively (Popa et al., 1991; Duda et al., 1995). A similar mode of capsid subunit processing buttresses the proposed relationship between these phages (Figure 12.1). On the other hand, several φC31 genes match genes in the

mycobacteriophages. For example, ORF's 9 and 10 of φC31 (Hartley et al., 1994) (now known to be a single gene called *9a*; Burns and Smith, in preparation) encodes a protein related to gp70 of mycobacteriophage TM4 (Ford et al., 1998b); φC31 gp11 is a DNA polymerase with greatest similarity to gp44 of L5 (43% identity); φC31 gp16 is related to gp48 of L5 (53% identity); and gp20 of φC31 is a dCMP deaminase closely related to gp36.1 of mycobacteriophage D29 (55% identity). While these are all early proteins, we note that the putative terminase of φC31 (gp33) shares similarity with the putative terminase (gp13) of L5 (27% identity), as it also does with the HK97 terminase, gp2, noted above. A pairwise alignment of the HK97gp2 and L5gp13 terminase sequences does not alone make a convincing case that they are related, but their mutual similarity to the φC31 protein argues that all three terminases are homologues. Taken together, the observations that these genes match across the phylogenetic chasms separating their hosts argue strongly that they – and perhaps most phage genes – are derived from a shared pool. While the modest levels of similarity that we see between the groups of phages infecting phylogenetically distant hosts argues against *direct* horizontal exchange of DNA in recent evolutionary time, we believe the observations are consistent with an ongoing pattern of exchanges among extensive chains of intermediates connecting the particular genomes we have examined.

A somewhat different set of examples of sequence relationships among "distant" phages comes from analysis of the genome sequence of mycobacteriophage D29, a close relative of phage L5 (Ford et al., 1998a). The left arms of the L5 and D29 genomes are very similar to each other (approximately 80% sequence identity) and contain closely related genes in collinear positions. A notable departure occurs in gene *10* of D29 which, while sharing sequence similarity with L5 gene *10*, is 603bp larger (1481bp vs. 878bp) and encodes a putative protein product (gp10) that is 201 amino acids bigger than its L5 counterpart (Figure 12.2 (see color plates)). Alignment of the coding sequences shows that D29 gp10 contains a contiguous block of 194 residues between codons 173 and 174, that is absent in L5.

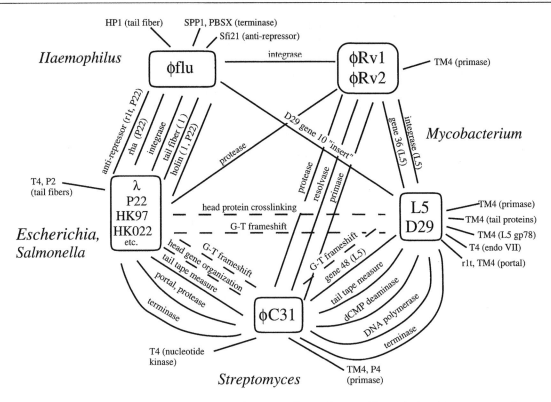

FIGURE 12.1 Sequence connections among phages and prophages. The relationships among phage and prophage sequences are indicated, with the solid lines representing sequence similarities and the dotted lines corresponding to commonalities of gene organization or gene function. Closely related phages are shown in boxes, and bacterial hosts are shown at the perimeter of the web. Sequence comparisons were performed using BLAST and GAPPED BLAST programs available at the National Center for Biotechnology Information web site (http://www.ncbi.nlm.nih.gov/) and the FASTA, TFASTA and BESTFIT programs within the Genetics Computer Groups Inc. (GCG) package. Protein sequences were considered to be related (that is, probable homologues) if they could be aligned over a substantial portion of their lengths with 20% or greater amino acid identities, or if the BLAST ouput reported a high probability of relatedness. A table of the protein similarities is available at http://www.pitt.edu/~gfh/table.html. Accession numbers for phage genome sequences are: HK97, AF069529; HK022, AF069308; φC31, AJ006589; L5, Z18946; D29, AF022214; TM4, AF068845.

Comparison of L5 gp10 with the protein databases fails to identify any close relatives (Hatfull and Sarkis, 1993). However, database searches with D29 gp10 reveals similarity to the putative 200 amino acid product of the HI1415 gene of *Haemophilus influenzae*, whose function is unknown (Fleischmann et al., 1995). The similarity is restricted to the region of D29 gp10 that is absent from L5 gp10, and this can be aligned with the HI1415 product with 34% identity (Figure 12.2). The evolutionary events that account for the relationship between HI1415 and gene *10* of L5 and D29 are not obvious, although

it seems unlikely that an intein is involved since the sequence features of known inteins (Perler et al., 1997) are not present.

A closer examination of the *H. influenzae* genome surrounding HI1415 suggests that this gene is part of a previously unidentified prophage which we propose to call φflu (Figure 12.3 (see color plates)). In particular, we note that several other genes in this region have sequence similarity to known phage genes; these are HI1403, HI1410, HI1411, HI1412, HI1415, HI1416, HI1418, HI1422, and HI1424 (Figure 12.3). The phages whose genes they match are associated with a wide diversity of bacterial hosts (Figure 12.3). Other noteworthy features are an integrase gene (HI1424) with a tRNA gene immediately upstream, a gene (HI1407) encoding a product related to *traN* of plasmid RP4, and the genes (HI1392 and HI1393) encoding the HindIII restriction/modification system (Figure 12.3). There are several open reading frames encoding proteins that do not match existing database entries. Since the only phage particles reported to be released from the sequenced strain *H. influenzae* Rd are those of a Mu-like prophage (Fleischmann et al., 1995) it seems likely that φflu is cryptic and unable to produce infectious viral particles.

The phage-related genes lie within a ~31.5 kb segment of the *H. influenzae* genome that is flanked by the *trpB* and *trpA* genes (HI1431 and HI1432 respectively) on one side, and *trpE*, *trpD* and *trpC* (HI1388, HI1389, and HI1389.1) on the other (Figure 12.3). This entire region contains approximately 40 genes, including several that are not obviously phage-related (e.g. HI1391, *valS*; HI1398, *fumC*; HI1401, *pyrD*; HI1425, *fnr*; HI1428, *purN*; HI1429, *purM*). A simple explanation accounting for this genome structure is that a defective transducing particle – generated by aberrant excision – integrated into the *trp* operon through illegitimate recombination.

The φflu cryptic prophage is remarkable in that the combination of identifiable phage genes has not previously been seen in genomes of infectious bacteriophages. Moreover, the phages that carry genes homologous to φflu genes (i.e. HP1, SF6, P22, D29, λ, Sfi21, and r1t) infect a wide range of bacterial hosts (*Haemophilus*, *Bacillus*, *Salmonella*, *Mycobacterium*, *Escherichia*, *Streptococcus* and *Lactococcus*) and have little or no sequence similarity with each other. As with the phage examples described above, these observations argue that there is substantial genetic exchange of phage genes across host phylogenetic boundaries.

The unusual combination of phage genes in φflu is not peculiar to this particular cryptic prophage but appears to be a feature of other, unrelated prophages. For example, the genome of *Mycobacterium tuberculosis* H37Rv (Philipp et al., 1996; Cole, 1998) contains at least two apparent cryptic phages (Figure 12.4 (see color plates)). Both of these are relatively small (~10 kb) but contain several phage-related genes. One of these prophages (φRv2) has at least two homologues of mycobacteriophage genes: integrase (Rv2659c, a member of the phage integrase-related family of recombinases but most closely related to integrases of phages L5 and D29), and a relative of L5 gene *36* (Rv2657c). In addition, it contains a second recombinase (Rv2647, related to the plasmid-encoded recombinase, *rci*), a homologue of the prohead protease genes of phages HK97 and φC31, and a homologue of a gene product of actinophage RP3 of unknown function (Figure 12.4). The second cryptic prophage (φRv1) is rather similar in overall structure to φRv2; it also contains homologues of the HK97/φC31 prohead protease and the actinophage RP3 gene. It has a gene similar to gene *70* of mycobacteriophage TM4 (Ford et al., 1998b) and to gene *9a* of *Streptomyces* phage φC31, encoding a probable DNA primase (Burns and Smith, in preparation). Finally, it has a recombinase (Rv1586c), that – like the integrase of φC31 (24) – is related to the transposon resolvase/DNA invertase family of site-specific recombinases (Stark et al., 1992). This cryptic prophage is flanked by two 12 bp direct repeats, which may represent recognition sites for this recombinase. Mahairas et al. (1996) noted previously that this region is only present in 16% of *M. tuberculosis* clinical isolates and is absent from bacille Calmette-Guérin (BCG) – an avirulent derivative of the *M. tuberculosis* complex of bacteria – which has only a single copy of the 12bp repeat. Presumably the recombination functions associated with this prophage are active and mediate integration and/or excision of this DNA segment. The φRv2 element may also be recombinationally active since the attachment junctions (*attL* and *attR*) are present and appear to derive from a phage attachment site (*attP*) that is structurally similar to L5 *attP* (Peña et al., 1997); the excisionase is provided by the homologue of L5 gene *36* (J. Lewis and G.F.Hatfull, unpublished observations). Although the two *M. tuberculosis* prophages appear too small and too deficient in virion structural genes to encode a complete virion, they could in fact be "complete" satellite phages, in the manner of coliphage P4, which uses the structural genes of another phage to package its genome (Lindqvist et al., 1993). In addition, they could be analogs of some other form of integrating plasmid, such as the *Streptomyces* integrating plasmid pSE211 (Brown et al., 1990), which has a gene with sequence similarity to a gene in φRv2 (Figure 12.4) as well as an indirect connection by sequence similarity through φC31 gene *9a* to the DNA primase gene of phage P4 (Burns and Smith, in preparation).

Regardless of the provenance and current function of these cryptic prophages, their unusual combinations of genes demonstrate that phages with apparently unrelated genomes (e.g. mycobacteriophage L5 and coliphage HK97) do not exist in genetic isolation. The bridge that φC31 provides between the L5 and HK97 sequences and the diverse combination of phage genes in φflu lead to similar conclusions, and we expect that many more such links will become apparent as new phage and prophage sequences become available. These connections are also consistent with previous observations that phages frequently have specific functional properties in common (for example, the crosslinked head proteins of L5 and HK97 (Duda et al., 1995) and ribosomal frameshifting in tail genes (Casjens et al., 1992)) even when sequence similarity is not evident (see Figure 12.1). The combination of these functional, organizational and now sequence similarities suggests that a significant number of dsDNA bacteriophages are in fact related, having had – and continuing to have – access to a large pool of functional genetic elements. The network of connections shown in Figure 12.1 is presumed to be representative of a vastly larger network of relationships among phage sequences that exists in nature. The sequence connections shown in the figure are unlikely to be the result of direct genetic exchanges between the indicated phage genomes; rather, we believe they are the visible end products of long chains of genetic interactions. We anticipate that these processes not only fuel the generation of new bacteriophages but that similar events are involved in the evolution of novel viruses of eukaryotic and archael hosts.

DISCUSSION

It has been estimated that there are $4–6 \times 10^{30}$ prokaryotic cells in the biosphere (Whitman et al., 1998), and direct counts on environmental samples typically show about 10-fold more tailed phage particles than cells (Bergh et al., 1989). Thus the total number of extant phages is enormous. We have no way of calibrating the age of phages as a group, but we presume them to be ancient – possibly comparable in age to their bacterial hosts. Against this very large and probably very old population, there are roughly 30 complete DNA sequences of contemporary phage genomes available, including those reported here, plus several prophage sequences contained in bacterial genome sequences. Here we discuss what our observations of this sample imply about the genetic structure and dynamics of the phage genome population as a whole, and more explicitly present our view of the most plausible interpretation of these data.

We take our observations of sequence similarities among genes found in phages and prophages of diverse hosts to imply common ancestry for those genes. We expect that the same is true for those genes for which sequence similarities are not yet apparent in the currently available data but which have similar functions, although this remains to be proven. Thus we suggest on the basis of the sequence similarities that many and probably most of the genes of contemporary phages derive from a common ancestral pool of genes. The view of common ancestry gains support as well from many striking similarities of phage gene organization and function, often in the absence of any remaining sequence similarity (Hatfull et al., 1994; Ford et al., 1998a,b; Hendrix et al., unpublished observations).

In comparing phage genomes it is clear that only some pairs of genes show significant sequence similarities. We believe that this argues strongly for horizontal exchange of sequences among the ancestors of the contemporary phage. Thus the juxtaposition of those genes that match with substantial sequence similarity between L5 and φC31 together with those that have no detectable similarity, seems most plausibly explained by past horizontal exchange of sequences. The same is true for the similarity of gene organization – and in some cases sequence similarity – between the late genes of HK97 and φC31 which are joined to the differently-organized and sequence-dissimilar early genes of these phages. The most extreme case, however, is in φFlu where genes with unequivocal sequence similarities to genes from a large group of phylogenetically diverse hosts are found, but for each homologous pair the adjacent genes have no sequence or functional resemblance. We find it very difficult to understand how such a set of relationships could have arisen without multiple horizontal exchanges. An alternative model that these phages have derived from a common ancestor in the absence of horizontal exchange cannot be formally ruled out. In this scenario differences in the degree of sequence similarity for different genes could be due to gene-specific differences in the rate of the mutational clock, but it is difficult to understand why, within different pairwise comparisons of genomes, different sets of genes would have slower mutational clocks than others.

If we allow that horizontal exchange has occurred, then we can ask when it happened. An extreme model is that most of the horizontal exchange giving rise to the current diversity occurred early in the history of phages and since some point early in evolutionary history – for example, the beginnings of bacterial speciation – has been passed down to contemporary phages largely by vertical transmission. If the phage gene pool was sufficiently diverse, the observed sequence similarities over large (host) phylogenetic distances between some phage genes, but not others, could simply reflect the assortment of the various combinations of genes at this early stage. Such a model might allow horizontal exchange to continue among phages infecting a single bacterial species. A less extreme model and one that we strongly prefer, is one in which horizontal exchange has continued up to the present. This is in part because we doubt that the degree of sequence similarity we see in some genes could have persisted over a period of time when other genes – for example, the virion structural genes, which are arguably homologous on the basis of shared gene organization and function – have often diverged past the point of

recognizable similarity. More importantly, a model with widespread contemporary horizontal exchange is more consistent with our knowledge of the biology of phages and their hosts.

It has been clear for some time that there is vigorous and ongoing horizontal exchange among the well studied "lambdoid" phages of *E. coli* and *Salmonella* – originally so named because they are able to form viable genetic hybrids with the prototype phage lambda (Campbell, 1994; R. Juhala et al., unpublished observations). Thus, for example, we can find pairs of genes in two lambdoid phages (such as HK97 and P22) that have nearly identical sequences, juxtaposed with genes with little or no sequence similarity, implying horizontal exchange in the very recent evolutionary past. This exchange presumably happens most often when two phage genomes find themselves in the same cell, either as two co-infecting phages or perhaps more importantly as a single phage infecting a cell that carries one or more prophages. Sequence comparisons by us and others (Brussow et al., 1998; Desiere et al., 1998; Ford et al., 1998a,b; Lucchini et al., 1998; Neve et al., 1998) of genomes of phages that infect other single or closely related groups of host species are beginning to make a case for similar clusters of intensely exchanging phages centered around these hosts. However, we believe it is very unlikely that horizontal exchange is confined to these species-related groups of phages. Phage host ranges are often not confined to a single bacterial species, and phages with overlapping host ranges should be able to exchange sequences, whether or not their host ranges are otherwise similar. Other mechanisms exist for transfer of bacterial DNA sequences, potentially including phage or prophage sequences, between bacterial species and it has been estimated that *E. coli* replaces approximately 16 kb of its genome by horizontal exchange from outside sources every million years (Lawrence and Ochman, 1998). Thus we believe there are ample opportunities for phage DNA sequences to travel among host species in the local phylogenetic neighborhood. However, our data and similar observations place an important constraint on how this process happens on a larger scale. Thus all the sequence similarities we see between phages or prophages associated with phylogenetically distant hosts are of only a modest level of sequence identity; this argues against *direct* horizontal exchange between these phage in recent evolutionary time. This most likely means that there is no mechanism available for direct exchange between two such phage – for example, that there is no common host for coliphage HK97 and actinophage φC31 – but it might also mean that a sequence adapted to function in one host might find itself at a disadvantage after leaping directly into a very different host. In either case, this argues that for a phage sequence to travel to a phylogenetically distant host requires a journey of many steps – a sort of random walk through phylogenetic space.

What then is our view of the genetic structure of the global phage population? We favor a model in which all the dsDNA phage and prophage genomes are mosaics with access by horizontal exchange to a large common gene pool. However, access is clearly not uniform. There are phylogenetically local areas of free and intense exchange of genetic information, as for example the lambdoid phages of enteric hosts or the phage L5 family of the mycobacteria. In addition, there is exchange beyond the confines of the local neighborhood, but only with reduced frequency. Thus we imagine that any given phage has access to all the sequences in the global pool, but that the frequency of that access depends strongly on the number of barriers (e.g. host ranges) between any particular sequence and that phage and, therefore, how many individual steps of genetic exchange are required to bring them together. The veracity of this view of bacteriophage population genetics and evolution, and the quantitative nature of the relationships implied, will only be determined, we believe, after substantially more data are determined of sequences and genetic organization of phages and their hosts.

ACKNOWLEDGMENTS

We thank members of our laboratories and Jeffrey Lawrence for useful discussions. This work was supported in part by NIH grant GM51975 to R.W.H. and G.F.H. and grants from the MRC (G9301410MB) and Royal Society to M.C.M.S. and in part by a grant from the Pittsburgh Supercomputing Center through the NIH National Center for Research Resources resource grant 2 P41 RR06009.

REFERENCES

Ackermann, H.-W. (1996) Frequency of morphological phage descriptions in 1995. *Arch. Virol.* **141**: 209–218.

Bergh, Ø., Børsheim, Y., Bratbak, G. and Heldal, M. (1989) High abundance of viruses found in aquatic environments. *Nature* **340**: 467–468.

Brown, D.P., Idler K.B. and Katz, L. (1990) Characterization of the genetic elements required for site-specific integration plasmid pSE211 in *Saccharopolyspora erythraea*. *J. Bacteriol.* **172**: 1877–1888.

Brussow, H., Bruttin, A., Desiere, F. et al. (1998) Molecular ecology and evolution of *Streptococcus thermophilus* bacteriophages, a review. *Virus Genes* **16**: 95–109.

Campbell, A. (1994) Comparative molecular biology of lambdoid phages. *Annu. Rev. Microbiol.* **48**: 193–222.

Campbell, A. and Botstein, D. (1988) Phage evolution and speciation. In *The Bacteriophages* (ed., R. Calenda), vol. 1, pp. 1–14, Plenum Press, New York.

Casjens, S.R., Hatfull, G.F. and Hendrix, R.W. (1992) Evolution of dsDNA tailed bacteriophage genomes. In *Seminars in Virology* (ed., E. Koonin), vol. 3, pp. 383–397, Academic Press, London.

Coetzee, J. (1987) In *Phage Ecology* (eds, S. M.Goval, C. Gerba and G. Bitton), pp. 45–85, John Wiley, New York.

Cole, S.T. (1998) Deciphering the biology of *Mycobacterium tuberculosis* from the complete genome sequence. *Nature* **393**: 537–544.

Desiere, F., Lucchini, S. and Brüssow, H. (1998) Evolution of *Streptococcus thermophilus* bacteriophage genomes by molecular exchanges followed by point mutations and small deletions and insertions. *Virology* **241**: 345–356.

Duda, R.L., Martincic, K., Xie, Z. and Hendrix. R.W. (1995) Bacteriophage HK97 head assembly. *FEMS Microbiol. Rev.* **17**: 41–46.

Fleischmann, R.D., Adams, M. D., White, O. et al. (1995) Whole-genome random sequencing and assembly of *Haemophilus influenzae*. *Science* **269**: 496–512.

Ford, M.E., Sarkis, G.J., Belanger, A.E. et al. (1998a) Genome structure of mycobacteriophage D29: implications for phage evolution. *J. Mol. Biol.* **279**: 143–164.

Ford, M.E., Hendrix, R.W. and Hatfull, G.F. (1998b) Mycobacteriophage TM4: genome structure and gene expression. *Tubercle Lung Dis.* **79**: 63–73.

Haggard-Ljungquist, E., Halling, C. and Calendar, R. (1992) DNA sequences of the tail fiber genes of bacteriophage P2: evidence for transfer of tail fiber genes among unrelated bacteriophages. *J. Bacteriol.* **174**: 1462–1477.

Hartley, N.M., Murphy, G.O., Bruton, C.J. and Chater, K.F. (1994) Sequence of the essential early region of phi C31, a temperate phage of *Streptomyces* spp. with unusual features in its lytic development. *Gene* **147**: 29–40.

Hatfull, G.F. and Sarkis, G. (1993) DNA sequence, structure and gene expression of mycobacteriophage L5: a phage system for mycobacterial genetics. *Mol. Microbiol.* **7**: 395–406.

Hatfull, G.F. and Jacobs, J.R. Jr (1994) In *Tuberculosis: Pathogenesis, Protection and Control* (ed., B.R. Bloom), pp. 165–183, ASM, Washington, DC.

Highton, P., Chang, Y. and Myers, R. (1990) Evidence for the exchange of segments between genomes during the evolution of lambdoid bacteriophages. *Mol. Microbiol.* **4**: 1329–1340.

Kuhstoss, S. and Rao, N.J. (1991) Analysis of the integration function of the streptomycete bacteriophage phi C31. *J. Mol. Biol.* **222**: 897–908.

Lawrence, J.G. and Ochman, H. (1998) Molecular archaeology of the *Escherichia coli* genome. *Proc. Natl Acad. Sci. USA* **95**: 9413–9417.

Lindqvist, B.H., Deho, G. and Calendar, R. (1993) Mechanisms of genome propagation and helper exploitation by satellite phage P4. *Microbiol. Rev.* **57**: 683–702.

Lucchini, S., Desiere, F. and Brüssow, H. (1998) The structural gene module in *Streptococcus thermophilus* bacteriophage phi Sfi11 shows a hierarchy of relatedness to Siphoviridae from a wide range of bacterial hosts. *Virology* **246**: 63–73.

Mahairas, G.E., Sabo, P. J. Hickey, M.J. et al. (1996) Molecular analysis of genetic differences between *Mycobacterium bovis* BCG and virulent *M. bovis*. *J. Bacteriol.* **178**: 1274–1282.

Monod, C., Repoila, F., Kutateladze, M. et al. (1997) The genome of the pseudo T-even bacteriophages, a diverse group that resembles T4. *J. Mol. Biol.* **267**: 237–249.

Neve, H., Zenz, K. I., Desiere, F. et al. (1998) Comparison of the lysogeny modules from the temperate *Streptococcus thermophilus* bacteriophages TP-J34 and Sfi21: implications for the modular theory of phage evolution. *Virology* **241**: 61–72.

Peña, C.E.A., Lee, M.H., Pedulla, M.L. and Hatfull, G.F. (1997) Characterization of the mycobacteriophage L5 attachment site, attP. *J. Mol. Biol.* **266**: 76–92.

Perler, F.B., Olsen, G.J. and Adam, E. (1997) Compilation and analysis of intein sequences. *Nucleic Acids Res.* **25**: 1087–1093.

Philipp, W.J., Poulet, S., Eiglmeier, K. et al. (1996) An integrated map of the genome of the tubercle bacillus, *Mycobacterium tumberculosis* H37Rv, and comparison with *Mycobacterium leprae*. *Proc. Natl Acad. Sci. USA* **93**: 3132–3137.

Popa, M.P., McKelvey, T.A., Hempel, J. and Hendrix, R.W. (1991) Bacteriophage HK97 structure: wholesale covalent cross-linking between the major head shell subunits. *J. Virol.* **65**: 3227–3237.

Sandmeier, H., Iida, S. and Arber, W. (1992) DNA inversion regions Min of plasmid p15B and Cin of bacteriophage P1: evolution of bacteriophage tail fiber genes. *J. Bacteriol.* **174**: 3936–3944.

Simon, M.N., Davis, R.W. and Davidson, N. (1971) Heteroduplexes of DNA molecules of lambdoid phages: physical mapping of their base sequence relationships by electron microscopy. In *The Bacteriophage Lambda* (ed., A.D. Hershey), pp. 313–328, Cold Spring Harbor Laboratory, Cold Spring Harbor, NY.

Stark, W.M., Boocock, M.R. and Sherratt, D.J. (1992) Catalysis by site-specific recombinases. *Trends Genet.* **12**: 432–439.

Tetart, F., Repoila, F., Monod, C. and Krisch, H.M. (1996) Bacteriophage T4 host range is expanded by duplications of a small domain of the tail fiber adhesin. *J. Mol. Biol.* **258**: 726–731.

Whitman, W.B., Coleman, D.C. and Wiebe, W.J. (1998) Prokaryotes: the unseen majority. *Proc. Natl Acad. Sci. USA* **95**: 6578–6583.

CHAPTER 13

Horizontal Gene Transfer in Bacteriophages

Gisela Mosig and Richard Calendar

Horizontal gene transfer in bacteriophages was demonstrated in 1959, when Denise Cohen infected *E. coli* strain B with temperate phage P2 and recovered a phage with a different immunity specificity, derived from a defective prophage (Cohen, 1959). Subsequently, many hybrid phages were created in the laboratory. In 1974 Botstein and Herskowitz constructed hybrids between coliphage lambda and *Salmonella* phage P22 and concluded that phages often exchanged modules of genes of related functions by illegitimate recombination. Recently, such exchange between phages has been noted within genes. For example, the tail fiber genes of *E. coli* phages are mosaics of small modules that have undergone extensive exchange (Haggard-Ljungquist et al., 1992; Sandmeier et al., 1992; Sandmeier, 1994). Another example of horizontal gene exchange is seen in some non-essential genes of coliphage P2. Several of these genes have an AT content of 67%, in contrast to the essential genes of P2, whose AT content is about 50%. Three of these non-essential genes with high AT content protect lysogenic cells from attack by heterologous phages, thus providing a selective advantage to the lysogenized host bacterium. The AT content of these non-essential genes argues for their acquisition by horizontal transfer from a bacterium whose DNA has a high AT content (Calendar et al., 1998; G. Christie, personal communication). These P2 non-essential genes must have been acquired relatively recently, because the related phage 186 does not have them. Hendrix et al. (1999) compared the sequences of lamdoid phage HK97 from *E. coli*, φC31 from *Streptomyces lividans* and phage L5 from *Mycobacterium tuberculosis*. They found that φC31 forms a bridge between the apparently unrelated phages HK97 and L5. For example, HK97 and φC31 exhibit more than 20% amino acid sequence identity in their capsid synthesis genes; φC31 and L5 have 43% identity of their DNA polymerase genes. Based on these and other observations, Hendrix et al. proposed that all phages are mosaics with access by horizontal exchange to a large common gene pool.

Specifically, they proposed that phage genomes can acquire additional genes, so-called "morons," by horizontal transfer (Juhala et al., 2000). Consistent with this idea, genomes of many related phages have unique, essential genes arranged in the same order, but in different members of the family different "non-essential," auxiliary, genes are interspersed. Prominent examples are the P2-like phages, mentioned above, and the family of T-even phages that are related to T2, T4 and T6 (Kim and Davidson, 1974). The mechanisms by which such morons are acquired, are, however, poorly understood. Illegitimate site-specific recombination and transposition have been proposed to account for transfer of entire gene modules, coding for interacting proteins. However, mosaic patterns of DNA sequences within single genes, particularly tail fiber genes, led to the conclusion that "current concepts cannot account for the formation of such chimeric genes, and the recombination mechanisms responsible are not known" (Kutter et al., 1995). Similar

141

Copyright © 2002 by Academic Press.
All rights of reproduction in any form reserved.

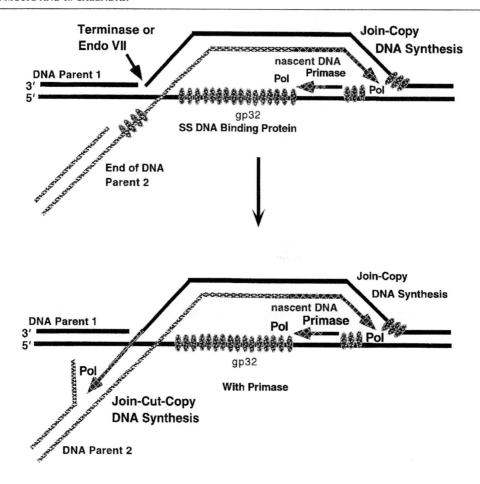

FIGURE 13.1 A recombinational intermediate (D-loop) formed by invasion of double-stranded DNA of parent 1 (solid line) by a single-stranded tail of the homologous DNA of parent 2 (patterned lines). Join-copy DNA synthesis is primed by the 3' end of the invading DNA. Join-cut-copy DNA synthesis is primed from a nick in the invaded DNA, inflicted by T4 terminase or endo VII. Except for occasional potential mutations, parents 1 and 2 are completely homologous.

mosaic patterns have been found in other bacteriophage families (Lucchini et al., 1999).

Our recent results have implicated homologous recombination and recombination-induced DNA replication in acquisition by T-even phages of morons (Gary et al., 1998; Mosig, 1998). The simplest interpretation of these results is that two recombination/replication pathways that we have called "join-copy" and "join-cut-copy," together with partial heteroduplex repair (Figures 13.1 and 13.2 (see color plates)) are responsible for horizontal gene transfer. The proposed mechanism readily explains the mosaic sequence variations near borders of many other suspected morons.

A MODEL FOR HORIZONTAL GENE TRANSFER BY HOMOLOGOUS RECOMBINATION PATHWAYS

In the now classical example of recombination-induced DNA replication, that we have dubbed "join-copy," a 3' end of a single-stranded DNA segment invading a homologous double-stranded DNA primes DNA synthesis in the rightward direction. This mechanism starts as soon as an origin-initiated replication fork has reached a chromosomal end or a double-stranded break, which can thereby be repaired (Mosig, 1998).

Another pathway, which we have dubbed "join-cut-copy," uses a single-stranded nick in a D-loop recombination intermediate to prime DNA replication in the leftward direction. In this case, priming of DNA synthesis on the *invading* DNA requires that a specific strand of the *invaded* DNA is nicked. In T4 this pathway depends on at least two DNA packaging genes which are expressed mainly late after infection (Figure 13.1), (Mosig, 1998). If both join-copy and join-cut-copy recombination occur, the intermediates appear like bidirectional replication forks initiated from an origin. A modification of this mechanism can accommodate the insertion of a foreign DNA segment containing moron X between two genes, A and B, as indicated in Figure 13.2 (see color plates).

Specifically, the model proposes that very limited sequence identity between the foreign DNA fragment and a resident gene A allows single-strand DNA invasion by a fragment's end (Figure 13.2a). This recombinational D-loop initiates join-copy replication from the 3'-invading end (in the rightward direction in Figure 13.2), which stabilizes the D-loop. Branch migration in the opposite direction (into a heterologous region) generates heteroduplexes with multiple mismatches and loops (Figure 13.2b). When certain enzymes (e.g. the T4 packaging enzymes Xsolvase endo VII (Kemper, 1998) or terminase (Bhattacharyya and Rao, 1994; Franklin et al., 1998)) cut the *invaded* strand of the recombinational intermediate (Figure 13.2b), the foreign DNA fragment can be replicated by the join-cut-copy pathway (Figure 13.2c–e). Limited homology of the foreign DNA fragment with another short sequence of gene B allows another recombinational invasion (Figure 13.2d) to complete the acquisition of the foreign fragment. Because of the limited homology of the foreign DNA to be acquired, branch migration generates many mismatched and looped-out bases in the heteroduplex region, and partial mismatch repair within the heteroduplex region (Figure 13.2c), including excision of loops and synthesis across looped-out DNA, is bound to lead to sequence divergence that appears like multiple mutations. If they occur within auxiliary genes these processes might inactivate or change the specificities of the

corresponding gene products, without serious consequences for phage survival. However, in essential genes only those combinations of repaired mismatches that regenerate a functional gene can survive.

This model most readily explains analyses of the gene *56* region of several T-even phages. Gene *56* encodes a dCTPase/dCDPase/dUTPase/dUDPase (Kutter and Wiberg, 1968), briefly called dCTPase, which is essential for growth of T-even phages, because it prevents incorporation of dCTP or dUTP into T-even DNA and allows incorporation of deoxy-hydroxymethyl-cytosine (from dHMCTP). This modification is essential for preventing degradation of T-even DNA by phage and host restriction enzymes (Carlson et al., 1994).

The dCTPase genes of several T-even phages have similar sizes, and their products complement each other (Gary et al., 1998) although the predicted amino acid sequences of the proteins differ considerably (up to 28%, Figure 13.3 (see color plates)). Most significantly in terms of our model, PCR and sequence analyses using primers corresponding to sequences shared by T2 and T4 in genes *56* and *soc* (small outer capsid protein) show that T4 and a close relative (class A) have a single gene, called *69*, between *56* and *soc*, T2, T6 and LZ5 (class B) have instead two small ORFs, and RB15 and several relatives (class C) have no gene between genes *56* and *soc*. The gene *56* DNA sequences of phages from the same class are very similar (less than 3% differences), but those belonging to different classes differ significantly by single mismatches and what appears as multiple compensating frameshift mutations (Mosig et al., 2001).

In terms of the model in Figure 13.2, we surmise that the large gene *56* differences of different classes were established concomitantly with acquisition of different adjacent morons by a mechanism depicted in Figure 13.2. Gene *56* corresponding to gene A, and *soc* corresponding to gene B, are adjacent in RB 15. We postulate that in T4, moron X (corresponding to gene *69*), and in T2 moron Y, corresponding to two short ORFs have been independently acquired by horizontal gene transfer. Consistent with our model, different functional gene *56* sequences were selected, which now appear like multiple base substitutions and compensating frameshifts. In protein segments whose amino acid sequences are not critical, these mutations can result in changes of an entire sequence block (Figure 13.3). Because invasion by different DNA fragments generated different viable combinations (e.g. in T4 and T2), it now appears as if multiple mutations were generated simultaneously.

The model predicts, and we found, that the present-day extensive sequence divergence between different T-even phages inhibits formation of viable recombinants. Because heteroduplexes of the dCTPase region between T-even phages belonging to different classes contain multiple mismatches and loops, partial heteroduplex repair is less likely to reconstitute a functional dCTPase gene, except when the dCTPase and adjacent DNA sequences are made similar (e.g. in chimeric phages). In the latter case viable recombinants are formed with frequencies expected from the size of the genes (Gary et al., 1998).

POTENTIAL EXCHANGES BETWEEN P2 AND T-EVEN GENES

As mentioned above, the tail fiber genes of P2, the T-evens and many other phages show considerable similarity (Haggard-Ljungquist et al., 1992; Hendrix and Duda, 1992; Henning and Hashemol-Hosseini, 1994; Sandmeier, 1994). The corresponding proteins are important for the host range of different phages, and the rapid emergence of new host specificities has been explained by exchanges of DNA of prophages and superinfecting phages (Henning and Hashemol-Hosseini, 1994; Tetart et al., 1998), as well as by partial duplications (Tetart et al., 1996). The mosaic pattern within tail fiber genes of different T-even phages is best explained by the mechanism depicted in Figure 13.2, assuming that relatively long hybrid DNA segments were formed and that the frequent apparent exchanges are due to partial heteroduplex repair, instead of frequent independent exchanges.

Is there other evidence for exchanges between P2 and T-even phages?

Both in P2 and T4 the tail fibers are attached to the baseplate at the bottom of the tail. Thus, the similarity of the T4 base plate outer wedge protein gp25 (Gruidl et al., 1988) with the base plate protein gpW of P2 (Haggård-Ljungquist et al., 1995) may have some functional significance, and we have considered the possibility that it may be related to horizontal gene transfer by the mechanism depicted in Figure 13.2. The amino acid similarity of these two proteins is limited to the central segment of the protein. Alignments of the corresponding DNA sequences show a similar pattern of apparent mismatches and compensating frameshift mutations (Figure 13.4 (see color plates)) as in the dCTPase genes 56 of the T-even phages (Gary et al., 1998; Mosig et al., 2001). Thus, we consider the possibility that one can use such patterns to deduce horizontal gene transfer in situations when neither the average AT content, nor the nucleotide signature (Karlin et al., 1998) of the putative transferred segment are significantly different.

IS HOMOLOGOUS RECOMBINATION A COMMON MECHANISM FOR HORIZONTAL GENE TRANSFER?

Three common aspects of the T-even phages are especially conducive to the acquisition of foreign genes by homologous recombination.

(1) A high recombination potential is essential for continuous DNA replication and successful progeny production of the T-even phages, and all DNA replication, recombination and repair enzymes are encoded in the phage genome (Mosig, 1998).

(2) The T-even enzymes endonuclease II and IV attack cytosine-containing DNA, i.e. foreign or host DNA including resident prophage DNA. Particularly, the sequence preference of endonuclease II qualifies it as a restriction enzyme that can process host DNA to fragments suitable for horizontal transfer (Carlson et al., 1994; Carlson and Kosturko, 1998).

(3) Large terminal redundancies (~5 kb) at the ends of packaged T-even chromosomes (Streisinger, 1966), can be reduced, or auxiliary genes can be deleted (Homyk and Weil, 1974), without loss of viability, when new genes ("morons") are acquired by horizontal transfer.

These properties, taken together, may enhance the role of homologous recombination (ascompared with site-specific recombination or transposition) in horizontal gene transfer during evolution of the T-even phages. However, there is no reason not to surmise a considerable role of homologous recombination during evolution of other organisms.

ACKNOWLEDGMENTS

Supported by NSF grant MCB 9983568 to GM and by NIH grant RO1 AI 08722 to RC.

REFERENCES

Bhattacharyya, S.P. and Rao, V.B. (1994) Structural analysis of DNA cleaved *in vivo* by bacteriophage T4 terminase. *Gene* 146: 67–72.

Botstein, D. and Herskowitz, I. (1974) Properties of hybrids between Salmonella phage P22 and coliphage lambda. *Nature* 251: 584–589.

Calendar, R., Yu, S., Myung, H. et al. (1998) The lysogenic conversion genes of coliphage P2 have unusually high AT content. In *Horizontal Gene Transfer* (ed., M. Syvanen), pp. 241–252, Chapman and Hall, London.

Carlson, K. and Kosturko, L.D. (1998) Endonuclease II of coliphage T4: a recombinase disguised as a restriction endonuclease? *Mol. Microbiol.* 27: 671–676.

Carlson, K., Raleigh, E.A. and Hattman, S. (1994) Restriction and modification. In *Molecular Biology of Bacteriophage T4* (eds, J. Karam, J.W. Drake, K.N. Kreuzer et al.), pp. 369–381, American Society for Microbiology, Wahington, DC.

Cohen, D. (1959) A variant of phage P2 originating in *Escherichia coli*, strain B. *Virology* 7: 112–126.

Franklin, J.L., Haseltine, D., Davenport, L. and Mosig, G. (1998) The largest (70kDa) product of the bacteriophage T4 terminase gene *17* binds to single-stranded DNA segments and digests them towards junctions with double-stranded DNA. *J. Mol. Biol.* 277: 541–557.

Gary, T.P., Colowick, N.E. and Mosig, G. (1998) A species barrier between bacteriophages T2 and T4: exclusion, join-copy and join-cut-copy recombination and mutagenesis in the dCTPase genes. *Genetics* 148: 1461–1473.

Gruidl, M.E., Canan, N.C. and Mosig, G. (1988) Bacteriophage T4 gene *25*. *Nucleic Acids Res.* 16: 9862.

Haggard-Ljungquist, E.C., Halling, C. and Calendar, R. (1992) DNA sequence of the tail fiber genes of bacteriophage P2: evidence for horizontal gene transfer of tail fiber genes among unrelated bacteriophages. *J. Bacteriol.* 174: 1462–1477.

Haggard-Ljungquist, E., Jacobsen, E., Rishovd, S. et al. (1995) Bacteriophage P2: genes involved in baseplate assembly. *Virology* 213: 109–121.

Hendrix, R.W. and Duda, R.L. (1992) Bacteriophage lambda PaPa: not the mother of all lambda phages. *Science* 258: 1145–1148.

Hendrix, R.W., Smith, M.C., Burns, R.N. et al. (1999) Evolutionary relationships among diverse bacteriophages and prophages: all the world's a phage. *Proc. Natl Acad. Sci. USA* 96: 2192–2197.

Henning, U. and Hashemol-Hosseini, S. (1994) Receptor recognition by T-even-type coliphages. In *Molecular Biology of Bacteriophage T4* (eds, J.D. Karam, J.W. Drake, K.N. Kreuzer et al.), pp. 291–298, American Society for Microbiology, Washington, DC.

Homyk, T. and Weil, J. (1974) Deletion analysis of two non-essential regions of the T4 genome. *Virology* 61: 505–523.

Juhala, R.J., Ford, M.E., Duda, R.L. et al. (2000) Genomic sequences of bacteriophages HK97 and HK022: pervasive genetic mosaicism in the lambdoid bacteriophages. *J. Mol. Biol.* 299: 27–51.

Karlin, S., Campbell, A.M. and Mra'zek, J. (1998) Comparative DNA analysis across diverse genomes. *Annu. Rev. Genet.* 32: 185–225.

Kemper, B. (1998) Branched DNA resolving enzymes (X-solvases). In *DNA Damage and Repair. Vol.1: DNA Repair in Prokaryotes and Lower Eukaryotes* (eds, J.A. Nickoloff and M. Hoekstra), pp. 179–204, Humana Press, Totowa.

Kim, J.-S. and Davidson, N. (1974) Electron microscope heteroduplex study of sequence relations of T2, T4, and T6 bacteriophage DNAs. *Virology* 57: 93–111.

Kutter, E., Gachechiladze, K., Poglazov, A. et al. (1995) Evolution of T4-related phages. *Virus Genes* 11: 285–297.

Kutter, E.M. and Wiberg, J.S. (1968) Degradation of cytosin-containing bacterial and bacteriophage DNA after infection of *Escherichia coli* B with bacteriophage T4D wild type and with mutants defective in genes *46, 47* and *56*. *J. Mol. Biol.* 38: 395–411.

Lucchini, S., Desiere, F. and Brüssow, H. (1999) The genetic relationship between virulent and temperate *Streptococcus thermophilus cos*-site phages Sfi19 and Sfi21. *Virology* 260: 232–243.

Mosig, G. (1998) Recombination and recombination-dependent DNA replication in bacteriophage T4. *Annu. Rev. Genet.* 32: 379–413.

Mosig, G., Gewin, J., Luder, A. et al. (2001) Two recombination-dependent DNA replication pathways of bacteriophage T4, and their roles in mutagenesis and horizontal gene transfer. *Proc. Natl Acad. Sci. USA* 98: 8306–8311.

Sandmeier, H. (1994) Acquisition and rearrangement of sequence motifs in the evolution of bacteriophage tail fibres. *Mol. Microbiol.* 12: 343–350.

Sandmeier, H., Iida, S. and Arber, W. (1992) DNA inversion regions Min of plasmid p15B and Cin of bacteriophage P1: evolution of bacteriophage tail fiber genes. *J. Bacteriol.* **174**: 3936–3944.

Streisinger, G. (1966) Terminal redundancy, or all's well that ends well. In *Phage and the Origins of Molecular Biology* (eds, J. Cairns, G.S. Stent and J.D. Watson), pp. 335–340, Cold Spring Harbor Laboratory, Cold Spring Harbor, NY.

Tetart, F., Monod, C. and Krisch, H.M. (1996) Bacteriophage T4 host range is expanded by duplications of a small domain of the tail fiber adhesin. *J. Mol. Biol.* **258**: 726–731.

Tetart, F., Desplats, C. and Krisch, H.M. (1998) Genome plasticity in the distal tail fiber locus of the T-even bacteriophage: recombination between conserved motifs swaps adhesin specificity. *J. Mol. Biol.* **282**: 543–556.

Horizontal Transfer of Mismatch Repair Genes and the Variable Speed of Bacterial Evolution

Ivan Matic, Olivier Tenaillon, Guillaume Lecointre, Pierre Darlu,
Miroslav Radman, François Taddei and Erick Denamur

During adaptation, methyl directed-mismatch repair (MMR) deficient bacterial mutators can be selected for by their association with favorable mutations they generate. However, once adaptation is achieved, their high rate of deleterious mutations creates selective pressure favoring lower mutation rates. Because defects in MMR genes stimulate homologous recombination, they could also facilitate MMR restoration by horizontal gene transfer from related strains. Indeed, the hyper-recombination phenotype of MMR deficient alleles, measured in the laboratory, correlates with the number of inferred horizontal transfers within these genes. Therefore, the mosaic structure of MMR genes might be the hallmark of recurrent losses and reacquisitions of MMR functions. Such events signal alternations between states of high and low rates of generation of genetic variability in the evolutionary past of bacteria.

INTRODUCTION

The level of genetic variability that maximizes the evolvability of population varies with the degree of its adaptation to the environment. It is low when environment is stable and high when environment is unstable and hostile. Since environmental conditions are changing, the adaptation is never permanent. Consequently, because the genetic variability in bacteria is first generated by mutagenesis, it could be expected that populations with high mutation rates would have better chance for successful evolution. Indeed, it was observed that significant fraction of *Escherichia coli*, *Salmonella enterica* and *Pseudomonas aeruginosa* natural populations exhibit high mutation rates (Treffers et al., 1954; Jyssum, 1960; Gross and Siegel, 1981; Tröbner and Piechocki, 1984; LeClerc et al., 1996; Matic et al., 1997; Oliver et al., 2000). Experimental observations (Mao et al., 1997; LeClerc et al., 1998) and theoretical calculations (Johnson, 1999; Boe et al., 2000) on the frequency of strains with high mutation rates (mutators) in bacterial populations at the mutation/selection equilibrium indicate that the frequency of mutators observed among natural isolates is quite high, suggesting the presence of conditions selecting for higher mutation rate in nature.

It has been experimentally demonstrated that the fraction of *E. coli* mutator cells can increase in bacterial populations under very strong or prolonged selection (antibiotic treatments (Mao et al., 1997) and adaptation to a new environment (Sniegowski et al., 1997)). Oliver et al. (2000) reported very high

Copyright © 2002 by Academic Press.
All rights of reproduction in any form reserved.

frequency (20%) of strong mutator *P. aeruginosa* strains isolated from lungs of cystic fibrosis (CF) patients. Those *P. aeruginosa* populations have been exposed to the challenges of the host immune defenses and antibiotic therapy for a long time. However, no mutators have been found among *P. aeruginosa* strains isolated from acutely infected non-CF patients. This last observation confirms the prediction that in the absence of a prolonged positive selective pressure, mutator cells should not accumulate. Mutators rise to a high frequency in bacterial populations under strong selective pressure through their association with the favorable mutations they generate, despite the cost caused by the load of deleterious mutations (Taddei et al., 1997; Tenaillon et al., 1999).

When adaptation is achieved, the load of deleterious mutations counterselects high-mutation rate mutations (Taddei et al., 1997). Moreover, continuous production of deleterious mutations by mutator cells leads to a fast genome erosion when populations pass through severe bottlenecks (Funchain et al., 2000). However, this phenomenon may be slow in big populations in a constant environment (Cooper and Lenski, 2000). It can be concluded that the strength of selection for lower mutation rates depends on numerous parameters. Therefore, adaptive alleles generated in mutator background will not necessarily be rapidly lost from population. They can be saved either through their transfer to non-mutator background or when the mutation rate of adapted mutator strain is reduced before the load of deleterious mutations became too high. The reduction of mutation rate might be achieved by several mechanisms: a reversion of the mutator mutation (Taddei et al., 1997), acquisition of suppressor mutations (e.g. in population of *mutT⁻* cells (Tröbner and Piechocki, 1984)), or reacquisition of the functional anti-mutator gene *via* horizontal gene exchange. We undertook to test the latter hypothesis (Denamur et al., 2000).

Most frequently found mutators in nature are methyl-directed mismatch repair (MMR) deficient (LeClerc et al., 1996; Matic et al., 1997; Oliver et al., 2000) which, unlike other mutators, have also increased capacity for recombination

between diverged DNA's (Rayssiguier et al., 1989). It follows that the very defect in a given MMR gene, besides increasing mutation rate, may increase the probability of horizontal transfer of any gene including reacquisition of its functional allele from a related but diverged bacterium. By studying the sequence polymorphism, phylogeny and mosaic patterns of MMR genes from *E. coli* natural isolates, we tested the hypothesis that, in the course of evolution, bacteria alternate between mutator and non-mutator states through loss and reacquisition of MMR genes.

METHYL-DIRECTED MISMATCH REPAIR SYSTEM

The primary function of MMR system is the control of genetic variability. Probably the most important role of MMR is control of the fidelity of DNA replication. Four genes have been shown to be required specifically for MMR in *E. coli*: *mutS*, *mutL*, mut*H* and *mutU* (*uvrD*) (see Friedberg et al. (1995) and references therein). MutS protein recognizes and binds to unpaired and mispaired nucleotides in duplex DNA. MutL associates with MutS-mismatch complex and activates MutH, which incises the newly synthesized strand at a hemimethylated 5'-GATC-3' sites. GATC sites are hemimethylated because methylation of adenine by Dam methylase lags behind replication by several minutes. MutH thus directs excision and resynthesis to the newly synthesized strand. Neither MutH nor a non-methylated strand is required if a single strand break is present in the substrate DNA. The helicase II (MutU), unwinds the DNA allowing excision and resynthesis. The inactivation of any of four MMR genes gives identical strong mutator phenotypes (Glickman and Radman, 1980; Schaaper and Dunn, 1987). MMR⁻ mutants have 10^2–10^3-fold increased rate of transition (both G:C $\rightarrow$ A:T and A:T $\rightarrow$ G:C) and frameshift (insertion/deletion of a few nucleotides) mutations, compared with MMR proficient bacteria.

Mismatched base pairs in DNA can arise also during recombination between nonidentical DNA sequences. MMR recognizes such structures and acts as an inhibitor of recombination between diverged DNA sequences as it has been shown by *in vitro* and *in vivo* experiments (for review see Matic et al. (1996) and references therein). This function of MMR may play role in maintenance of structural integrity of chromosomes. For example, the inactivation of *mutS* genes increases 10–15-fold the frequency of chromosomal duplications resulting from recombination between *E. coli rhsA* and *rhsB* loci (0.9% divergent) (Petit et al., 1991) and about 10^3-fold gene conversion between *S. enterica* (serovar Typhimurium) *tufA* and *tufB* genes (about 1% divergent) (Abdulkarim and Hughes, 1996).

MMR system controls also conjugational and transductional recombination between divergent strains and species (Rayssiguier et al., 1989; Matic et al., 1995; Vulic et al., 1997; Zahrt and Maloy, 1997). Even low divergence is sufficient to impede recombination. For example, transductional recombination between two serovars (Typhimurium and Typhi) of *S. enterica*, whose genomes differ only 1–2% at DNA sequence level, increase 10^2–10^3-fold in MMR deficient genetic backgrounds (Zahrt and Maloy, 1997). However, unlike identical effect on mutagenesis, the inactivation of different MMR genes has distinct and characteristic effect on the interspecies recombination (Rayssiguier et al., 1989; Stambuk and Radman, 1998; Denamur et al., 2000). Very strong hyper-recombination effect is observed upon inactivation of *mutS* and *mutL* genes. On the contrary, the overproduction of MutS and MutL proteins severely reduces recombination frequency even between bacterial strains with very low genomic divergence (Vulic et al., 1997). The hyper-recombinagenic effect of inactivation of *mutU* and *mutH* genes is relatively weak. The presence of single strand breaks in recombination intermediates can explain the weak effect of *mutH* gene inactivation, while helicase II is probably replaced by other enzymes that are involved in recombination, e.g. RecG. The extensive knowledge of the role of MMR enzymes in control of recombination efficiency allowed us to test for their impact on gene flow in natural populations (Denamur et al., 2000).

MISMATCH REPAIR DEFICIENT MUTANTS IN NATURE

MMR deficient strains are most frequently found mutators in bacterial natural populations (LeClerc et al., 1996; Matic et al., 1997; Oliver et al., 2000). Such abundance can be explained either by: 1) the high frequency of generation of MMR⁻ alleles, 2) the hypothesis that MMR⁻ alleles are close to neutral and therefore enriched in some bacterial populations through genetic drift or 3) positive selection for high mutation rates (Taddei et al., 1997). The first hypothesis seems not plausible since the rate of mutations that inactivate MMR genes is not higher than that inactivating housekeeping genes such as *hisD* and *lacI* (measured in the laboratory) (Drake, 1991 (and references therein); Boe et al., 2000). In short-term cultures of *E. coli* and *S. enterica* laboratory strains grown without particular selective pressure, the frequency of MMR⁻ cells is about 10^{-5} and 10^{-6}, respectively (Mao et al., 1997; LeClerc et al., 1998). In addition, it has been demonstrated in the laboratory that the fitness of MMR⁻ is lower than that of wild-type cells (Chao and Cox, 1983; Boe et al., 2000), indicating that MMR⁻ alleles cannot be considered as neutral. This leaves the third hypothesis.

Inactivation of over 20 different *E. coli* genes can confer mutator phenotypes of different strength (Miller, 1996; Horst et al., 1999), but among strong mutators, only MMR deficient alleles appear to be present at high frequency in nature. One explanation for this phenomenon is that inactivation of the other genes involved in important aspects of DNA metabolism (e.g. replication) may have too high a cost to be compensated by advantageous mutations. Even different MMR defective alleles are not found with the same frequency. An example is provided by *mutL*⁻ alleles which are less abundant in nature than *mutS*⁻ populations (LeClerc et al., 1996; Matic et al., 1997). Yet, inactivation of these genes has indistinguishable effect on spontaneous mutagenesis and homologous recombination (Glickman and Radman, 1980; Schaaper and Dunn, 1987; Rayssiguier et al., 1989). In contrast to the under-representation of *mutL*⁻ strains in nature, *mutL*⁻ alleles can be isolated in the laboratory as readily as *mutS⁻* in the absence of natural selection and competition (Brégeon et al., 1999). The genomic environment of the *mutL* gene may provide a clue for this disparity. The *mutL* gene is part of a superoperon with a complicated arrangement of seven genes that mediate several important cellular processes, e.g. DNA repair, tRNA modification, proteolysis and pleiotropic regulation (Tsui et al., 1994). Mutations within the *mutL* gene may be counterselected in nature due to interference with the transcription of other genes of this complex operon.

The specific advantage of MMR deficient strains over other mutators (for example *mutT⁻*), which might also explain their abundance in nature, is their hyper-recombination phenotype. Recombination can increase adaptability by increasing genetic variability (Crow and Kimura, 1965). Consequently, it might be expected that genotypes with increased recombination rates might be selected for by virtue of the favorable genotypes they generate by associating beneficial mutations that have appeared, in different individuals (Otto and Michalakis, 1998). Moreover, recombination can alleviate the load of deleterious mutations by separating the rare favorable mutations from deleterious ones (Peck, 1994). Therefore, it may not be pure coincidence that some genetic systems, e.g. SOS and MMR, simultaneously increase or decrease mutagenesis and recombination, in relation to the degree of adaptation to environmental conditions.

HORIZONTAL TRANSFER OF MISMATCH REPAIR GENES

The frequency of detectable horizontal gene transfers in natural populations is dependent on the rate of transfer from donor to recipient cell, recombination rate and the fitness of the recombinants. Although it is likely that all chromosomal regions have a similar probability of being exchanged, the selection acting on a given recombinant gene depends on its function and the interactions of its product with other cellular components. For example, it has been observed that genes encoding proteins participating in

transcription and translation are less likely to be successfully transferred than other housekeeping genes (Jain et al., 1999). Housekeeping genes are exchanged at lower rates than genes encoding surface antigens and antibiotic resistance targets whose variability provides an immediate selective advantage to bacteria exposed to the immune system and antibiotics (Maynard-Smith et al., 1991; Marklund et al., 1992; Li et al., 1994). It has been reported previously that some antimutator genes like *mutD* (*dnaQ*), *mutH* and *mutT* may have been acquired by *E. coli* K12 from an organism with a different codon preference (Médigue et al., 1991). However, prior to our study (Denamur et al., 2000), the rate of horizontal transfer of MMR genes has been unknown.

Phylogeny of mismatch repair genes

Horizontal transfers of chromosomal genes between strains can be revealed by incongruence between individual gene and strain phylogenies (Dykhuizen and Green, 1991; Bull et al., 1993). We have looked for such incongruence by comparing the phylogenies of MMR genes with the strain phylogeny in a representative set of 30 *E. coli* natural isolates from the ECOR collection (Ochman and Selander, 1984). In addition, as a control, two more genes have been analysed: *mutT* gene whose inactivation increases spontaneous mutagenesis and does not stimulate recombination (Fowler and Schaaper, 1997; Denamur et al., 2000) and *recD* gene whose inactivation does not increase spontaneous mutagenesis but stimulates recombination (Lovett et al., 1988; Zahrt and Maloy, 1997). Phylogenies obtained by the maximum likelihood method showed significant incongruence (ILD test of Farris et al. (1995)) of *mutS*, *mutL* and *mutU* genes with the whole genome reference dataset (Desjardins et al., 1995) it was nearly significant for *mutH* gene, but it was not significant for *mutT* and *recD* genes (Denamur et al., 2000). To estimate the extent of incongruence due to recombination, single strains or combinations of strains were progressively removed, and the analysis repeated, until the incongruence was no longer significant. For *mutS*, *mutL* and *mutU*, at least eight, three, and three strains respectively have to be

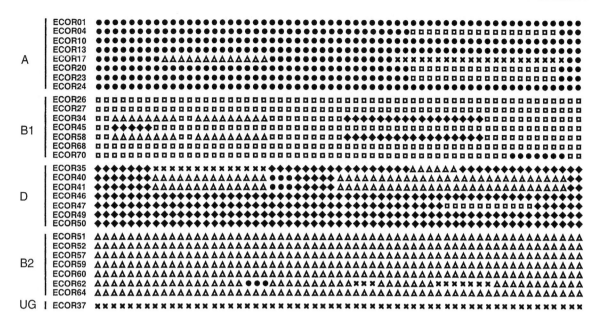

FIGURE 14.1 Mosaicism of *E. coli mutS* genes. Alignment of the informative sites of *E. coli mutS* gene segments from ECOR reference collection based on the strain phylogeny. Each symbol represents one nucleotide from genes belonging to A (●), B1 (□), D (◆) and B2 (△) phylogenetic groups, as well as one ungrouped strain (✘). Transferred segments among different genes were defined by at least three bases delineating a stretch of DNA which is different from that of the sequences of the other strains belonging to the same phylogenetic group, but which is identical to those of another phylogenetic group. It was considered that one transfer event is more parsimonious than at least three independent mutational events occurring within the corresponding DNA stretch. Ungrouped strain has been considered only as a donor and not as a recipient due to its unresolved phylogenetic position.

removed to raise the P-value above the significance threshold.

However, the incongruence observed for MMR genes is not as high as those observed for highly diversified genes like those coding for surface antigens. For example, to relieve incongruence of the *gnd* locus, which reflects a strong selective pressure exerted by the host immune system on the closely linked O antigen complex, 28 out of 34 strains must be removed to eliminate incongruence (data not shown). The reason for this difference might be that, in the case of MMR genes, the selective pressure is on restoration of MMR functions rather than on diversification of MMR genes and protein sequences. Indeed, most of the sequence polymorphism found in this phylogenetic analysis of MMR genes is neutral (Denamur et al., 2000). Only 1–12% of base substitutions in MMR genes result in an amino acid change, which is similar to the average of 11.5% for housekeeping genes analyzed in this study. This is very low compared with approximately 80% amino acid changing mutations observed in the natural polymorphism of genes encoding major outer membrane protein in *Chlamydia trachomatis*, which is subjected to selective pressure by the immune system (Kaltenboeck et al., 1993).

Quantification of mosaicism of mismatch repair genes

To determine whether recombination at the MMR loci was significantly higher than that observed for housekeeping genes, we have quantified the transferred segments among genes from different phylogenetic lineages (Denamur et al., 2000). Transferred segments were identified by a computer algorithm based on aligned

DNA sequences and the previously established phylogeny (Lecointre et al., 1998). They were defined by at least three nucleotides, contiguous or not, delineating a fragment of DNA whose sequence is different from those of the strains belonging to the same phylogenetic group, but which is identical to those of another phylogenetic group. The datasets analyzed here exhibit particular properties allowing to infer horizontal transfers. Some positions along the DNA sequences show nucleotide substitutions, which do not fit correctly the grouping inferred from the phylogeny. They are rare and, when observed, usually not randomly distributed along the sequences but clustered together in a stretch favoring another contradictory grouping. This distinctive pattern leads us to retain the far more parsimonious hypothesis of transfer of part of a gene rather than the hypotheses of multiple correlated substitutions (Figure 14.1). Moreover, since the sequences differ more between main strain groups A, B1, D, B2 than within each group, the donor group can often be unambiguously identified.

All the incongruences detected by the ILD test in four MMR, as well as in *mutT* and *recD* genes, were identified as recombinational events. Some additional recombination events were identified that were not detected by the ILD test due to insufficient lengths of transferred stretches of DNA. Different MMR genes show different levels of intragenic recombination, but all of them have significantly higher level of mosaicism than the control group (*aceK, crr, gapA, icd, mdh, mutT, pabB, putP, recD, trpA, trpB* and *trpC* genes) even when more stringent criteria (five and four instead of three nucleotides) for detection of recombined fragments were used.

The majority of strains have detectable traces of recombination of at least one of these antimutator genes. Mosaic MMR alleles have been observed in strains belonging to all phylogenetic groups. No obvious correlation was established between the degree of MMR gene mosaicism and the presence of pathogenic determinants (Bingen et al., 1998; Boyd and Hartl, 1998), geography or the bacterial host (Ochman and Selander, 1984). Out of 30 analyzed strains, 18 showed detectable traces of recombination within MMR genes. Moreover, seven, five and five strains showed recombination in one, two and three different MMR genes, respectively. ECOR 37 strain, with the highest level of mosaicism within MMR genes (*mutS*: four transfers, *mutL*: two transfers) seems to be a special case. This strain belongs to ungroup (UG) strains that do not have a clear phylogenetic position within the *E. coli* species, probably because they have highly mosaic genomes (Lecointre et al., 1998). These observations suggest that the high mosaicism of MMR genes is a general characteristic of the *E. coli* species, and is not limited to a group of strains.

CONCLUSIONS

Different MMR genes show different numbers of inferred intragenic recombination events: *mutS* and *mutL* exhibit higher levels of sequence mosaicism than *mutU* and *mutH* (Denamur et al., 2000). Similary, the inactivation of *mutS* and *mutL* genes result in much stronger hyper-recombination phenotype than the inactivation of *mutU* and *mutH*. The highly significant correlation between these two variables strongly supports the hypothesis that the inferred recombination events accounting for the mosaic structure of the MMR genes occurred under MMR deficient conditions. The restoration of MMR functions by recombination with an identical gene (which is expected to be the most efficient recombination) leaves no detectable trace at the DNA sequence level and may also often erase a mosaic gene by replacing it. Thus, the rates of loss and reacquisition of MMR functions inferred from this study must be greatly underestimated.

Different strains have different evolutionary histories, but most of them apparently repeatedly pass through the periods of loss and reacquisition of MMR functions. Therefore, brief episodes of accelerated adaptive evolution driven by hypermutation and hyper-recombination are probably periodically experienced by all bacterial populations. High mutation and recombination rates are subject to indirect selection via direct natural selection of the favorable genetic modifications they create.

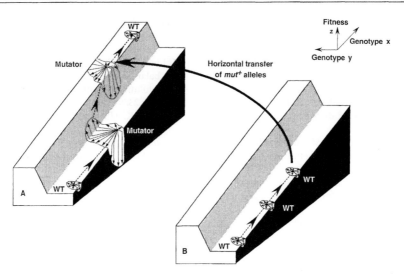

FIGURE 14.2 The role of mutators in adaptive evolution of bacteria. These figures represent bacterial populations (presented as "amoebas") climbing a fitness landscape. The genotypes can be modified along x and y axis and the fitness is represented on the z axis. The trajectory of the most common genotypes is represented as a dashed line. The sizes of the "amoebas" reflect the mutation rate of the populations and are proportional to the sequence space explored. In this simplified drawing the drift along the y axis, which may allow the population to get closer to the grey hill and then climb it, has been neglected because it is a slow and stochastic process. The selective pressure acting on the wild-type (WT) non-mutator population can select a mutator cells (e.g. mismatch repair defective mutants) which, by generating more mutations than WT cells along the y axis, have higher probability to generate mutations giving a higher fitness and thus start climbing the grey hill (A). However, at the end of the adaptive process, when no more adaptive mutations can be found, the mutator has a cost because it generates lots of deleterious mutations. Such mutator cells, as well as adaptive alleles they carry, can be "saved" by restoration of antimutator functions through reacquisition of the functional antimutator gene *via* horizontal gene exchange from WT cells (B). The hyper-rec phenotype of mismatch repair deficient mutators facilitates this process.

Hence, the fate of mutator cells depends on their adaptation to environmental conditions. They can facilitate and accelerate the adaptation of bacterial populations to new and/or changing environments, such as coloniszation of new hosts, acquisition of pathogenic determinants and antibiotic resistance. However, once the bacterial population is adapted, the ever increasing genetic load of deleterious mutations will favor restoration of the antimutator functions often acquired through genetic exchange (Figure 14.2). Consequently, the evolutionary success of bacteria may largely result from an optimized interplay of mutation and recombination. Therefore, it may be that the global mosaic structure of *E. coli* genome (Milkman and McCane, 1995; Guttman, 1997) represents a structural testimony of such overlapping vertical and horizontal genetic variation.

ACKNOWLEDGMENTS

This work was supported by grants from the Programme de Recherche Fondamentale en Microbiologie et Maladies Infectieuses et Parasitaires – MENRT, the Ligue contre le Cancer, the Programme Environnement et Santé – MATE, and the Association pour la Recherche sur le Cancer.

REFERENCES

Abdulkarim, F. and Hughes, D. (1996) Homologous recombination between the *tuf* genes of *Salmonella typhimurium*. *J. Mol. Biol.* **260**: 506–522.

Bingen, E., Picard, B., Brahimi, N. et al. (1998) Phylogenetic analysis of *Escherichia coli* strains causing neonatal meningitis suggests horizontal gene transfer from a predominant pool of highly virulent B2 group strains. *J. Infect. Dis.* **177**: 642–650.

Boe, L., Danielsen, M., Knudsen, S. et al. (2000) The frequency of mutators in populations of *Escherichia coli*. *Mutat. Res.* **448**: 47–55.

Boyd, E.F. and Hartl, D.L. (1998) Chromosomal regions specific to pathogenic isolates of *Escherichia coli* have a phylogenetically clustered distribution. *J. Bacteriol.* **180**: 1159–1165.

Brégeon, D., Matic, I., Radman, M. and Taddei, F. (1999) Inefficient mismatch repair: genetic defects and down regulation. *J. Genet.* **78**: 21–28.

Bull, J.J., Huelsenbeck, J.P., Cunningham, C.W. et al. (1993) Partitioning and combining data in phylogenetic analysis. *Syst. Biol.* **42**: 384–397.

Chao, L. and Cox, E.C. (1983) Competition between high and low mutating strains of *Escherichia coli*. *Evolution* **37**: 125–134.

Cooper, V.S. and Lenski, R.E. (2000) The population genetics of ecological specialization in evolving *Escherichia coli* populations. *Nature* **407**: 736–739.

Crow, J.F. and Kimura, M. (1965) Evolution in sexual and asexual populations. *Am. Nat.* **99**: 439–450.

Denamur, E., Lecointre, G., Darlu, P. et al. (2000) Evolutionary implications of the frequent horizontal transfer of mismatch repair genes. *Cell* **103**: 711–721.

Desjardins, P., Picard, B., Kaltenbock, B. et al. (1995) Sex in *Escherichia coli* does not disrupt the clonal structure of the population: evidence from random amplified polymorphic DNA and restriction-fragment-length polymorphism. *J. Mol. Evol.* **41**: 440–448.

Drake, J.W. (1991) A constant rate of spontaneous mutation in DNA-based microbes. *Proc. Natl Acad. Sci. USA* **88**: 7160–7164.

Dykhuizen, D.E. and Green, L. (1991) Recombination in *Escherichia coli* and the definition of biological species. *J. Bact.* **173**: 7257–7268.

Farris, J.S., Källersjö, M., Kluge, A.G. and Bult, C. (1995) Testing significance of incongruence. *Cladistics* **10**: 315–319.

Fowler, R.G. and Schaaper, R.M. (1997) The role of *mutT* gene of *E. coli* in maintaining replication fidelity. *FEMS Microbiol. Rev.* **21**: 43–54.

Friedberg, E.C., Walker, G.C. and Siede, W. (1995) *DNA Repair and Mutagenesis*, ASM Press, Washington, DC.

Funchain, P., Yeung, A., Stewart, J.L. et al. (2000) The consequences of growth of a mutator strain of *Escherichia coli* as measured by loss of function among multiple gene targets and loss of fitness. *Genetics* **154**: 959–970.

Glickman, B.W. and Radman, M. (1980) *Escherichia coli* mutator mutants deficient in methylation-instructed DNA mismatch correction. *Proc. Natl Acad. Sci. USA* **77**: 1063–1067.

Gross, M.D. and Siegel, E.C. (1981) Incidence of mutator strains in *Escherichia coli* and coliforms in nature. *Mutat. Res.* **91**: 107–110.

Guttman, D.S. (1997) Recombination and clonality in natural populations of *Escherichia coli*. *Trends Ecol. Evol.* **12**: 16–22.

Horst, J.P., Wu, T.H. and Marinus, M.G. (1999) *Escherichia coli* mutator genes. *Trends Microbiol.* **7**: 29–36.

Jain, R., Rivera, M.C. and Lake, J.A. (1999) Horizontal gene transfer among genomes: the complexity hypothesis. *Proc. Natl Acad. Sci. USA* **96**: 3801–3806.

Johnson, T. (1999) The approach to mutation-selection balance in an infinite asexual population, and the evolution of mutation rates. *Proc. R. Soc. Lond. B. Biol. Sci.* **266**: 2389–2397.

Jyssum, K. (1960) Observation of two types of genetic instability in *Escherichia coli*. *Acta Pathol. Microbiol. Immunol. Scand.* **48**: 113–120.

Kaltenboeck, B., Kousoulas, K.G. and Storz, J. (1993) Structures of and allelic diversity and relationships among the major outer membrane protein (*ompA*) genes of the four chlamydial species. *J. Bacteriol.* **175**: 487–502.

LeClerc, J.E., Li, B., Payne, W.L. and Cebula, T.A. (1996) High mutation frequencies among *Escherichia coli* and *Salmonella* pathogens. *Science* **274**: 1208–1211.

LeClerc, J.E., Payne, W.L., Kupchella, E. and Cebula, T.A. (1998) Detection of mutator subpopulations in *Salmonella typhimurium* LT2 by reversion of his alleles. *Mutat. Res.* **400**: 89–97.

Lecointre, G., Rachdi, L., Darlu, P. and Denamur, E. (1998) *Escherichia coli* molecular phylogeny using the incongruence length difference test. *Mol. Biol. Evol.* **15**: 1685–1695.

Li, J., Nelson, K., McWhorter, A.C. et al. (1994) Recombinational basis of serovar diverity in *Salmonella enterica*. *Proc. Natl Acad. Sci. USA* **91**: 2552–2556.

Lovett, S.T., Luisi-DeLuca, C. and Kolodner, R.D. (1988) The genetic dependence of recombination in *recD* mutants of *Escherichia coli*. *Genetics* **120**: 37–45.

Mao, E.F., Lane, L., Lee, J. and Miller, J.H. (1997) Proliferation of mutators in a cell population. *J. Bacteriol.* **179**: 417–422.

Marklund, B., Tennent, J.M., Garcia, E. et al. (1992) Horizontal gene transfer of the *Escherichia coli pap* and *prs* pili operons as a mechanism for the development of tissue-specific adhesive properties. *Mol. Microbiol.* **6**: 2225–2242.

Matic, I., Rayssiguier, C. and Radman, M. (1995) Interspecies gene exchange in bacteria: the role of SOS and mismatch repair systems in evolution of species. *Cell* **80**: 507–515.

Matic, I., Taddei, F. and Radman, M. (1996) Genetic barriers among bacteria. *Trends Microbiol.* **4**: 69–73.

Matic, I., Radman, M., Taddei, F. et al. (1997) Highly variable mutation rates in commensal and pathogenic *E. coli*. *Science* **277**: 1833–1834.

Maynard-Smith, J., Dawson, C.G. and Spratt, B.G. (1991) Localized sex in bacteria. *Nature* **349**: 29–31.

Médigue, C., Rouxel, T., Vigier, P. and Hénaut, A. (1991) Evidence for horizontal gene transfer in *Escherichia coli* speciation. *J. Mol. Biol.* **222**: 851–856.

Milkman, R. and McCane, M. (1995) DNA sequence variation and recombination in *E. coli*. In *Population Genetics of Bacteria* (eds, S. Baumberg, J.P.W. Young, E.M.H. Wellington and J.R. Saunders), pp. 127–142, Cambridge University Press, Leicester.

Miller, J.H. (1996) Spontaneous mutators in bacteria: insight into pathways of mutagenesis and repair. *Annu. Rev. Microbiol.* **50**: 625–643.

Ochman, H. and Selander, R.K. (1984) Standard reference strains of *Escherichia coli* from natural population. *J. Bacteriol.* **157**: 690–693.

Oliver, A., Canton, R., Campo, P. et al. (2000) High frequency of hypermutable *Pseudomonas aeruginosa* in cystic fibrosis lung infection. *Science* **288**: 1251–1254.

Otto, S.P. and Michalakis, Y. (1998) The evolution of recombination in changing environments, *Trends Ecol. Evol.* **13**: 145–151.

Peck, J.R. (1994) A ruby in the rubbish: beneficial mutations, deleterious mutations and the evolution of sex. *Genetics* **137**: 597–606.

Petit, M.A., Dimpfl, J., Radman, M. and Echols, H. (1991) Control of chromosomal rearrangements in *E. coli* by the mismatch repair system. *Genetics* **129**: 327–332.

Rayssiguier, C., Thaler, D.S. and Radman, M. (1989) The barrier to recombination between *Escherichia coli* and *Salmonella typhimurium* is disrupted in mismatch-repair mutants. *Nature* **342**: 396–401.

Schaaper, R.M. and Dunn, R.L. (1987) Spectra of spontaneous mutations in *Escherichia coli* strains defective in mismatch correction: the nature of *in vivo* DNA replication errors. *Proc. Natl Acad. Sci. USA* **84**: 6220–6224.

Sniegowski, P.D., Gerrish, P.J. and Lenski, R.E. (1997) Evolution of high mutation rates in experimental populations of *E. coli. Nature* **387**: 703–705.

Stambuk, S. and Radman, M. (1998) Mechanism and control of interspecies recombination in *Escherichia coli*. I. Mismatch repair, methylation, recombination and replication functions. *Genetics* **150**: 533–542.

Taddei, F., Radman, M., Maynard-Smith, J. et al. (1997) Role of mutators in adaptive evolution. *Nature* **387**: 700–702.

Tenaillon, O., Toupance, B., Le Nagard, H. et al. (1999) Mutators, population size, adaptive landscape and the adaptation of asexual populations of bacteria. *Genetics* **152**: 485–493.

Treffers, H.P., Spinelli, V. and Belser, N.O. (1954) A factor (or mutator gene) influencing mutation rates in *E. coli. Proc. Natl Acad. Sci. USA* **40**: 1064–1071.

Tröbner, W. and Piechocki, R. (1984) Selection against hypermutability in *Escherichia coli* during long term evolution. *Mol. Gen. Genet.* **198**: 177–178.

Tsui, H.-C.T., Zhao, G., Feng, G. et al. (1994) The *mutL* repair gene of *Escherichia coli* K-12 forms a superoperon with a gene encoding a new cell-wall amidase. *Mol. Microbiol.* **11**: 189–202.

Vulic, M., Dionisio, F., Taddei, F. and Radman, M. (1997) Molecular keys to speciation: DNA polymorphism and the control of genetic exchange in enterobacteria. *Proc. Natl Acad. Sci. USA* **94**: 9763–9767.

Zahrt, T.C. and Maloy, S. (1997) Barriers to recombination between closely related bacteria: MutS and RecBCD inhibit recombination between *Salmonella typhimurium* and *Salmonella typhi. Proc. Natl Acad. Sci. USA* **94**: 9786–9791.

Eukaryotic Mobile Elements

Horizontal gene transfer is common in bacteria and is increasingly recognized to have been important in the evolution of the eukaryotic cell. But we still hear the assertion that it does not occur or is not important among eukaryotes. That is, at the very least, debatable. Some of the best examples of horizontal gene transfer have been seen with the transposable elements in metazoans. This section contains a detailed review by Clark et al. in Chapter 15 of the classic example of horizontal gene transfer observed with the DNA transposon *P* factor among the drosophilids. The case for the role of *P* factors in hybrid dysgenisis has resulted in the most rigorously documented example of horizontal gene transfer in a metazoan. The invasion of these elements into *Drosophila melanogaster* occurred in historical times, after geneticists began studying this organism. Thus we have been able to observe the invasion and spread of this element in real time, much as we have observed the invasion and spread of antibiotic resistance genes among pathogenic bacteria. Another excellent example of the likely horizontal spread of a transposable element involves *mariner* elements. *Mariner* is another DNA transposon that was found originally in *Drosophila* but was subsequently uncovered in many other metazoan phyla (see Chapter 16 by Robertson et al.). Phylogeny based on *Mariner* gene sequences is highly incongruent with the phylogeny of the underlying species. One of the more striking similarities, from a human perspective, is the close affinity between a class of human *mariners* and those found in nematodes.

There exists a second *Drosophila* element known as *hobo* that is responsible for a hybrid dysgenisis-like phenomenon and apparently represents another recent invasion into *Drosophila* (Simons et al., 1976). *Hobo* is a member of a larger group of DNA transposons known as hAT elements. These are found in plants, fungi and metazoa. Indeed, there are 195,000 copies of hAT members in the human genome (Lander et al., 2000). These DNA transposon elements are widely suspected to sustain themselves over time through horizontal transfer events because although they tend to decay once established in a genome, they nevertheless have managed to maintain a presence. There is one well-documented case of horizontal gene transfer of an hAT transposon in vertebrates. Koga et al. (2000) identified the Tol2 transposon as a hAT member in the medaka fish and concludes that this is the result of a recent gene transfer event. Tol2 was found in two species, but not eight other species, of the medaka fish, and the *Tol12* sequences found in the two species showed near identity. Thus, Koga et al. (2000) conclude that, in recent years, Tol2 invaded those two species well after their speciation separation.

The example of the transposons brings up an issue that has not been discussed up to now in this book. *Mariner* as well as *hobo* and *P* factor are members of multigene families, and these multiple copies arose from many gene duplication events. This raises the problem of comparing parologous genes with orthologous genes and mistaking a gene duplication event for a horizontal gene transfer event. An explanation of this problem will be helpful.

PAROLOGY AND ORTHOLOGY

A number of the premature reports of possible horizontal transfers resulted from comparing parologous genes and treating them as if they were orthologous. Two genes are defined as parologous if they diverged after gene duplication, whereas the two genes are defined as orthologous if they diverged after a speciation event. Let us construct the following scenario to illustrate this problem. Consider a gene in an ancestral strain that underwent a duplication to give rise to gene A and gene B, which subsequently diverged from one another. Further consider the evolution of that ancestral strain into two major branches. In the extant species, all descendants of gene A will belong to one orthologous set, and descendants of gene B will belong to another. Members within the gene A group are said to be parologous to members of the gene B group. If both sets can be identified in the extant species, this does not pose a problem. However, consider what happens if, during the evolution of these two major branches, the duplicated pair is reduced back to a single copy, and A survives in some extant species while B survives in others. This could result in both gene A and gene B being located in the two major clades and their parologous nature not being recognized. If a single tree is constructed, then the A–B gene tree would be incongruent with the species tree and hence a horizontal transfer incorrectly inferred. This scenario can also arise from what is essentially a sampling error, when too few genes and their incompletely characterized products are compared.

Comparisons of parologous genes were responsible for a number of early reports, and likely erroneous reports, of horizontal gene transfer. Therefore, when we are dealing with multigene families, a simple incongruency between a gene tree and an underlying species tree does not provide an adequate basis to conclude that a horizontal gene transfer event occurred. Additional evidence must be supplied. In the cases presented in the following three chapters, that evidence is based on molecular clock estimates. That is, two genes are just too closely related to one another to have diverged as long ago as the last speciation event. This type of argument makes a very strong case for both *P* factor and *mariner* involvement in numerous horizontal gene transfer events.

RETROTRANSPOSONS

There are other examples of transposable elements that have likely been involved in horizontal gene transfer events; these can be mentioned here since they are not covered in any systematic review. A major class of transposable elements in eukaryotes is known as retroposons. The name "retroposon" reflects on the molecular mechanism by which this class of elements moves. A DNA copy of the elements is first transcribed into RNA and these are then reverse transcribed into a DNA copy, which inserts into some chromosomal target. The mechanism of movement is related to the cycle of the chromosome replication observed for the retroviruses (and many of the retroposons are related to this important class of retroviruses and it is suspected that they probably serve as the transpecies carriers for these elements).

An important group of retroposons is called "*SINES*." This is a large group of short (i.e. ~300 bp) mobile elements that are often related to tRNA genes. They can be present in extremely high copy numbers per genome. For example, there are 1.5 million *SINES* found in the human genome, of which 1 million belong to the Alu group (Lander et al., 2001). In general *SINES* can serve as a useful genealogical marker for the underlying species because their inheritance is mostly vertical. A particularly instructive example was the finding of highly similar *SINES* in both whales and the hippopotamus that links these two species to the exclusion of the other artiodactyls (Nomura et al., 1998; Shimamura et al., 1999). If other molecular evidence were not available showing the cetaceans evolved from an ancestor shared with the hippopotamus, this close resemblance of these *SINES* could have been erroneously attributed to a horizontal gene transfer event. There are, however, at least two other cases where horizontal movement still remains the most parsimonious explanation for observed similarities. Bucci et al. (1999) has postulated at least one

SINE gene transfer between species in the frog genus *Rana*. And Ohshima et al. (1993) postulated that an unexpected similarity between an squid *SINE* and those found in vertebrates might be evidence for a trans-phyla gene movement. The distribution of *SINES* among different salmonid species and the related white fish has been postulated to arise from horizontal gene transfer events (Murata et al., 1993; Takasaki et al., 1997; Hamada et al., 1997). These possible cases of horizontal transfer by *SINES* are not strongly supported by rate arguments, but mostly by intriguing incongruities.

Another group of retroposons that appear to have leapt species boundaries are the *LINES*. This group of elements is about 2.5 kb in size and encodes its own enzyme that promotes integration of the *LINE* DNA element into new chromosomal locations. The one documented case shows that cows and snakes share a *LINE*; it is suggested that an artiodactyl ancestor received a reptilian *LINE* 40 Myr ago (Kordis and Gubensek, 1997, 1998).

REFERENCES

Bucci, S., Ragghianti, M., Mancino, G., Petroni, G., Guerrini. F. and Giampaoli, S. (1999) Rana/Pol III: a family of *SINE*-like sequences in the genomes of western Palearctic water frogs. *Genome* **42**: 504–411.

Hamada, M., Kido, Y., Himberg, M., Reist, J.D., Ying, C., Hasegawa, M. and Okada, N. (1997) A newly isolated family of short interspersed repetitive elements (*SINEs*) in coregonid fishes (whitefish) with sequences that are almost identical to those of the SmaI family of repeats: possible evidence for the horizontal transfer of *SINEs*. *Genetics*, **146**: 355–367.

Koga, A., Shimada, A., Shima, A., Sakaizumi, M., Tachida, H. and Hori H. (2000) Evidence for recent invasion of the medaka fish genome by the Tol2 transposable element. *Genetics* **155**: 273–281.

Kordis, D. and Gubensek, F. (1997) Bov-B long interspersed repeated DNA (LINE) sequences are present in *Vipera ammodytes* phospholipase A2 genes and in genomes of Viperidae snakes. *Eur. J. Biochem.* **246**: 772–779.

Kordis, D. and Gubensek, F. (1998) The Bov-B lines found in *Vipera ammodytes* toxic PLA2 genes are widespread in snake genomes. *Toxicon* **36**: 1585–1590.

Lander et al., International Human Genome Sequencing Consortium (2001) Initial sequencing and analysis of the human genome. *Nature* **409**: 860–921.

Murata, S., Takasaki, N., Saitoh, M. and Okada N. (1993) Determination of the phylogenetic relationships among Pacific salmonids by using short interspersed elements (*SINEs*) as temporal landmarks of evolution. *Proc. Natl Acad. Sci. USA* **90**: 6995–6999.

Nomura, O., Lin, Z.H., Muladno, Wada, Y. and Yasue, H. (1998) A *SINE* species from hippopotamus and its distribution among animal species. *Mammal. Genome* **9**: 550–555.

Ohshima, K., Koishi, R., Matsuo, M. and Okada, N. (1993) Several short interspersed repetitive elements (*SINEs*) in distant species may have originated from a common ancestral retrovirus: characterization of a squid *SINE* and a possible mechanism for generation of tRNA-derived retroposons. *Proc. Natl Acad. Sci. USA* **90**: 6260–6264.

Shedlock, A.M. and Okada, N. (2000) *SINE* insertions: powerful tools for molecular systematics. *Bioessays* **22**: 148–160.

Shimamura, M., Abe, H., Nikaido, M., Ohshima, K. and Okada N. (1999) Genealogy of families of *SINEs* in cetaceans and artiodactyls: the presence of a huge superfamily of tRNA(Glu)-derived families of *SINEs*. *Mol. Biol. Evol.* **16**: 1046–1060.

Takasaki, N., Yamaki, T., Hamada, M., Park, L. and Okada, N. (1997) The salmon SmaI family of short interspersed repetitive elements (*SINEs*): interspecific and intraspecific variation of the insertion of *SINEs* in the genomes of chum and pink salmon. *Genetics* **146**: 369–380.

Evidence for Horizontal Transfer of *P* Transposable Elements

Jonathan B. Clark, Joana C. Silva and Margaret G. Kidwell

Evidence is reviewed here for multiple instances of horizontal transfer of the *P* transposable element in *Drosophila* and related genera. More than a decade ago, the *Drosophila melanogaster P* element provided one of the earliest, and best documented, examples of eukaryotic horizontal transfer and subsequent invasion of a new species. Later, additional likely cases of *P* element horizontal transfer were inferred from incongruities between *P* element and species phylogenetic trees. The results of recent studies suggest that *P* element interspecies transfer has been rampant among some *Drosophila* species endemic to South America. Although interspecific hybridization may account for some fraction of transfers among closely related species, the mechanism(s) involved in more phylogenetically distant transfers is unknown.

INTRODUCTION

Although difficult to document rigorously, there is some evidence that eukaryotic transposable genetic elements may be more susceptible to horizontal gene transfer than eukaryotic host-derived genes that are transmitted strictly by Mendelian inheritance (Kidwell, 1993). This is not surprising because transposable elements have the obvious advantage of possessing the molecular machinery that facilitates such transfer. Furthermore, it has been argued that horizontal transfer to new hosts may be essential for the survival of some transposable elements over long periods of evolutionary time (Hurst et al., 1992, Kidwell, 1992, Lohe et al., 1995).

Evidence for the horizontal transfer of a number of transposable element families has accumulated during the last two decades. Prominent among these are several members of the Class II elements, which transpose by means of a DNA intermediate. This group of elements includes *P* elements (described in this chapter), *mariner* elements (see Chapter 16) and *hobo* elements. Recent evidence indicates that *copia* (Jordan et al., 1999), *Penelope* (Evgen'ev et al., 2000), and other Class I elements which use reverse transcriptase for transposition, have also been transferred horizontally between species.

Males from strains of *D. melanogaster* that carry full-sized *P*, *I*, or *hobo* elements produce a very striking phenotype when mated with females lacking the respective element. A number of abnormal traits are often seen in the F1 or F2 hybrids of these matings, including mutations, increased recombination, chromosomal rearrangements, and reduced fertility and viability. The phenotypes of the offspring from the reciprocal cross are usually normal, but the paternal chromosomes of these normal looking hybrids will frequently be observed to pick up elements as a result of transposition. This phenomenon is called hybriddysgenesis. The high rate of *P*-element transposition in dysgenic hybrids of this outcrossing species probably accounts for the world-wide conservation of all *D. melanogaster*

Copyright © 2002 by Academic Press.
All rights of reproduction in any form reserved.

populations to the P+ condition within the last century.

The purpose of this chapter is to review the evidence for *P* element horizontal transfer that has accumulated during the last two decades. There are two main advantages to using *P* elements to study horizontal transfer. First, the *P* element is one of the best described eukaryotic transposable elements (Engels, 1989), and provided one of the earliest reports of transposable element horizontal transfer (Daniels et al., 1990) and subsequent invasion of a naive contemporary species (Kidwell, 1979; Anxolabéhère et al., 1988). Second, the evolutionary history of the genus *Drosophila* has been well-studied and reasonable estimates exist for species divergence times. This provides a context in which to evaluate the timing of horizontal transfer events.

Overview of *P* element distribution in *Drosophila* and related genera

In contrast to *mariner* elements, the distribution of the *P* family of elements is restricted to the order Diptera. This restricted host range may be due to the requirement of host factors for transposition (Beall et al., 1994). *P*-homologous sequences identified to date are mostly concentrated in species of the subgenus *Sophophora* in the genus *Drosophila* which includes four principal species groups (Figure 15.1). The ancestral sophophoran arose in tropical Africa and subsequently diverged in the Old World tropics, leading to the *melanogaster* and *obscura* species groups (Throckmorton, 1975). The two New World species groups, *saltans* and *willistoni*, probably arose in tropical North America from a single lineage and subsequently diversified in South America (Throckmorton, 1975).

Within the subgenus *Sophophora*, *P* element sequences have been detected in all species of the *willistoni* and *obscura* species groups and in most species of the *saltans* species group. Their distribution in the *melanogaster* species group is somewhat patchy, and in species outside of the subgenus *Sophophora*, it is both sparse and patchy (Anxolabéhère et al., 1985; Lansman et al., 1985; Daniels et al., 1990). Weak hybridization signals in Southern blots have been

observed in two species of the *quinaria* group and three species from the *immigrans* radiation (subgenus *Drosophila*). A *P* element with very high similarity to elements in species of *Sophophora* has been found in one species of the *tripunctata* group (subgenus *Drosophila*) (Loreto et al., 1998), *P* elements have also been detected in a few species of the subgenus *Scaptodrosophila*, in *Drosophila busckii* (subgenus *Dorsilopha*), and in drosophilids of the genera *Liodrosophila*, *Lordiphosa*, and *Scaptomyza* (Anxolabéhère et al., 1985; Lansman et al., 1985; Daniels et al., 1990; Haring et al., 2000). To date, *P*-homologous sequences have been reported in only two non-drosophilids, the blow fly, *Lucilia cuprina* (Perkins and Howells, 1992) and the house fly, *Musca domestica* (Lee et al., 1999). Since these two sequences are quite divergent from drosophilid *P* element sequences and since sampling has been limited, it is difficult to determine the origin of these two elements. They may represent ancient elements that were at one time found in several Dipteran species, or they may have been transferred to these taxa horizontally at some time in the past.

Overview of *P* element structures

P elements are members of the Class II transposable elements (Finnegan, 1989), which replicate by means of a DNA intermediate. Like many other Class II elements they can be divided into two structural types, autonomous elements and non-autonomous, usually non-functional, elements. Non-autonomous elements are derived from autonomous elements, often following internal deletions. The structure of *P* elements has been most thoroughly studied in *Drosophila melanogaster* from which the so-called canonical *P* element was first characterized (O'Hare and Rubin, 1983). Canonical *P* elements are 2.9 kb in length and include flanking 31-bp inverted repeats. Four exons are designated as open reading frames (ORFs) 0, 1, 2, and 3. They encode two polypeptides involved in *P* element transposition: an 87 kDa transposase, encoded by all four exons, and a 66 kDa repressor protein encoded by the first three exons (Rio et al., 1986; Misra and Rio, 1990). The majority of *P* elements in the *D. melanogaster* genome are, however, non-

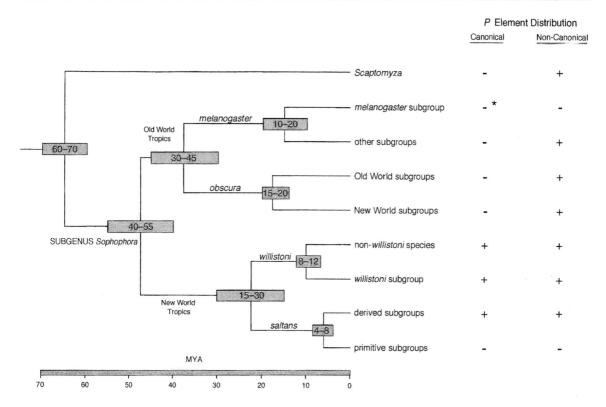

FIGURE 15.1 A phylogeny of *Sophophora* showing the relationship of this subgenus to the genus *Scaptomyza*. Relationships and divergence times for the four principle species groups are indicated and are based on a number of morphological, cytogenetic, biogeographical, and molecular studies. Within each species group a major split among subgroups is indicated. The presence (+) or absence (–) of both canonical and non-canonical *P* element sequences is indicated; the asterisk (*) identifies the canonical *P* element transferred horizontally to *D. melanogaster*.

autonomous as a result of internal deletions, or other mutations, that disrupt coding sequences. These non-autonomous elements are unable to transpose unless the genome in which they reside carries at least one autonomous element. Some deleted non-autonomous elements may also act as repressors of transposition. Many species in which *P* elements have resided for long periods of time carry only diverged non-autonomous elements.

With the exceptions of *D. melanogaster* and *Drosophila mediopunctata* (subgenus *Drosophila*), canonical *P* elements are restricted to the *saltans* and *willistoni* species groups. Non-canonical elements are any sequences that are phylogenetically distinct from the canonical *P* element (Clark et al., 1995). Non-canonical *P* elements have been detected in most of the species of the subgenus

Sophophora. Exceptions are *Drosophila emarginata* and *Drosophila neocordata* of the *saltans* species group, and several species of the *melanogaster* species group (Clark et al., 1998; Daniels et al., 1990). It is intriguing that none of the close relatives of *D. melanogaster* (in the *melanogaster* subgroup) possesses detectable *P* elements of any kind.

EVIDENCE FOR HORIZONTAL TRANSFER OF CANONICAL *P* ELEMENTS

Recent *P* element horizontal transfer into *D. melanogaster*

In earlier work on hybrid dysgenesis, we observed that samples of *D. melanogaster* natural populations collected in the late 1970s and early 1980s invariably carried *P* elements. In contrast, *P* elements were absent from strains collected more than 20 years earlier from around the world (Kidwell, 1979, 1983; Anxolabéhère et al., 1988). This was consistent with a recent *P* element invasion of the cosmopolitan species, *D. melanogaster* (Kidwell, 1979, 1983). Recent invasion was supported by a number of observations. Among these, it was observed that complete *P* elements from geographically dispersed populations of *D. melanogaster* were virtually identical (O'Hare and Rubin, 1983; Nitasaka et al., 1987) suggesting a recent common evolutionary origin.

The *P* element involved in the *D. melanogaster* horizontal transfer belongs to the canonical *P* element subfamily described above. Although characteristic of this species, the canonical *P* element is not found in any other species in the *melanogaster* species group of the genus *Drosophila*, to which *D. melanogaster* belongs (see Figure 15.1). Subsequent sequence analysis demonstrated that the canonical *P* element was almost certainly transferred horizontally to *D. melanogaster* from *Drosophila willistoni* of the *willistoni* species group (Daniels et al., 1990; Clark et al., 1998)

It is not a coincidence that the horizontal transfer that introduced the *P* element into *D. melanogaster* has occurred only relatively recently. This species is a relative newcomer to the Americas having probably been introduced to the West Indies from Africa some time during the last 200 years (Engels, 1992). Consequently, only during the last century or so has there actually been an opportunity for *D. melanogaster* to share the host range of *D. willistoni* which is limited to southern Florida, and Central and South America. Therefore, horizontal transfer of the *P* element from *D. willistoni* to *D. melanogaster* would not have been possible in nature until only very recently.

Origin of the canonical P element clade

A summary of the phylogeny of *P* elements from the *saltans* and *willistoni* species groups is presented in Figure 15.2. There are two main features of *P* element evolution in these two groups. First, there are at least four different subfamilies, or clades, of *P* elements in these New World lineages, differing from each other by as much as 44% in nucleotide sequence (Clark and Kidwell, 1997). Thus each subfamily of *P* elements – including the canonical elements – has a distinct evolutionary origin. Second, in contrast to the other three subfamilies, canonical *P* elements are relatively homogeneous, differing from each other by at most 10% in nucleotide sequence (Clark et al., 1995).

Based on the molecular evolution studies described below, the age of the most recent common ancestor of the canonical *P* elements has been estimated to be 2–3 Myr (Silva and Kidwell, 2000). Since the ages of the *saltans* and *willistoni* species groups themselves are much greater than this estimate (Figure 15.1), it can be concluded that most of the species in these two groups existed prior to the invasion of the New World *Sophophora* by the canonical *P* element. This conclusion is consistent with several observations concerning the distribution of the canonical *P* element in these two groups. For example canonical *P* elements are absent in *Drosophila insularis* of the *willistoni* subgroup (Daniels et al., 1990; Clark et al., 1995). This species is apparently the oldest in this subgroup (Gleason et al., 1998) and has the most restricted geographic distribution, confined to the Lesser Antilles (Spassky et al., 1971). Thus, a plausible explanation for the lack of canonical *P* elements in this species is its geographic isolation from other members of the *willistoni* group prior to the invasion of its sister species by the canonical element.

It appears that the canonical *P* element has only recently entered the New World lineages of *Sophophora*. The original source of canonical *P* elements is unknown. While *P* element sequences have been detected in species outside of the family Drosophilidae (Perkins and Howells, 1992; Lee et al., 1999), these are neither canonical, nor autonomous, *P* elements. The existence of canonical *P* elements in *Drosophila mediopunctata* (subgenus *Drosophila*) ((Loreto et al., 2001) and see below) is best explained by horizontal transfer from a species of the *willistoni* or *saltans* group, rather than suggesting the existence of an independent reservoir of canonical *P* elements

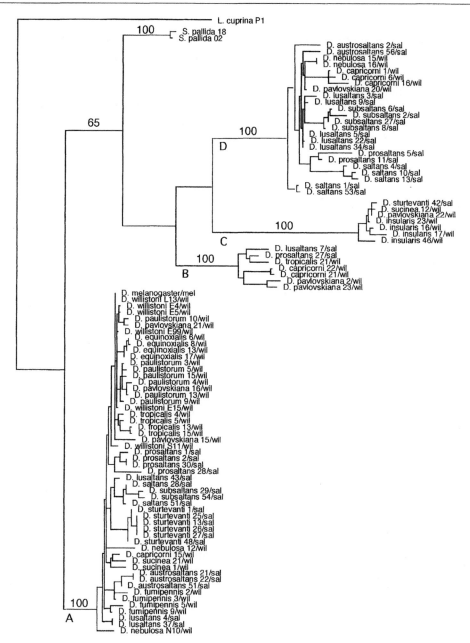

FIGURE 15.2 Phylogenetic analysis of *P* element sequences from the *saltans* and *willistoni* species groups. Comparisons were limited to a 449 bp fragment amplified from exon 3 of the transposase gene by PCR. The *P* element from the blow fly, *Lucilia cuprina*, was used as an outgroup and two *P* element sequences from *Scaptomyza pallida* are included for reference. This tree was generated by parsimony analysis as implemented by PAUP 3.1.1 (Swofford, 1993) using the heuristic search algorithm, TBR branch swapping and random stepwise addition of taxa. This is an arbitrarily chosen representative of 1000 equally parsimonious trees, each requiring 1158 steps. Letters A, B, C, D refer to the four distinct subfamilies of *P* elements in these two species groups; subfamily A represents the canonical *P* elements. Bootstrap values are shown above the branches and represent percentages for 100 replicates. Values of 50% or greater are shown for only the major clades. Species names are given followed by a clone number; species group designations are given after the slash: sal: *saltans* species group; wil: *willistoni* species group. A single canonical sequence from *D. melanogaster* is included for reference. See Clark et al. (1995) for more details.

outside of the *Sophophora*. In light of the evidence for recent invasion by canonical *P* elements, each of the distinct subfamilies of *P* elements in the *saltans* and *willistoni* species groups may have a distinct evolutionary origin. Thus *P* element evolution in these two groups may be characterized by successive waves of horizontal transfer that have occurred at various times in the past. The relatively extreme sequence divergence seen in the non-canonical subfamilies compared with the relatively homogeneous canonical *P* element subfamily suggests that such non-canonical element invasions occurred in the distant past.

Multiple horizontal transfers in species of the *willistoni* and *saltans* species groups

Horizontal transfer can be inferred whenever species that are only distantly related carry identical, or extremely similar, transposable element sequences. This kind of inference cannot be used in the case of the *willistoni* and *saltans* species groups, as they are sister groups and relatively young in age (Figure 15.1). Instead, in the analysis of closely related taxa, a quantitative method to assess horizontal transfer has to be used; divergence in synonymous sites among taxa is compared for transposable elements and host genes (Silva and Kidwell, 2000). For this rationale to work reliably, transposable element synonymous sites should be evolving under no selective constraints, such as in *P* canonical elements. When this is the case, element synonymous sites evolve at least as fast as the synonymous sites of host genes. Consequently, a lower divergence between transposable element sequences than between host genes is likely to result from horizontal transfer of transposable elements. This method was used (Silva and Kidwell, 2000) to estimate the number of horizontal transfer events within and between the *saltans* and *willistoni* groups of *Sophophora*. A horizontal transfer event between a pair of taxa was only inferred whenever *P* element divergence in synonymous sites was significantly lower than that of all host genes compared between those same taxa. In addition, only the most similar pair of *P* elements within each pair of taxa was analyzed. This

analysis lead to the inference of a minimum of 11 horizontal transfer events among 18 species surveyed. Seven of these are within the *willistoni* group and four within the *saltans* group (Figure 15.3). Incomplete reproductive isolation has been documented between some of these species (Bock, 1984). It can then be suggested that some of the horizontal transfer events mentioned could have been elicited by interspecific hybridization. At least one more event has occurred between one of these taxa and *D. mediopunctata* of the subgenus *Drosophila* ((Loreto et al., 2001) and see below).

The method used by Silva and Kidwell (2000) is conservative for at least two reasons. First, no horizontal transfer event is inferred in cases in which the divergence of *P* element synonymous sites are found to be intermediate between those of multiple host genes analyzed, even though horizontal transfer is likely to have occurred. It is also possible that in a fraction of the cases in which *P* element divergence is larger than that of host genes of the same taxa, *P* elements could still have been transmitted horizontally between them. Second, because only a pair of *P* elements is analyzed for each pair of taxa, multiple horizontal events are ruled out, even though such occurrences are hypothetically possible.

The divergence among *P* elements in synonymous sites was also used to estimate the age of the most recent common ancestor of these element (Silva and Kidwell, 2000). This estimate is dependent on the fact that no selective constraints are detected on synonymous sites of the canonical *P* elements. *P* element divergence was calibrated in three ways: (1) using the rate of synonymous substitution in *Drosophila* low-codon bias genes (Sharp and Li, 1989); (2) using the rate of synonymous substitutions in *R1* and *R2* transposable elements relative to the *Adh* locus (Eickbush et al., 1995); and (3) using the relationship between codon bias and synonymous substitution rate in the host species. In this way the age of the most recent common ancestor of these elements was estimated to be approximately 2–3 Myr (Silva and Kidwell, 2000).

The minimum rate of horizontal transfer of canonical *P* elements among these taxa was therefore estimated to be 12 events per 3 Myr,

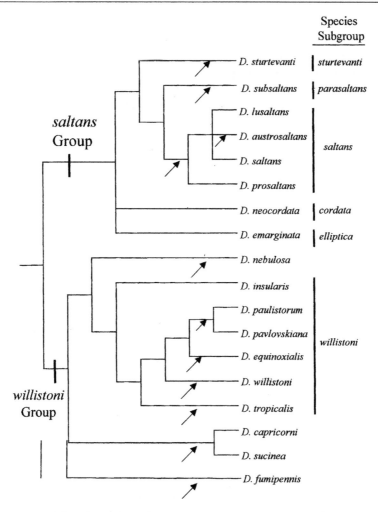

FIGURE 15.3 Phylogeny of flies from the *willistoni* and *saltans* species groups. A minimum of 11 horizontal transfer events of *P* elements, as indicated by arrows, have been detected among species of the *willistoni* and *saltans* species groups (Silva and Kidwell 2000). The phylogeny of these groups is based, respectively, on Gleason et al. (1998) and on Bicudo (1973), Throckmorton (1975), and O'Grady et al. (1998).

or one event per 250 000 years. For the reasons stated above this is very likely to be an underestimate. It can be concluded that canonical *P* elements have transferred horizontally between species at a rate higher than anticipated previously.

Canonical *P* elements in *Drosophila mediopunctata*

As described above, canonical *P* elements appeared at one time to be restricted to New World species of the subgenus *Sophophora*. The only previous exception was *D. melanogaster* which acquired its *P* elements very recently by horizontal transfer from *D. willistoni* when it extended its range to include South and Central America. Recently, *P* elements have been identified in *D. mediopunctata*, a species of the *tripunctata* group in the subgenus *Drosophila* (Loreto et al., 1998). Furthermore, these *P* elements were identified as belonging to the canonical subfamily with nucleotide divergence between 2 and 6% from other members of this subfamily (Loreto et al., 2001). On the basis of the following evidence a strong case can be made for horizontal transfer of a canonical P element to *D. mediopunctata* from a member of the *willistoni* or *saltans* groups: (1) P elements are completely absent from other species of the *D. tripunctata* group that have been examined; (2) the nucleotide divergence of the *D. mediopunctata* P elements is in the same range of high similarity (less than 10% sequence divergence) reported among canonical P elements in species of the *willistoni* and *saltans* groups (Loreto et al., 2001) and is thus much lower than expected, given the taxonomic distance between species of the *Drosophila* and *Sophophora* subgenera; (3) the geographical range of *D. mediopunctata* overlaps with the geographical range of the *willistoni* and *saltans* group species, as do their choice of feeding/breeding habitat. Thus the requirement of horizontal transfer for ecological and spatial overlap is satisfied.

EVIDENCE FOR HORIZONTAL TRANSFER OF NON-CANONICAL P ELEMENTS

In contrast to the extensively studied closely related canonical P elements that have a maximum age of 2–3 Myr in *Sophophora* (Silva and Kidwell, 2000), non-canonical elements have an ancient origin with nucleotide divergences among subfamilies exceeding 40% (Clark and Kidwell, 1997). Because of the great age of many of these elements, horizontal transfer events are more difficult to identify and document than for younger elements. However, a few cases of horizontal transfer involving non-canonical P elements have been identified. The most prominent examples involve two out of three of the P element subfamilies that have been identified in species of the *obscura* group, the M-type and O-type subfamilies (Hagemann et al., 1998; Haring et al., 2000) that are described below.

The M-type subfamily of non-canonical P elements

The M-type subfamily of non-canonical P elements has been identified in species of two subgroups (*montium* and *rhopala*) of the *melanogaster* species group and in about one-third of species of the *obscura* group that have been examined (Hagemann et al., 1998). These elements appear to have been present prior to speciation, and thus may have been present for a relatively long time in these lineages. M-type elements are also found in the genera *Scaptomyza* and *Lordiphosa* (Haring et al., 2000). The phylogeny of the M-type elements based on partial nucleotide sequences is not congruent with that of the host species suggesting horizontal transfer (Hagemann et al., 1998; Haring et al., 2000). The most striking example of horizontal transfer involving the M-type subfamily occurred between *Drosophila bifasciata* of the *obscura* species group and *Scaptomyza pallida*. In spite of a divergence time of at least 60 Myr between these two taxa, their P element sequences differ by only about 7% (Hagemann et al., 1992).

The O-type subfamily of non-canonical P elements

A second subfamily of P elements, known as the O-type, has been identified in *D. bifasciata* and these sequences differ from the M-type elements by over 30% (Hagemann et al., 1994). These elements are also found in the genera *Scaptomyza* and *Lordiphosa* (Haring et al., 2000). It was shown that P element sequences highly similar to the O-type elements are common in the *saltans* and *willistoni* species groups (Clark and Kidwell, 1997). To explain these observations a minimum of two horizontal transfer events are required (Haring et al., 2000). A plausible scenario is an initial transfer from a fly in the *saltans* or *willistoni* group to the cosmopolitan *S. pallida* and a second transfer from *S. pallida* to *D. bifasciata*. This would be consistent with what is known about the geographic distribution of these flies.

P elements from *Drosophila sucinea* and two New World species of the *obscura* group

In a previous phylogenetic analysis of 92 P element sequences from the *saltans* and *willistoni* species groups, a single sequence (*D. sucinea* 10) from *Drosophila sucinea* of the *willistoni* group showed no affinity to any of the other 91 sequences (Clark et al., 1995). Whereas nucleotide sequence divergence between this element and other sequences from the *saltans* and *willistoni* groups is about 35%, this sequence differs by only 10% from sequences isolated from two species of the *obscura* species group, *D. affinis* and *D. azteca* (Clark and Kidwell, 1997). These are two of the many species of the *obscura* group that colonized the New World at some time in the past (see Figure 15.1). It is intriguing that there is considerable overlap of the geographic distributions of *D. sucinea* and *D. azteca* in central Mexico. Thus, both ecology and sequence comparisons are consistent with a horizontal transfer event involving these two species.

CONCLUSIONS

The phylogenetic analyses presented here provide strong support for frequent horizontal transfer in the evolutionary history of the *P* transposable element. These conclusions are based primarily on the incongruence between species and *P* element phylogenies, and on the high degree of sequence similarity between elements isolated from distant taxa. In several instances, the case for horizontal transfer is strengthened when the ecology of individual species is examined more closely. A striking example is the recent entry of *D. melanogaster* into the New World tropics, a prerequisite for the transfer of the canonical *P* element to this species from *D. willistoni* as described above. The case of M-type *P* elements from *D. bifasciata* of the *obscura* species group and *Scaptomyza pallida*, which share 93% sequence identity, is a second example. *S. pallida* is a cosmopolitan species whereas *D. bifasciata* is the most widespread member of the *obscura* species group. Thus these two species share considerable habitat overlap in the palearctic temperate region. A third example of horizontal transfer that is consistent with ecology is found in *D. mediopunctata*, which has been recently shown to possess canonical *P* elements and is sympatric with many of the flies from the *saltans* and *willistoni* species groups (Loreto et al., 2001).

Alternative explanations to horizontal transfer, such as the comparison of paralogous sequences, or retention of ancestral polymorphisms, are more difficult to exclude when dealing with closely related species. However, the examination of sequence evolution described above allows conclusions that are consistent with both horizontal transfer and the molecular behavior of these transposable elements.

What is the role of horizontal transfer in the long-term survival of *P* elements? Simulation studies suggest that the eventual fate of a transposable element system like *P* is extinction, as elements accumulate mutations that render them incapable of producing transposase, and deletions render other copies non-functional, at a higher rate than new functional elements arise by transposition (Kaplan et al., 1985). Horizontal transfer of a functional element in to a naive host, with the concomitant production of large number of functional copies by transposition, provides a chance to escape this extinction (Hurst et al., 1992; Kidwell, 1992; Lohe et al., 1995). Our results suggest that horizontal transfer is the main selective constraint on the canonical *P* element transposase in the species surveyed here (Silva and Kidwell, 2000). This agrees with other studies of *P* element evolution (Witherspoon, 1999) and of the *mariner* transposable element (Robertson and Lampe, 1995). When viewed together with the phylogenetic analyses, these conclusions suggest that horizontal transfer of the *P* element has been much more widespread than was previously realized.

REFERENCES

Anxolabéhère, D., Kidwell, M.G. and Périquet, G. (1988) Molecular characteristics of diverse populations are consistent with the hypothesis of a recent invasion of *Drosophila melanogaster* by mobile *P* elements. *Mol. Biol. Evol.* **5**: 252–269.

Anxolabéhère, D., Nouaud, D. and Périquet, G. (1985) Séquences homologues a l'élément *P* chez des espèces de *Drosophila* du groupe *obscura* et chez *Scaptomyza pallida* (Drosophilidae). *Génét. Sél. Evol.* **17**: 579–584.

Beall, E.L., Admon, A. and Rio, D.C (1994) A *Drosophila* protein homologous to the human p70 Ku autoimmune antigen interacts with the P transposable element inverted repeats. *Proc. Natl Acad. Sci. USA* **91**: 12681–12685.

Bicudo, H.E.M.C. (1973) Chromosomal polymorphism in the *saltans* group of *Drosophila*. I. The *saltans* subgroup. *Genetica* **44**: 520–552.

Bock, I.R. (1984) Interspecific hybridization in the genus *Drosophila*. In *Evolutionary Biology* (eds, M.K. Hecht, B. Wallace and G.T. Prance), vol. 18, pp. 41–70, Plenum, New York.

Clark, J.B. and Kidwell, M.G. (1997) A phylogenetic perspective on *P* transposable element evolution in *Drosophila*. *Proc. Natl Acad. Sci. USA* **94**: 11428–11433.

Clark, J.B., Kim, P. and Kidwell, M.G. (1998) Molecular evolution of *P* transposable elements in the genus *Drosophila*. III. The *melanogaster* species group. *Mol. Biol. Evol.* **15**: 746–755.

Clark, J.B., Altheide, T.K., Schlosser, M.J. and Kidwell, M.G. (1995) Molecular evolution of *P* transposable elements in the genus *Drosophila*. I. The *saltans* and *willistoni* species groups. *Mol. Biol. Evol.* **12**: 902–913.

Daniels, S.B., Peterson, K.R., Strausbaugh, L.D. et al. (1990) Evidence for horizontal transmission of the *P* transposable element between *Drosophila* species. *Genetics* **124**: 339–355.

Eickbush, D.G., Lathe III, W.C., Francino, M.P. and Eickbush, T.H. (1995) *R1* and *R2* retrotransposable elements of *Drosophila* evolve at rates similar to those of nuclear genes. *Genetics* **139**: 685–695.

Engels, W.R. (1989) *P* elements in *Drosophila*. In *Mobile DNA* (eds, D. Berg, and M. Howe), pp. 437–484, American Society for Microbiology, Washington, DC.

Engels, W.R. (1992) The origin of *P* elements in *Drosophila melanogaster*. *Bioessays* **14**: 681–686.

Evgen'ev, M., Zelentsova, H., Mnjoian, L. et al. (2000) Invasion of *Drosophila virilis* by the *Penelope* transposable element. *Chromosoma* **109**: 350–357.

Finnegan, D.J. (1989) Eukaryotic transposable elements and genome evolution. *Trends Genet.* **5**: 103–107.

Gleason, J.M., Griffith, E.C. and Powell, J.R. (1998) A molecular phylogeny of the *Drosophila willistoni* group: conflicts between species concepts? *Evolution* **52**: 1093–1103.

Hagemann, S., Miller, W.J. and Pinsker, W. (1992) Identification of a complete *P* element in the genome of *Drosophila bifasciata*. *Nucleic Acids Res.* **20**: 409–413.

Hagemann, S., Miller, W.J. and Pinsker, W. (1994) Two distinct *P* element subfamilies in the genome of *Drosophila bifasciata*. *Mol. Gen. Genet.* **244**: 168–175

Hagemann, S., Haring, E. and Pinsker, W. (1998) Horizontal transmission vs. vertical inheritance of *P* elements in *Drosophila* and *Scaptomyza*: has the M-type subfamily spread from East Asia *J. Zool. Syst. Evol. Res.*, in press.

Haring, E., Hagemann, S. and Pinsker, W. (2000) Ancient and recent horizontal invasions of Drosophilids by *P* elements. *J. Mol. Evol.* **51**: 577–586.

Hurst, G.D.D., Hurst, L.D. and Majerus, M.E.N. (1992) Selfish genes move sideways. *Nature* **356**: 659–660.

Jordan, I.K., Matyunina, L.V. and McDonald, J.F. (1999) Evidence for the recent horizontal transfer of long

terminal repeat retrotransposon. *Proc. Natl Acad. Sci. USA* **96**: 12621–12625.

Kaplan, N., Darden, T. and Langley, C. (1985) Evolution and extinction of transposable elements in Mendelian populations. *Genetics* **109**: 459–480.

Kidwell, M.G. (1979) Hybrid dysgenesis in *Drosophila melanogaster*: The relationship between the P-M and I-R interaction systems. *Genet. Res.* **33**: 105–117.

Kidwell, M.G. (1983) Evolution of hybrid dysgenesis determinants in *Drosophila melanogaster*. *Proc. Natl Acad. Sci. USA* **80**: 1655–1659.

Kidwell, M.G. (1992) Horizontal transfer. *Curr. Opin. Genet. Dev.* **2**: 868–873.

Kidwell, M.G. (1993) Lateral transfer in natural populations of eukaryotes. *Annu. Rev. Genet.* **27**: 235–256.

Lansman, R.A., Stacey, S.N., Grigliatti, T.A. and Brock, H.W. (1985) Sequences homologous to the P mobile element of *Drosophila melanogaster* are widely distributed in the subgenus Sophophora. *Nature* **318**: 561–563.

Lee, S.-H., Clark, J.B. and Kidwell, M.G. (1999) A *P* element-homologous sequence in the house fly, *Musca domestica*. *Insect Mol. Biol.* **8**: 491–500.

Lohe, A.R., Moriyama, E.N., Lidholm, D.A. and Hartl, D.L. (1995) Horizontal transmission, vertical inactivation, and stochastic loss of *mariner*-like transposable elements. *Mol. Biol. Evol.* **12**: 62–72.

Loreto, E.L., da Silva, L.B., Zaha, A. and Valente, V.L. (1998) Distribution of transposable elements in neotropical species of *Drosophila*. *Genetica* **101**: 153–165.

Loreto, E.L. d.S., Valente, V.L. d.S., Zaha, A. et al. (2001) *Drosophila mediopunctata P* elements: a new example of horizontal transfer. *J. Heredity*. In press.

Misra, S. and Rio, D.C. (1990) Cytotype control of *Drosophila P* element transposition: the 66 kd protein is a repressor of transposase activity. *Cell* **62**: 269–284.

Nitasaka, E., Mukai, T. and Yamazaki, T. (1987) Repressor of *P* elements in *Drosophila melanogaster*: cytotype determination by a defective *P* element with only open reading frames 0 through 2. *Proc. Natl Acad. Sci. USA* **84**: 7605–7608.

O'Grady, P.M. Clark J.B. and Kidwell, M.G. (1998) Phylogeny of the *Drosophila saltans* species group based on combined analysis of nuclear and mitochondrial DNA sequences. *Mol. Biol. Evol.* **15**: 656–664.

O'Hare, K. and Rubin, G.M. (1983) Structures of *P* transposable elements and their sites of insertion and excision in the *Drosophila melanogaster* genome. *Cell* **34**: 25–35.

Perkins, H.D. and Howells, A.J. (1992) Genomic sequences with homology to the *P* element of *Drosophila melanogaster* occur in the blowfly *Lucilia cuprina*. *Proc. Natl Acad. Sci. USA* **89**: 10753–10757.

Rio, D.C., Laski, F.A. and Rubin, G.M. (1986) Identification and immunochemical analysis of biologically active *Drosophila P* element transposase. *Cell* **44**: 21–32.

Robertson, H.M. and Lampe, D J. (1995) Distribution of transposable elements in arthropods. *Annu. Rev. Entomol.* **40**: 333–357.

Sharp, P.M. and Li, W.-H. (1989) On the rate of DNA

sequence evolution in *Drosophila*. *Mol. Biol. Evol.* **28**: 398–402.

Silva, J.C. and Kidwell, M.G. (2000) Selection and horizontal transfer in the evolution of *P* elements. *Mol. Biol. Evol.* **17**: 1542–1557.

Spassky, B.S., Richmond, R.C., Peréz-Salas, S. et al. (1971) Geography of sibling species related to *D. willistoni* and semispecies of the *D. paulistorum* complex. *Evolution* **25**: 129–143.

Swofford, D. (1993) *PAUP: Phylogenetic Analysis Using Parsimony*, Version 3.1.1, Smithsonian Institution, Washington, DC.

Throckmorton, L.H. (1975) The phylogeny, ecology and geography of *Drosophila*. In *Handbook of Genetics* (ed., R.C. King), pp. 421–469, vol. 3, Plenum, New York.

Witherspoon, D.J. (1999) Selective constraints on *P* element evolution. *Mol. Biol. Evol.* **16**: 472–478.

The *mariner* Transposons of Animals: Horizontally Jumping Genes

Hugh M. Robertson, Felipe N. Soto-Adames, Kimberly K. O. Walden, Rita M. P. Avancini and David J. Lampe

The *mariner* transposons of insects and other animals are now known to comprise a large family of small (±1300 bp) transposable elements characterized by a D,D34D catalytic domain in their encoded transposases. They transpose by a DNA-mediated cut-and-paste mechanism, and form one branch of the D,D35E transposase/integrase megafamily of proteins, being most closely related to the *Tc1* family that is also widespread in animals. Within the *mariner* family at least 16 distinct subfamilies can be recognized, five of which have numerous representatives in the genomes of diverse insects, other arthropods, nematodes, flatworms, hydras, and mammals, including humans. Phylogenetic analysis of the relationships of *mariners* within particular subfamilies clearly demonstrate that they have undergone multiple horizontal transfers between hosts, sometimes across animal phyla. We have studied three particular instances of relatively recent horizontal transfers in detail, involving transfer of irritans subfamily *mariners* across two orders of insects, mellifera subfamily *mariners* across four orders of insects, and cecropia subfamily *mariners* across three phyla of animals. In most cases these *mariners* evolve neutrally and accumulate incapacitating mutations within particular hosts, whereas comparisons between hosts indicate that most of the evolutionary conservation of their transposase genes occurs in conjunction with horizontal transfers between hosts. The ability of *mariner* transposase to catalyze transposition without species-specific host factors appears to allow this unusual evolutionary pattern. *Mariners* and the related *Tc1* family of transposons, which evidence many of these same characteristics, have thereby affected the composition of most animal genomes.

INTRODUCTION

The first *mariner* element was discovered in the fruit fly *Drosophila mauritiana* (Hartl, 1989, 2001) and a large body of work has characterized the distribution of this element in related drosophilids, including a likely horizontal transfer event (Maruyama and Hartl, 1991), and its functionality (Lohe and Hartl, 1996; Lohe et al., 1997; Hartl et al., 1997). Following the serendipitous discovery of a distantly related *mariner* in the genome of the giant silk moth, *Hyalophora cecropia* (Lidholm et al., 1991), we employed a PCR assay to detect and preliminarily characterize *mariner* transposons with a widespread but patchy distribution in insects (Robertson, 1993) and related arthropods (Robertson and MacLeod, 1993). Since then we have used this assay, as well as additional primers designed from even more conserved amino acid blocks, to detect members of this family of transposons in the genomes of mites, flatworms, hydras, and mammals (Robertson, 1997, 1999; Robertson

173

Copyright © 2002 by Academic Press.
All rights of reproduction in any form reserved.

and Zumpano, 1997). Several additional *mariners* have been identified by us and others in the genomes of various insects as well as the nematode *Caenorhabditis elegans* and humans (Robertson and Asplund, 1996; Robertson, 1997).

In our initial phylogenetic analyses of *mariner* relationships it was clear that horizontal transfers across large host phylogenetic distances, such as orders of insects, must have occurred. Since then work with hosts from other phyla has shown that transfers across phyla have also occurred. Here we describe an additional dataset of PCR fragments from diverse insects, and analyze these within individual subfamilies, rather than across the entire family. For the central most conserved region of their transposase genes that are amplified with our PCR assay, sequences within subfamilies are colinear with each other and generally share at least 40% encoded amino acid identity, indicating that they diverged relatively recently (see Robertson and MacLeod, 1993, for definitions of the subfamilies). Although the rates of evolution of *mariners* are unclear, most evidence suggests that within particular hosts they evolve neutrally and hence relatively rapidly (for example, the neutral rate for DNA sequence divergence in *Drosophila* flies is about 1% per Myr). Thus phylogenetic analyses of *mariners* that reveal major incongruencies with host phylogenies, for example, closely related within-subfamily *mariners* in hosts from different insect orders or animal phyla, must imply horizontal transfers because these transposons could not have evolved for 200–600 Myr as separate lineages within their host genomes and remained so similar to each other.

METHODS

Two hundred and sixty-seven additional species of insects have been examined for the presence of *mariners* using our PCR assay since our initial reports (Table 16.1). These reflect a similar diversity of insects to our previous study of over 400 species (Robertson and MacLeod, 1993), but instead of a quick DNA extraction method, DNA was extracted by conventional methods involving denaturation of proteins using heat and

TABLE 16.1 Numbers of species tested and positive for mariners from various orders of insects (following nomenclature of Borror et al., 1989)

Order	Tested	Positive
Diplura	2	1
Collembola	8	0
Microcoryphia	2	2
Thysanura	2	1
Ephemeroptera	1	0
Grylloblattaria	1	1
Zoraptera	1	0
Embiidina	1	0
Isoptera	1	0
Orthoptera	5	2
Hemiptera	12	5
Mantodea	1	1
Homoptera	24	13
Blattaria	5	3
Dermaptera	4	1
Psocoptera	2	0
Neuroptera	16	8
Mecoptera	2	1
Coleoptera	46	24
Strepsiptera	2	0
Diptera	79	21
Trichoptera	2	1
Lepidoptera	23	10
Hymenoptera	39	18
Total	267	113

SDS, precipitation of proteins using high salt, extraction of proteins using phenol/chloroform, and precipitation of the DNA with ethanol. Over 100 of these species tested positive using our original PCR assay that yields a ±500 bp fragment (Robertson, 1993). Most of the 267 samples were also examined with two additional primers designed to regions encoding conserved amino acid blocks inside the original PCR fragment region (Robertson, 1997), and in each case this additional assay was also positive. A few samples only yielded amplification with the second set of primers, and one example, the grape phylloxera *Daktulosphaira vitifoliae*, was examined further and is included below. Twenty-five species (Table 16.2) were examined further by sequencing of a sample of 2–15 PCR fragment clones from each. These were conceptually translated, with judicious introduction of frameshifts to maintain aligned reading frames

TABLE 16.2 Species from which mariner sequences were obtained. Common names are used herein, abbreviated where necessary, except for *Drosophila ananassae*

Common name	Scientific name	Family	Order
Surinam cockroach	*Pycnocelus surinamensis*	Blaberidae	Blattaria
Backswimmer	*Buenoa* sp.	Notonectidae	Hemiptera
Damsel bug	*Nabis* sp.	Nabidae	Hemiptera
Grape phylloxera	*Daktulosphaira vitifoliae*	Phylloxeridae	Homoptera
Hangingfly	*Bittacus strigosus*	Bittacidae	Mecoptera
Mantidfly	*Mantispa pulchella*	Mantispidae	Neuroptera
Blister beetle	*Epicauta funebris* (= *pestifera*)	Meloidae	Coleoptera
Locust borer	*Megacyllene robiniae*	Cerambycidae	Coleoptera
Flour beetle	*Tribolium madens*	Tenebrionidae	Coleoptera
Rusty grain beetle	*Cryptolestes ferrugineus*	Cucujidae	Coleoptera
Mosquito	*Culex restuans*	Culicidae	Diptera
Tsetse fly	*Glossina palpalis*	Muscidae	Diptera
Stable fly	*Stomoxys uruma*	Muscidae	Diptera
False stable fly	*Muscina stabulans*	Muscidae	Diptera
Soybean nodule fly	*Rivellia quadrifasciata*	Platystomatidae	Diptera
Otitid fly	*Delphinia picta*	Otitidae	Diptera
–	*Drosophila ananassae*	Drosophilidae	Diptera
Microcaddisfly	*Orthotrechia* cf. *cristata*	Hydroptilidae	Trichoptera
Parsnip webworm	*Depressaria pastinacella*	Oecophoridae	Lepidoptera
Ailanthus webworm	*Atteva punctella*	Yponomeutidae	Lepidoptera
Diamondback moth	*Plutella xylostella*	Plutellidae	Lepidoptera
Silkworm moth	*Bombyx mori*	Bombycidae	Lepidoptera
Stingless bee 1	*Plebia frontalis*	Apidae	Hymenoptera
Stingless bee 2	*Plebia jatiformis*	Apidae	Hymenoptera
Andrenid bee	*Andrena erigenia*	Andrenidae	Hymenoptera

in some cases, and aligned with the available dataset of PCR fragments (Robertson, 1993, 1997; Robertson and MacLeod, 1993; Robertson and Lampe, 1995), as well as the equivalent regions of several full-length clones and consensus sequences now available (Robertson and Asplund, 1996; Robertson, 1997). Representative sequences, essentially all of those shown in bold in the figures, have been submitted to GenBank (accession No. U91342–U91393).

There are many reasons to be confident that these sequences originate from the insects examined (Robertson, 1993, 1997). PCR contamination of samples was avoided by performing pre- and post-PCR work in different rooms with completely separate equipment and reagents. In any case contaminants would appear as identical sequences from different species, and while the remarkable aspect of our results is the extreme similarity of *mariners* from different hosts, no instances of identical sequences were obtained. Furthermore, as described below, we have cloned and sequenced copies of many particularly recently horizontally transferred *mariners* from genomic libraries of their hosts, thereby confirming that they originate from these genomes.

Phylogenetic analysis was conducted on the aligned ±170 amino acid sequences using maximum parsimony as implemented by the PAUP v3.1.1 software for the Macintosh (Swofford, 1993), employing the heuristic search algorithm with random addition of sequences, 10 iterations, and tree-bisection-and-reconnection branch swapping. All positions and amino acid changes were weighted equally. Where sets of multiple clones of similar kind were obtained (for example in Robertson (1993) there were six clones representing the canonical *Drosophila mauritiana mar1* element) only one was used in the phylogenetic analysis to reduce the large numbers of equally parsimonious trees typically obtained in these analyses, and to allow boot-strap analysis. In each case the most intact clone, that is with the fewest frameshifts and/or encoded stop codons was employed to represent that type (an indication of the divergence of the available clones is given after the name in the figures). The consensus sequences of genomic clones were employed where possible. The four largest subfamilies (mauritiana, mellifera, cecropia, and irritans) were analyzed separately, while the small capitata and lineata subfamilies were analyzed together with single sequence types representing apparently novel subfamilies. The outgroup for analyses of the four large subfamilies was chosen to be all of the sequences from Robertson and Asplund (1996), that is, three divergent representatives from each of the mauritiana, mellifera, cecropia, and irritans subfamilies, as well as two from the nematode *Caenorhabditis elegans* (for each subfamily analyzed that subfamily's three sequences were removed from the outgroup). The basal *mariner* lineage from *Bombyx mori* (Robertson and Asplund, 1996) was not included in the outgroup because it is so divergent it only complicates the analyses (it shares just 20% amino acid identity with other *mariners*). Sequences from different subfamilies generally share 25–40% amino acid identity with each other. The outgroup for the capitata, lineata, and smaller subfamilies was the three irritans subfamily sequences from Robertson and Asplund (1996), because they are a relatively basal lineage within the *mariner* family. Bootstrap analyses of at least 100 replications were performed on each dataset.

RESULTS

The overall rate of positive insect species found in this survey (42%; Table 16.1) is considerably higher than in our earlier surveys (around 15%). In large part we ascribe this difference to the use of highly purified genomic DNA. The effect is particularly strong for certain groups, e.g. the Coleoptera, which are hard to extract using our earlier quick extraction method. Bigot et al. (1994) reported that the great majority of Hymenoptera they examined had evidence of

mauritiana subfamily *mariners* alone, so we may still be underestimating the proportion of insects with *mariner* transposons in their genomes.

The alignment of these ±500 bp fragments encoding ±170 amino acids is largely unambiguous (see for example Robertson, 1997), and entirely colinear within subfamilies, one defining feature of the subfamilies (Robertson and MacLeod, 1993). This fragment of *mariners* includes most of the D,D34D region homologous to the catalytic site of the D,D35E megafamily of transposases and integrases (Doak et al., 1994; Robertson, 1995), and represents fully half of the transposase protein sequence. It therefore constitutes a good sample of each *mariner* that provides considerable phylogenetic information. A phylogenetic analysis of this region for the entire family, using the *Tc1* family as an outgroup and a sample of *mariner* sequences, is available in Robertson (1997), so here we focus on analyses of all available distinct *mariner* sequences within individual subfamilies. The 54 new types of *mariners* described here from 25 additional species reinforce and extend previous conclusions about *mariners*, for example, we again find that many insects have multiple different kinds of *mariners*, and their phylogenetic relationships indicate additional horizontal transfers across orders of insects and phyla of animals.

The mauritiana subfamily (Figure 16.1) now includes *mariners* in three phyla of animals (arthropods, platyhelminths, and cnidaria) and the new sequences reveal additional examples of horizontal transfers. For example, sequences very similar to the Deer.fly.9.6 clone were found in a fruit fly, *Drosophila ananassae*, a mantidfly, *Mantispa pulchella*, and a hangingfly, *Bittacus strigosus*, the latter two species belonging to the orders Neuroptera and Mecoptera respectively. This relationship is strongly supported by bootstrapping. These three *mariners* share 89–95% amino acid identity, and as is typical for closely related *mariners*, their per cent DNA identity is similar (91–96%). These orders are at least 265 Myr old (Robertson and MacLeod, 1993). In contrast, there appears to be a basal lineage within this subfamily found to date only in various Hymenoptera, although it is not supported by bootstrapping.

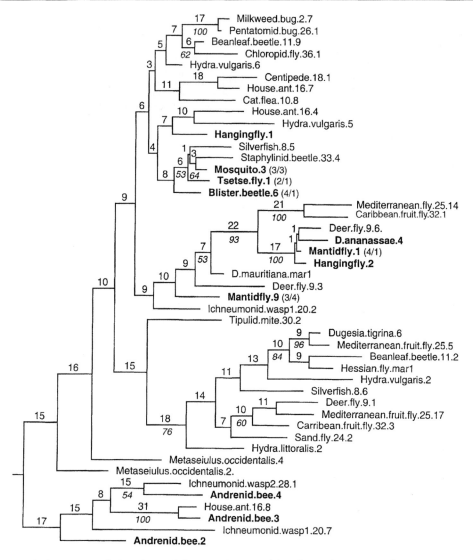

FIGURE 16.1 Phylogenetic relationships of the mauritiana subfamily *mariners*. A single most parsimonious tree was obtained. Branch lengths in number of amino acid changes are shown above all branches supporting nodes. Bootstrap percentages are shown in italics below branches supporting nodes with more than 50% support. Names of clone sequences reported here for the first time are in bold. Where appropriate the average percent amino acid divergence between the encoded transposase sequences of multiple similar clones are shown in parentheses after these bold names.

The mellifera subfamily (Figure 16.2) is now the largest subfamily of *mariners* with over 50 different kinds. The placement of the flatworm Dugesia.tigrina.1 and the mite Tetranychus.urticae.1 sequences at the base of this family is tentative (Robertson, 1997, 1999) because while they have length variants that place them in this or the cecropia subfamilies, they have sequence features that in analyses of the entire family cause them to cluster at the base of the mauritiana subfamily. The new sequences provide many additional examples of horizontal transfers in this subfamily in that almost all of their relationships are highly incongruent with those of their hosts. For example, several clones from the blister beetle

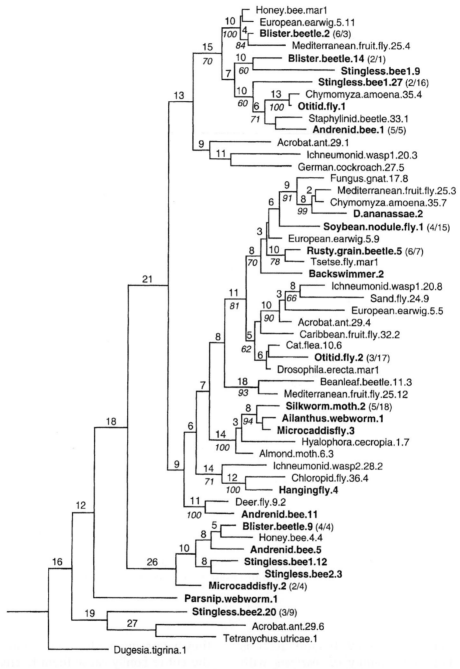

FIGURE 16.2 Phylogenetic relationships of the mellifera subfamily *mariners*. This tree is an arbitrary representative of the 72 equally parsimonious trees obtained. Branch lengths are shown only for those branches supporting nodes present in 100% of the trees, according to the semi-strict consensus option of PAUP. See the caption to Figure 16.1 for other features.

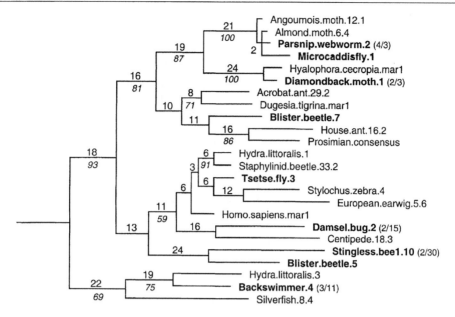

FIGURE 16.3 Phylogenetic relationships of the cecropia subfamily *mariners*. See the caption to Figure 16.1 for other features.

Epicauta funebris cluster confidently with those from the European honey bee, European earwig, and Mediterranean fruit fly (the first three of these *mariners* share at least 90% amino acid and DNA identity). Clones from the rusty grain beetle, *Cryptolestes ferrugineus*, cluster with the full length *mariner* described from the tsetse fly *Glossina palpalis* (Blanchetot and Gooding, 1995). Some sequences from a microcaddisfly cluster with those from several moths, while others cluster with those of several bees, and the Hangingfly.4 clone is the closest relative of the Chloropid.fly.36.4 clone.

Cecropia subfamily *mariners* (Figure 16.3) have been found in the most diverse hosts, including many insects, two flatworms, a hydra, and primates. Again a microcaddisfly clone clusters confidently with several moth clones, perhaps suggesting that transfers among Lepidoptera and Trichoptera are common. A tsetse fly clone provides yet another insect with *mariners* closely related to the first known human *mariner*, Hsmar1, and *mariners* from *Hydra littoralis* and the marine flatworm *Stylochus zebra* (Robertson, 1997; Robertson and Zumpano, 1997). The staphylinid beetle, Hsmar1 consensus, and *Hydra littoralis*

mariners share at least 85% amino acid and DNA identity, which is extraordinary for sequences from three phyla that last shared a common ancestor at least 600 Myr ago (Robertson, 1997). Another cluster of sequences from extremely diverse hosts is the original flatworm *Dugesia tigrina* mar1 (Garcia-Fernàndez et al., 1995), the insect Acrobat.ant.29.2, House.ant.16.2, and Blister.beetle.7 sequences, and the consensus of multiple sequences found in diverse prosimians (Robertson and Zumpano, 1997).

The best studied example of recent horizontal transfer of *mariner* elements is in the irritans subfamily (Figure 16.4) involving a green lacewing, the horn fly, *Anopheles gambiae*, and *Drosophila ananassae* (Robertson and Lampe, 1995), to which sequences from a stable fly, *Stomoxys uruma*, can now be added (they all share at least 88% amino acid and DNA identity). The other new sequences in this subfamily provide additional examples of *mariner* phylogenetic relationships incongruent with those of their hosts. The second human *mariner*, Hsmar2, has been tentatively placed as a basal lineage of this subfamily (Oosumi et al., 1995; Reiter et al., 1996; Robertson et al., 1996; Robertson and Asplund, 1996; Robertson, 1997; Robertson and

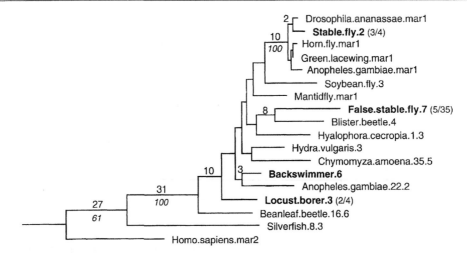

FIGURE 16.4 Phylogenetic relationships of the irritans subfamily *mariners*. This tree is an arbitrary representative of the 22 equally parsimonious trees obtained. Branch lengths are shown only for those branches supporting nodes present in 100% of the trees, according to the semi-strict consensus option of PAUP. The Soybean.fly.3 and Blister.beetle.4 sequences were reported in Robertson and Lampe (1995) and hence are not in bold. See the caption to Figure 16.1 for other features.

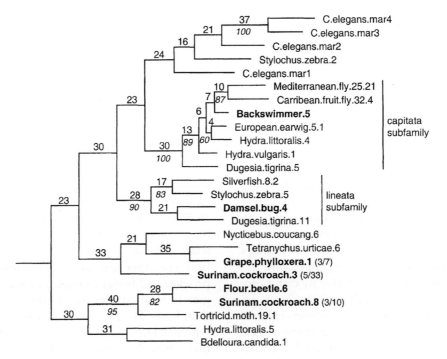

FIGURE 16.5 Phylogenetic relationships of the smaller subfamilies of *mariners*. See the caption to Figure 16.1 for other features.

Martos, 1997), and it is not particularly closely related to any other *mariner*.

Finally, in Figure 16.5 the relationships within the small capitata and lineata subfamilies, as well as single *mariner* types that might represent novel subfamilies, are shown. These include sequences from all four phyla, as well as four from the nematode *Caenorhabditis elegans* (which also has three more *mariners* similar to the extremely basal *Bombyx mori mariner1* (Robertson and Asplund, 1996; Robertson, 1997)). Several of the new sequences might represent novel subfamilies, for example, those from the grape phylloxera, the surinam cockroach, and the flour beetle. In each case these sequences have length differences from even their closest relatives in the tree, and generally share with them and each other only 20–40% amino acid identity (values typical of subfamily relationships). It is not yet known whether these are very rare subfamilies, or are common in animals not yet sampled, or are so divergent in sequence that they do not readily amplify with the PCR primers employed to date (which might explain why the grape phylloxera sequences were only amplified with the internal primers).

DISCUSSION

The phylogenetic relationships of *mariners* revealed by our work and that of others (Garcia-Fernàndez et al., 1995; Lohe et al., 1995) are clearly highly incongruent with those of their hosts. These incongruencies can only be reasonably explained by repeated, indeed seemingly fairly frequent, horizontal transfers between hosts. The alternative explanation of vertical inheritance of particular *mariner* lineages within only certain host lineages, and their loss from all others examined, is untenable, because it is inconceivable that these *mariners* could have remained so similar to each other (commonly at least 85% identity) for such long periods in such divergent hosts (all over 200 Myr old). To provide additional data to support this claim we have examined in detail three particularly clear examples of relatively recent horizontal transfers across large host phylogenetic gaps.

Robertson and Lampe (1995) describe the remarkably similar irritans subfamily *mariners* (Figure 16.4) found in the green lacewing, *Chrysoperla plorabunda* (order Neuroptera), the horn fly, *Haematobia irritans* and a drosophilid fly *D. ananassae* (order Diptera, suborder Brachycera), and the African malaria mosquito *Anopheles gambiae* (order Diptera, suborder Nematocera). For example, the consensus DNA sequences of the 1044 bp transposase genes of the lacewing and horn fly *mariners* differ by just 2 bp, leading to the 345 amino acid sequences of their encoded transposases differing by just one. Given the age of these insect lineages (200 Myr in the case of fly suborders, and 265 Myr for the orders) it is inconceivable that these *mariners* have evolved vertically within these lineages, been conserved to such an extent, including third codon positions and non-coding sequence, and have been lost from most other neuropteran and fly lineages examined. Indeed, when the evolution of the individual copies from their consensus sequences within each host lineage was examined, most copies appeared to be evolving neutrally, and therefore rapidly, because the rate of synonymous changes in the transposase genes usually equaled that of non-synonymous changes.

These results, and similar results obtained by others (Lohe et al., 1995), have led to the view that horizontal transfers to new hosts are not simply an evolutionary curiosity of these transposons, but rather are central to their evolutionary persistence (Robertson and Lampe, 1995). Thus it appears that once such a transfer has occurred, presumably by a single active copy of a *mariner*, the copy number increases and the element spreads throughout the host species. The individual copies in most cases appear to evolve independently and neutrally from each other thereafter, with their transposition activity either repressed by the host, by themselves, or by accumulation of internally deleted copies that titrate transposase (Lohe and Hartl, 1996). They will eventually be lost from the host genome or become an unrecognizable part of the DNA of the host. Occasionally a still active copy might undergo a horizontal transfer to a new host, repeating the cycle. This is probably the only stage at which significant selection for activity takes place, because while this copy may have acquired mutations, none of these can compromise activity for it to be able to achieve the horizontal transfer and establish a new population of *mariners* in its new host. From the example of the hornfly and green lacewing *mariners* above, we believe that in fact most such horizontal transfers to new hosts occur reasonably soon after a *mariner* invades a host, with at most only a few mutations being acquired, and that therefore each comparison described above of relatively closely related (>85% identity) *mariners* in highly divergent hosts in fact reflects tens to hundreds of horizontal transfers to species we have not examined. In no case do we believe we have identified direct donor and recipient hosts, in contrast to the example of extremely recent horizontal transfer of the *P* element among *Drosophila* species (see Chapter 15).

Additional support for this model of *mariner* molecular evolution is provided by our unpublished analyses of two other sets of relatively recent horizontal transfers of *mariners*. One is the set of closely related mellifera subfamily *mariners* found in insects from four different orders (Figure 16.2); the European honey bee *Apis mellifera* (Hymenoptera), the European earwig *Forficula auricularia* (Dermaptera), the Mediterranean fruit fly *Ceratitis capitata* (Diptera), and a blister beetle *Epicauta funebris* (= *pestifera*) (Coleoptera). Again, multiple full-length copies of these *mariners* have been cloned from genomic libraries of these species, with the exception of the blister beetle, where for reasons unexplained we cannot obtain them from a genomic library, but can by PCR. Comparison of these sequences reveals that they have generally evolved neutrally within their hosts, but show evidence of conservation when compared across hosts.

Similarly, we have cloned and sequenced multiple copies of the cecropia subfamily *mariners* (Figure 16.3) found in the human genome, *Hydra littoralis*, and a staphylinid beetle *Carpelimus* sp. The latter two *mariners* are relatively recent invaders of their hosts, as judged by the high similarity of the copies within their genomes (e.g. 99% DNA identity for the beetle copies), however this human *mariner*, *Hsmar1*, is an old constituent of our genome, having

invaded about 50 Myr ago (most copies are highly mutated and defective, differing from their consensus by ±7.5% in DNA sequence) (Robertson and Zumpano, 1997).

These results from genomic libraries abundantly confirm the observations based on PCR fragments. That such horizontal transfers occur is also supported by the findings of Lohe et al. (1995) of a mellifera subfamily *mariner* in *Drosophila erecta* that is very similar to a PCR fragment we reported from the cat flea (Figure 16.2). Furthermore, Garcia-Fernàndez et al. (1995) describe a cecropia subfamily *mariner* in the flatworm *Dugesia tigrina* that shares 75% amino acid identity with an acrobat ant *mariner* PCR fragment (Figure 16.3). The copies within this flatworm differ from each other by less than 1% DNA divergence, and no congeners examined had this *mariner*, strongly supporting the argument for a recent horizontal transfer into the flatworm, but not of course directly from the acrobat ant. That animals generally acquire *mariners* by horizontal transfer is further supported by a recent study of the anciently asexual bdelloid rotifers, whose genomes appear to be devoid of vertically inherited RNA or Class I transposons, but have acquired at least four lineages of *mariner* transposons, presumably by horizontal transfer (Arkhipova and Meselson, 2000). We can therefore be fairly confident that all of the unusual relationships in the phylogenetic trees of *mariners* shown herein are genuine examples of horizontal transfers. Indeed, it seems inescapable that each example in fact reflects many horizontal transfers, because we have only examined a very small sample of animals in detail.

The implications of these observations for *mariner* transposons, and likely also the closely related *Tc1* family of transposons, are twofold. First, they must be capable of functioning in diverse host cellular environments. This inference has been confirmed for two *mariners*, the horn fly irritans subfamily element *Himar1* (Lampe et al., 1996) and the canonical *Mos1* element (Tosi and Beverley, 2000), and the *Tc1* element from *C. elegans* (Vos et al., 1996). In each case purified transposase was able to catalyze transposition of a marked cognate transposon from one plasmid to another, with the products recovered in bacteria. Therefore these transposons are unlikely to require species-specific host factors for mobility, beyond the common host enzymes involved in repair of excision sites and new integrations. This transposon lifestyle probably represents one extreme on a continuum including elements like the *P* elements which reveal reasonably frequent horizontal transfers among a phylogenetically restricted set of hosts, the family Drosophilidae (Silva and Kidwell, 2000; see Chapter 15), to retrotransposons such as the *R* elements that are widespread in insects, yet appear to persist primarily by vertical and conserved evolution within their hosts (Lathe et al., 1995).

Secondly, there must be some mechanism by which these transposons are able to move from one host to another. This movement has to be into the germline of the new host and across orders of insects and phyla of animals. It seems unlikely that there will be a single mechanism for such transfers (Kidwell, 1992), however the best current candidates are perhaps various DNA viruses, some of which are known to be suitable targets for transposons from their hosts (Fraser et al., 1985), including two members of the *Tc1* family (Jehle et al., 1995, 1998).

Consequences of such horizontal transfers for the hosts are that essentially all animals are probably vulnerable to invasion by these transposons, indeed it appears that while the *mariner* family is fairly widespread, the *Tc1* family has multiple representatives in the genomes of most animals (Avancini et al., 1996; Grossman et al., 1999). Given the antiquity of these two transposon families, this is a process animals have been subjected to for a very long time. In some cases the hosts appear to be particularly vulnerable to such invasions, for example, the horn fly has a relatively enormous genome of 2.2×10^9 bp, nearly the size of ours, and 10 times the size of *Drosophila melanogaster*. Fully 1% of this DNA consists of 17 000 copies of the particular *mariner* described above (Robertson and Lampe, 1995), while at least 17 different kinds of *Tc1* family elements constitute an undetermined portion (Avancini et al., 1996).

The discovery of the two human *mariners* (Auge-Gouillou et al., 1995; Morgan 1995; Oosumi et al., 1995; Reiter et al., 1996; Robertson et al., 1996; Smit and Riggs, 1996; Robertson and Martos, 1997; Robertson and Zumpano, 1997; DeMattei et al., 2000) demonstrated that these consequences have applied to the evolution of our genome as well. In each case these are very old constituents and comprise a very small portion of our genome. There are now known to be many other Class II or DNA-mediated transposons in our genome (Smit, 1999), including a series called *Tiggers* that are members of the *pogo* family of elements distantly related to *mariners* and the *Tc1* family (Robertson, 1996; Smit and Riggs, 1996), and Smit (1999) inferred that altogether remnants of DNA-mediated transposons constitute at least 1% of our genome. It seems unlikely that any of these elements are currently active in our genome, however they likely had some mutagenic and perhaps recombinational influence on its evolution.

ACKNOWLEDGMENTS

We thank the many colleagues who contributed insects and/or DNA samples for this work, particularly Alain Blanchetot, Raul Cano, Jan Conn, Thomas Coudron, Jeffrey Granett, Kostas Iatrou, Karen McClellan, Barbara Stay, Durdica Ugarkovic, Lisa Vawter, and David Weaver. Matthew Sharkey, Karen Zumpano, Michelle Lepkowitz, and Paul White provided technical assistance. This work was supported by NSF grant MCB 93–17586.

REFERENCES

Arkhipova, I. and Meselson, M. (2000) Transposable elements in sexual and ancient asexual taxa. *Proc. Natl Acad. Sci. USA* **97**: 14473–14477.

Auge-Gouillou, C., Bigot, Y., Pollet, N. et al. (1995) Human and other mammalian genomes contain transposons of the *mariner* family. *FEBS Letters* **368**: 541–546.

Avancini, R.M.P., Walden, K.K.O. and Robertson, H.M. (1996) The genomes of most animals have multiple members of the *Tc1* family of transposable elements. *Genetica* **98**: 131–140.

Bigot, Y., Hamelin, M-H., Capy, P. and Periquet, G. (1994) *Mariner*-like elements in hymenopteran species: Insertion site and distribution. *Proc. Natl Acad. Sci. USA* **91**: 3408–3412.

Blanchetot, A. and Gooding, R.H. (1995) Identification of a *mariner* element from the tsetse fly, *Glossina palpalis palpalis*. *Insect Mol. Biol.* **4**: 89–96.

Borror, D.J., Triplehorn, C.A. and Johnson, N.E. (1989) *An Introduction to the Study of Insects*, 6th edn, Saunders College Publishing.

DeMattei, M.V, Auge-Gouillou, C., Pollet, N. et al. (2000) Features of the mammal mar1 transposons in the human, sheep, cow, and mouse genomes and implications for their evolution. *Mammal. Genome* **11**: 1111–1116.

Doak, T.G., Doerder, F.P., Jahn, C.L. and Herrick, G. (1994) A proposed superfamily of transposase-related genes: new members in transposon-like elements of ciliated protozoa and a common "D35E" motif. *Proc. Natl Acad. Sci. USA* **91**: 942–946.

Garcia-Fernàndez, J., Bayascas-Ramírez, J.R., Marfany, G. et al. (1995) High copy number of highly similar *mariner*-like transposons in planarian (Platyhelminthe): Evidence for a trans-phyla horizontal transfer. *Mol. Biol. Evol.* **12**: 421–431.

Grossman, G.L., Cornel, A.J., Rafferty, C.S. et al. (1999) Tsessebe, Topi and Tiang: three distinct Tc1-like transposable elements in the malaria vector, *Anopheles gambiae*. *Genetica* **105**: 69–80.

Fraser, M.J., Brusca, J.S., Smith, G.E. and Summers, M.D. (1985) Transposon-mediated mutagenesis of a baculovirus. *Virology* **145**: 356–361.

Hartl, D.L. (1989) Transposable element *mariner* in *Drosophila* species. In *Mobile DNA* (eds, D.E. Berg and M.M. Howe), pp. 5531–5536, American Society for Microbiology, Washington, DC.

Hartl, D.L. (2001) Discovery of the transposable element *mariner*. *Genetics* **157**: 471–476.

Hartl, D.L., Lohe, A.R. and Lozovskaya, E.R. (1997) Modern thoughts on an ancyent marinere: function, evolution, regulation. *Annu. Rev. Genet.* **31**: 337–358.

Jehle, J.A., Fritsch, E., Nickel, A. et al. (1995) TCI4.7: A novel lepidopteran transposon found in *Cydia pomonella* granulosis virus. *Virology* **207**: 369–379.

Jehle, J.A., Nickel, A., Vlak, J.M. and Backhaus, H. (1998) Horizontal escape of the novel Tc1-like lepidopteran transposon TCp3.2 into *Cydia pomonella* granulovirus. *J. Mol. Evol.* **46**: 215–224.

Kidwell, M.G. (1992) Horizontal transfer. *Curr. Opin. Genet. Dev.* **2**: 868–873.

Lampe, D.J., Churchill, M.E.A. and Robertson, H.M. (1996) A purified *mariner* transposase is sufficient to mediate transposition *in vitro*. *EMBO J.* **15**: 5470–5479.

Lathe, W.C. III, Burke, W.D., Eickbush, D.G. and Eickbush, T.H. (1995) Evolutionary stability of the *R1* retrotransposable element in the genus *Drosophila*. *Mol. Biol. Evol.* **12**: 1094–1105.

Lidholm, D-A., Gudmundsson, G.H. and Boman, H.G. (1991) A highly repetitive, *mariner*-like element in the genome of *Hyalophora cecropia*. *J. Biol. Chem.* **266**: 11518–11521.

Lohe, A.R. and Hartl, D.L. (1996) Autoregulation of *mariner* transposase activity by overproduction and dominant-negative complementation. *Mol. Biol. Evol.* **13**: 549–555.

Lohe, A.R., Moriyama E.N., Lidholm D-A. and Hartl, D.L. (1995) Horizontal transmission, vertical inactivation, and stochastic loss of *mariner*-like transposable elements. *Mol. Biol. Evol.* **12**: 62–72.

Lohe, A.R., De Aguinar, D. and Hartl, D.L. (1997) Mutations in the *mariner* transposase: the D,D(35)E consensus sequence is nonfunctional. *Proc. Natl Acad. Sci. USA* **94**: 1293–1297.

Maruyama, K. and Hartl, D.L. (1991) Interspecific transfer of the transposable element *mariner* between *Drosophila* and *Zaprionus*. *J. Mol. Evol.* **33**: 514–524.

Morgan, G.T. (1995) Identification in the human genome of mobile elements spread by DNA-mediated transposition. *J. Mol. Biol.* **254**: 1–5.

Oosumi, T., Belknap, W.R. and Garlick, B. (1995) *Mariner* transposons in humans. *Nature* **378**: 672.

Reiter, L.T., Murakami, T., Koeuth, T. et al. (1996) A recombination hotspot responsible for two inherited peripheral neuropathies is located near a *mariner* transposon-like element. *Nature Genet.* **12**: 288–297.

Robertson, H.M. (1993) The *mariner* transposable element is widespread in insects. *Nature* **362**: 241–245.

Robertson, H.M. (1995) The *Tc1-mariner* superfamily of transposons in animals. *J. Insect Physiol.* **41**: 99–105.

Robertson, H.M. (1996) Members of the pogo superfamily of DNA-mediated transposons in the human genome. *Mol. Gen. Genet.* **252**: 761–766.

Robertson, H.M. (1997) Multiple *mariners* in flatworms and hydras are related to those of insects. *J. Heredity* **88**: 195–201.

Robertson, H.M. (1999) *Mariner* transposable elements in mites. In *Acarology IX: Volume 2, Symposia* (eds, G.R. Needham, R. Mitchell, D.J. Horn and W.C. Welbourn), pp. 1–6, Ohio Biological Survey, Columbus, OH.

Robertson, H.M. and Asplund, M.L. (1996) *Bmmar1*: a basal lineage of *mariner* family of transposable elements in the silkmoth, *Bombyx mori*. *Insect Biochem. Mol. Biol.* **26**: 945–954.

Robertson, H.M. and Lampe, D.J. (1995) Recent horizontal transfer of a *mariner* element between Diptera and Neuroptera. *Mol. Biol. Evol.* **12**: 850–862.

Robertson, H.M. and MacLeod, E.G. (1993) Five major subfamilies of *mariner* transposable elements in insects, including the Mediterranean fruit fly, and related arthropods. *Insect Mol. Biol.* **2**: 125–139.

Robertson, H.M. and Martos, R. (1997) Molecular evolution of the second ancient human *mariner* transposon, *Hsmar2*, illustrates patterns of neutral evolution in the human genome lineage. *Gene* **205**: 219–228.

Robertson, H.M. and Zumpano, K.L. (1997) Molecular evolution of an ancient *mariner* transposon, *Hsmar1*, in the human genome. *Gene* **205**: 203–217.

Robertson, H.M., Zumpano, K.L., Lohe A.R. and Hartl, D.L. (1996) Reconstruction of the ancient *mariners* of humans. *Nature Genet.* **12**: 360–361.

Silva, J.C. and Kidwell, M.G. (2000) Horizontal transfer and selection in the evolution of *P* elements. *Mol. Biol. Evol.* **17**: 1542–1557.

Smit, A.F. (1999) Interspersed repeats and other mementos of transposable elements in mammalian genomes. *Curr. Opin. Genet. Dev.* **9**: 657–663.

Smit, A.F.A. and Riggs, A.D. (1996) *Tiggers* and other DNA transposon fossils in the human genome. *Proc. Natl Acad. Sci. USA* **93**: 1443–1448.

Swofford, D.L. (1993) *PAUP: Phylogenetic Analysis Using Parsimony* Version 3.1.1, Smithsonian Institution, Washington, DC.

Tosi, L.R. and Beverley, S.M. (2000) cis and trans factors affecting Mos1 mariner evolution and transposition in vitro, and its potential for functional genomics. *Nucl. Acids Res.* **28**: 784–790.

Vos, J.C., De Baere, I. and Plasterk, R.H.A. (1996) Transposase is the only nematode protein required for *in vitro* transposition of Tc1. *Genes Dev.* **10**: 755–761.

CHAPTER 17

The Splicing of Transposable Elements: Evolution of a Nuclear Defense Against Genomic Invaders?

Michael D. Purugganan

In the last decade, it has become apparent that many eukaryotic transposable elements possess the ability to act as introns and be spliced from pre-mRNA. The ability to act as introns has associated fitness consequences for both the transposable element as well as the host organism. Splicing may represent a mechanism to circumvent negative selection on the element that may arise from insertions into gene coding regions. Based on the details of the molecular features of transposable element splicing, it is possible that the intron-like features of many transposable element insertions may result from the coevolution of element-encoded splicing signals and the host cellular splicing machinery. The cellular splicing machinery may thus act as a general defense against foreign insertions into genes, including those elements that may be transposing at high frequencies as a result of recent horizontal transfer into new genomes.

INTRODUCTION

Five decades ago, Barbara McClintock observed the genetic effects of a series of loci that changed their position in the maize genome (McClintock, 1950, 1984). These mobile loci are now referred to as transposable elements, and over the last few years molecular geneticists have managed to isolate and characterize hundreds of these elements in both prokaryotic and eukaryotic systems (Berg and Howe, 1989; Wessler, 1989; Robertson and Lampe, 1995a,b). These mobile sequences are ubiquitous components of organismal genomes, and they have been shown to exist in almost every organism where their presence has been sought (Berg and Howe, 1989). The widespread distribution and mobility of these sequences have led to suggestions that transposable elements serve as agents of evolutionary change by increasing genetic variability within populations (Syvanen, 1984; McDonald, 1990, 1995). Several others have suggested, however, that transposable elements are essentially genomic parasites that confer no advantage to the host and may even be deleterious (Hickey, 1980; Doolittle and Sapienza, 1984; Orgel and Crick, 1984).

Transposable elements are not only persistent components of organismal genomes, but they also appear capable of moving between species via horizontal transfer events. *Drosophila melanogaster* P elements provide a classic example of the lateral movement between species of a transposable element (Anxolabehere et al., 1988; Clark et al., 1994). P elements in the *melanogaster* genome, which were introduced in this species in the 1950s, appears to have originated from *Drosophila willistoni*. It is believed

Copyright © 2002 by Academic Press.
All rights of reproduction in any form reserved.

that this horizontal transfer of elements occurred through the agency of the mite *Proctolaelaps regalis* (Houck et al., 1991). Phylogenetic analysis of several other transposable elements suggests that the evolution of several element families are characterized by horizontal transfer of these mobile sequences between species genomes (Flavell, 1992; Voytas et al., 1992; Robertson, 1993; Robertson and Lampe, 1995a,b; Robertson, this volume). Among plants, the prolific ability of species to undergo interspecific hybridization presents yet another, albeit sexual, mechanism by which elements can invade the genomes of other taxa (Purugganan and Wessler, 1994).

What are the consequences of the movement of transposable elements between species genomes? The establishment of invading elements in new host genomes by horizontal transfer events may present a whole suite of challenges for both invader and new host. Unrestricted movement of transposable elements in new host genomes promotes the spread of these mobile loci. In interspecific hybrids between *Drosophila buzzatii and D. koepferae*, high rates of germline transposition of several elements appear to occur (Labrador and Fontdevila, 1994). Elevated rates of transposition could have severe consequences for the host organism, resulting in higher mutation rates, chromosomal aberrations and dysgenesis (Kidwell, 1985; Mackay et al., 1992). The transposition of these mobile sequences have been shown to be a significant source of spontaneous insertion mutations in maize, yeast and *Drosophila*. In the maize *waxy* locus, five of 17 spontaneous mutant alleles are the result of transposable element insertions (Wessler and Varagona, 1985; Varagona et al., 1992). Cytogenetic effects, such as chromosomal breakage (Weil and Wessler, 1993) and inversions, are also observed to accompany transposon movement. Finally, hybrid dysgenesis, with accompanying lost of fertility, provides another example of the effects of element activity on organismal viability (Kidwell, 1985). The dynamics of element spread in natural populations thus reflects the balance between the fitness of the transposable element and the possible negative impact the element may have towards the new host organism (Hickey, 1980;

Charlesworth and Charlesworth, 1983; Charlesworth and Langley, 1989).

Transposable elements which have been resident in a genome over a protracted period of evolutionary time are likely to co-evolve with their host genomes to prevent serious disruption of host gene activity that could lead to an excessive reduction in host fitness (Hickey, 1980). Charlesworth and co-workers have also suggested that transposable elements evolve mechanisms to inhibit high rates of transposition within host genomes. Indeed, under normal conditions in the wild the activity of many transposable elements are low or genetically suppressed (McDonald, 1990). There are numerous examples that illustrate a variety of mechanisms that control transposition behaviour (Fedoroff, 1995; Lozovskaya et al., 1995). The maize *Ac* element, for example, exhibits a negative dosage effect in which increasing copies of the element reduces the rate by which *Ac* transposes (McClintock, 1951). *P* elements introduced into *D. simulans* have also been shown to evolve towards weak P or M' types that transpose at reduced rates (Kimura and Kidwell, 1994). Finally, the *Tc1* element in *Caenorhabditis elegans* appear to be under host control in nematode somatic tissues (Emmons and Yesner, 1984).

These co-evolved mechanisms, however, may not initially be operational when a new transposable element enters a host genome via a horizontal transfer event. Horizontal transfer events place transposable elements in new genomic environments where host controls on the activity of these specific invading elements are inefficient or absent (Kidwell, 1993). The maize *Ac* system again provides an illustrative example: when *Ac* is introduced into new plant species as a transgenic construct, it loses the negative dosage effect that partially controls element transposition rate in its nominal host *Zea mays* (Scofield et al. 1993). A new series of co-evolutionary changes must then occur over time to control the number and activity of these invading transposable sequences if a new host organism is to remain viable after a horizontal transfer event. Until these new controls evolve, however, there remains a risk that the activities of the new invading elements are so severe as to cause widespread genetic damage to the host.

TABLE 17.1 Some transposable elements that are spliced from pre-mRNA

Species	Transposable elements	Host genes
Zea mays	*Ds family, dSpm*	*waxy, adh, bronze, a2, shrunken*
Drosophila melanogaster	*P, 412*	*vermillion, yellow*
Caenorhabditis elegans	*Tc1*	*unc-54, hlh-1, mlc-2*

In these instances, mechanisms to prevent transposable elements from damaging the genetic constitution of new hosts may operate as a short-term defense against the new genomic invaders. The ability of transposable elements to be spliced from pre-mRNA appears to provide one such mechanism that ameliorates some of the mutational impact of element insertions in eukaryotic genes (Purugganan and Wessler, 1992; Purugganan, 1993).

THE SPLICING OF TRANSPOSABLE ELEMENTS

Leaky phenotypes exhibited by transposable element insertion alleles provided early clues as to mechanisms that could mitigate the impact of element insertions into eukaryotic genes. In several model species, it has been known that not all element-induced mutant alleles result in null phenotypes (Wessler, 1989; Weil and Wessler, 1990). In the maize *waxy* locus, for example, the allele *wx-m9* displays residual levels of gene expression despite the insertion of a 4.37 kb *Ds* element in a *waxy* exon (Wessler et al., 1987). In *Drosophila*, the *vermillion* allele v^k also displays appreciable levels of gene activity despite the presence of a large *412* element insertion into the *v* coding region (Fridell et al., 1990).

The molecular basis for these leaky allele phenotypes was elucidated in 1987, when Wessler and her co-workers demonstrated that the partial restoration of gene activity in the *Ds* transposable element allele *wx-m9* was due to the splicing of the element insertion from the gene transcript (Wessler et al., 1987). In this allele, the 4.37 kb *Ds* element insertion was located in exon 10 near the 3' end of the *waxy* coding region. The allele encodes a wild type-sized protein and mRNA despite this large insertion in a translated exon. In *wx-m9*, it appears that the *Ds* element behaves as a nuclear intron, and the element sequence is processed out of pre-mRNA by the nuclear splicing machinery resulting in the partial restoration of gene function. Three cryptic splice donor sites at the termini of the element are utilized by the splicing machinery, as well as one of two cryptic splice acceptor sites – one near the 3' site of the *Ds* element and downstream of the *Ds* insertion in *wx* exon 10 (Figure 17.1). Some of these splicing combinations result in a mutant transcript in the correct reading frame, permitting the translation of a wild type-sized protein with mutational alterations at the sequence surrounding the site of element insertion.

The splicing of transposable elements has turned out to be a fairly widespread phenomenon and is apparently an integral part of transposon biology (Table 17.1). In maize, transposable element splicing is not confined to alleles containing insertions of the *Ac/Ds* family. The *a2-m1* allele of the anthocyanin biosynthetic gene *A2*, for example, contains a *dSpm* transposon that is spliced from pre-mRNA (Menssen et al., 1990). The *a2-m1* allele possesses a 2.2 kb *dSpm* insertion in the *A2* coding region, and displays low-level anthocyanin pigmentation in the kernel. The *dSpm* element is spliced is this allele using splice sites within the element itself (Figure 17.1). A derivative of *a2-m1*, the allele *a2-m1 (Class II)*, possesses an inserted *dSpm* element with a 900 basepair deletion; this appears to increase splicing efficiency and the allele displays even higher levels of pigmentation. This allele also shows decreased responsiveness to transpose in the presence of the autonomous *Spm* element.

Numerous cases of transposable element splicing has since been identified not only in maize but in other species as well. In *Drosophila melanogaster*, the hypomorphic v^k allele of the

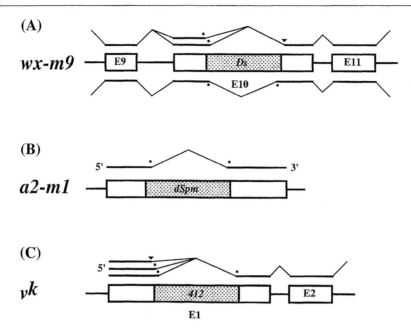

FIGURE 17.1 Schematic of three examples of transposable element splicing. (A) The maize *Ds* insertion in *wx-m9*, (B) the maize *dSpm* insertion in *a2-m1*, and (C) the *Drosophila 412* retrotransposon insertion in *v^k*. Boxes indicate exons while horizontal lines are introns. The shaded boxes represent the transposable element insertions. Splicing events are indicated above and below by horizontal lines for exons and slanted lines for spliced introns or element sequences. The dots indicate splice signals within the element while arrows denote splice sites in the host sequences used to process transposable element sequences.

vermillion gene contains a 7.5 kb *412* retrotransposon insertion in the untranslated leader region of exon 1 (Fridell et al., 1990). As in *wx-m9*, the *v^k* allele conditions an intermediate phenotype and transcribes wild type-sized mRNAs. Analysis of *v^k* shows that the *412* retrotransposon sequences are processed from the *vermillion* pre-mRNA through splice sites located within the element's LTR (Figure 17.1). Three donor sites are clustered at the 5' end of the element sequence; an additional chimeric donor site containing both *vermillion* and *412* sequence is also utilized. These sites are then joined to a cryptic acceptor located just upstream of the element's 3' termini. Interestingly, an intragenic revertant of *v^k*, the allele *v^{+37}*, was found to contain an insertion of the *B104/roo* element within the original *412* insertion; the combined insertion now totals 11.1 kb in *vermillion* exon 1 (Pret and Searles, 1991). The *roo* element, located at the *412* 5' terminus, appears to provide better splice donor sites that increases the efficiency of element splicing.

Transposable element insertion alleles that display element splicing are identified in genetic screens as hypomorphic alleles with intermediate mutant phenotypes. This biases the number of insertion alleles identified – those alleles that are spliced efficiently and confer no mutant phenotype escape detection in these genetic screens. A relatively unbiased sampling of *Tc1* insertions in the *C. elegans hlh-1*, *mlc-2* and *unc-54* loci by molecular (as opposed to genetic) screening suggests that the fraction of element insertions that are spliced but give no phenotype may be high (Rushforth and Anderson, 1996). Although null alleles of these genes have clear mutant phenotypes, eight independent molecularly-identified insertion alleles of these loci display a wild-type phenotype. The *Tc1* insertions in all these alleles are spliced from pre-mRNA, resulting in the production of in-frame messages. In many cases, the splicing machinery recognized cryptic splice sites both in the element and the surrounding exon.

These are just a few examples of element splicing patterns that have thus far been characterized (Kim et al., 1987; Dennis et al., 1988; Geyer et al., 1991). Splicing of transposable elements in all these cases rely on cryptic splice sites encoded either within the element or in the host gene's exon (Purugganan and Wessler, 1992). These sites are either at the element termini or, if located in the host gene, close to the insertion junction; utilizing these splice sites thus leads to the removal of most of the inserted element sequences. The splicing is imprecise, however, since these splice sites are not usually found at the exact insertion junctions. Splicing can delete host gene sequence, or, more often, leads to the addition of nucleotide sequences at the insertion junction in the mature transcript, resulting in translational frameshifts or altered amino acid sequences. Expressed proteins from these genes, although mutant, may still be capable of activity. Thus, spliced insertion alleles can lead to leaky or even wild-type phenotypes (Wessler, 1989; Weil and Wessler, 1990), rather than the null phenotypes commonly exhibited by transposon insertion alleles that do not exhibit splicing.

THE EVOLUTION OF TRANSPOSABLE ELEMENT SPLICING: MECHANISMS AND CONSEQUENCES

If transposable elements are viewed as genomic parasites, then splicing appears to represent a host nuclear defense against element insertions. Moreover, the ability to be spliced provides unique advantages to the new, invading mobile elements as they move throughout the genome (Kidwell, 1993; Purugganan, 1993). Splicing circumvents some of the mutational effects associated with element insertions. The data on *Ds* and *dSpm* element insertions in maize, *P* and *412* insertions in *Drosophila* and *Tc1* insertions in *C. elegans*, demonstrate that splicing of elements may result in partial to full restoration of gene expression (for a review, see Purugganan and Wessler, 1992). By permitting even limited levels of gene expression, splicing can reduce

the negative fitness consequences of transposable element activity within eukaryotic genomes (Gierl, 1990; Purugganan, 1993). Transposable elements that evolve the ability to be spliced would thus be able to escape negative selection in spite of their insertional effects into host genes.

There are several reasons to believe that cryptic splice signals within the termini of transposable elements may have evolved to allow for the removal of the element from exons (Purugganan and Wessler, 1992). First, most element-encoded terminal splice sites are positioned fairly close to element ends, thus ensuring that most of the element sequences are removed by splicing. Secondly, the ability to be spliced can occur among differing members of specific transposable element families. The maize *Ds* elements, for example, have diverse internal sequences but all share conserved terminal ends that are required for element transposition in the presence of an autonomous *Ac* element. It also appears that the splice sites that mediate the removal of different *Ds* elements from pre-mRNA are conserved between these disparate elements, suggesting that the presence of signals for splicing are under positive selection (Wessler, 1991).

It has also been suggested that the cellular splicing machinery itself may have evolved in part to deal with transposable elements that insert themselves into eukaryotic genes. Even before the demonstration that transposable elements could be spliced, Crick speculated that splicing evolved as a defense against transposons within genomes (Crick, 1979). If this is indeed the case, then the mechanism by which this is accomplished appears to involve the ability of the eukaryotic splicing apparatus to differentiate between exon and intron sequences in a process referred to as "exon definition" (Robberson et al., 1992; Berget, 1995)

Recent results suggest that during nuclear splicing, cellular factors initially search for and recognize a pair of splice sites across an exon in the pre-mRNA (Berget, 1995). When such a pair is found, the binding of U1 and U2 snRNPs, and other associated splicing factors occurs. These associated factors include the 3' splice site recognition factor U2AF, and the 5' splice site

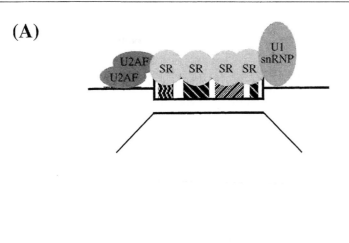

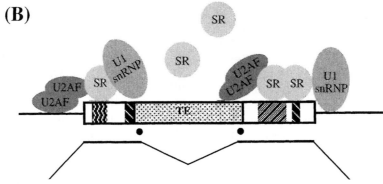

FIGURE 17.2 Possible mechanism for transposable element splicing. (A) Early steps in normal spliceosome assembly. The boxes represent the exons, with patterned areas denoting exonic splicing enhancers. The binding of SR proteins to these enhancers as well as their interactions with U1 snRNP and U2AF factors define the exon. (B) Hypothetical mechanism for element splicing. The lack of SR binding to the inserted transposable element sequence forces weak binding of U1 snRNP and U2AF factors at the junctions of the exon and element insertion. The weak binding is facilitated by strong SR protein binding to exonic enhancers not disrupted by the transposon. Positioning of the splicing factors may also be facilitated by a mechanism to recognize the presence of non-exon sequences.

recognition factor ASF/SF2; together, these then define the exon as a unit in the pre-mRNA. Neighboring defined exons are then brought together and ligated to form the mature processed transcript.

Exon definition is facilitated by a group of serine/arginine-rich SR proteins that bind to pre-mRNA exons and facilitate the assembly of the spliceosome (Fu, 1995; Adams et al., 1996; Reed, 1996). These SR proteins bind to exonic splicing enhancers, which contain purine-rich enhancer elements that promote the splicing of adjacent 5' and 3' splice sites (Kohtz et al., 1994; Staknis and Reed, 1994) (Figure 17.2A). SR proteins bound to

exons also promote the interaction between U2 snRNP and the intron branchpoint sequence. In general, the SR proteins appear to regulate the initial steps in the recruitment of snRNPs in the formation of functional spliceosomes (Adams et al., 1996; Reed, 1996).

Based on studies of transposable element splicing, it may be that the mechanism of exon definition is supplemented by the ability of the splicing machinery to recognize transposable element insertions as "non-exon" sequences. One scenario could be that in alleles with transposable element insertions, the SR proteins initially recognize constitutive exon sequences by binding to exonic enhancers. The presence of non-exon sequences (the transposable element) is then detected by the splicing machinery, possibly by the lack of SR binding to the insertion sequence (Figure 17.2B). The splicing machinery then positions splicing factors to recognize cryptic splice sites at the boundary of the exon and transposable element insertion; these include element-encoded splice sites as well as cryptic sites in exon sequences. The assembled spliceosomes then proceed to process the element insertion from the pre-mRNA as an intron.

There are several lines of evidence that suggest that the splicing of transposable elements may be a generalized mechanism of the cell to remove insertions in pre-mRNA. First, the ability to be spliced appears to be a widespread trait, occurring in various species and types of transposons (both IR transposons and retrotransposons) (Purugganan and Wessler, 1992; Purugganan, 1993). The elements themselves are of diverse origin, and the terminal sequences harboring the splice sites are unrelated to each other. Thus, the splice sites within these elements evolved independently several times in diverse transposable element evolutionary lineages, and it appears likely that these splice sites were responding to the general ability of the splicing machinery to remove these element insertions. Secondly, the use of cryptic sites not only in the element but in adjacent host sequences lends support to the notion that the splicing machinery can recognize the presence of transposable element insertions. The use of these host exon-encoded cryptic splice sites suggests that transposable element splicing does not rely solely on element encoded splice signals (Purugganan and Wessler, 1992). Rather, the splicing machinery itself may distinguish between exon and non-exon (transposable element) sequences in the pre-mRNA, and utilize appropriate cryptic splice signals near the insertion junction to remove element sequences.

Selection may thus operate on both the transposable elements and the eukaryotic splicing machinery – the splicing of transposable elements can then be viewed as a co-evolved cellular system as a genomic defense against these mobile sequences. The splicing machinery develops the capability to splice

element insertions, while the element itself facilitates the process by evolving more intron-like features, such as terminal splice signals. The organism can minimize the deleterious effects of transposable element insertions and thus increase its fitness.

From the standpoint of the transposable element, there are distinct advantages in remaining camouflaged within the genome as an intron. By mitigating the negative selection pressure that accompanies deleterious insertion into host genes, the transposable element can increase its probability of survival within a genome. For the continued persistence of the transposon, however, the element must retain the ability to spread within populations via transposition. Although transposons can act as introns, they cannot evolve to be simply introns; the result would be insertions that are spliced but cannot move. Indeed, mutations within element insertions that decrease transposition rates while increasing splicing efficiency have been observed (Mennsen et al., 1990). These element insertions effectively become introns, and their ability to transpose and subsequently replicate is lost. For splicing to be effective as a strategy for increasing element fitness, it must function in concert with the continued ability of elements to move within genomes.

It is clear that there exists alternative strategies for transposable elements to escape negative selection, including the modulation of transposition activity by both host-encoded and transposon-encoded mechanisms (Emmons and Yesner, 1984; Kimura and Kidwell, 1994; Fedoroff, 1995). The importance of splicing as a strategy towards enhancing fitness, however, has yet to be explored in a systematic fashion. It would be interesting to determine the conditions in which the ability to be spliced has real selective advantages. Empirical and theoretical studies to address this question can be undertaken to assess whether transposable elements that possess the ability to be spliced can compete effectively against elements that cannot. Together, these approaches may begin to address the question of whether splicing truly represents a cellular defense against transposable element invasions, and the subsequent mutagenic insertions they create.

ACKNOWLEDGMENTS

The author wishes to thank the anonymous reviewers for critical comments that improved the manuscript. The author would also like to express his gratitude to Sue Wessler and John McDonald for numerous interesting discussions on the subject of transposable element splicing. This work was supported by the North Carolina Agricultural Service and grants from the Alfred P. Sloan and National Science Foundations.

REFERENCES

Adams, M.D., Rudner, D.Z. and Rio, D.C. (1996) Biochemistry and regulation of pre-mRNA splicing. *Curr. Opin. Cell Biol.* **8**: 331–339.

Anxolabehere, D., Kidwell, M.G. and Periquet, G. (1988) Molecular characterization of diverse populations are consistent with the hypothesis of a recent invasion of *D. melanogaster* by mobile *P* elements. *Mol. Biol. Evol.* **5**: 252–269.

Berg, D. and Howe, M. (1989) *Mobile DNA,* American Society for Microbiology, Washington, DC.

Berget, S.M. (1995) Exon recognition in vertebrate splicing. *J. Biol. Chem.* **270**: 2411–2414.

Charlesworth, B. and Charlesworth, D. (1983) The population dynamics of transposable elements. *Genetical Res.* **42**: 1–27.

Charlesworth, B. and Langley, C.H. (1989) The population genetics of *Drosophila* transposable elements. *Annu. Rev. Genetics* **23**: 251–287.

Clark, J.B., Maddison, W.P. and Kidwell, M.G. (1994) Phylogenetic analysis supports horizontal transfer of *P* transposable elements. *Mol. Biol. Evol.* **11**: 40–50.

Crick, F. (1979) Split genes and RNA splicing. *Science* **204**: 264–271.

Dennis, E., Sachs, M., Gehrlach, W. et al. (1988) The *Ds1* transposable element acts as an intron in the mutant allele *adh1-Fm335* and is spliced from the message. *Nucl. Acids Res.* **16**: 3315–3328.

Doolittle, W.F. and Sapienza, C. (1984) Selfish genes: the phenotype paradigm and genome evolution. *Nature* **284**: 601–603.

Emmons, S. and Yesner, L. (1984) High frequency excision of transposable element *Tc1* in the nematode *C. elegans. Cell* **36**: 599–605.

Fedoroff, N. (1995) Maize transposable element regulation. *Maydica* **40**: 7–12.

Flavell, A.J. (1992) *Ty1-copia* group retrotransposons and the evolution of retroelements in the eukaryotes. *Genetica* **86**: 203–214.

Fridell, R., Pret, A. and Searles, L. (1990) A retrotransposon *412* insertion within an exon of the *D. melanogaster vermilion* gene is spliced from precursor RNA. *Genes and Dev.* **4**: 559–566.

Fu, X.D. (1995) The superfamily of arginine/serine-rich splicing factors. *RNA* **1**: 663–680.

Geyer, P., Chien, A., Corces, V. and Green, M. (1991) Mutations in the *su(s)* gene affects RNA processing in *Drosophila melanogaster. Proc. Natl Acad. Sci. USA* **88**: 7116–7120.

Gierl, A. (1990) How maize transposable elements escape negative selection. *Trends Genet.* **6**: 155–158.

Hickey, D. (1980) Selfish DNA, a sexually-transmitted nuclear parasite. *Genetics* **101**: 519–531.

Houck, M.A., Clark, J.B., Peterson, K.R. and Kidwell, M.G. (1991) Possible horizontal transfer of *Drosophila* genes by the mite *Proctolaelaps regalis. Science* **253**: 1125–1129.

Kidwell, M.G. (1985) Hybrid dysgenesis in *D. melanogaster*: Nature and inheritance of *P* element regulation. *Genetics* **111**: 337–350.

Kidwell, M.G. (1993) Lateral transfer in natural populations of eukaryotes. *Annu. Rev. Genetics.* **27**: 235–256.

Kim, H., Schiefelbein, J., Raboy, V. et al. (1987) RNA splicing permits expression of a maize gene with a defective *Spm* transposable element insertion in an exon. *Proc. Natl Acad. Sci. USA* **84**: 5863–5867.

Kimura, K. and Kidwell, M.G. (1994) Differences in *P* element population dynamics between the sibling species *D. melanogaster* and *D. simulans. Genetical Res.* **63**: 27–38.

Kohtz, J.D., Jamison, S.F., Will, C.L. et al. (1994) Protein–protein interaction and 5′ splice site recognition in mammalian mRNA precursors. *Nature* **368**: 119–124.

Labrador, M. and Fontdevila, A. (1994) High transposition rates of *Osvaldo*, a new *Drosophila buzzatii* retrotransposon. *Mol. Gen. Genet.* **245**: 661–674.

Lozovskaya, E., Hartl, D.L. and Petrov, D.A. (1995) Genomic regulation of transposable elements in *Drosophila. Curr. Opin. Genetics Dev.* **5**: 768–773.

Mackay, T.F.C., Lyman, R.F. and Jackson, M.S. (1992) Effects of *P* element insertions on quantitative traits in *Drosophila melanogaster. Genetics* **130**: 315–332.

McClintock, B. (1950) The origin and behaviour of mutable loci in maize. *Proc. Natl Acad. Sci. USA* **36**: 344–355.

McClintock, B. (1951) Chromosome organization and genic expression. *Cold Spring Harb. Symp. Quant. Biol.* **16**: 13–47.

McClintock, B. (1984) The significance of responses of the genome to challenges. *Science* **226**: 792–801.

McDonald, J.F. (1990) Macroevolution and retroviral evolution. *BioScience* **40**: 183–191.

McDonald, J.F. (1995) Transposable elements – possible catalysts of organismic evolution. *Trends Ecol. Evol.* **10**: 123–126.

Menssen, A., Hohmann, W.M., Schnable, P. et al. (1990) The *En/Spm* transposable element of *Z. mays* contains splice sites at the termini generating a novel intron from a *dSpm* element in the *A2* gene. *EMBO J.* **9**: 3051–3057.

Orgel, L.E. and Crick, F. (1984) Selfish DNA: The ultimate parasite. *Nature* **284**: 604–607.

Pret, A. and Searles, L. (1991) Splicing of retrotransposon insertion from transcripts of the *Drosophila melanogaster vermilion* gene in a revertant. *Genetics* **129**: 1137–1145.

Purugganan, M.D. (1993) Transposable elements as introns: evolutionary connections. *Trends Ecol. Evol.* **8**: 239–243.

Purugganan, M.D. and Wessler, S.R. (1992) The splicing of transposable elements and its role in intron evolution. *Genetica* **86**: 295–303.

Purugganan, M.D. and Wessler, S.R. (1994) Molecular evolution of *magellan*, a maize *Ty3/gypsy*-like retrotransposon. *Proc. Natl Acad. Sci. USA* **91**: 11674–11678.

Reed, R. (1996) Initial splice-site recognition and pairind during pre-mRNA splicing. *Curr. Opin. Genetics Dev.* **6**: 215–220.

Robberson, B.L., Cote, G.J. and Berget, S.M. (1992) Are vertebrate exons scanned during splice site selection? *Nature* **360**: 277–280.

Robertson, H.M. (1993) The mariner transposable element is widespread in insects. *Nature* **362**: 241–245.

Robertson, H.M. and Lampe, D.J. (1995a) Recent horizontal transfer of a *mariner* transposable element among and between Diptera and Neuroptera. *Mol. Biol. Evol.* **12**: 850–862.

Robertson, H.M. and Lampe, D.J. (1995b) Distribution of transposable elements in arthropods. *Annu. Rev. Entomol.* **40**: 333–357.

Rushforth, A.M. and Anderson, P. (1996) Splicing removes the *C. elegans* transposon *Tc1* from most mutant pre-mRNAs. *Mol. Cell. Biol.* **16**: 442–429.

Scofield, S.R., English, J.J. and Jones, J.D. (1993) High level expression of the Ac transposase gene inhibits the excision of Ds in tobacco cotyledons. *Cell* **75**: 507–517.

Staknis, D. and Reed, R. (1994) SR proteins promote the first specific recognition of pre-mRNA and are present together with the U1 small nuclear ribonucleoprotein particle in a general splicing enhancer complex. *Mol. Cell. Biol.* **14**: 7670–7682.

Syvanen, M. (1984) The evolutionary implications of mobile genetic elements. *Annu. Rev. Genetics* **18**: 271–293.

Varagona, M.J., Purugganan, M.D. and Wessler, S.R. (1992) Alternative splicing induced by insertions of retrotransposons into the maize *waxy* gene. *Plant Cell* **4**: 811–820.

Voytas, D.F., Cummings, M.P., Konieczny, A. et al. (1992) *Copia*-like retrotransposons are ubiquitous among plants. *Proc. Natl Acad. Sci. USA* **86**: 7124–7128.

Weil, C.F. and Wessler, S.R. (1990) The effects of plant transposable element insertions on transcription initiation and RNA processing. *Annu. Rev. Plant. Physiol. Plant Mol. Biol.* **41**: 527–552.

Weil, C.F. and Wessler, S.R. (1993) Molecular evidence that chromosome breakage by Ds elements is caused by aberrant transposition. *Plant Cell* **5**: 515–522.

Wessler, S. R. (1989) Phenotypic diversity mediated by the maize transposable elements Ac and Spm. *Science* **242**: 399–405.

Wessler, S. (1991) The maize transposable Ds1 element is alternatively spliced from exon sequences. *Mol. Cell. Biol.* **11**: 6192–6196.

Wessler, S.R., Baran, G. and Varagona, M.J. (1987) The maize transposable element Ds is spliced from RNA. *Science* **237**: 916–918.

Wessler, S.R. and Varagona, M.J. (1985) Molecular basis of mutations at the *waxy* locus of maize: Correlation with the fine-structure genetic map. *Proc. Natl Acad. Sci. USA* **82**: 4177–4181.

SECTION IV

Transfer Mechanisms Involving Plants and Microbes

Phylogenetic evidence is available suggesting that horizontal gene transfer has occurred in the past. Less clear are the mechanisms that facilitate these transfers, especially among eukaryotes. This section is composed of a series of reviews on cross-species transfer mechanisms involving eukaryotes that occur, or possibly occur, in nature. Riesberg and Welch (Chapter 18) and Ellstrand et al. (Chapter 19) describe evidence for introgressive hybridization in plants. Though this phenomenon has been documented for centuries, recent concerns about the safety of bioengineered crops have stimulated interest in new investigations. This concern has also stimulated research into transfer of genes from transgenic plants to microbes, a topic that is reviewed in Chapter 20 by Syvanen. Wöstemeyer et al in Chapter 21 describe an interesting phenomenon that could be possibly facilitate transfer among fungi. This example of parasitica, as well as *Agrobacterium tumefaciens*, reviewed in Section I of this book, serve as a model for vectors that could possibly move DNA among many different eukaryotes.

Chisholm et al. (Chapter 22) and Grillot-Courvalin et al. (Chapter 23) deal with genetic engineering technologies, but are included here to point out possible mechanisms of horizontal gene transfer. They discuss bacterial vectors that can act as donors to fungal and metazoan hosts after intracellular localization of the bacteria. Given the large number of intracellular bacterial pathogens and symbiotes (such as the *Rickettsia* and *Wollbachia*), these mechanisms could prove to be important.

Finally, the first vectors suggested to be of importance for horizontal gene transfer among eukaryotes were viruses, either DNA or RNA. There has been little new information on this topic since it was last reviewed (Syvanen, 1986, 1987).

REFERENCES

Syvanen, M. (1986) Cross-species gene transfer: a major factor in evolution. *Trends Genet.* **2**: 63–66.

Syvanen M. (1987) Molecular clocks and evolutionary relationships: possible distortions due to horizontal gene flow. *J. Mol. Evol.* **26**: 16–23.

Gene Transfer Through Introgressive Hybridization: History, Evolutionary Significance, and Phylogenetic Consequences

Loren H. Rieseberg and Mark E. Welch

The movement of genetic material across species barriers can be achieved either horizontally through the various mechanisms described in this book or vertically through introgressive hybridization. This chapter focuses on the history, evolutionary role, and phylogenetic impact of gene transfer through introgressive hybridization. The rudimentary understanding of hybridization and heredity demonstrated by ancient humans was largely lost until late in the 17th century. Over the next two centuries, however, students of hybridization such as Linnaeus, Kölreuter, and Mendel elucidated the principles of sexual reproduction, hybridization, and inheritance. The possibility of gene transfer through introgression was not recognized until early in the 20th century, but its evolutionary importance has been debated ever since. We argue that hybridization frequencies are high enough for introgression to play a significant evolutionary role in plants and animals and show that hybrid zones generally do not represent significant barriers to the flow of neutral or advantageous alleles. The role of introgression in adaptive evolution is more difficult to assess, however, because advantageous alleles spread quickly, making it difficult to catch the introgression of favorable alleles in action. Nonetheless, molecular marker surveys indicate that introgression is widespread in both plants and animals. This body of evidence poses a significant challenge to (1) theories of speciation that require geographic isolation and *ad hoc* mechanisms to prevent gene flow (Dobzhansky, 1941; Mayr, 1942), and (2) current phylogenetic methods, which assume purely dichotomous patterns of evolution.

INTRODUCTION

Why should hybridization be discussed in a book devoted to horizontal gene transfer? Horizontal transfer refers to the movement of genetic material between related or unrelated neighbors (Amábile-Cuevas and Chicurel, 1992; Ochman et al., 2000), whereas hybridization refers to vertical gene transfer between distinct evolutionary lineages through sexual reproduction. Both processes may result in the movement of genetic material across species barriers. Thus, horizontal gene transfer and hybridization represent viable alternative explanations for discordance between gene lineages and organismal pedigrees.

Horizontal Gene Transfer
ISBN: 0-12-680126-6

Copyright © 2002 by Academic Press.
All rights of reproduction in any form reserved.

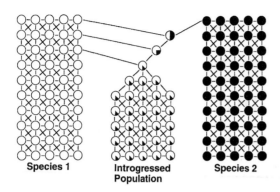

FIGURE 18.1 Gene transfer through introgressive hybridization. Open circles = populations of species 1; closed circles = populations of species 2; black lines = crosses between populations. (After Rieseberg and Wendel, 1993.)

The process by which genes move between hybridizing lineages is referred to as introgression (Anderson and Hubricht, 1938). Briefly, hybridization between two taxa is followed by backcrossing between the hybrids and one or both parental lineages. After several generations of backcrossing, parental-like individuals are recovered that contain only a small proportion of the alien genome (Figure 18.1). If frequent, introgression represents a challenge to cherished concepts such as the biological species and the hierarchical nature of phylogeny. Here we review the historical developments that have led to our current understanding of introgression, and its evolutionary importance.

HISTORY

Knowledge of hybridization in ancient times

Exhaustive treatments of the early history of hybridization are provided by Roberts (1929), Zirkle (1935), and Stubbe (1972). Here we summarize the highlights from the early literature on hybridization and attempt to make connections with modern theory.

Hybridization was probably first recognized by humans shortly after the first animals were domesticated more than 10 000 years ago

(Olsen, 1985). These early cases, which may have been promoted by people, involved the hybridization of domesticated animal stocks with wild populations for the purpose of enhancing domesticated blood lines, a process which was eventually recorded by Romans (Stubbe, 1972). The oldest documentary evidence of hybridization appears to come from Homer's Odyssey (800 BC), in which mules were recognized as hybrids between the horse and the ass (Zirkle, 1935). That mules were the earliest documented hybrids is not surprising. Of the early animal domesticates, horses and asses will most readily mate and produce viable offspring (Rife, 1965).

The mule (female horse × male ass) and hinny (female ass × male horse) were perhaps the first hybrid combinations to be studied in detail. Several generalizations (or stereotypes) about hybrids appear to have arisen in part from these early observations. First, mules and hinnies are morphologically intermediate relative to their parents, exemplifying the aphorism of hybrid intermediacy (Rieseberg and Ellstrand, 1993). Secondly, male mules and male hinnies are always sterile, illustrating the principle of hybrid sterility. Thirdly, female mules or female hinnies are sometimes fertile (Rife, 1965), providing the earliest example of Haldane's Rule (1922), which states that, if only one hybrid sex is fertile or viable, it is nearly always the homogametic (XX) sex.

Because of the more cryptic nature of sex in plants it is unlikely that early humans recognized the role of heredity in plant reproduction (reviewed in Zirkle, 1935). However, there are two lines of evidence suggesting that early farmers may have had some knowledge of the sexual nature of plants and of plant hybrids. The first line of evidence involves the intentional cross-pollination of date palms and figs for agricultural purposes. Archaeological evidence indicates that date palms, which have distinct male and female forms, were cultivated as early as the 4th millennium BC (Zohary and Hopf, 1988). The species is wind-pollinated in nature, and populations have a 1:1 ratio of males and females. With hand pollination, however, the proportion of female fruit-bearing trees can be increased at the expense of non-productive male trees. Evidence from orchard layout indicates that Babylonians had discovered the efficacy of hand pollination by 2400 BC (Zirkle, 1935), and Assyrian bas-reliefs from 800 BC illustrate actual pollination. Farmers from the same region fertilized cultivated figs with wild pollen to prevent immature fruit from falling to the ground (Zirkle, 1935). There is little evidence, however, to suggest that the knowledge of pollination in date palms and figs was transferred to other species.

The hybrid origin of many cultivated plants provides a second line of evidence suggesting that ancient humans may have had some knowledge of hybridization in plants. As a result of spreading agriculture, species that were previously geographically isolated were incidentally brought into reproductive contact in the fields of early farmers. Hybridization between species and varieties occurred, allowing hybrids to compete and interbreed with parental genotypes (Zohary and Hopf, 1988). As a result, some hybrid forms replaced their parental lineages completely. The most notable of these is common bread wheat, *Triticum aestivus*, which is an allohexaploid (Zohary and Hopf, 1988). However, the hybrid origin of certain crops does not necessarily mean that early farmers understood plant sexuality and hybridization. Hybrid crop plants likely increased gradually in number and may have escaped the notice of early farmers. In contrast, the hand-pollination of dates and figs required a conscious effort. Whatever knowledge early farmers may have had of plant sexuality and hybridization was largely lost until late in the 17th century (Roberts, 1929; Zirkle, 1935).

The first recorded theories of heredity were developed by Greek philosophers. These early theories were diverse and had little influence on the work of early hybridizers or on the development of modern theories of inheritance. However, the arguments of one group, the atomic and materialist philosophers, bear a striking resemblance to modern views (Darlington, 1937). The conclusions of atomic theory in relation to heredity are summarized by Lucretius (98–55 BC) in a description of evolution by natural selection. Atomic theory proposed that inheritance was achieved by transmitting material

bodies from one generation to the next. Because this material was believed to be atomic (or particulate), inheritance was reasoned to be particulate as well (this was later demonstrated by Mendel). Offspring were believed to receive more or less equal contributions of these particles from both parents, a principle that some primitive people have not yet learned. Finally, atomic theory correctly proposed that these particles of inheritance would separate and recombine through sexual reproduction (Darlington, 1937).

The many theories developed by Greek and Roman philosophers were debated well into the Middle Ages, but experimentation was rare and poorly documented. Thus, it is perhaps not surprising that atomic theory was rejected or at least forgotten. Not until the 17th century did Western science begin to unravel the mysteries of heredity and reproduction in plants and animals. Hybridization experiments played a key role in this process.

Early hybridizers

The sexuality of plants was not recognized by modern man until 1694 when Camerarius demonstrated that pollen was required before female flowers would produce viable seed (Olby, 1985). Hence, it was observations of spontaneous animal hybrids that provided the impetus for early hybridization experiments. Unfortunately, the animal stocks available for study were large domesticated species with long generation times. As a result, animal hybridization experiments were slow and progeny sizes were small. The French biologist, Georges Louis Buffon, is perhaps the best known of the early animal hybridizers (Stubbe, 1972). In the early 1700s, Buffon made several crosses among wild and domesticated stocks of long-lived animals such as species of goats and wolves (Olby, 1985). Buffon had hoped that the capacity to hybridize could be used as a diagnostic for species status, but he abandoned this notion when fertile offspring were reported between what were otherwise good species of birds (Mayr, 1982).

Studies of plant hybridization were also initiated during this period. Spontaneous hybrids were first reported in a letter written by Cotton

Mather in 1716, in which he describes hybrids between Indian and yellow corn and between gourds and squash (Zirkle, 1935). The first intentional plant hybrid (carnation × sweet William) may have been generated in the same year by the English horticulturist, Thomas Fairchild (Zirkle, 1935). Many garden plants have features that are favorable for hybridization studies, such as ease of propagation, short generation time, and easily circumvented barriers to interspecific mating. As a result, hybridization studies developed much more rapidly in plants than animals.

Scientific study of plant hybrids was initiated by Linnaeus, the "father of taxonomy." Linnaeus was burdened with the doctrine of special creation, which considered species to be immutable. Thus, he initially considered hybridization to be unimportant and viewed it only as a means of species determination. However, the successful synthesis of hybrids between what were considered "good" species of plants and animals by Linnaeus and his contemporaries slowly undermined the notion of fixity. In fact, by 1751 Linnaeus considered most species to be of hybrid origin (Olby, 1985), and in an essay on the sexes of plants in 1760, he asks whether all members of a genus might not have a common ancestor (Roberts, 1929). These views of Linnaeus, while foreshadowing Lamarck and Darwin, do not constitute a full-blown theory of evolution. Moreover, his views on species formation through hybridization were based on a few first generation hybrids; he apparently was not aware of potential difficulties with the theory of hybrid speciation such as hybrid sterility and segregation.

Linnaeus' work was followed by publication of a much more comprehensive set of hybridization experiments by Kölreuter (1893; cited in Mayr, 1986). These experiments, which involved more than 100 plant species, resulted in several major discoveries about hybrids that are widely accepted today: (1) first generation hybrids are uniformly intermediate and often more vigorous than either parent; (2) the fertility expressed by interspecific hybrids varies from full fertility to complete sterility, but favors sterility; (3) unlike first generation hybrids, second generation hybrids are highly variable;

and (4) later generation hybrids often revert to one parental form or the other. Kölreuter viewed his findings as proof of the fixity of species, and as inconsistent with Linnaeus' claims regarding the establishment of new true-breeding species through hybridization. While Kölreuter's hybridization experiments were far more sophisticated than those of Linnaeus, his work was much less popular. It would be more than 50 years before Kölreuter would finally receive due credit and recognition (Olby, 1985).

In 1826, Sagaret replicated and confirmed several of Kölreuter's experiments and helped redeem the latter's reputation (Roberts, 1929). Shortly thereafter, Gärtner, the most thorough of the early plant hybridizers, initiated the experiments that finally settled many issues concerning plant reproduction and heredity (Olby, 1985). However, with one exception, no real extension of Kölreuter's work was made. Over a long career, Gärtner conducted nearly 1000 hybridization experiments and gained many insights through his experience. He recognized that there are fundamental differences among interspecific hybrids in their expression of parental characters from generation to generation. Kölreuter considered the F_1 to be intermediate. Gärtner also saw this blending of traits. However, Gärtner found it much more common for hybrids to favor one parent or the other in specific traits and yet be generally balanced in their expression of parental traits. Rieseberg and Ellstrand (1993) later showed that this "mosaic" expression of parental characters in hybrids was the rule rather than the exception. Gärtner also recognized that certain interspecific hybrids were biased. That is, the hybrids favored one parental species but still showed characteristics indicative of the second parental species. And lastly, he recognized that some hybrids were constant in their transmission of hybrid character combinations to successive generations and did not show signs of reversion or segregation. Today we know these constant hybrids to be polyploids, products of apomixis (asexual reproduction via seed), or permanent interchange hybrids (Grant, 1981).

In addition to the confusion caused by constant hybrids, most of Gärtner's contemporaries failed to distinguish among different classes of sexual, homoploid hybrids (Roberts, 1929). Instead, they lumped interspecific hybrids into one category, and attempted to find laws that governed them all. This practice not only muddied the hybridization literature, but also misled early students of heredity (Olby, 1985). For example, Darwin's failure to elucidate the principles governing heredity can be traced to his attempts to reconcile all data concerning hybrids (Roberts, 1929). In contrast, Mendel's focus on segregating hybrid classes, and his willingness to ignore observations and reports that were inconsistent with his theories, were major contributors to his success (Roberts, 1929).

During this period, there was little discussion of the evolutionary role of hybrids, although Mendel commented that "... constant hybrids attain the status of new species." (Mendel, 1865; quoted in Roberts, 1929). Likewise, Naudin (1863; cited in Roberts, 1929) pointed out that hybrid characters might become fixed in later generation hybrids and that this might facilitate species formation. However, there was no evidence that any of the early hybridizers had considered the possibility of natural interspecific gene transfer through introgression. Rather, interest focused on the development of hybrid varieties that might be of horticultural interest or on the validity of hybrid sterility as a species criterion (Roberts, 1929). In addition, there were increasing numbers of reports documenting spontaneous natural hybrids. The implications of these reports were summarized by Focke (1881), who noted that (1) natural hybridization has a very uneven taxonomic distribution; (2) species with zygomorphic flowers are more likely to hybridize than those with actinomorphic symmetry; and (3) hybridization is most likely when the ratios of the parental species are strongly biased. All of these observations/predictions have been confirmed by later studies (Stebbins, 1959; Arnold, 1997).

Cytology and hybrid speciation

Chromosomes were first observed by Carl W. Von Nägeli in a study of pollen cells in 1842 (Stubbe, 1972). However, it would not be until Theodor Boveri produced haploid sea urchins in 1889 that chromosomes began gaining

acceptance as the vehicles of heredity. Shortly thereafter, cytologists such as Strasburger and Fleming initiated studies of chromosome number variation within and among species (Stubbe, 1972). They found that related species tended to have similar numbers of chromosomes or multiples of the same number of chromosomes. This latter observation was explained correctly as a "reduplication" of the genome, or polyploidy. The evolutionary significance of this discovery was recognized by Wingë (1917), who suggested that genome doubling could instantaneously produce new true breeding and reproductively isolated forms and might contribute to apomixis. Wingë's polyploid speciation hypothesis was confirmed by Clausen and Goodspeed (1925), and allopolyploidy is now recognized as perhaps the most significant evolutionary outcome of natural hybridization (Soltis and Soltis, 1993; Leitch and Bennett, 1997; Ramsey and Schemske, 1998).

The discovery of allopolyploidy provided an explanation for the constant hybrids reported in the early hybridization literature and thus reconciled these observations with the Mendelian theory. It also forced students of hybridization to distinguish between polyploid and diploid or "homoploid" hybrids and to explore the evolutionary consequences of hybridization at different ploidal levels.

With respect to homoploid hybrids, an obvious question emerged. Could new species arise via hybridization, but without a change in ploidy? Homoploid hybrid speciation is theoretically difficult because, in the absence of some type of hybrid founder event, reproductive isolation must be developed in sympatry (Rieseberg, 1997). Otherwise, the new hybrid lineage will be overcome by gene flow with its parents. Müntzing (1930) suggested that reproductive isolation might be achieved through the sorting of chromosomal rearrangements in later generation hybrids. In this "recombinational" model, fertility selection could lead to the development of hybrid lineages that were homozygous for a new karyotype and thus partially isolated from their parental species. The recombinational model was later expanded to incorporate genic sterility (Stebbins, 1957), ecological differentiation (Grant, 1981), and hybrid founder events (Charlesworth, 1995). Exploration of the recombinational model through artificial hybridization experiments (Stebbins, 1957; Grant, 1966; Rieseberg et al., 1996b) and computer simulation (McCarthy et al., 1995; Buerkle et al., 2000) indicate that it is feasible, but that the requirements are stringent. Thus, it is perhaps unsurprising that homoploid hybrid speciation appears to be rare in nature, with only a handful of confirmed examples (Rieseberg, 1997; Dowling and Secor, 1997).

Introgression

Another possible outcome of hybridization was proposed by Ostenfeld (1927). He reasoned that natural hybridization followed by several generations of backcrossing might lead to the gradual infiltration of the germplasm of one species into another. Empirical evidence to support this hypothesis quickly followed. For example, Du Rietz (1930) noted that sympatric populations of *Dacrophyllum*, *Coprosma*, and *Salix* species tend to converge in certain morphological features. Du Rietz attributed this convergence to the "infection" of one species by genes from another. Likewise, backcrossing experiments by Marsden-Jones (1930) showed that, as predicted by Ostenfeld, parental-like individuals could be recovered after only a few generations of backcrossing.

The process of gene transfer through backcrossing was termed introgressive hybridization by Edgar Anderson and co-workers (reviewed in Anderson, 1949, 1953), who also collected substantial morphological evidence for introgression, developed methods for its analysis, recognized the link between habitat disturbance and hybridization, and suggested that foreign alleles acquired through introgression might contribute to adaptive evolution. In fact, Anderson went so far as to suggest that the "raw material for evolution brought about by introgression must greatly exceed the new genes produced directly by mutation" (Anderson, 1949).

However, these claims were received with a certain degree of skepticism by the scientific community. The primary problem was that the supporting evidence often had alternative explanations (Gottlieb, 1972; Heiser, 1973; Rieseberg and Wendel, 1993). For example, Baker (1947) was skeptical about the use of hybrid indices and other biometric tools in the absence of knowledge concerning the genetic basis of the studied characters. Dobzhansky (1941) worried that the morphological convergence attributed to introgression could be due instead to parallel selection or to the joint retention of ancestral character states (symplesiomorphy). Mayr (1942) noted that clines, which were typically viewed as evidence for hybridization following secondary contact, might often arise *in situ*.

There also were theoretical concerns that hampered the widespread acceptance of introgression as an important evolutionary force. Early zoological workers, for example, envisioned hybrid zones as being ephemeral (Dobzhansky, 1941; Mayr, 1942; Wilson, 1965). Hybridizing species either merged due to extensive gene flow or reproductive barriers were reinforced, leading to the cessation of hybridization. It was difficult to envision how introgression could contribute significantly to adaptive evolution under either scenario. Another theoretical problem concerned the genetics of reproductive barriers, which were thought to be controlled by very many genes. Under this infinitesimal model (Fisher, 1930), it was difficult to see how foreign alleles could move across a reproductive barrier, because they would inevitably be tightly linked to negatively selected alleles. Moreover, for polygenic traits, alleles contributing to the trait would have to be individually advantageous to transgress the species' barrier (Barton and Bengtsson, 1986).

The acceptance of Anderson's ideas was also slowed by the view that hybridization was too rare in many animal groups for introgression to play a major role (Mayr, 1942). In addition, the notion that species could evolve and/or be maintained in the presence of gene flow represented a challenge to the biological species concept and to prevailing views about the requirement of allopatry for speciation (Mayr, 1963).

Both theoretical and empirical advances have been made over the past several decades that have allayed some of these concerns and have begun to provide reliable evidence regarding

TABLE 18.1 Hybrid zone models

Model	Selection	Hybrid fitness	Clinal variation	Reference
Ecological selection gradient	Exogenous	>Parents	Yes	Moore and Price (1993)
Dynamic equilibrium	Endogenous	<Parents	Yes	Barton (1979)
Mosaic	Exogenous and endogenous	<Parents	Spatial scale dependent	Harrison (1986)
Evolutionary novelty	Exogenous and endogenous	Variable	Yes, if along ecotone	Arnold (1997)

the frequency of introgression and its likely consequences (reviewed in Rieseberg and Wendel, 1993; Arnold, 1997; Dowling and Secor, 1997).

THEORY

Stability of hybrid zones

The assumption that hybrid zones were ephemeral was widely held until the mid-1970s, when a compilation of evidence from empirical hybrid zone studies was used to argue that hybrid zones could be stable over long periods (Hewitt, 1975; Endler, 1977; Moore, 1977). These empirical conclusions spurred the development of theoretical models that could account for the observed stability. One such model proposed that hybrid zones could be stable if the fitness of hybrids was at least equal to that of the parental species in some portion of the hybrid zone (Table 18.1). In this "ecological selection gradient" or "bounded hybrid superiority" model (Endler, 1977; Moore, 1977; Moore and Price, 1993), adaptation to different environments by the parental species and their hybrids prevents homogenization of gene pools. The model therefore predicts that selection in the hybrid zone will be largely exogenous (i.e. environment dependent), that diagnostic characters should exhibit clinal variation if the zone occurs along an ecotone, and that only a subset of favored hybrid genotypes are likely to be common in the hybrid zone.

Shortly after publication of the hybrid superiority model, Barton (1979) and Barton and Hewitt (1981, 1985, 1989) demonstrated mathematically that hybrid zones could be

maintained through a balance of dispersal of parental individuals into the hybrid zones and selection against hybrids as a result of their mixed ancestry. In this "dynamic equilibrium" or "tension zone" model, selection in the hybrid zone is endogenous (environment independent) and should not vary over geographic space. Like the ecological selection gradient model, clinal variation is predicted for diagnostic traits (Table 18.1). However, a much wider diversity of hybrid genotypes are likely to be present in dynamic equilibrium zones than in zones maintained by an ecological selection gradient. In addition, dynamic equilibrium zones are predicted to be fairly uniform in width, whereas the width of zones maintained by ecological selection are expected to vary according to the steepness of the environmental gradient.

The major differences between the ecological selection gradient and dynamic equilibrium models are the emphasis on equality or superiority of hybrids and exogenous selection in the former and hybrid inferiority and endogenous selection in the latter (Table 18.1). Of course, in natural hybrid zones, both kinds of selection may operate and relative fitnesses may vary (Bert and Arnold, 1995; Burke et al., 1998a,b). The recognition of this potential complexity has led to two additional hybrid zone models. One of these, the "mosaic" model (Harrison, 1986; Howard, 1986), assumes that hybrid zones are maintained through the same dispersal/negative selection balance previously described for the dynamic equilibrium model (Table 18.1). However, like the ecological selection gradient model, it recognizes that species are often adapted to different, patchily distributed habitats. This results in a patchy distribution of parental and hybrid genotypes. As a result of this structure, a mosaic pattern of variation for diagnostic characters is predicted, although clinal variation may be observed at either very local or very broad geographic scales (Rand and Harrison, 1989).

The final model was developed by Arnold (1997). This "evolutionary novelty" model is the most pluralistic of the models in that both exogenous and endogenous selection are assumed to be important in hybrid zones and that hybrid fitness is predicted to vary (from less than to more than one or both parents) depending on both genotype and habitat (Table 18.1). As in the ecological selection gradient and mosaic models, "evolutionary novelty" zones are predicted to be structured according to habitat. However, clinal variation is not predicted unless the hybrid zone happens to occur along an ecotone.

So which of these models most accurately describes natural hybrid zones? Barton and Hewitt (1985, 1989) argue that many of the most intensively studied animal hybrid zones are best explained by the dynamic equilibrium model. Evidence in favor of this view includes observations of inviability or sterility in many animal hybrids, the deficit of hybrid forms in many zones, the uniform width of most zones, and the fact that hybrid zones can often be shown to have moved, suggesting that they are environment independent. However, as these and other zones have been analyzed in greater depth, their general fit to the dynamic equilibrium model has been questioned (reviewed in Arnold, 1997). Two irregularities have been observed with disconcerting frequency. First, hybrid fitnesses often are more variable than expected under the dynamic equilibrium model (Howard et al., 1993; Bert and Arnold, 1995; Nürnberger et al., 1995; Reed and Sites, 1995; Rolan-Alvarez et al., 1997; Semlitsch et al., 1999). Second, a substantial number of these zones show evidence of habitat-dependent selection (Harrison, 1986; Bert and Arnold, 1995; Rolan-Alvarez et al., 1997; MacCallum et al., 1998; Semlitsch et al., 1999). Although the

accumulating evidence is consistent with a dispersal/negative selection balance, it also suggests a greater role for exogenous selection in animal hybrid zones than was originally envisioned, supporting the mosaic model.

There is less evidence for the dynamic equilibrium model in plants, primarily because plants tend to be more closely tied to their environment than are animals (Arnold, 1997; Rieseberg, 1998). In general, plant hybrid zones between genetically or chromosomally divergent taxa seem best explained by a mosaic model (Heiser, 1947, 1979; Bloom and Lewis, 1972; Hauber and Bloom, 1983; Potts, 1986; Meyn and Emboden, 1987; Keim et al., 1989; Burke et al., 1998b; Rieseberg et al., 1998; Carney et al., 2000). In contrast, zones between closely related and interfertile taxa often appear to be maintained via an ecological selection gradient (Levin and Schmidt, 1985; H. Wang et al., 1997; Campbell et al., 1998). Of course, many of the examples listed above could also be explained by the evolutionary novelty model, given its pluralistic nature (Arnold, 1997).

Although we are not yet sure which of these models explains the largest number of natural hybrid zones, all four provide reasonable explanations for the maintenance and stability of hybrid zones. Thus, one of the major objections to Anderson's hypothesis of introgression and its role in evolution can be discarded.

Introgression

The four hybrid zone models described above also provide a framework for addressing another critical theoretical concern. That is, can hybrid zones serve as bridges for the movement of advantageous alleles between species and thus contribute to adaptive evolution as postulated by Anderson (1949)? This question has been addressed most thoroughly within the framework of the dynamic equilibrium model (Barton, 1979, 1986; Barton and Bengtsson, 1986; Barton and Hewitt, 1989; Pialek and Barton, 1997). However, conclusions from this work should apply to mosaic, ecological selection gradient, and evolutionary novelty zones as well (Petry, 1983; Gavrilets and Cruzan, 1998).

Theoretical work indicates that successful introgression will depend both on the selective value of individual alleles in a hybrid genetic background, and on the number and distribution of loci that contribute to the species barrier (Barton and Hewitt, 1989). In the middle of the zone, alleles derived from the same parental species will tend to be associated and selection will act on these correlated sets of alleles. However, as alleles move away from the center of the zone, they will recombine into a new genetic background and selection will act on their individual effects. With respect to introgression, this means that alleles that contribute to reproductive isolation and thus are negatively selected in hybrids will fail to move beyond the center of the hybrid zone. Likewise, hybrid zones will slow the introgression of neutral alleles due to linkage. However, advantageous alleles are unlikely to be significantly impeded by most hybrid zones because once a few alleles successfully transgress the barrier, they will rapidly increase in frequency.

It is possible that very tight linkage between advantageous and negatively selected alleles could prevent the flow of the former (R.L. Wang et al., 1997). However, linkage must be very tight to be effective in reducing gene flow at nearby loci. For example, Ting et al. (2000) have shown that regions of DNA separated by just 2 kb appear to act independently with respect to the retention of shared polymorphisms and/or history of introgression.

The important conclusion from this work is that hybrid zones should not provide a significant barrier to the movement of advantageous alleles, which of course are the kinds of alleles that are most likely to contribute significantly to adaptive evolution. Thus, hybrid zone theory is completely consistent with Anderson's (1949) hypothesis of a possible role for introgression in adaptive evolution. Moreover, the fact that hybrid zones will block the movement of negatively selected alleles indicates that the genetic coherence of a species' gene pool is unlikely to be seriously threatened by interspecific hybridization (Barton and Bengtsson, 1986). However, this theory does not invalidate the requirement that for polygenic traits, alleles must be individually advantageous to transgress the species' barrier.

A final comment relates to whether the divergence of populations will be halted if they are connected by gene flow. For example, Barton and Bengtsson (1986) suggest that hybridizing taxa will "evolve in consort, in that an advantageous mutation that occurs in one population will be able to pass through even a strong barrier and continue to spread into the other population." However, this need not be the case for all mutations. If the selective value of new mutations is sensitive to genetic background or habitat, divergence may continue even in the presence of substantial gene flow. Thus, there is good theoretical support for the old botanical concept of the syngameon (Grant, 1981), in which groups of related species are able to evolve and diverge while interconnected by gene flow.

EMPIRICAL DATA

While theory is consistent with the Andersonian hypothesis, this does not necessarily mean that introgression has been important in evolution. Remember, some authors consider hybridization to be too rare for introgression to have played a major role. In addition, there are the documentary issues that disturbed Dobzhansky (1941), Baker (1947), Barber and Jackson (1957), Gottlieb (1972), Heiser (1973), and others. The detection of adaptive trait introgression represents a special problem. If an allele is advantageous in a pair of species connected by a hybrid zone, then it should move rapidly to fixation in both taxa, making it difficult to catch adaptive trait introgression in action. We discuss these issues below.

Frequency of hybridization

In the botanical literature, it is often stated that up to 80% of flowering plant species are of hybrid origin (Whitham et al., 1991; Masterson, 1994). However, the cited estimates of hybridization frequencies actually refer to the proportion of plant species that have chromosome numbers in excess of what is considered to be the basal number for angiosperms (Grant, 1981). Since most polyploids are thought to be of hybrid origin (i.e. allopolyploids), the chromosome number data are then extrapolated to provide an estimate of hybridization frequencies. Of course, there are several problems with this basic approach. First, if an allopolyploid founds a speciose lineage, a single hybridization event could account for the so-called hybrid origins of hundreds of species. Secondly, the ubiquity of allopolyploids vs. autopolyploids has recently been called into question (Soltis and Soltis, 1993). Thirdly, chromosome number is highly labile in many groups (Stebbins, 1971), further undermining these estimates. Finally, there is considerable disagreement about the base chromosome number in flowering plants, leading to wide variances in the estimates provided above.

Better estimates of hybridization frequencies in plants come from a recent study of five biosystematic floras (Ellstrand et al., 1996). The frequency of spontaneous hybridization was found to range from approximately 22% for the British flora to 5.8% for the intermountain flora of North America (Ellstrand et al., 1996). The occurrence of hybridization was non-randomly distributed among taxa, with the majority of reported hybrids coming from a small fraction of families and genera. These may represent underestimates of hybridization frequencies, however, since several of the studied floras remain poorly known.

Hybridization is thought to occur less frequently in animals than in plants, but quantitative estimates are only available for birds and fish. Between 9% and 10% of bird species hybridize (Grant and Grant, 1992) – a percentage similar to that reported for plants. Likewise, hybridization frequencies in families of North American freshwater fish range from 3% to 17% (Hubbs, 1955). If other groups of animals have similar hybridization frequencies, introgression should not be limited by a low incidence of hybridization. In this context, it is noteworthy that Gray (1972) lists several thousand hybrid combinations involving 573 mammal species. Although most of the mammal hybrids were produced in captivity, the list does suggest the possibility of more widespread hybridization in nature than is typically suspected for mammals.

It also should be pointed out that the estimates provided above are of current

hybridization. Phylogenetic data suggest many species that are no longer hybridizing may have done so in the past, particularly during the early stages of species divergence (Wendel and Albert, 1992; Soltis and Kuzoff, 1995; Song et al., 1995; Setoguchi and Watanabe, 2000).

Documentary issues

The most important modern contribution to the documentation of introgression stems from advances in molecular marker technology (Rieseberg and Ellstrand, 1993; Cruzan, 1998). We now have an abundance of molecular markers that can be applied to essentially any group of organisms. The large number of markers available increases the sensitivity with which introgression can be detected. In addition, molecular markers appear less prone than morphological characters to parallel or convergent evolution and offer the ability to monitor both nuclear and cytoplasmic gene flow. Of course, molecular markers are subject to ancestral sorting just like morphological traits. However, the detection of linkage disequilibrium among nuclear markers or between cytoplasmic and nuclear markers (cytonuclear disequilibrium) can serve as a diagnostic for introgression (Sites et al., 1996; Linder et al., 1998). Note that linkage disequilibrium is expected for both linked and unlinked markers, but it should be strongest for tightly linked markers (Rieseberg et al., 1999a). By contrast, character coherence among morphological traits may have a variety of causes (Arnold, 1994), reducing its utility as a diagnostic for introgression. Finally, many molecular markers differ by multiple nucleotide substitutions. If variation at these markers is analyzed in a phylogenetic fashion, it becomes feasible to detect ancient introgression events (Wendel and Albert, 1992; Soltis and Kuzoff, 1995; Setoguchi and Watanabe, 2000), and to provide rough estimates of the timing of introgression at a given locus (R.L. Wang et al., 1997).

Application of molecular marker data to natural plant and animal populations has been fruitful and numerous good examples of introgression have been documented in all manner of organisms (Arnold, 1997). For plants,

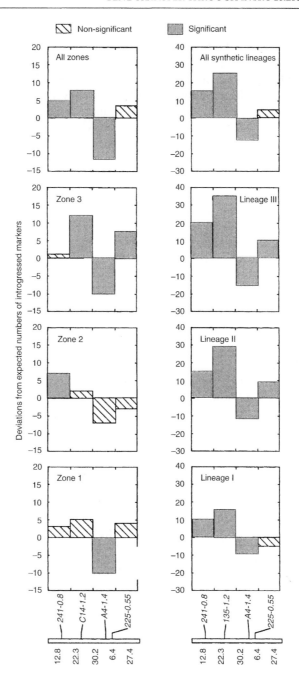

FIGURE 18.2 Direct count deviations from the expected numbers of introgressed markers in three natural hybrid zones (n = 50, 44, 45; Rieseberg et al., 1996a) and three synthetic hybrid lineages (n = 58, 56, 56; Rieseberg et al., 1999b) between *H. annuus* and *H. petiolaris*, as well as from pooled datasets. Markers or chromosomal blocks that introgressed at higher than expected rates presumably contain alleles that are advantageous in a heterospecific genetic background. In contrast, those that are found at lower than expected frequencies are hypothesized to contribute in some way to reduced hybrid fitness. Map distances are given below each linkage group and represent averages across three genetic maps for wild sunflowers (Rieseberg et al., 1995; Ungerer et al. 1998). Markers are given above each linkage group and are shown in the same order as found in *H. annuus*. Note that marker *135–1.2*, which was assayed in the synthetic lineages, was replaced by a closely linked taxon-specific marker (*C14–1.2*) in the natural hybrid zones.

Rieseberg and Wendel (1993) compiled a list of 165 proposed cases of introgression, of which 65 were considered well documented. We were unable to find a comparable summary in animals, although numerous examples of introgression are described in Hewitt (1989), Avise (1994) and Dowling and Secor (1997). In addition, Arnold (1997) provides a useful summary of some of the best case histories in both plants and animals. There is probably little to be gained by combing the literature for additional examples, given that introgression appears to be an inevitable result of hybridization (except in those few cases in which first generation hybrids are completely sterile or inviable or the reproductive barrier is so complex as to eliminate all gene flow).

Introgression of advantageous alleles in the wild

As was alluded to earlier, the rapid escape and spread of alleles or traits that are advantageous makes detection difficult. That is, there will be a fairly short window after secondary contact during which it will be feasible to detect the introgression of advantageous alleles. Once an allele introgresses and spreads to fixation, continued introgression at the locus will no longer be detectable because of a lack of taxon-specific polymorphisms. An understanding of the role of introgression in evolution requires measurements of the frequency with which alleles arise that are advantageous in pairs of hybridizing taxa along with empirical evidence for the fixation of these alleles in both members of the pair.

One possible method for detecting the introgression of advantageous alleles is to focus on hybrid zones that are of very recent origin. This is feasible because many hybrid zones in plants and animals have arisen as a result of recent human disturbance or of human-mediated translocations (Anderson, 1948; Grant and Grant, 1992; Avise, 1994; Rhymer and Simberloff, 1996). A possible example of the utility of this approach comes from a study of hybridization between the golden-collared and white-collared manakin (Parsons et al., 1993). Parsons et al. were able to show that plumage

characteristics of the former taxon had moved further across a recently formed hybrid zone than either mitochondrial or nuclear markers, possibly as a result of positive sexual selection. More recently, Rieseberg et al. (1999b) examined patterns of introgression across three wild sunflower hybrid zones that were thought to have formed when western Nebraska was homesteaded nearly 100 years ago. Nine chromosomal blocks (as assayed with linked molecular markers) introgressed at a rate significantly greater than expected under neutrality, suggesting that these blocks conferred a fitness advantage in a heterospecific genetic background (Figure 18.2). However, it is possible that the number of advantageous chromosomal blocks was overestimated in this study because of uncertainty associated with the estimation of neutral expectations (Rieseberg et al., 1999b).

Although recently formed hybrid zones may provide a means for detecting adaptive trait introgression, the successful implementation of this approach may be prevented by the history of the parental taxa. For example, many species, particularly those native to temperate floras and faunas, appear to have undergone range contraction-expansion cycles throughout much of their evolutionary histories (Hewitt, 1999). As a result, recently formed hybrid zones may simply represent the last of a long series of contacts between the taxa. Advantageous alleles may therefore have flowed across the species barrier during an earlier episode of contact.

Fortunately, several other approaches may allow the introgression of advantageous alleles to be detected. Thus, studies of the introgression of adaptive traits need not be restricted to species with long histories of allopatric divergence followed by recent contact. Consider, for example, alleles that offer a habitat-specific fitness advantage. If the recipient species (the taxon that receives an alien allele through introgression) occurs in multiple habitats, the allele will only be advantageous in a portion of the recipient's geographic range, producing a detectable polymorphism. Just such a scenario appears to have taken place in the annual sunflowers of the genus *Helianthus* (Heiser, 1951; Kim and Rieseberg, 1999, unpublished data). Briefly, Heiser (1951) postulated that the common sunflower (*H. annuus*) was able to colonize Texas by acquiring advantageous alleles from *H. debilis* ssp. *cucumerifolius*, which was native to the area. The introgressive form that was created (*H. annuus* ssp. *texanus*) is fully compatible with other forms of *H. annuus*, but approaches *H. debilis* in several morphological features. Mapping of quantitative trait loci (QTL) responsible for these trait differences, combined with analysis of flanking markers in natural populations, suggests that several *H. debilis* QTL are now essentially fixed in populations of ssp. *texanus*. This result accords well with expectations for the adaptive trait introgression hypothesis.

Another approach that holds considerable promise for the study of introgression over a historical time frame involves the analysis of multiple loci within a coalescence framework. For example, Wang and Hey (1996) and R.L. Wang et al. (1997) analyzed sequence variation at three genes (*Adh*, *Hsp82* and *Period*) in *Drosophila pseudoobscura* and *D. persimilis*. They showed that (1) gene exchange apparently ceased at *Hsp82* at the time of species formation, (2) introgression was limited and relatively ancient at the *Period* locus, and (3) gene flow has continued until the present at *Adh*. Thus, it appears that with polymorphic gene sequences, it may be feasible to estimate gene flow rates at different loci during and after species divergence. Of course, this is far from inferring that the observed gene flow at *Adh* and *Period* was adaptive. Variation at these two genes may be neutral to introgression and/or the patterns reported may be governed primarily by selection at linked loci. Nonetheless, this example does raise the possibility that by reconstructing genealogies for candidate genes we may be able to assess the history of gene flow and selection at the loci in question.

Experimental studies of introgression

Thus far, we have limited our discussion to the analysis of patterns of molecular marker variation in natural hybrid zones or in species with a history of hybridization. However, this "pattern analysis" approach is perhaps most powerful when combined with experimental data. The latter can provide a useful framework in which to interpret the observed patterns. For example, five of the nine blocks that introgressed at higher than expected rates in three natural hybrid zones of sunflower (Rieseberg et al., 1999b) did so in three experimental backcross lineages as well (Figure 18.2; Rieseberg et al., 1996a). Thus, the evidence that this subset of blocks is advantageous in a heterospecific background is quite compelling. Similar results have been reported in cotton (Stephens, 1949; Wang et al., 1995) and iris (Burke et al., 1998b).

Another kind of experiment that may be helpful are estimates of hybrid fitnesses in natural populations. The hybrids analyzed may be synthetic (Burke et al., 1998a; Hauser et al., 1998; Parris et al., 1999), spontaneous (Levin and Schmidt, 1985; H. Wang et al., 1997; Campbell et al., 1998), or both (Daehler and Strong, 1997). If some subset of hybrids have a fitness advantage, the alleles conferring this advantage seem likely to spread in nature.

For organisms with short generation times, it may be feasible to perform selection experiments in the laboratory or nature to assess the effects of hybridization on rates of evolution. If hybrid lines adapt more quickly to environmental changes, this would imply the existence of interspecific gene combinations that were favored by selection. For example, the Australian fruit fly (*Dacus tryoni*) is thought to have recently expanded its range by acquiring alleles from a related species (*D. humeralis*) that make it less sensitive to heat stress (Lewontin and Birch, 1966). This hypothesis was tested by measuring the response of parental and hybrid populations to temperature stress over a two-year period (Lewontin and Birch, 1966). At the end of the two years, the hybrid lineages were able to outperform both parental species at higher temperatures, providing evidence for the existence of favorable hybrid gene combinations. Of course, these results do not prove that *D. tryoni* actually did expand its range through introgression, but they do demonstrate the plausibility of the hypothesis. In contrast to the Lewontin and Birch (1966) study, Hercus and Hoffman (1999) found no evidence that hybridization increased the rate of adaptation in hybrid versus parental lineages of *Drosophila serrata* and *D. birchii*. Rather, the hybrid lineages tended to converge on one or the other parental species in their performance and phenotype.

Ultimately, the best kind of experiment may be to introgress individual chromosomal blocks into the genetic background of the recipient taxon. This would allow the fitness consequences of individual blocks to be tested in a heterospecific genetic background and might be the best predictor of introgression in nature.

PHYLOGENETIC CONSEQUENCES OF INTROGRESSION

Whether introgression is adaptive or not, it may cause problems in phylogenetic tree construction, particularly when trees are based on single genes. The theory behind this is straightforward. Loci that contribute in some way to reproductive isolation or species differentiation should faithfully track species relationships when used for phylogenetic reconstruction (R.L. Wang et al., 1997; Ting et al., 2000). In contrast, those that represent shared ancestral polymorphisms, or that have a history of gene flow after the onset of speciation, may not be consistent with species relationships. As we have discussed earlier, it appears that only very small blocks containing "speciation" genes are likely to be impermeable to gene flow (Ting et al., 2000). Thus, introgression has the potential to affect phylogenetic reconstruction even if reproductive barriers are strong and genetically complex. Note that discrepancies among tree topologies are likely even if introgression acts very early in species formation.

Empirical data are consistent with theory in suggesting that introgression has impacted phylogenetic reconstruction in many organismal groups. For example, one of the most striking results from early phylogenetic studies in plants was the widespread incongruence observed between trees based on chloroplast DNA and those inferred from nuclear characters (Rieseberg and Soltis, 1991; Rieseberg et al., 1996c). Even in the infancy of chloroplast DNA systematics, Rieseberg and Soltis (1991) were able to compile a list of 37 examples of phylogenetic incongruence, apparently due to cytoplasmic introgression. Of

these, 28 were deemed conclusive. An additional 61 cases of phylogenetic incongruence due to cytoplasmic introgression were added to this list by Rieseberg et al. (1996c), as were several cases in which discordance was attributable to the introgression of nuclear markers. There seems little reason to expand on this list with reference to more recent literature except to note that many species of plants appear to have continued exchanging genes after the onset of speciation and that reports of conflicting gene trees are likely to increase in the future as additional nuclear genes are sampled.

Data from animals are less striking, and as far as we are aware a comprehensive treatment of phylogenetic incongruence in animals has not been published. Nonetheless, many examples of incongruence possibly resulting from introgression have been reported in the animal literature (Solignac and Monnerot, 1986; Wayne and Jenks, 1991; Dowling and DeMarais, 1993; Roy et al., 1994; R. L. Wang et al., 1997; Ting et al., 2000), and excellent case studies are reviewed by Avise (1994), Arnold (1997), and Dowling and Secor (1997). One of the notable differences between the plant and animal literature is that phylogenetic incongruence in animals is often assumed to result from the sorting of ancestral gene lineages (Avise, 1994), whereas botanists are more likely to attribute the same observations to introgression (Rieseberg and Soltis, 1991). Unfortunately, it is difficult to differentiate between these two hypotheses, although evidence from patterns of multilocus linkage disequilibrium may ultimately allow them to be distinguished. Presently, the literature may be best summarized as showing that the genomes of many species appear to be mosaics of chromosomal blocks that vary in genealogy due to introgression, the sorting of ancestral polymorphisms, and horizontal transfer (see elsewhere in this volume).

CONCLUSIONS

It is clear that species evolution may be complicated by the continued movement of genes after the onset of species formation. Species appear able to diverge at some loci while evolving in concert at others due to introgression. This process has several implications for evolutionary theory. First, it indicates that microevolutionary forces such as selection, drift, migration, and recombination must be considered at both the intra- and interspecific levels. In this context, hybridization may be viewed as a kind of wide recombination (Stebbins, 1959) and introgression as wide migration of alleles. Secondly, these observations challenge theories of species and speciation that require species to diverge in allopatry and that emphasize *ad hoc* mechanisms that prevent gene flow (Mayr, 1942; Dobzhansky, 1941). The accumulating data on introgression suggests instead that species are more correctly viewed as groups of populations at different adaptive peaks, which are primarily maintained by selection (Carson, 1985). Thirdly, the view of species genomes as mosaics of chromosomal blocks with different pedigrees, represents a substantial challenge to current methods of phylogenetic reconstruction, which generate trees that are exclusively dichotomous and branching. Given the very small size of chromosomal blocks that appear likely to share the same genealogy, the tree-building algorithms that dominate the literature are poorly designed to detect or portray accurately the complex, reticulate nature of genome and species relationships in many organismal groups. Of course, horizontal transfer can only make this problem worse.

ACKNOWLEDGMENTS

The authors' research on hybridization has been supported by grants from the NSF, the USDA, and the NIH. We thank John Burke for helpful comments on an earlier version of this manuscript.

REFERENCES

Amábile-Cuevas, C.F. and Chicurel, M.E. (1992) Bacterial plasmids and gene flux. *Cell* **70**: 189–199.

Anderson, E. (1948) Hybridization of the habitat. *Evolution* **2**: 1–9.

Anderson, E. (1949) *Introgressive Hybridization*, Wiley, New York.

Anderson, E. (1953) Introgressive hybridization. *Biol. Rev.* **28**: 280–307.

Anderson, E. and Hubricht, L. (1938) Hybridization in *Tradescantia*. III. The evidence for introgressive hybridization. *Am. J. Bot.* **25**: 396–402.

Arnold, M.L. (1997) *Natural Hybridization and Evolution*, Oxford University Press, New York.

Arnold, S.J. (1994) Multivariate inheritance and evolution: a review of concepts. In *Quantitative Genetic Studies of Behavioral Evolution* (ed., C.R.B. Boake), pp. 17–48, Chicago Press, Chicago.

Avise, J.C. (1994) *Molecular Markers, Natural History, and Evolution*, Chapman & Hall, New York.

Baker, H.G. (1947) Criteria for hybridity. *Nature* **159**: 221–223.

Barber, H.N. and Jackson, W.D. (1957) Natural selection in action in *Eucalyptus*. *Nature* **179**: 1267–1269.

Barton, N.H. (1979) Gene flow past a cline. *Heredity* **43**: 333–339.

Barton, N.H. (1986) The effects of linkage and density-dependent regulation on gene flow. *Heredity* **57**: 415–426.

Barton, N.H. and Bengtsson, B.O. (1986) The barrier to genetic exchange between hybridising populations. *Heredity* **56**: 357–376.

Barton, N.H. and Hewitt, G.M. (1981) A chromosomal cline in the grasshopper *Podisma pedestris*. *Evolution* **35**: 1008–1018.

Barton, N.H. and Hewitt, G.M. (1985) Analysis of hybrid zones. *Annu. Rev. Ecol. Syst.* **16**: 113–148.

Barton, N.H. and Hewitt, G.M. (1989) Adaptation, speciation and hybrid zones. *Nature* **341**: 497–503.

Bert, T.M. and Arnold, W.S. (1995) An empirical-test of predictions of 2 competing models for the maintenance and fate of hybrid zones-both models are supported in a hard-clam zone. *Evolution* **49**: 276–289.

Bloom, W. and Lewis, H. (1972) Interchanges and interpopulational gene exchange in *Clarkia speciosa*. *Chromosomes Today* **3**: 268–284.

Buerkle, C.A., Morris, R.J., Asmussen, M.A. and Rieseberg, L.H. (2000) The likelihood of homoploid hybrid speciation. *Heredity* **84**: 441–451.

Burke, J.M., Carney, S. E. and Arnold, M. L. (1998a) Hybrid fitness in the Louisiana irises: analysis of parental and F1 performance. *Evolution* **52**: 37–43.

Burke, J.M., Voss, T.J. and Arnold, M.L. (1998b) Genetic interactions and natural selection in Louisiana iris hybrids. *Evolution* **52**: 1304–1310.

Campbell, D.R., Waser, N.M. and Wolf, P.G. (1998) Pollen transfer by natural hybrids and parental species in an *Ipomopsis* hybrid zone. *Evolution* **52**: 1602–1611.

Carney, S.E., Gardner, K.A. and Rieseberg, L.H. (2000) Evolutionary changes over the fifty-year history of a hybrid population of sunflowers (*Helianthus*). *Evolution* **54**: 462–474.

Carson, H.L. (1985) Unification of speciation theory in plants and animals. *Syst. Bot.* **10**: 380–390.

Charlesworth, D. (1995) Evolution under the microscope. *Curr. Biol.* **5**: 835–836.

Clausen, R.E. and Goodspeed, T.H. (1925) Interspecific hybridization in *Nicotiana*. II. A tetraploid *glutinosa-*

tabacum hybrid, an experimental verification of Winge's hypothesis. *Genetics* **10**: 278–284.

Cruzan, M.B. (1998) Genetic markers in plant evolutionary ecology. *Ecology* **79**: 400–412.

Daehler, C.C. and Strong, D.R. (1997) Hybridization between introduced smooth cordgrass (*Spartina alterniflora*, Poaceae) and native California cordgrass (*S. foliosa*) in San Francisco Bay, California, USA. *Am. J. Bot.* **84**: 607–611.

Darlington, C.D. (1937) The early hybridizers and the origin of genetics. *Herbertia* **4**: 63–69.

Dobzhansky, T.H. (1941) *Genetics and the Origin of Species*, Columbia University Press, New York.

Dowling, T.E. and DeMarais, B.D. (1993) Evolutionary significance of introgressive hybridization in cyprinid fishes. *Nature* **362**: 444–446.

Dowling, T.E. and Secor, C.L. (1997) The role of hybridization and introgression in the diversification of animals. *Annu. Rev. Ecol. Syst.* **28**: 593–619.

Du Rietz, G.E. (1930) The fundamental units of biological taxonomy. *Sven. Bot. Tidskr.* **24**: 333–428.

Ellstrand, N.C., Whitkus, R. and Rieseberg, L.H. (1996) Distribution of spontaneous plant hybrids. *Proc. Natl Acad. Sci. USA* **93**: 5090–5093.

Endler, J.A. (1977) *Geographic Variation, Speciation, and Clines*, Princeton University Press, Princeton.

Fisher, R.A. (1930) *The Genetical Theory of Natural Selection*, Clarendon Press, Oxford.

Focke, W.O. (1881) *Die Pflanzen-Mischlinge*, Borntraeger, Berlin.

Gavrilets, S. and Cruzan, M.B. (1998) Neutral gene flow across single locus clines. *Evolution* **52**: 1277–1284.

Gottlieb, L.D. (1972) Levels of confidence in the analysis of hybridization in plants. *Ann. Mo. Bot. Gard.* **59**: 435–446.

Grant, P.R. and Grant, B.R. (1992) Hybridization of bird species. *Science* **256**: 193–197.

Grant, V. (1966) The origin of a new species of *Gilia* in a hybridization experiment. *Genetics* **54**: 1189–1199.

Grant, V. (1981) *Plant Speciation*, Columbia University Press, New York.

Gray, A.P. (1972) *Mammalian Hybrids: A Check-list with Bibliography*, Commonwealth Agricultural Bureaux, Slough.

Haldane, J.B.S. (1922) Sex ratio and unisexual sterility in hybrid animals. *J. Genet.* **12**: 101–109.

Harrison, R.G. (1986) Pattern and process in a narrow hybrid zone. *Heredity* **56**: 337–349.

Hauber, D.P. and Bloom, W.L. (1983) Stability of a chromosomal hybrid zone in *Clarkia speciosa* ssp. *polyantha* complex (Onagracea). *Am. J. Bot.* **70**: 1454–1459.

Hauser, T.P., Jorgensen R.B. and Ostergard, H. (1998) Fitness of backcross and F-2 hybrids between weedy *Brassica rapa* and oilseed rape (*B. napus*). *Heredity* **81**: 436–443.

Heiser, C.B., Jr. (1947) Hybridization between the sunflower species *Helianthus annuus* and *H. petiolaris*. *Evolution* **1**: 249–262.

Heiser, C.B., Jr. (1951) Hybridization in the annual sunflowers: *Helianthus annuus* × *H. debilis* ssp. *cucumerifolius*. *Evolution* **5**: 42–51.

Heiser, C.B., Jr. (1973) Introgression re-examined. *Bot. Rev.* **39**: 347–366.

Heiser, C.B., Jr. (1979) Hybrid populations of *Helianthus divaricatus* and *H. microcephalus* after 22 years. *Taxon* **28**: 71–75.

Hercus, M.J. and Hoffman, A.A. (1999) Does interspecific hybridization influence evolutionary rates? An experimental study of laboratory adaptation in hybrids between *Drosophila serrata* and *Drosophila birchii*. *Proc. R. Soc. Lond. Ser. B Biol. Sci.* **266**: 2195–2200.

Hewitt, G.M. (1975) A sex chromosome hybrid zone in the grasshopper *Podisma pedestris* (Orthoptera: Acrididae). *Heredity* **35**: 375–387.

Hewitt, G.M. (1989) The subdivision of species by hybrid zones. In *Speciation and Its Consequences* (eds, D. Otte and J.A. Endler), pp. 85–110, Sinauer Associates, Inc., Sunderland.

Hewitt, G.M. (1999) Post-glacial re-colonization of European biota. *Biol. J. Linn. Soc.* **68**: 87–112.

Howard, D.J. (1986) A zone of overlap and hybridization between two ground cricket species. *Evolution* **40**: 34–43.

Howard, D.J., Waring, G.L., Tibbets, C.A. and Gregory, P.G. (1993) Survival of hybrids in a mosaic hybrid zone. *Evolution* **47**: 789–800.

Hubbs, C.L. (1955) Hybridization between fish species in nature. *Syst. Zool.* **4**: 1–20.

Keim, P., Paige, K.N., Whitham, T.G. and Lark, K.G. (1989) Genetic analysis of an interspecific hybrid swarm of *Populus*: occurrence of unidirectional introgression. *Genetics* **123**: 557–565.

Kim, S.C. and Rieseberg, L.H. (1999) Genetic architecture of species differences in annual sunflowers: implications for adaptive trait introgression. *Genetics* **153**: 965–977.

Kölreuter, J.G. (1893) *Voläufige Nachricht von einigen das Geschlecht der Pflanzen betreffenden und Beobachtungen, nebst Fortsetzungen 1,2, und 3*. In *Ostwald's Klassiker der exacten Wissenschaften* No. 41 (ed. W. Pfeffer), Wilhelm Engelmann, Leipzig.

Leitch, I.J. and Bennett, M.D. (1997) Polyploidy in angiosperms. *Trends Plant Sci.* **2**: 470–476.

Levin, D.A. and Schmidt, K.P. (1985) Dynamics of a hybrid zone in *Phlox*: an experimental demographic investigation. *Am. J. Bot.* **72**: 1404–1409.

Lewontin, R.C. and Birch, L.C. (1966) Hybridization as a source of variation for adaptation to new environments. *Evolution* **20**: 315–336.

Linder, C.R., Taha, I., Seiler, G.J. et al. (1998) Long-term introgression of crop genes into wild sunflower populations. *Theor. Appl. Genet.* **96**: 339–347.

MacCallum, C.J., Nürnberger, B., Barton, N.H. and Szymura, J.M. (1998) Habitat preference in the *Bombina* hybrid zone in Croatia. *Evolution* **52**: 227–239.

Marsden-Jones, E.M. (1930) The genetics of *Geum intermedium* Willd. haud Ehrh. and its back-crosses. *J. Genet.* **23**: 377–395.

Masterson, J. (1994) Stomatal size in fossil plants–evidence for polyploidy in the majority of angiosperms. *Science* **264**: 421–424.

Mayr, E. (1942) *Systematics and the Origin of Species*, Columbia University Press, New York.

Mayr, E. (1963) *Animal Species and Evolution*, Harvard University Press, Cambridge, MA.

Mayr, E. (1982) *The Growth of Biological Thought*, The Belknap Press of Harvard University Press, Cambridge, MA.

Mayr, E. (1986) Joseph Gottlieb Kölreuter's contributions to biology. *OSIRIS, 2nd series* **2**: 135–176.

McCarthy, E.M., Asmussen, M.A. and Anderson, W.W. (1995) A theoretical assessment of recombinational speciation. *Heredity* **74**: 502–509.

Mendel, G. (1865) Versuche über Pflanzenhybriden. *Verhandlungen Naturforschenden Vereines in Brünn* **10**: 1–62.

Meyn, O. and Emboden, W.A. (1987) Parameters and consequences of introgression in *Salvia apiana* × *S. mellifera* (Lamiaceae). *Syst. Bot.* **12**: 390–399.

Moore, W.S. (1977) An evaluation of narrow hybrid zones in vertebrates. *Q. Rev. Biol.* **52**: 263–267.

Moore, W.S. and Price, J.T. (1993) Nature of selection in the northern flicker hybrid zone and its implications for speciation theory. In *Hybrid Zones and the Evolutionary Process* (ed., R.G. Harrison), pp. 196–225, Oxford University Press, Oxford.

Müntzing, A. (1930) Outlines to a genetic monograph of the genus *Galeopsis* with special reference to the nature and inheritance of partial sterility. *Hereditas* **13**: 185–341.

Naudin, C. (1863) De l'hybridité considérée comme cause de variabilité dans les végétaux. *C. R. Acad. Sci.* **59**: 837–845.

Nürnberger, B., Barton, N.H., MacCallum, C. et al. (1995) Natural selection on quantitative traits in the *Bombina* hybrid zone. *Evolution* **49**: 1224–1238.

Ochman, H., Lawrence, J.G. and Groisman, E.A. (2000) Lateral gene transfer and the nature of bacterial inovation. *Nature* **405**: 299–304.

Olby, R. (1985) *Origins of Mendelism*, The University of Chicago Press, Chicago.

Olsen, S.J. (1985) *Origins of the Domestic Dog*, University of Arizona Press, Tucson.

Ostenfeld, C.H. (1927) The present state of knowledge on hybrids between species of flowering plants. *J. R. Hortic. Soc.* **53**: 31–44.

Parris, M.J., Semlitsch, R.D. and Sage, R.D. (1999) Experimental analysis of the evolutionary potential of hybridization in leopard frogs (Anura: Ranidae). *J. Evol. Biol.* **12**: 662–671.

Parsons, T.J., Olson, S.L. and Braun, M.J. (1993) Unidirectional spread of secondary sexual plumage traits across an avian hybrid zone. *Science* **260**: 1643–1646.

Petry, D. (1983) The effect on neutral gene flow of selection at a linked locus. *Theor. Popul. Biol.* **23**: 300–313.

Pialek, J. and Barton, N.H. (1997) The spread of an advantageous allele across a barrier: the effects of random drift and selection against heterozygotes. *Genetics* **145**: 493–504.

Potts, B.M. (1986) Population dynamics and regeneration of a hybrid zone between *Eucalyptus risdonii* and *Eucalyptus amygdalina*. *Aust. J. Bot.* **34**: 305–330.

Ramsey, J. and Schemske, D.W. (1998) Pathways, mechanisms, and rates of polyploid formation in flowering plants. *Annu. Rev. Ecol. Syst.* **29**: 467–501.

Rand, D.M. and Harrison, R.G. (1989) Ecological genetics of a mosaic hybrid zone: mitochondrial, nuclear, and reproductive differentiation of crickets by soil type. *Evolution* **43**: 432–439.

Reed, K.M. and Sites, J.W. (1995) Female fecundity in a hybrid zone between 2 chromosome races of the *Sceloporus grammicus* complex (Sauria, Phrynosomatidae). *Evolution* **49**: 61–69.

Rhymer, J.M. and Simberloff, D. (1996) Extinction by hybridization and introgression. *Annu. Rev. Ecol. Syst.* **27**: 83–109.

Rieseberg, L.H. (1997) Hybrid origins of plant species. *Annu. Rev. Ecol. Syst.* **28**: 359–389.

Rieseberg, L.H. (1998) Molecular ecology of hybridization. In *Advances in Molecular Ecology* (ed., G.R. Carvalho), pp. 243–265, IOS Press, Amsterdam.

Rieseberg, L.H. and Ellstrand, N.C. (1993) What can morphological and molecular markers tell us about plant hybridization? *Crit. Rev. Plant Sci.* **12**: 213–241.

Rieseberg, L.H. and Soltis, D.E. (1991) Phylogenetic consequences of cytoplasmic gene flow in plants. *Evol. Trends Plants* **5**: 65–84.

Rieseberg, L.H. and Wendel, J. (1993) Introgression and its consequences in plants. In *Hybrid Zones and the Evolutionary Process* (ed., R. Harrison), pp. 70–109, Oxford University Press, New York.

Rieseberg, L.H., Van Fossen, C. and Desrochers, A. (1995) Hybrid speciation accompanied by genomic reorganization in wild sunflowers. *Nature* **375**: 313–316.

Rieseberg, L.H., Arias, D.M., Ungerer, M.C. et al. (1996a) The effects of mating design on introgression between chromosomally divergent sunflower species. *Theor. Appl. Genet.* **93**: 633–644.

Rieseberg, L.H., Sinervo, B., Linder, C.R. et al. (1996b) Role of gene interactions in hybrid speciation: evidence from ancient and experimental hybrids. *Science* **272**: 741–745.

Rieseberg, L.H., Whitton, J. and Linder, C.R. (1996c) Molecular marker incongruence in plant hybrid zones and phylogenetic trees. *Acta Bot. Neerl.* **45**: 243–262.

Rieseberg, L.H., Baird, S.J. and Desrochers, A.M. (1998) Patterns of mating in wild sunflower zones. *Evolution* **52**: 713–726.

Rieseberg, L.H., Kim, M.J. and Seiler, G.J. (1999a) Introgression between cultivated sunflowers and a sympatric wild relative, *Helianthus petiolaris* (Asteraceae). *Int. J. Plant Sci.* **160**: 102–108.

Rieseberg, L.H., Whitton, J. and Gardner, K. (1999b) Hybrid zones and the genetic architecture of a barrier to gene flow between two sunflower species. *Genetics* **152**: 713–727.

Rife, D.C. (1965) *Hybrids*, Public Affairs Press, Washington, DC.

Roberts, H.F. (1929) *Plant Hybridization Before Mendel*, Princeton University Press, Princeton.

Rolan-Alvarez, E., Johannesson, K. and Erlandsson, J. (1997) The maintenance of a cline in the marine snail *Littorina saxatilis*: the role of home site advantage and hybrid fitness. *Evolution* **51**: 1838–1847.

Roy, M.S., Geffen, E., Smith, D. et al. (1994) Patterns of differentiation and hybridization in North American wolflike canids revealed by analysis of microsatellite loci. *Mol. Biol. Evol.* **11**: 553–570.

Semlitsch, R.D., Pickle, J., Parris, M.J. and Sage, R.D. (1999) Jumping performance and short-term repeatability of newly metamorphosed hybrid and parental leopard frogs (*Rana spenocephala* and *Rana blairi*). *Can. J. Zool.* **77**: 748–754.

Setoguchi, H. and Watanabe, I. (2000) Intersectional gene flow between insular endemics of *Ilex* (Aquifoliaceae) on the Bonin Islands and the Ryukyu Islands. *Am. J. Bot.* **87**: 793–810.

Sites, J.W., Jr., Basten, C.J. and Asmussen, M.A. (1996) Cytonuclear stucture of a hybrid zone in lizards of the *Sceloporus grammicus* complex (Sauria, Phrynosomatidae). *Mol. Ecol.* **5**: 379–392.

Solignac, M. and Monnerot, M. (1986) Race formation, speciation, and introgression within *Drosophila simulans*, *D. mauritania*, *D. sechellia* inferred from mitochondrial DNA analysis. *Evolution* **40**: 531–539.

Soltis, D.E. and Soltis, P.S. (1993) Molecular data and the dynamic nature of polyploidy. *Crit. Rev. Plant Sci.* **12**: 243–273.

Soltis, D.E. and Kuzoff, R.K. (1995) Discordance between nuclear and chloroplast phylogenies in the *Heuchera* group (Saxifragaceae). *Evolution* **49**: 727–742.

Song, K., Lu, P., Tang, K. and Osborn, T. (1995) Rapid genome change in synthetic polyploids of *Brassica* and its implications for polyploid evolution. *Proc. Natl Acad. Sci. USA* **92**: 7719–7723.

Stebbins, G.L. (1957) The hybrid origin of microspecies in the *Elymus glaucus* complex. *Cytologia*, Suppl. vol., 336–340.

Stebbins, G.L. (1959) The role of hybridization in evolution. *Proc. Am. Philos. Soc.* **103**: 231–251.

Stebbins, G.L. (1971) *Chromosomal Evolution in Higher Plants*, Edward Arnold, London.

Stephens, S.G. (1949) The cytogenetics of spciation in *Gossypium*: I. Selective elimination of the donor parent genotype in interspecefic backcrosses. *Genetics* **34**: 627–637.

Stubbe, H. (1972) *History of Genetics, from Prehistoric Times to the Rediscovery of Mendel's Laws*, MIT Press, Cambridge, MA.

Ting, C.T., Tsaur, S.C. and Wu, C.I. (2000) The phylogeny of closely related species as revealed by the genealogy of a speciation gene, *Odysseus. Proc. Natl Acad. Sci. USA* **97**: 5313–5316.

Ungerer, M.C., Baird, S. Pan, J. and Rieseberg, L.H. (1998) Rapid hybrid speciation in wild sunflowers. *Proc. Natl Acad. Sci. USA* **95**: 11757–11762.

Wang, G.L., Dong, J. M. and Paterson, A.H. (1995) The distribution of *Gossypium hirsutum* chromatin in *G. barbadense* germplasm: molecular analysis of introgressive hybridization. *Theor. Appl. Genet.* **91**: 1153–1161.

Wang, H., McArthur, E.D., Sanderson, S.C. et al. (1997) Narrow hybrid zone between two subspecies of big sagebrush (*Artemisia tridentata*: Asteraceae). IV. Reciprocal transplant experiments. *Evolution* **51**: 95–102.

Wang, R.L. and Hey, J. (1996) The speciation history of *Drosophila pseudoobscura* and close relatives: inferences from DNA sequence variation at the period locus. *Genetics* **144**: 1113–1126.

Wang, R.L., Wakely, J. and Hey, J. (1997) Gene flow and natural selection in the origin of *Drosophila pseudoobscura* and close relatives. *Genetics* **147**: 1091–1106.

Wayne, R.K. and Jenks, S.M. (1991) Mitochondrial-DNA analysis implying extensive hybridization of the endangered red wolf *Canis rufus. Nature* **351**: 565–568.

Wendel, J.F. and Albert, V.A. (1992) Phylogenetics of the cotton genus (*Gossypium*)-character-state weighted parsimony analysis of chloroplast-DNA restriction site data and its systematic and biogeographic implications. *Syst. Bot.* **17**: 115–143.

Whitham, T.G., Morrow, P.A. and Potts, B.M. (1991) Conservation of hybrid plants. *Science* **254**: 779–780.

Wilson, E.O. (1965) The challenge from related species. In *The Genetics of Colonizing Species* (eds, H.G. Baker and G.L. Stebbins), pp. 7–24, Academic Press, Orlando.

Wïngë, Ö. (1917) The chromosomes: their number and general importance. *C. R. Trav. Lab. Carlsberg* **13**: 131–275.

Zirkle, C. (1935) *The Beginnings of Plant Hybridization*, University of Pennsylvania Press, Philadelphia.

Zohary, D. and Hopf, M. (1988) *Domestication of Plants in the Old World: the Origin and Spread of Cultivated Plants in West Asia, Europe and the Nile Valley*, Oxford University Press, Oxford.

CHAPTER 19

Gene Flow and Introgression from Domesticated Plants into their Wild Relatives[*]

Norman C. Ellstrand, Honor C. Prentice and James F. Hancock

Domesticated plants cannot be regarded as evolutionarily discrete from their wild relatives. Most domesticated plants mate with wild relatives somewhere in the world, and gene flow from crops may have a substantial impact on the evolution of wild populations. In a literature review of the world's 13 most important food crops, we show 12 of these crops hybridize with wild relatives in some part of their agricultural distribution. Likewise, experiments have demonstrated that crop transgenes will introgress into wild populations as easily as other alleles. We use population genetic theory to predict the evolutionary consequences of gene flow from crops to wild plants and discuss two applied consequences of crop-to-wild gene flow – the evolution of aggressive weeds and the extinction of rare species. We suggest ways of assessing the likelihood of hybridization, introgression, and the potential for undesirable gene flow from crops into weeds or rare species.

INTRODUCTION

Planting, growing, and harvesting plants for food, fiber, fodder, pharmaceuticals, and many other uses have profound effects on the ecology and evolution of organisms in the surrounding habitats. Some of these effects may be direct and obvious. For example, tilling affects the composition of the local plant and microbial communities; likewise, replacing pre-existing vegetation with sunflower fields alters nectar and pollen sources for local insect communities. Evolutionary effects may be less obvious and less immediate. For example, weeds may evolve crop mimicry under directional selection from constant mechanical weed control (Barrett, 1983), and the continued use of herbicides may lead to evolution of herbicide resistance (Jasieniuk et al., 1996).

More subtle, but no less important, are evolutionary effects that arise from spontaneous mating of domesticated plants (those that have evolved under human selection and management) with their wild relatives. Such hybridization may lead to gene flow – "the incorporation of genes into the gene pool of one population from one or more populations" (Futuyma, 1998). If new or locally rare alleles from the domesticate persist in wild populations, gene flow may lead to significant evolutionary change in the recipient populations (Anderson, 1949). Crop-to-weed gene flow will have important practical and economic consequences if it promotes the evolution of more aggressive weeds (Anderson, 1949; Barrett, 1983). Hybridization with domesticated species has also been implicated in the extinction of certain wild crop

[*]Adapted with permission from the *Annual Review of Ecology and Systematics*, Volume 30, © 1999, by Annual Reviews (www.AnnualReviews.org).

Copyright © 2002 by Academic Press.
All rights of reproduction in any form reserved.

relatives (Small, 1984; Ellstrand and Elam, 1993). Over the last decade, much attention has been focused on crop-to-weed hybridization as a potential avenue for the escape of crop transgenes into natural populations (Colwell et al., 1985; Dale, 1994; Hancock et al., 1996). Although that narrow issue has been the topic of numerous theoretical, empirical, and synthetic publications, the general issue of the consequences of the flow of alleles from domesticated plants (whether genetically engineered or not) to their wild relatives has received scant attention (deWet and Harlan, 1975; Small, 1984). Indeed, a few scientists have questioned the validity of reports of spontaneous hybridization between crops and their wild relatives, asserting that "evidence for...hybridization between crops and their wild relatives has often been only circumstantial" (National Research Council, 1989). Our review addresses the general question of gene flow from domesticated plants to their wild relatives; we find that evidence for this phenomenon is typically much more than "circumstantial." Indeed, cases of gene flow from crops to their wild relatives provide examples of contemporary micro-evolution that also have important applied implications.

Our review addresses the following questions: (1) "What do we know about spontaneous hybridization between domesticated plants and their wild relatives?" and (2) "How and when does gene flow from domesticated plants to their wild relatives play a role in the evolution of wild populations?" Our specific focus is the flow of domesticated alleles into the populations of their wild relatives. The pollen parent in the initial cross may involve a plant in cultivation, a volunteer left from a previous planting or from seed spillage into a natural population, or a recent escape from cultivation. An alternate pathway may involve a wild plant that acts as the pollen parent in a cross with a plant in cultivation, giving rise to a hybrid that spontaneously backcrosses with wild plants.

While we recognize that gene flow in the reverse direction (into the crop) may have important evolutionary consequences for crop taxa, that topic is beyond the scope of our review. Also beyond our review's scope is the topic of artificial hybridization – the production of hybrids by hand-crossing or under laboratory conditions. Finally, although we discuss a few cases where hybridization has resulted in the evolution of new taxa, we do not address the evolutionary origins of wild crop relatives *per se*. We recognize that wild crop relatives may be the descendants of feral crops, plants with both wild and domesticated ancestors, or plants with no domesticated ancestors in their genealogy. We define "wild" plants as those that grow and reproduce without being deliberately planted. We define "weeds" as wild plants that interfere with human objectives.

First, we present an overview of gene flow by hybridization in plants. Next, we provide some examples of spontaneous hybridization between the world's most important crops and their wild relatives and use these to illustrate the generality of the phenomenon. We then examine the theoretical consequences of gene flow. We describe two applied implications – the evolution of aggressive weeds and the extinction of vulnerable species. Finally, we suggest ways of assessing the likelihood of hybridization, introgression, and the potential for undesirable gene flow from crops into weeds or rare species.

GENE FLOW BY HYBRIDIZATION IN PLANTS

"Hybridization is a frequent and important component of plant evolution and speciation" (Rieseberg and Ellstrand, 1993). More than 70% of plant species may be descended from hybrids (Grant, 1981). Natural interspecific and, in certain families, even intergeneric hybridization is not rare; well-studied examples number well over 1000 (Grant, 1981; Arnold, 1997). Nonetheless, even if hybridization is common, it is not ubiquitous. The incidence of natural hybridization varies substantially among plant genera and families (Ellstrand et al., 1996). In certain cases, two taxa naturally form hybrid swarms in some areas of contact but not in others (Meyn and Emboden, 1987).

Hybridization in plants depends on a variety of factors (Grant, 1981). Cross-pollination must occur. For this to happen, both taxa must be in flower at the same time. The plants must be close enough in space to allow a vector to carry pollen between them. They must be cross-compatible so that the pollen is able to germinate and effect fertilization. If the resulting embryos develop into viable seeds and germinate, the F_1 plants typically have some reduced fertility but are rarely fully sterile (Stace, 1975). In fact, many spontaneous plant hybrids do not suffer reduced fitness and, in certain cases, exhibit increased fitness relative to either or both of the parental taxa (Klinger and Ellstrand, 1994; Arnold, 1997; Hauser et al., 1998a).

The frequent occurrence of fertile hybrids increases the chances of introgression – the incorporation of alleles from one taxon into another (Richards, 1986). Note that this definition is essentially equivalent to that of gene flow at the intertaxon level. Studies employing allozymes and DNA-based genetic markers have revealed dozens of instances of natural introgression in plants (Rieseberg and Ellstrand, 1993). In many cases, morphological intermediacy and molecular confirmation of introgression go hand in hand. But, in other cases, one or few introgressed genetic markers may be found in otherwise morphologically pure individuals, even far beyond the morphologically defined limits of a hybrid zone of contact (Rieseberg and Wendel, 1993; Runyeon and Prentice, 1997). (For a more thorough discussion of hybridization see Chapter 18.)

GENE FLOW FROM DOMESTICATED PLANTS TO THEIR WILD RELATIVES

The hundreds of well-studied cases of natural hybridization and introgression involving wild plants suggest that most domesticated plants will hybridize naturally with their cross-compatible wild relatives when they come into contact. A growing number of both experimental and descriptive studies, using genetically-based markers, have demonstrated domesticated alleles can and do enter and persist in natural populations. The domesticated species involved are amazingly diverse, ranging from mushrooms (Xu et al., 1997) and raspberries (Luby and McNichol, 1995) to ornamental shrubs (Sanders, 1987) and forage crops (Stace, 1975). The accumulating evidence suggests these examples are probably the rule rather than the exception.

The risk of hybridization as an avenue for the escape of crop transgenes has stimulated researchers to evaluate opportunities for spontaneous gene flow from major cultivated species to wild plants. Some reviews have been regional in scope, addressing a country's major domesticated plants and their wild taxa in the local flora. For example, for the United Kingdom, about one-third of the 31 domesticated species reviewed spontaneously hybridize with one or more elements of the local flora (Raybould and Gray, 1993); for the Netherlands, that fraction was about one-quarter of the 42 reviewed species (deVries et al., 1992). These proportions are surprisingly high, given that (1) the vast majority of the cultivated species of the United Kingdom and the Netherlands were domesticated elsewhere and (2) both countries are small in relation to the world-wide area covered by most of the cultivated species reviewed.

Other reviews are taxon-specific in scope, addressing whether a crop is known to hybridize with wild plants over all or part of its range. Examples of crops examined by such reviews include rapeseed (Scheffler and Dale, 1994), oat (Burdon et al., 1992), and potato (Love, 1994). These reviews typically report evidence for spontaneous mating between the crop in question and some subset of its wild relatives. Rather than focusing on a single region or a single crop, we present the results of a literature review of a suite of different crops, and ask whether these crops hybridize with wild taxa in any part of their range of cultivation.

Case studies: the world's 13 most important food crops

We chose the 13 most important crops (in terms of area harvested) grown for human consumption (including oil crops) for our case studies because their impact on human well-being makes them likely to be among the best studied

TABLE 19.1 Spontaneous hybridization between the world's most important food crops and their wild relatives

Rank	Crop	Scientific name	Kilohectares[a]	Evidence for hybridization[b]	Implicated in Weed evolution	Implicated in Extinction risk
1	Wheat	*Triticum aestivum*[c]	228 131	+	No	No
		T. turgidum[c]		+	Yes	No
2	Rice	*Oryza sativa*	149 555	+	Yes	Yes
		O. glaberrima		+	No	No
3	Maize, including sweet and field corn	*Zea mays mays*	143 633	+	No	No
4	Soybean	*Glycine max*	67 450	+	Yes	No
5	Barley	*Hordeum vulgare*	65 310	m	No	No
6	Cottonseed	*Gossypium hirsutum*	51 290	+	No	Yes
		G. barbadense		+	No	No
7	Sorghum	*Sorghum bicolor*	45 249	+	Yes	No
8	Millet	*Eleusine coracana*[c]	38 077	m	No	No
		Pennisetum glaucum[c]		+	Yes	No
9	Beans, dry, green, and string	*Phaseolus vulgaris*[c]	28 671	+	Yes	No
10	Rapeseed	*Brassica napus*	24 044	+	No	No
		B. rapa		+	No	No
11	Groundnut	*Arachis hypogaea*	23 647	None	–	–
12	Sunflower seed	*Helianthus annuus*	19 628	+	Yes	?
13	Sugar cane	*Saccharum officinarum*[c]	19 619	+	No	No

[a]Estimated area of production for 1997 from the 1998 FAOSTAT website, 6/15/98.
[b]m, morphological intermediacy only; +, more substantial evidence for hybridization.
[c]Other taxa account for only a small portion of world production of this crop.

domesticated plants. We used the FAOSTAT Statistics Database (http://apps.fao.org/) to obtain recent (1997) estimates of area harvested for each crop. Some crops comprise more than one species; we reviewed as many species as were necessary to account for a substantial majority of the area harvested (Table 19.1). The number of crops reviewed was limited by considerations of time and space. The case studies represent a relatively heterogeneous group of 18 tropical, subtropical, and temperate species belonging to five different families.

We used the AGRICOLA and BIOSIS Previews bibliographic databases points of departure for our literature search for information on whether the reviewed crops naturally hybridize with wild species. After compiling our conclusions for each crop, we consulted one or more experts (see Acknowledgments) to pre-review the conclusions, to suggest other experts, and to

identify gaps in our treatment. We, however, take final responsibility for interpretation of the data.

For each crop below, we start with its name, its rank, and the name(s) of the taxa involved, following the nomenclature of Smartt and Simmonds (1995). We then present evidence for natural hybridization and introgression (summarized in Table 19.1). In some cases, the only evidence for hybridization is the presence of morphologically-intermediate plants in localities where the crop and wild relative are sympatric. Such inferred evidence of hybridization is relatively weak, because morphological intermediacy may result from phenotypic plasticity or from convergent evolution for crop mimicry resulting from selection alone rather than from selection and gene flow in concert (Barrett, 1983; National Research Council, 1989). In many cases, however, sympatric intermediates have been genetically evaluated and compared with reference populations (putatively pure allopatric crop and wild populations). The presence of crop-specific alleles in morphologically-intermediate populations provides strong evidence for a history of hybridization. Another approach is to examine experimentally whether a crop will mate spontaneously with a wild relative under field conditions. Progeny testing of seed set by wild plants grown adjacent to plantations of their crop relative can be used to identify whether the crop is the paternal father. Gene flow rates can be obtained directly by this method.

1. Wheat: *Triticum aestivum* (L.) Thell. (bread wheat) and *T. turgidum* ssp. *turgidum* (L.) Thell. (durum wheat) (Poaceae).

Spontaneous intermediates between cultivated wheats and their wild relatives occur frequently on margins of wheat fields when wild *T. turgidum* subspecies or certain species of *Aegilops* are present (Popova, 1923; Kimber and Feldman, 1987; Ladizinsky, 1992; van Slageren, 1994). Hybrids are reported from the Middle East, Africa, Europe, Asia, and North America. At least 11 different *Aegilops* species apparently hybridize naturally with bread wheat: *Ae. biuncialis*, *Ae. crassa*, *Ae. cylindrica*, *Ae. geniculata*, *Ae. juvenalis*, *Ae. neglecta*, *Ae. speltoides* var. *ligustica*, *Ae. tauschii*, *Ae. triuncialis*, *Ae. umbellulata*, and *Ae. ventricosa* (van Slageren, 1994). At least four *Aegilops* species (*Ae. columnaris*, *Ae. geniculata*, *Ae. neglecta*, and *Ae. ventricosa*) are thought to hybridize spontaneously with durum (van Slageren, 1994), but the list is not likely to be exhaustive (van Slageren, personal communication). In addition to their morphological intermediacy, the presumed hybridity of these plants is supported by their sterility. Although breeders have produced fertile hybrids between wheat and its wild relatives (Jiang et al., 1994; Feldman et al., 1995), "all natural hybrids ... are highly sterile, although seeds may occasionally be found" (Popova, 1923; van Slageren, 1994). This hybrid sterility may explain why hybridization generally appears to be restricted to F_1s with little evidence for subsequent introgression.

However, introgression from cultivated durum into wild emmer, *T. dicoccoides*, has been implicated in the evolution of a distinct race of wild emmer in the Upper Jordan Valley. That race of wild emmer is native to a region with a history of durum cultivation. At present, wild emmer in the Upper Jordan Valley has a number of morphological, physiological, and isozyme traits that are common in durum and rare or absent in other wild emmer populations (Blumler, 1998).

Gene flow between crops and their wild relatives is generally considered to occur with the wild plant as the seed parent (Ladizinsky, 1985). However, two herbicide-resistant hybrids of *T. aestivum* × *Ae. cylindrica* appeared in seed collected from herbicide-resistant wheat plants grown as a field trial in the United States (Seefeldt et al., 1998), the result of gene flow into the wheat from nearby stands of wild goatgrass. In this case, the domesticated plant is the seed parent.

Clearly, cultivated wheat can and does occasionally naturally hybridize with certain wild relatives. The extent of introgression is presumed to be limited to highly sterile hybrids. A thorough, population genetic analysis of some of these hybrid swarms would reveal whether that is indeed the case.

2. Rice: *Oryza sativa* L. and *O. glaberrima* Steud. (Poaceae).

Both cultivated rice species are interfertile with certain close wild relatives; hand crosses are easily accomplished (Chu and Oka, 1970; Morishima et al., 1992). F_1 fitnesses from crosses between wild and cultivated rice are generally high, but hybrids from certain crosses show reduced fertility (Chu et al., 1969). In contrast, when the two cultivated species are intercrossed, the hybrids are highly sterile (Chang, 1995).

Spontaneous intermediates between cultivated rice species and their wild relatives occur frequently in and near rice fields when wild taxa are present. In the case of *O. glaberrima*, which is cultivated in west Africa, the sympatric wild relatives are weedy forms of the same species as well as *O. barthii* and *O. longistamina*. For the more widespread *O. sativa*, they are weedy conspecifics (*O. sativa* f. *spontanea*), as well as *O. nivara* and *O. rufipogon*. The intermediate plants usually appear as hybrid swarms. In the case of *O. sativa*, the intermediates may sometimes be relatively uniform and behave as stabilized races. Genetic analysis of the wild intermediates has been conducted using isozymes, RAPDs, and progeny segregation studies of morphological and physiological traits. Intermediate plants have a combination of alleles specific to both the pure crop and the pure wild taxon (Oka and Chang, 1961; Chu and Oka, 1970; Oka, 1988; Tang and Morishima, 1988; Majumder et al., 1997; Suh et al., 1997), supporting the hypothesis that the intermediates are products of hybridization or introgression.

Natural rates of hybridization can be substantial. Allozyme progeny analysis of experimental mixed stands of cultivated *O. sativa* and the weed, red rice (*O. sativa* f. *spontanea*), in Louisiana revealed rates of natural hybridization ranging from 1% to 52%, depending on the cultivar acting as pollen parent (Langevin et al., 1990). Hybridization rate tended to increase with phenological overlap. The hybrids demonstrated heterosis; they were generally taller and produced more tillers than either parental type (Langevin et al., 1990).

A notable example of crop-to-wild gene flow involving rice occurred during the present century. To facilitate weeding, a red-pigmented rice cultivar was planted in India to distinguish its seedlings from those of the unpigmented weed. The strategy was thwarted by introgression. After a few seasons of gene flow and selection, the weed populations had accumulated the pigmentation allele at a high frequency (Oka and Chang, 1959).

Natural hybridization with cultivated rice has been implicated in the near extinction of the endemic Taiwanese taxon, *O. rufipogon* ssp. *formosana*. Collections of this wild rice over the last century show a progressive shift towards characters of the cultivated species and a coincidental decrease in fertility of seed and pollen (Kiang et al., 1979). Indeed, throughout Asia, typical specimens of other subspecies of *O. rufipogon* and the wild *O. nivara* are now rarely found because of extensive hybridization with the crop (Chang, 1995).

3. Maize: *Zea mays* ssp. *mays* L. (Poaceae).

Morphologically-intermediate plants between maize and teosintes (various wild *Zea*) often occur spontaneously in and near Mexican maize fields when teosinte, particularly *Z. m.* ssp. *mexicana*, is abundant (Wilkes, 1977). At present, the genetic data do not offer a clear view of the extent of hybridization and introgression that have occurred. Allozyme analysis of accessions of the teosintes *Z. luxurians*, *Z. diploperennis*, and *Z. m.* ssp. *mexicana* revealed that alleles that are otherwise maize-specific occurred at extremely low frequencies suggesting a very low level of introgression from maize into these teosinte taxa (Doebley, 1990).

However, cytogenetic analyses "offer no evidence of … maize-teosinte introgression in either direction" (Kato, 1997). Allozyme analyses of accessions or bulk seed collections of teosinte growing sympatric with maize showed no evidence of introgression from the crop into the wild taxa (Doebley, 1990). Such comparisons are not available for *Z. m.* ssp. *parviglumis* for which no crop-specific allozyme alleles are apparently available. To our knowledge, no one has genetically analyzed spontaneous, morphologically-intermediate plants to test for hybridity in the same way as in the thorough rice studies discussed above. Such analyses would identify whether the maize-teosinte intermediates are true hybrids, introgressants, or crop mimics.

Strong reproductive barriers exist between maize and one teosinte taxon, *Z. perennis* (Doebley, 1990). No evidence suggests that natural hybridization occurs between these taxa.

4. Soybean: *Glycine max* (L.) Merr. (Fabaceae).

Wild plants morphologically-intermediate to *G. max* and *G. soja* often occur spontaneously near Chinese, Korean, and Japanese soybean fields where *G. soja* is present (Abe et al., 1999). An apparently stabilized hybrid taxon in northeastern China, *G. gracilis* (= *G. max* f. *gracilis*), is intermediate to and interfertile with both *G. max* and *G. soja*. (The three are sometimes considered the same species (J. Doyle, personal communication).) *Glycine gracilis* accessions in germplasm collections have been genetically characterized with morphological, chemical, and RLFP markers; all studies have shown the accessions have a mixture of alleles diagnostic for both putative parents (Broich and Palmer, 1980, 1981; Keim et al., 1989). However, population-level analysis has not yet been conducted to determine the extent to which crop alleles have moved into natural populations. It is not yet clear whether *G. gracilis* represents a stabilized hybrid-derived taxon or a subset of the variation found within hybrid swarms between *G. max* and *G. soja*, as self-fertilizing accessions maintained in collections.

5. Barley: *Hordeum vulgare* L. (Poaceae).

Wild plants morphologically-intermediate to cultivated barley and its wild, weedy relative *H. spontaneum* (= *H. vulgare* ssp. *spontaneum*) often "occur where the two species are found together" (Harlan, 1995) "in less intensively cultivated fields or areas adjacent to cultivated fields" in the Middle East (von Bothmer et al., 1991). Despite morphological evidence for hybridization, we are not aware of any genetic analysis of the spontaneous intermediates. Barley can be crossed with *Hordeum* species other than *H. spontaneum*, but the resulting hybrids are highly sterile (Harlan, 1995).

6. Cottonseed: *Gossypium hirsutum* L. and *G. barbadense* L. (Malvaceae).

Allozyme and DNA analyses have demonstrated that limited interspecific introgression has occurred from the cultivated cotton species to these wild relatives. *Gossypium barbadense*-specific alleles were found in low frequency in wild or feral populations of *G. hirsutum* that are sympatric with the crop in meso-America and the Caribbean. In the same regions, *G. hirsutum*-specific alleles occur in wild *G. barbadense* populations that are sympatric with cultivated *G. hirsutum* (Wendel et al., 1992; Brubaker et al., 1993; Brubaker and Wendel, 1994). Population-level analysis has not yet been conducted to determine the extent to which crop alleles have moved into natural populations. Intraspecific gene flow from these crops to their wild forms has apparently not been investigated.

Fryxell (1979) suggested, on the basis of morphology, that both *G. darwinii* of the Galapagos Islands and *G. tomentosum* of the Hawaiian Islands were at risk of extinction as a result of hybridization with *G. hirsutum*. The presence of *G. hirsutum*-specific allozyme alleles in wild populations has confirmed *G. darwinii* has experienced substantial introgression from the crop (Wendel and Percy, 1990). No evidence of interspecific hybridization between *G. hirsutum* and *G. tomentosum* has yet been detected; however, attempts to determine whether it occurs have been hampered by the lack of species-specific markers (DeJoode and Wendel, 1992). Allozyme analysis has revealed limited introgression of *G. hirsutum* alleles into the Brazilian endemic *G. mustelinum* (Wendel et al., 1994).

7. Sorghum: *Sorghum bicolor* (L.) Moench (Poaceae).

Spontaneous, morphologically-intermediate, plants between cultivated sorghum and its wild relatives (both conspecifics and congenerics) occur frequently in and near sorghum fields when wild taxa are present, in both the Old World and the New World (Baker, 1972; Doggett and Majisu, 1968). Analyses of progeny segregation, allozymes and RFLPs all reveal crop-specific alleles in wild *S. bicolor* when it co-occurs with the crop in Africa, suggesting that intraspecific hybridization and introgression are common (Doggett and Majisu, 1968; Aldrich and Doebley, 1992; Aldrich et al., 1992).

Allozyme progeny analysis of the tetraploid weed *S. halepense*, planted around cultivated sorghum fields in California, detected spontaneous hybridization as far as 100 m from the crop (Arriola and Ellstrand, 1996). The resulting, presumably triploid, hybrids showed no fitness differences from the weed under field conditions (Arriola and Ellstrand, 1997).

Hybridization between crop sorghum and *S. propinquum* has been implicated in the origin of the weedy *S. almum*. Likewise, introgression from crop sorghum has been implicated in the evolution of enhanced weediness in one of the world's worst weeds, *S. halepense* (Holm et al., 1977). RFLP analysis has shown that both *S. almum* and *S. halepense* contain a combination of alleles specific to both putative parent species (Paterson et al., 1995).

8. Millets: *Eleusine coracana* (L.) Gaertn. (finger millet) and *Pennisetum glaucum* (L.) R. Br. (= *P. americanum* = *P. typhoides*) (pearl millet) (both in Poaceae). (Millets are a collection of unrelated, tropical small grain species; these two species

account for most of the world's production (Hancock, 1992).)

Plants with a range of morphologies between finger millet and the wild *E. coracana* ssp. *africana* are often found along roadsides as well as in, and at the edges of, finger millet fields where the two taxa co-occur in Africa (Mehra, 1962; Phillips, 1972). These spontaneous intermediates sometimes behave as noxious weeds (deWet, 1995a). Despite morphological evidence for hybridization, we are not aware of any genetic analysis of the spontaneous intermediates. Natural hybrids between finger millet and other wild relatives have not been reported.

Spontaneous intermediates between pearl millet and the wild species, *P. violaceum* (= *P. g.* ssp. *monodii*) and *P. sieberanum* (= *P. g.* ssp. *stenostachyum*), often occur in and near pearl millet fields in West Africa and northern Namibia when the wild taxa are present (Brunken et al., 1977). Allozyme analysis of *P. violaceum* accessions revealed that populations sympatric with the crop were genetically more similar to the crop than allopatric populations (Tostain, 1992), supporting the hypothesis of crop-to-wild introgression in areas of sympatry.

Hybridization between pearl millet and *P. violaceum* may have been involved in the evolution of *P. sieberanum*, "shibra," which is morphologically-intermediate to the two species. "In Africa, shibras are found throughout much of the area of pearl millet cultivation" (Brunken et al., 1977), where, as "obligate weeds of cultivation," they "do not persist for more than one generation after cultivated fields have been abandoned" (deWet, 1995b).

Natural hybridization between wild and cultivated *Pennisetum* can occur at a substantial rate. Genetic markers (allozyme and morphological) were used to measure hybridization between naturally adjacent wild and cultivated populations of millet in Niger (Marchais, 1994). The crop sired about 35% of the progeny of both *P. violaceum* and *P. sieberanum*. Also, a plot of interplanted wild and cultivated millet was genetically structured so that allozyme progeny analysis would identify hybrids (Renno et al., 1997). In this experiment, the crop sired 8% of

the progeny of *P. violaceum* and 39% of the progeny of *P. sieberanum*.

9. Dry, string, and green beans: *Phaseolus vulgaris* L. (Fabaceae). (This species, common bean, comprises most of the world's production of the crop.)

Plants appearing to be hybrids between cultivated bean and wild *P. vulgaris* species often occur in South America at the margins of bean fields when the wild beans are present (Beebe et al., 1997). These intermediates often persist for years (Freyre et al., 1996). Plants appearing to be hybrids between common bean and its wild relatives, *P. aborigineus* and *P. mexicanus*, are known from Mexico (Vanderborght, 1983; Delgado Salinas et al., 1988; Acosta et al., 1994). However, morphological intermediates are not always present where cultivated and wild taxa co-occur (Debouck et al., 1993; Freyre et al., 1996).

Genetic analysis of the wild intermediates in South America has been conducted using phaseolin seed proteins and progeny segregation studies of morphological traits. Intermediate plants have a combination of alleles specific to both the pure crop and the pure wild taxon (Beebe et al., 1997), supporting the hypothesis that the intermediates have a hybrid ancestry. In fact, some of the weedy populations are fixed for both crop-specific and wild-specific characters, suggesting that they are genetically stabilized lineages arising from past hybridization. We are not aware of similar genetic analysis of the putative hybrids in Mexico.

10. Rapeseed (canola): *Brassica napus* L. and *B. campestris* L. (= *B. rapa*) (Brassicaceae). (Other cultivars of *B. napus* and *B. campestris* are grown as vegetable and fodder crops.)

Spontaneous hybridization between a *B. campestris* vegetable cultivar and wild *B. campestris* ssp. *eu-campestris* has been experimentally measured (Manasse, 1992). Wild plants homozygous for a recessive allele were planted at varying distances around stands of a cultivar homozygous for a dominant anthocyanin marker allele. Progeny testing from the wild plants revealed that *B. campestris* readily hybridizes with its weedy conspecific under field conditions. We are not aware of any descriptive or experimental study addressing whether oilseed *B. campestris* can spontaneously hybridize with wild *B. campestris*. However, hybridization rates involving *B. campestris* oilseed cultivars probably will be similar to those for the vegetable cultivar.

"It is uncertain whether or not *B. napus* exists in truly wild form," except for escapes from cultivation (McNaughton, 1995), but there are unlikely to be reproductive barriers that would prevent spontaneous hybridization between wild and cultivated *B. napus*. The hybrid between cultivated *B. napus*, a tetraploid, and wild *B. campestris*, a diploid, (*B* × *harmsiana*) occurs sporadically in the British Isles in crops of *B. napus* that are adjacent to or sympatric with *B. campestris* (Harberd, 1975). Jørgensen and Andersen (1994) found *B. napus*-specific allozyme alleles in two wild *B. campestris*-like plants in Denmark – evidence of past hybridization and introgression. Most recently, Jørgensen et al. (1998) analyzed a population of *B. campestris* growing in a field of volunteer oilseed rape and identified a number of hybrids and backcrosses based on a combination of morphology, chromosome counts, isozymes, and RAPDs.

Field-based experiments using genetically-based morphological, cytogenetic, allozyme, and DNA markers have measured spontaneous hybridization rates between *B. napus* and wild *B. campestris*. *Brassica campestris* growing within stands of the crop produced anywhere from 9% to 93% hybrid progeny, depending on the experimental design (Jørgensen et al., 1996). Likewise, stands of *B. napus* oilseed rape transgenic for herbicide resistance intermixed with wild *B. campestris* spontaneously produced transgenic interspecific hybrids on both parental species, "the frequency of hybrids 3% and 0.3% with the weedy species and oilseed rape as female, respectively" (Jørgensen et al., 1998).

The transgenic hybrids were grown in plots with wild *B. campestris* to assess the potential for backcrossing to the wild parent. A total of 865 herbicide-resistant progeny of the hybrids were grown to flowering. Among 44 analyzed progeny with a *B. campestris*-like morphology, a few plants were found with the chromosome

number of *B. campestris*, high pollen fertility, and cross-compatibility with pure *B. campestris* (Mikkelsen et al., 1996).

When wild *B. campestris* × cultivated *B. napus* hybrids were grown with their parents under field conditions, the hybrids were significantly more fit than wild *B. campestris*, producing on average fewer seeds per fruit but many more fruits per plant (Hauser et al., 1998a). Backcrosses and F_2s were also compared with wild *B. campestris* and cultivated *B. napus* under field conditions. These genotypes had, on average, reduced fitness relative to their wild grandparent (Hauser et al., 1998b). The authors of the study note, however, that some of the individuals were as fit as their parents, and conclude that the introgression of crop alleles from oilseed rape to wild *B. campestris* will be slowed, but not stopped, by the low fitness of the second generation hybrids.

Spontaneous hybridization between *B. napus* and wild *B. juncea* was investigated by planting 12 individuals of the latter in a field of the former (Jørgensen et al., 1996). Progeny from the wild plants were analyzed for hybridity with RAPDs, isozymes, chromosome counts, and morphological measurements. Three percent of the *B. juncea* progeny tested were identified as interspecific hybrids; Bing et al. (1996) reported the same hybridization rate for a *B. juncea* cultivar interplanted with *B. napus* oilseed rape. In a subsequent experiment, *B. juncea* was planted at a range of frequencies into stands of cultivated *B. napus* (Jørgensen et al., 1998). When *B. napus* was the maternal parent, the hybridization rate was constant, about 1%; when *B. juncea* was the maternal parent, the hybridization rate decreased from 2.3% to 0.3% as *B. juncea* became more frequent in the stand (increasing from 75% of the planting to over 90%).

Field-based experiments have also demonstrated that *B. napus* can act as a successful pollen parent in spontaneous intergeneric crosses, albeit at a very low rate. Herbicide-resistant *B. napus* was interplanted with the pantemperate weed, hoary mustard (*Hirschfeldia incana* = *B. adpressa*) in the fields at a low density (Lefol et al., 1996b). About 2% of seedlings from the wild plants were herbicide resistant. Isozyme, morphological, and cytogenetic analysis confirmed their hybridity.

The *Hirschfeldia-Brassica* hybrids were all triploids, producing no fertile pollen grains and less than one seed per plant under greenhouse conditions. In a similar experiment, herbicide-resistant *B. napus* was interplanted with the serious pan-temperate weed, jointed charlock (*Raphanus raphanistrum*), under equal frequencies (1:1) and low frequency for the weed (1:600 charlock:rape) (Darmency et al., 1998). Progeny from the wild plants were tested for herbicide resistance and had their hybridity double-checked by isozyme and cytogenetic analysis. Only a few hybrids were detected in the low frequency treatment in only one of the three years of study. The number of hybrid progeny per maternal plant averaged 0.03 over the whole experiment. No hybrids were detected in the equal frequency treatment. *Raphanus–Brassica* hybrids have very low fertility under field conditions, averaging less than one seed per plant (Chèvre et al., 1998). Both paternal and maternal fertility in the field were observed to increase in first- and second-generation backcrosses to jointed charlock (Chèvre et al., 1998), although it is not clear from the data presented whether fitness in the second generation backcrosses had recovered to the level typical of the wild species. A third experiment involved a transgenic herbicide-resistant *B. napus* cultivar interplanted with the serious weed, wild mustard (*Sinapis arvensis*) (Lefol et al., 1996a); despite progeny testing of millions of seeds from the wild species for herbicide resistance, no hybridization was detected.

Although *B. napus* can successfully cross spontaneously with certain species in other genera, it is apparently reproductively isolated from certain other *Brassica*. For example, a field experiment in Saskatchewan failed to detect any spontaneous hybridization between *B. napus* and the weed *B. nigra* in hundreds of seeds that were tested (Bing et al., 1996).

11. Groundnut (peanut): *Arachis hypogea* L. (Fabaceae).

Experimental crosses between groundnut and the wild South American species, *A. monticola*, readily produce fertile hybrids, but the crop does not cross freely with other wild relatives (Singh, 1995). We did not find any report suggesting that spontaneous hybridization occurs between groundnut and any wild relative. Although *A. hypogea* may show limited natural cross-pollination (S. Hegde, personal communication), all *Arachis* species are geocarpic and are presumed to be predominantly self-pollinated (Singh, 1995), a situation which is likely severely to limit opportunities for spontaneous interspecific hybridization, even in sympatry.

12. Sunflower seed: *Helianthus annuus* L. (Asteraceae).

Cultivated sunflower is cross-compatible with wild *H. annuus* and other species in the genus (Rogers et al., 1982). Both morphological and molecular evidence suggest that wild *H. annuus* naturally hybridizes with other wild sunflower species in areas of sympatry in western North America (Heiser, 1978; Rogers et al., 1982; Dorado et al., 1992). Indeed, certain rare sunflower species naturally hybridize with wild (or possibly feral) *H. annuus*, which may increase their risk of extinction (Rogers et al., 1982). Molecular markers have confirmed the hybrid origin of certain *Helianthus* species that involves *H. annuus* as one of the parent species (Rieseberg et al., 1990). However, the existing data on hybridization between wild taxa are of limited value in predicting gene flow dynamics between cultivated *H. annuus* and wild congeners.

Spontaneous hybridization between cultivated sunflower and wild *H. annuus* has been the subject of considerable research. Allozyme progeny analysis of wild *H. annuus*, planted around cultivated sunflower fields in Mexico, detected spontaneous hybridization at substantial rates and over distances of up to 1000 m from the crop (Arias and Rieseberg, 1994). The fitness of hybrids between the wild and cultivated sunflower has been compared with that of wild plants under field conditions. "In general, hybrid plants had fewer branches, flower heads, and seeds than wild plants, but in two crosses fecundity of the hybrids was not significantly different from that of purely wild plants" (Snow et al., 1998). Sunflowers also provide one of the few cases of the relative field fitness of crop-wild hybrids under attack from biological enemies. Cummings et al. (1999) compared the seed produced by wild plants and wild × crop F_1s at three experimental field sites in eastern

Kansas. "The average hybrid plant has 36.5% of its seeds … eaten by insect larvae while the average wild plant lost only 1.8% … to seed predators" (Cummings et al., 1999). Introgression of crop alleles into wild sunflower populations is likely to be impeded rather than prevented by the lowered fitness of the hybrids.

Indeed, introgression from cultivated to wild *H. annuus* appears to occur with little impediment. Whitton et al. (1997) found crop-specific RAPD markers persisting in a California population of wild sunflowers for five generations following a single season of hybridization with a nearby crop plantation. Linder et al. (1998) analyzed three wild *H. annuus* populations from the northern Great Plains that had had long-term (20–40 years) contact with the crop. They found substantial introgression of crop-specific RAPD markers, with an "average overall frequency of cultivar markers greater than 35%" and "every individual…tested contained at least" one cultivar-specific allele. Given the substantial gene flow from the crop to its wild form, it is not surprising that introgression from the crop has been implicated in the evolution of increased weediness in wild *H. annuus* (Heiser, 1978).

One study has addressed introgression of cultivated sunflower alleles into natural populations of another *Helianthus* species. Rieseberg et al. (1999) genetically analyzed four Great Plains populations of *H. petiolaris* that were distant from wild *H. annuus* populations but adjacent to fields of cultivated sunflowers. They found cultivar-specific AFLP alleles in a relatively small number of individuals and at relatively low frequencies within each population. "All hybrids appeared to represent later-generation backcrosses" (Rieseberg et al., 1999).

13. Sugarcane: *Saccharum officinarum* L. (This species, noble cane, comprises most of the world's production of the crop.)

Sugarcane does not need to flower to produce its crop. Nonetheless, spontaneous hybridization between cultivated and wild *Saccharum* in Australasia and islands of the Indian Ocean has been implicated in the origin of many cultivars (Roach, 1995). While most of the evidence for the hybrid origin of these cultivars comes from morphology, some supporting evidence also comes

from genetically based traits. For example, certain wild cases from New Guinea that are morphologically classified as *S. spontaneum* "have leaf flavonoids and triterpenoids common to … *S. officinarum*" (Daniels and Roach, 1987). Likewise, chromosome data indicate that certain wild canes from Java and Mauritius are hybrids between the cultivated *S. officinarum* and the wild *S. spontaneum* (Stevenson, 1965).

Spontaneous intergeneric hybridization has been implicated in the origin of particular accessions of wild sugarcane relatives (Daniels and Roach, 1987). Chromosomal evidence supports the suggestion, based on morphology, that *Erianthus maximus* is derived from a natural cross of the cultivated *S. officinarum* with the wild genus, *Miscanthus* (Daniels and Roach, 1987). Likewise, a *Miscanthus* clone was found to have the same chloroplast restriction fragment sites as cultivated *Saccharum* species, suggesting a history of introgression between the two genera in the lineage of that genotype (Sobral et al., 1994). Population-level analysis has not yet been conducted to determine the extent to which crop alleles have moved into natural populations.

Summary of case studies

All but one of the reviewed crops hybridize naturally with wild realtives in some part of their agricultural distribution. These results make sense in an evolutionary context. Domesticated plants represent lineages that diverged from their progenitors no more than a few thousand generations ago. There is no reason to assume that reproductive isolation should be absolute. Whether the evidence is reviewed on a regional basis or a crop-by-crop basis, it is clear that spontaneous hybridization and introgression of genes from domesticated plants to wild relatives is a common characteristic of domesticated plant taxa.

EVOLUTIONARY CONSEQUENCES OF GENE FLOW

Gene flow can be a potent evolutionary force. A small amount of gene flow is capable of counteracting the other evolutionary forces of mutation, drift, and selection (Slatkin, 1987). Gene flow's impact depends on its magnitude. The magnitude of gene flow among natural plant populations is idiosyncratic, varying among species, populations, individuals, and even years (Ellstrand, 1992). In the same way, levels of gene flow from domesticated plants to their wild relatives are expected to be highly variable and to depend on a variety of spatio-temporal factors. Gene flow may be effectively zero for plants that are cross-incompatible, spatially-isolated, or do not overlap in flowering time. Even when compatible species with similar flowering phenologies grow in spatial proximity, levels of gene flow may vary. For example, differences in the relative sizes of the source and sink populations may result in different rates of hybridization and gene flow (Ellstrand and Elam, 1993). The cultivation of one or a few plants in a backyard or dooryard garden may result in low levels of crop-to-weed gene flow. In contrast, a stand of an agronomic crop grown according to the practices of modern industrial agriculture may contain millions of plants. In such situations, weed populations hundreds of meters away may show high levels of hybridization and introgression.

Experimentally-measured hybridization rates between crops and cross-compatible wild relatives typically exceed 1%, sometimes even over distances of 100 m or more (Langevin et al., 1990; Arias and Rieseberg, 1994; Arriola and Ellstrand, 1996; Renno et al., 1997). The rate of incorporation of foreign alleles under such levels of hybridization is likely to be orders of magnitude higher than typical mutation rates (Grant, 1975), and we would expect gene flow in these systems to be much more important than mutation.

The best known evolutionary consequence of gene flow is its tendency to homogenize population structure (Slatkin, 1987). The conditions for homogenization will vary depending on whether immigrant alleles are neutral, detrimental, or beneficial in the ecological and genomic environment of the population that receives them.

Neutral gene flow

When gene flow is absent, the evolution of neutral alleles in a population depends on stochastic processes (genetic drift). Genetic drift will lead to genetic differentiation among populations, especially when effective population sizes are small (Wright, 1969). When gene flow is present, the rule of thumb is that one immigrant every other generation, or one interpopulation mating per generation, will be sufficient to prevent strong interpopulation differentiation of neutral alleles (Slatkin, 1987). Interestingly, this relationship is independent of the size of the recipient population. Conservation geneticists often conclude one migrant per generation will be sufficient to counteract the effects of drift, preventing divergence between populations and buffering against the loss of within-population variation (Allendorf, 1983).

The homogenization of genetic variation by gene flow does not necessarily result in the enhancement of local variation. Ultimately, changes in local diversity will depend on levels of allelic richness and genetic diversity of the source population (in our case, the crop) relative to that of the sink population. Crop cultivars typically contain substantially less genetic variation than populations of their wild relatives (Ladizinsky, 1985). Thus, we can predict that levels of neutral variation in a wild population will often decrease under gene flow from a domesticated relative.

Detrimental gene flow

Immigration of disadvantageous alleles can counteract local directional selection, reducing local fitness (Antonovics, 1976). Generally, detrimental alleles will swamp those that are locally adaptive when $m \geq s$, where m is the fraction of immigrants relative to the total population per generation, and s is the local selective coefficient against the immigrant alleles (Slatkin, 1987). That is, moderate rates of incoming gene flow ($\sim$1–5% per generation) are expected to be sufficient to introduce and maintain alleles when the local selective coefficient against them is of the same magnitude (i.e. 1–5%).

Data from natural systems support this expectation. Reciprocal transplant studies often reveal local adaptive differentiation in plant populations (Levin, 1984; Waser, 1993), but seldom at the microgeographic level at which substantial gene flow occurs (Waser and Price, 1985; Antonovics et al., 1988). However, such differentiation becomes apparent if selection is very strong, $s > 0.99$ (Antonovics and Bradshaw, 1970). Adaptive differences between populations (in our case, adaptation to growth and reproduction under cultivation versus adaptation to growth and reproduction in unmanaged ecosystems) may lead to outbreeding depression, "a fitness reduction following hybridization" between populations (Templeton, 1986). Outbreeding depression has been detected in many natural plant populations (Waser, 1993), and naturally occurring interspecific hybrids are often partially or completely sterile (Stace, 1975). Similarly, attempts by plant breeders to introduce commercially desirable alleles to crops from distant relatives (the secondary gene pool) are often constrained by the reduced fertility and viability of hybrids resulting from wide crosses (Simmonds, 1981).

Beneficial gene flow

When gene flow and selection work in concert, gene flow will accelerate the increasing frequency of favorable alleles in a sink population. Wright (1969) modeled several situations that involve gene flow in combination with favorable selection. His "Continent-Island" model, with one-way immigration from a large source population into a sink population, is the one most likely to be applicable to gene flow from crops to their wild relatives. Our primary concern is the movement of domesticated alleles into natural populations; these alleles are more-or-less fixed in the crop and absent from the wild population. With these assumptions, Wright's model can be simplified to predict that the fraction of crop-to-wild immigrants, m, will have the same effect on the speed of fixation of the favorable alleles as the local selective coefficient, s. That is, if the magnitudes of both migration and the local selective coefficient are equal, then the rate of allele frequency change per generation is doubled relative to the situation with selection alone. If m is twice as great as s, then the rate will be tripled compared with the situation without gene flow.

APPLIED IMPLICATIONS

Gene flow between domesticated plants and their wild relatives may have two potentially harmful consequences – the evolution of increased weediness and the increased likelihood of extinction of wild relatives. Applied plant scientists can collect data, from experimental or descriptive studies and (if available) from the pre-existing scientific literature, to evaluate the likelihood and potential importance of these gene flow risks. Such data should be used to guide decision-makers about the deployment of new cultivars or the introduction of a new crop into a region. Guidelines have been created for evaluating the gene flow risks of transgenic crops (Rissler and Mellon, 1996), but, to our knowledge, have never been extended to domesticated plants in general. Below, we examine two important risks of gene flow from a domesticated plant species to a wild taxon and address possible ways of evaluating those risks.

Weed evolution

The first potential problem is the transfer of crop alleles to a weedy taxon to create a more aggressive weed. While some crop alleles may be disadvantageous in the wild, others may confer a selective advantage for weedy populations. As noted above, the spread of alleles under favorable selection is accelerated by gene flow. The problem is far from hypothetical. Increased weediness resulting from gene flow from traditionally improved crops to weedy relatives has been well documented. Crop-to-weed gene flow has been implicated in the evolution of enhanced weediness in wild relatives of seven of the world's 13 most important crops (Table 19.1). An additional example is the recent evolution of a weedy rye derived from natural hybridization between cultivated rye, *Secale cereale*, and wild *S. montanum* in California. The weed's hybrid origin has been confirmed by genetic analysis (Sun and Corke, 1992). The weed has become such a serious problem that "farmers have abandoned efforts to grow cultivated rye for human consumption" (National Research Council, 1989).

Under what circumstances is hybridization with a crop most likely to result in the evolution of enhanced weediness? The evolution of enhanced weediness will depend on (1) whether hybridization can occur, (2) whether the hybrids can reproduce in the wild, and (3) whether one or more alleles from the domesticate confer an advantage on the weed.

Without hybridization, gene flow cannot occur. Problems with the evolution of enhanced weediness will only arise when a crop is grown in proximity to a cross-compatible relative. The appropriate way to test whether hybridization is likely to occur under field conditions is to create experimental stands of crop and weed under conditions that simulate those in which the crop and weed will co-occur after field release. Gene flow can then be measured by testing the progeny of the weed for crop-specific genetically based markers. Sample sizes must be large enough to detect gene flow at biologically important levels (sensitive enough to detect hybridization rates of at least 1%). Our case studies include several experimental studies of this type that measure crop-to-wild hybridization rates (Langevin et al., 1990; Arriola and Ellstrand, 1996; Arias and Rieseberg, 1999). Hand pollination and embryo rescue are inappropriate techniques for the investigation of crop-to-weed gene escape because they do not simulate natural vectors or natural seed production.

How these data on hybridization are interpreted is as important as the data themselves. Phrases such as "gene flow drops off rapidly with distance" are misleading. If any hybridization is detected within distances that are typical of those occurring between the crop and the weed, the appropriate interpretation is, "crop genes will move into weedy populations."

The next question is, "Will hybrids reproduce in the wild?". If hybrids reproduce, and if the weedy populations are exposed to a constant rain of compatible pollen from the crop, favorable and neutral crop alleles will persist, even if the hybrids do not show heterosis. Appropriate data on hybrid persistence are those that indicate whether first generation hybrids will reproduce (by seed and/or by vegetative reproduction) under field conditions. We are aware

of only a few studies that have measured the fitness of weed-crop hybrids compared with that of their weed progenitors under field conditions (Hauser et al., 1998a; Klinger and Ellstrand, 1994; Arriola and Ellstrand, 1997). Interestingly, in the majority of such studies, the fitness of the hybrid tends to be the same as or even higher than that of its wild parent.

An alternative to experimental approaches to the first two questions is to conduct descriptive population genetic studies that ask whether crop-specific genetic markers occur in natural populations that grow in the vicinity of the crop in question. The presence of and the frequency of crop-specific markers can then be used to determine levels of past introgression. Such studies must be based on sufficiently large population samples. If crop-specific alleles occur at biologically important frequencies ($>1\%$) in adjacent weedy populations, then it is clear that domesticated alleles can become incorporated into natural populations via hybridization and introgression. Numerous studies have already demonstrated introgression of crop alleles into natural populations; some of these are mentioned in our case studies (Jørgensen et al., 1996; Beebe et al., 1997; Linder et al., 1998).

The final question is, "Will the crop genes enhance weediness?". Unlike the preceding two questions, this question cannot be addressed without direct consideration of the biology of the new cultivar. Certain novel traits (e.g. herbicide tolerance, disease resistance, insect resistance) will enhance particular fitness components of a weed in particular environments but may (or may not) have attendant costs (Bergelson and Purrington, 1996). Pleiotropic fitness effects may also occur with alleles in a novel genetic context (Bergelson et al., 1998). Ideally, the fitness impact of the novel crop alleles should be examined experimentally in the context of the wild genome and the environment of the weed. The fitness of hybrids containing the novel allele could be compared with the fitness of those without the allele under field conditions. Alternatively, the field-measured fitness of first generation backcrosses to the weed could be compared with that of pure wild plants. Selective factors – herbivores, diseases, and herbicides – should not be excluded from the experiments if they are common features of the field environment. Few such experiments have been reported (Cummings et al., 1999) – largely motivated by the desire to estimate the fitness impact of a transgene in a hybrid genome. The interpretation of the fitness data is critical. We expect that a fitness boost of 5% will enhance weediness to the point that would have important practical consequences.

The issue of evolution of enhanced weediness has received considerable attention as a potential consequence of the field release of genetically engineered crops and their subsequent hybridization with wild relatives (Colwell et al., 1985; Rissler and Mellon, 1996). As noted above, hybridization involving conventionally bred crops has already resulted in the evolution of enhanced weediness, leading to human hardship. There is no reason to expect that engineered crops should behave any differently than their traditionally improved counterparts. Given the rapid increase in planting of transgenic crops, it would not be surprising to find transgenes in populations of wild relatives. We are not aware of any such reports outside of the controlled experimental studies mentioned in our case studies.

In summary, bringing a novel cultivar or a new crop taxon into contact with a cross-compatible weed may lead to evolution of enhanced weediness if hybridization occurs under field conditions, the hybrids persist to reproduce, and one or more crop alleles give a fitness boost to the weed populations. The economic and environmental impact of that enhanced weediness will depend on how much fitness increases and under what environmental conditions.

Extinction of wild relatives

The other potential problem with natural hybridization between domesticated plants and their wild relatives is the increased risk of extinction of wild taxa. Largely neglected as a general conservation issue, extinction by hybridization has recently begun to attract attention (Ellstrand and Elam, 1993; Levin et al., 1996; Rhymer and Simberloff, 1996). Rare taxa may be threatened by hybridization in two ways.

As noted above, outbreeding depression from detrimental gene flow will reduce the fitness of a locally rare species that is mating with a locally common one. An alternate route to extinction is by swamping, which occurs when a locally rare species loses its genetic integrity and becomes assimilated into a locally common species as a result of repeated bouts of hybridization and introgression (Ellstrand and Elam, 1993). We would expect swamping to result from gene flow that is largely neutral or beneficial.

Both outbreeding depression and swamping are frequency-dependent phenomena and show positive feedback. With each succeeding generation of hybridization and backcrossing, genetically pure individuals of the locally rare species become increasingly rare until extinction occurs (Ellstrand and Elam, 1993). Both phenomena can lead to extinction rapidly. If 900 individuals of a locally common species mate randomly with 100 individuals of the locally rare one, extinction by outbreeding depression and/or swamping can occur in two generations (N. Ellstrand, unpublished data).

In contrast to the problem of enhanced weediness, the problem of extinction by hybridization does not depend on relative fitness, but only on patterns of mating. Specifically, the risk decreases as intertaxon mating rates depart from random mating (that is, as assortative mating increases). Consider the example above, with 900 individuals of a locally common species mating with 100 individuals of the locally rare one. If we impose assortative mating, such that the rare species mates with its own kind five times more frequently than the rate expected under random mating, the time to extinction by outbreeding depression and/or swamping is expected to double relative to random mating (N. Ellstrand, unpublished data).

Extinction by hybridization with domesticated species has been implicated in the extinction or increased risk of extinction of several wild species, including wild relatives of two of the world's 13 most important crops (Table 19.1; also cf. Small, 1984). Indeed, hybridization between crops and their wild relatives has probably resulted in the extinction of many species since the Agricultural Revolution thousands of years ago (Small, 1984). The extraordinary speed of the extinction process outlined above may be one reason those extinctions have been overlooked.

The risk of extinction by hybridization will occur when a previously allopatric rare taxon becomes sympatric or peripatric with a recently introduced crop. That risk will increase as hybridization and introgression proceed over generations and will also increase with intertaxon hybridization rates. Measuring hybridization rates is critical for the assessment of the risk of extinction by hybridization. The appropriate way to assess hybridization rates under field conditions is to create experimental stands of the crop and wild taxon under conditions comparable to those in which the crop and the wild taxon will co-exist when field release occurs. Progeny testing of the wild taxon for crop-specific genetic markers can then used to measure gene flow. Sample sizes must be large enough to detect gene flow at biologically significant levels (sensitive enough to detect hybridization rates of at least 10%). We are not aware of any experiments measuring hybridization rates between a common plant species and a rare relative. However, the experimental studies mentioned above that measure crop-to-weed hybridization rates are appropriate models.

After assessing hybridization rates, the sizes of the populations of the rare taxon can be estimated by direct counts and adjusted to allow for future disturbance and fragmentation. These numbers, combined with the experimentally-assessed hybridization rates, can then be used to develop crude estimates of the numbers of individuals in each wild population that will be replaced by hybrids and introgressants with each generation of hybridization. The speed of replacement will indicate whether the risk of extinction is biologically important.

ACKNOWLEDGMENTS

This project grew out of projects funded with USDA grant support to NCE and SJFR (the Swedish Forestry and Agricultural Research Council) grant support to NCE and HCP. Helena Parrow helped with the initial compilation of case studies. The following scientists pre-reviewed our case studies: J. Antonovics, P. Arriola, L. Blancas, A.H.D. Brown, E. Buckler, J. Doebley, J. Doyle, D. Garvin, P. Gepts, W. Hanna, S. Hegde, K. Hilu, T. Holtsford, F. Ibarra-Perez, R. Jørgensen, P. Keim, J. Kohn, H. Morishima, H. Oka, R. Palmer, L. Rieseberg, Y. Sano, K. Schertz, A. Scholz, B. Sobral, H. Suh, M. van Slageren, J.G. Waines, and J. Wendel. This paper improved as a result of thoughtful suggestions from N. Barton, D. Bartsch, S. Baughman, J.M.J. deWet, J. Endler, S. Frank, D. Gessler, J. Hamrick, A. Rankin, and M. Syvanen.

REFERENCES

Abe, J., Hasegawa, A., Fukushi. H. et al. (1999) Introgression between wild and cultivated soybeans of Japan revealed by RFLP analysis for chloroplast DNAs. *Econ. Bot.* **53**: 285–291.

Acosta, J., Gepts, P. and Debouck, D.G. (1994) Observations on wild and weedy accessions of common beans in Oaxaca, Mexico. *Annu. Rept. Bean Improv. Coop* **37**: 137–138.

Aldrich, P.R. and Doebley, J. (1992) Restriction fragment variation in the nuclear and chloroplast genomes of cultivated and wild *Sorghum bicolor. Theor. Appl. Genet.* **85**: 293–302.

Aldrich, P.R., Doebley, J., Schertz. K.F. and Stec, A. (1992) Patterns of allozyme variation in cultivated and wild *Sorghum bicolor. Theor. Appl. Genet.* **85**: 451–460.

Allendorf, F.W. (1983) Isolation, gene flow, and genetic differentiation among populations. In *Genetics and Conservation: A Reference for Managing Wild Animal and Plant Populations* (eds, C.M. Schonewald-Cox, S.M. Chambers, B. MacBryde and W.L. Thomas), pp. 51–65, Benjamin-Cummings, Menlo Park, CA.

Anderson, E. (1949) *Introgressive Hybridization,* John Wiley & Sons, New York.

Antonovics, J. (1976) The nature of limits to natural selection. *Ann. Mo. Bot. Gard.* **63**: 224–247.

Antonovics, J. and Bradshaw, A.D. (1970) Evolution in closely adjacent populations. VIII. Clinal patterns at a mine boundary. *Heredity* **25**: 349–362.

Antonovics, J., Ellstrand, N.C. and Brandon, R.N. (1988) Genetic variation and environmental variation: expectations and experiments. In *Plant Evolutionary Biology: A Symposium Honoring G. Ledyard Stebbins* (eds, L. Gottlieb and S.K. Jain), pp. 275–303, Chapman & Hall, London.

Arias, D.M. and Rieseberg, L.H. (1994) Gene flow between cultivated and wild sunflowers. *Theor. Appl. Genet.* **89**: 655–660.

Arnold, M.L. (1997) *Natural Hybridization and Evolution* Oxford University Press, New York.

Arriola, P.E. and Ellstrand, N.C. (1996) Crop-to-weed gene flow in the genus *Sorghum* (Poaceae): spontaneous interspecific hybridization between johnsongrass, *Sorghum halepense*, and crop sorghum, *S. bicolor. Am. J. Bot.* **83**: 1153–1160.

Arriola, P.E. and Ellstrand, N.C. (1997) Fitness of interspecific hybrids in the genus *Sorghum*: persistence of crop genes in wild populations. *Ecol. Appl.* **7**: 512–518.

Baker, H.G. (1972) Human influences on plant evolution. *Econ. Bot.* **26**: 32–43.

Barrett, S.C.H. (1983) Crop mimicry in weeds. *Econ. Bot.* **37**: 255–282.

Beebe, S., Toro, O., González, A.V. et al. (1997) Wild-weed-crop complexes of common bean (*Phaseolus vulgaris* L., Fabaceae) in the Andes of Peru and Colombia, and their implications for conservation and breeding. *Genet. Resources Crop. Evol.* **44**: 73–91.

Bergelson, J. and Purrington, C.B. (1996) Surveying patterns in the cost of resistance in plants. *Am. Nat.* **148**: 536–558.

Bergelson, J., Purrington, C.B. and Wichmann, G. (1998) Promiscuity in transgenic plants. *Nature* **395**: 25.

Bing, D.J., Downey, R.K. and Rakow, G.F.W. (1996) Hybridizations among *Brassica napus, B. rapa*, and *B. juncea* and their two weedy relatives *B. nigra* and *Sinapis arvensis* under open pollination conditions in the field. *Plant Breeding* **115**: 470–473.

Blumler, M.A. (1998) Introgression of durum into wild emmer and the agricultural origin question. In *The Origins Of Agriculture and Crop Domestication* (eds, A.B. Damania, J. Valkoun, G. Willcox and C.O. Qualset), pp. 252–68, ICARDA, Aleppo.

Broich, S.L. and Palmer, R.G. (1980) A cluster analysis of wild and domesticated soybean phenotypes. *Euphytica* **29**: 23–32.

Broich, S.L. and Palmer, R.G. (1981) Evolutionary studies of the soybean: the frequency and distribution of alleles among collections of *Glycine max* and *G. soja* of various origin. *Euphytica* **30**: 55–64.

Brubaker, C.L., Koontz, J.A. and Wendel, J.F. (1993) Bidirectional cytoplasmic and nuclear introgression in the New World cottons, *Gossypium barbadense* and *G. hirsutum* (Malvaceae). *Am. J. Bot.* **80**: 1203–1208.

Brubaker, C.L. and Wendel, J.F. (1994) Reevaluating the origin of domesticated cotton (*Gossypium hirsutum*; (Malvaceae) using nuclear restriction fragment length polymorphisms (RFLPs). *Am. J. Bot.* **81**: 1309–1326.

Brunken, J., deWet, J.M.J. and Harlan, J.R. (1977) The morphology and domestication of pearl millet. *Econ. Bot.* **31**: 163–174.

Burdon, J.J., Marshall, D.R. and Oates, J.D. (1992) Interactions between wild and cultivated oats. In *Wild Oats in World Agriculture* (eds, A.R. Barr and R.W. Medd), *Proc. 4th Int. Oat Conf.* **2**: 82–87.

Chang, T.T. (1995) Rice. In *Evolution of Crop Plants* (eds, J. Smartt and N.W. Simmonds), 2nd edn, pp. 147–155, Longman, Harlow.

Chèvre, A.M., Eber, F., Baranger, A. et al. (1998) Characterization of backcross generation obtained under field conditions from oilseed rape – wild radish F_1

interspecific hybrids: an assessment of transgene dispersal. *Theor. Appl. Genet.* **97**: 90–98.

Chu, Y.E., Morishima, H. and Oka, H.I. (1969) Reproductive barriers distributed in cultivated rice species and their wild relatives. *Japan. J. Genet.* **4**: 207–223.

Chu, Y.E. and Oka, H.I. (1970) Introgression across isolating barriers in wild and cultivated *Oryza* species. *Evolution* **24**: 344–355.

Colwell, R.K., Norse, E.A., Pimentel, D. et al. (1985) Genetic engineering in agriculture. *Science* **229**: 111–112.

Cummings, C.L., Alexander, H.M. and Snow, A.A. (1999) Increased pre-dispersal seed predation in sunflower crop-wild hybrids. *Oecologia* **121**: 330–338.

Dale, P.J. (1994) The impact of hybrids between genetically modified crop plants and their related species: general considerations. *Molec. Ecol.* **3**: 31–36.

Daniels, J. and Roach, B.T. (1987) Taxonomy and evolution. In *Sugarcane Improvement through Breeding* (ed., D.J. Heinz), pp. 7–83, Elsevier, Amsterdam.

Darmency, H., Lefol, E. and Fleury, A. (1998) Spontaneous hybridizations between oilseed rapes and wild radish. *Molec. Ecol.* **7**: 1467–1473

Debouck, D.G., Toro, O., Paredes, O.M. et al. (1993) Genetic diversity and ecological distribution of *Phaseolus vulgaris* in northwestern South America. *Econ. Bot.* **47**: 408–423.

DeJoode, D.R. and Wendel, J.F. (1992) Genetic diversity and origin of the Hawaiian Islands cotton, *Gossypium tomentosum. Am. J. Bot.* **79**: 1311–1319.

Delgado Salinas, A., Bonet, A. and Gepts, P. (1988) The wild relative of *Phaseolus vulgaris* in Middle America. In *Genetic Resources of Phaseolus beans* (ed., P. Gepts), pp. 163–184, Kluwer, Dordrecht.

deVries, F.T., van der Meijden, R. and Brandenburg, W.A. (1992) Botanical files: a study of the real chances for spontaneous gene flow from cultivated plants to the wild flora of the Netherlands. *Gorteria* suppl. **1**: 1–100.

deWet, J.M.J. (1995a) Finger millet. In *Evolution of Crop Plants* (eds, J. Smartt and N.W. Simmonds), 2nd edn, Longman, Harlow.

deWet, J.M.J. (1995b) Pearl millet. In *Evolution of Crop Plants* (eds, J. Smartt and N.W. Simmonds), 2nd edn, Longman, Harlow.

deWet, J.M.J. and Harlan, J.R. (1975) Weeds and domesticates: evolution in the man-made habitat. *Econ. Bot.* **79**: 99–107.

Doebley, J. (1990) Molecular evidence for gene flow among *Zea* species. *BioScience* **40**: 443–448.

Doggett, H. and Majisu, B.N. (1968) Disruptive selection in crop development. *Heredity* **23**: 1–23.

Dorado, O., Rieseberg, L.H. and Arias, D.M. (1992) Chloroplast DNA introgression in southern California sunflowers. *Evolution* **46**: 566–572.

Ellstrand, N.C. (1992) Gene flow among seed plant populations. *New Forests.* **6**: 241–256.

Ellstrand, N.C. and Elam, D.R. (1993) Population genetic consequences of small population size: implications for plant conservation. *Annu. Rev. Ecol. Syst.* **24**: 217–242.

Ellstrand, N.C., Whitkus, R.W. and Rieseberg, L.H. (1996) Distribution of spontaneous plant hybrids. *Proc. Natl Acad. Sci. USA* **93**: 5090–5093.

Feldman, M., Lipton, F.G.H., Miller, T.E. (1995) Wheats. In *Evolution of Crop Plants* (eds, J. Smartt and N.W. Simmonds), pp. 184–192, Longman, Harlow.

Freyre, R., Ríos, R., Guzmán, L. et al. (1996) Ecogeographic distribution of *Phaseolus* ssp. (Fabaceae) in Bolivia. *Econ. Bot.* **50**: 195–215.

Fryxell, P.A. (1979) *The Natural History of the Cotton Tribe (Malvaceae, Tribe Gossypieae)*, Texas A&M University, College Station.

Futuyma, D.J. (1998) *Evolutionary Biology*, 3rd edn, Sinauer, Sunderland.

Grant, V. (1975) *Genetics of Flowering Plants*, Columbia University Press, New York.

Grant, V. (1981) *Plant Speciation*, 2nd edn, Columbia University, New York.

Hancock, J.F. (1992) *Plant Evolution and the Origin of Crop Species*, Prentice-Hall, Englewood Cliffs, NJ.

Hancock, J.F., Grumet, R. and Hokanson, S.C. (1996) The opportunity for escape of engineered genes from transgenic crops. *HortScience* **31**: 1080–1085.

Harberd, D.J. (1975) *Brassica* L. In *Hybridization and the Flora of the British Isles* (ed., C.A. Stace), pp. 137–139, Academic Press, London.

Harlan, J.R. (1995) Barley. In *Evolution of Crop Plants* (eds, J. Smartt and N.W. Simmonds), 2nd edn, pp. 140–147, Longman, Harlow.

Hauser, T.P., Shaw, R.G. and Østergård, H. (1998a) Fitness of F_1 hybrids between weedy *Brassica rapa* and oilseed rape (*B. napus*). *Heredity* **81**: 429–435.

Hauser, T.P., Jørgensen, R.B. and Østergård, H. (1998b) Fitness of backcross and F2 hybrids between weedy *Brassica rapa* and oilseed rape (*B. napus*). *Heredity* **81**: 436–443.

Heiser, C.B. (1978) Taxonomy of *Helianthus* and the origin of domesticated sunflower. In *Sunflower Science and Technology* (ed., J.F. Carter), pp. 31–54, Am. Soc. Agronomy, Crop Sci. Soc. & Soil Sci. Soc. Am, Madison.

Holm, L.G., Plucknett, D.L., Pancho, J.V. and Herberger, J.P. (1977) *The World's Worst Weeds: Distribution and Biology*, University Press of Hawaii, Honolulu.

Jasieniuk, M., Brûlé-Babel, A.L. and Morrison, I.N. (1996) The evolution and genetics of herbicide resistance in weeds. *Weed Sci.* **44**: 176–193.

Jiang, J., Freibe, B. and Gill, B.S. (1994) Recent advances in alien gene transfer in wheat. *Euphytica* **73**: 199–212.

Jørgensen, R.B. and Andersen, B. (1994) Spontaneous hybridization between oilseed rape (*Brassica napus*) and weedy *Brassica campestris* (Brassicaceae): a risk of growing genetically modified oilseed rape. *Am. J. Bot.* **81**: 1620–1626.

Jørgensen, R.B., Andersen, B., Landbo, L. and Mikkelsen, T. (1996) Spontaneous hybridization between oilseed rape (*Brassica napus*) and weedy relatives. In *Proceedings of the International Symposium on Brassicas/Ninth Crucifer Genetics Workshop* (eds, J.S. Dias, I, Crute and A.A. Monteiro), pp. 193–197, ISHS, Lisbon.

Jørgensen, R.B., Andersen, B., Hauser, T.P. et al. (1998) Introgression of crop genes from oilseed rape (*Brassica napus*) to related wild species – an avenue for the escape of engineered genes. In *Proceedings of the International Symposium on Brassicas* (eds, T. Grégoire and A.A. Monteiro), pp. 211–217, ISHS, Rennes.

Kato, T.A. (1997) Review of introgression between maize and teosinte. In *Gene Flow among Maize Landraces, Improved Maize Varieties, and Teosinte: Implications for Transgenic Maize* (eds, J.A. Serratos, M.C. Willcox and F. Castillo), pp. 44–53, DF: CIMMYT, Mexico.

Keim, P., Shoemaker, R.C. and Palmer, R.G. (1989) Restriction fragment length polymorphism diversity in soybean. *Theor. Appl. Genet.* **77**: 786–792.

Kiang, Y.T., Antonovics, J. and Wu, L. (1979) The extinction of wild rice (*Oryza perennis formosana*) in Taiwan. *J. Asian Ecol.* **1**: 1–9.

Kimber, G. and Feldman, M. (1987) *Wild Wheat*, Special Report 353, University of Missouri, Columbia.

Klinger, T. and Ellstrand, N.C. (1994) Engineered genes in wild populations: fitness of weed-crop hybrids of radish, *Raphanus sativus* L. *Ecol. Appl.* **4**: 117–120.

Ladizinsky, G. (1985) Founder effect in crop-plant evolution. *Econ. Bot.* **39**: 191–199.

Ladizinsky, G. (1992) Crossability relations. In *Distant Hybridization of Crop Plants* (eds, G. Kalloo and J.B. Chowdhury), pp. 15–31, Springer-Verlag, Berlin.

Langevin, S., Clay, K. and Grace, J.B. (1990) The incidence and effects of hybridization between cultivated rice and its related weed red rice (*Oryza sativa* L.). *Evolution* **44**: 1000–1008.

Lefol, E., Danielou, V. and Darmency, H. (1996a) Predicting hybridization between transgenic oilseed rape and mustard. *Field Crops Research* **45**: 153–161.

Lefol, E., Fleury, A. and Darmency, H. (1996b) Gene dispersal from transgenic crops. II. Hybridization between oilseed rape and the wild hoary mustard. *Sex. Plant Reprod.* **9**: 189–196.

Levin, D.A. (1984) Immigration in plants: an exercise in the subjunctive. In *Perspectives on Plant Population Ecology* (eds, R. Dirzo and J. Sarukhán), pp. 242–260, Sinauer, Sunderland.

Levin, D.A., Francis-Ortega, J. and Jansen, R.K. (1996) Hybridization and the extinction of rare plant species. *Conserv. Biol.* **10**: 10–16.

Linder, C.R., Taha, I., Seiler, G.J. et al. (1998) Long-term introgression of crop genes into wild sunflower populations. *Theor. Appl. Genet.* **96**: 339–347.

Love, S.L. (1994) Ecological risk of growing transgenic potatoes in the United States and Canada. *Am. Potato J.* **71**: 647–658.

Luby, J.J. and McNichol, R.J. (1995) Gene flow from cultivated to wild raspberries in Scotland: developing a basis for risk assessment for testing and deployment of transgenic cultivars. *Theor. Appl. Genet.* **90**: 1133–1137.

Majumder, N.D., Ram, T. and Sharma, A.C. (1997) Cytological and morphological variation in hybrid swarms and introgressed population of interspecific hybrids (*Oryza rufipogon* Griff. X *Oryza sativa* L.) and its impact on evolution of intermediate types. *Euphytica* **94**: 295–302.

Manasse, R.S. (1992) Ecological risks of transgenic plants: effects of spatial dispersion of gene flow. *Ecol. Appl.* **2**: 431–438.

Marchais, L. (1994) Wild pearl millet population (*Pennisetum glaucum*, Poaceae) integrity in agricultural Sahelian areas. An example from Keita (Niger). *Pl. Syst. Evol.* **189**: 233–245.

McNaughton, I.M. (1995) Swedes and rapes. In *Evolution of Crop Plants* (eds, J. Smartt and N.W. Simmonds), 2nd edn, pp. 68–75, Longman, Harlow.

Mehra, K.L. (1962) Natural hybridization between *Eleusine coracana* and *E. africana* in Uganda. *J. Indian Bot. Soc.* **41**: 531–539.

Meyn, O. and Emboden, W.A. (1987) Parameters and consequences of introgression of *Salvia apiana* X *S. mellifera* (Lamiaceae). *Syst. Bot.* **12**: 390–399.

Mikkelsen, T.R., Andersen, B. and Jørgensen, R.B. (1996) The risk of crop transgene spread. *Nature* **380**: 31.

Morishima, H., Sano, Y. and Oka, H.I. (1992) Evolutionary studies in rice and its wild relatives. *Oxford Surveys in Evolutionary Biology* **8**: 135–184.

National Research Council (1989) *Field Testing Genetically Modified Organisms: Framework for Decisions*, National Academy, Washington, DC.

Oka, H.I. (1988) *Origin of Cultivated Rice*, JSSP/Elsevier, Tokyo.

Oka, H.I. and Chang, W.T. (1959) The impact of cultivation on populations of wild rice, *Oryza sativa* f. *spontanea*. *Phyton* **13**: 105–117.

Oka, H.I. and Chang, W.T. (1961) Hybrid swarms between wild and cultivated rice species, *Oryza perennis* and *O. sativa*. *Evolution* **15**: 418–430.

Paterson, A.H., Schertz, K.F., Lin, Y.R. et al. (1995) The weediness of wild plants: molecular analysis of genes influencing dispersal and persistence of johnsongrass, *Sorghum halepense* (L.) Pers. *Proc. Natl Acad. Sci. USA* **92**: 6127–6131.

Phillips, S.M. (1972) A survey of the genus *Eleusine* Gaertn. (Gramineae) in Africa. *Kew. Bull.* **27**: 251–270.

Popova, G. (1923) Wild species of *Aegilops* and their mass-hybridisation with wheat in Turkestan. *Bull. Appl. Bot.* **13**: 475–482.

Raybould, A.F. and Gray, A.J. (1993) Genetically modified crops and hybridization with wild relatives: a UK perspective. *J. Appl. Ecol.* **30**: 199–219.

Renno, J.F., Winkel, T., Bonnefous, F. and Benzançon, G. (1997) Experimental study of gene flow between wild and cultivated *Pennisetum glaucum*. *Can. J. Bot.* **75**: 925–931.

Rhymer, J.M. and Simberloff, D. (1996) Extinction by hybridization and introgression. *Annu. Rev. Ecol. Syst.* **27**: 83–109.

Richards, A.J. (1986) *Plant Breeding Systems*, Allen & Unwin, Hemel Hempstead.

Rieseberg, L.H., Carter, R. and Zona, S. (1990) Molecular tests of the hypothesized hybrid origin of two diploid *Helianthus* species. *Evolution* **44**: 1498–1511.

Rieseberg, L.H. and Ellstrand, N.C. (1993) What can molecular and morphological markers tell us about plant hybridization? *Crit. Rev. Pl. Sci.* **12**: 213–241.

Rieseberg, L.H. and Wendel, J.F. (1993) Introgression and its evolutionary consequences in plants. In *Hybrid Zones and the Evolutionary Process* (ed., R. Harrison), pp. 70–109, Oxford University Press, New York.

Rieseberg, L.H., Kim, M.J. and Seiler, G.J. (1999) Introgression between the cultivated sunflower and a sympatric wild relative, *Helianthus petiolaris* (Asteraceae). *Inter. J. Pl. Sci.* **160**: 102–108.

Rissler, J. and Mellon, M. (1996) *The Ecological Risks of Engineered Crops*, MIT Press, Cambridge, MA.

Roach, B.T. (1995) Sugar canes. In *Evolution of Crop Plants* (eds, J. Smartt and N.W. Simmonds), 2nd edn, pp. 160–166, Longman, Harlow.

Rogers, C.E., Thompson, T.E. and Seiler, G.J. (1982) *Sunflower Species of the United States*, National Sunflower Association, Bismarck.

Runyeon, H. and Prentice, H.C. (1997) Genetic differentiation in the bladder campions, *Silene vulgaris* and *S. uniflora* (Caryophyllaceae), in Sweden. *Biol. J. Linn. Soc.* **61**: 559–584.

Sanders, R.W. (1987) Identity of *Lantana depressa* and *L. ovatifolia* (Verbenaceae) of Florida and the Bahamas. *Syst. Bot.* **12**: 44–60.

Scheffler, J.A. and Dale, P.J. (1994) Opportunities for gene transfer from transgenic oilseed rape (*Brassica napus*) to related species. *Transgenic Research* **3**: 263–278.

Seefeldt, S.S., Zemetra, R., Young, F.L. and Jones, S.S. (1998) Production of herbicide-resistant jointed goatgrass (*Aegilops cylindrica*) × wheat (*Triticum aestivum*) hybrids in the field by natural hybridization. *Weed Sci.* **46**: 632–634.

Simmonds, N.W. (1981) *Principles of Crop Improvement*, Longman, London.

Singh, A.K. (1995) Groundnut. In *Evolution of Crop Plants* (eds, J. Smartt and N.W. Simmonds), 2nd edn, pp. 246–250, Longman, Harlow.

Slatkin, M. (1987) Gene flow and the geographic structure of natural populations. *Science* **236**: 787–792.

Small, E. (1984) Hybridization in the domesticated-weed-wild complex. In *Plant Biosystematics* (ed., W.F. Grant), pp. 195–210, Academic, Toronto.

Smartt, J. and Simmonds, N.W. (1995) *Evolution of Crop Plants* 2nd edn, Longman, Harlow.

Snow, A.A., Moran-Palma, P., Rieseberg, L.H. et al. (1998) Fecundity, phenology, and seed dormancy of F_1 wild-crop hybrids in sunflower (*Helianthus annuus*, Asteraceae). *Am. J. Bot.* **85**: 794–801.

Sobral, B.W.S., Braga, D.P.V., LaHood, E.S. and Keim, P. (1994) Phylogenetic analysis of chloroplast restriction enzyme site mutations in the *Saccharinae* Griesb. subtribe of the *Andropogoneae* Dumort. tribe. *Theor. Appl. Genet.* **87**: 843–853.

Stace, C.A. (1975) *Hybridization and the Flora of the British Isles*, Academic Press, London.

Stevenson, G.C. (1965) *Genetics and Breeding of Sugar Cane*, Longman, London.

Suh, H.S., Sato, Y.I. and Morishima, H. (1997) Genetic characterization of weedy rice (*Oryza sativa* L.) based on morpho-physiology, isozymes and RAPD markers. *Theor. Appl. Genet.* **94**: 316–321.

Sun, M. and Corke, H. (1992) Population genetics of colonizing success of weedy rye in northern California. *Theor. Appl. Genet.* **83**: 321–329.

Tang, L.H. and Morishima, H. (1988) Characteristics of weed rice strains. *Rice Genet. Newsl.* **5**: 70–72.

Templeton, A.R. (1986) Coadaptation and outbreeding depression. In *Conservation Biology: The Science of Scarcity and Diversity* (ed., M. Soulé), pp. 105–16, Sinauer, Sunderland.

Tostain, S. (1992) Enzyme diversity in pearl millet (*Pennisetum glaucum* L.). 3. Wild millet. *Theor. Appl. Genet.* **83**: 733–742.

Vanderborght, T. (1983) Evaluation of *Phaseolus vulgaris* wild and weedy forms. *Plant Genet. Res. Newslett.* **54**: 18–25.

van Slageren, M.W. (1994) *Wild Wheats: a Monograph of Aegilops L. and Amblyopyrum (Jaub. & Spach) Eig (Poaceae)*, Veenman Druckers, Wageningen.

von Bothmer, R., Jacobsen, N., Baden, C. et al. (1991) *An Ecogeographical Study of the Genus*, Hordeum IBPGR, Rome.

Waser, N.M. (1993) Population structure, optimal outbreeding, and assortative mating in angiosperms. In *The Natural History of Inbreeding and Outbreeding, Theoretical and Empirical Perspectives* (ed., N.W. Thornhill), pp. 173–199, University of Chicago, Chicago.

Waser, N.M. and Price, M.V. (1985) Reciprocal transplants with *Delphinium nelsonii* (Ranunculaceae): evidence for local adaptation. *Am. J. Bot.* **72**: 1726–1732.

Wendel, J.F., Brubaker, C.L. and Percival, A.E. (1992) Genetic diversity in *Gossypium hirsutum* and the origin of upland cotton. *Am. J. Bot.* **79**: 1291–1310.

Wendel, J.F. and Percy, R.G. (1990) Allozyme diversity and introgression in the Galapagos Islands endemic *Gossypium darwinii* and its relationship to continental *G. barbadense*. *Biochem. Ecol. Syst.* **18**: 517–528.

Wendel, J.F., Rowley, R. and Stewart, J.M.cD. (1994) Genetic diversity in and phylogenetic relationships of the Brazilian endemic cotton *Gossypium mustelinum Pl. Syst. Evol.* **192**: 49–59.

Whitton, J., Wolf, D.E., Arias, D.M. et al. (1997) The persistence of cultivar alleles in wild populations of sunflowers five generations after hybridization. *Theor. Appl. Genet.* **95**: 33–40.

Wilkes, H.G. (1977) Hybridization of maize and teosinte, in Mexico and Guatemala and the improvement of maize. *Econ. Bot.* **31**: 254–293.

Wright, S. (1969) *Evolution and the Genetics of Populations. Volume 2. The Theory of Gene Frequencies*, University of Chicago, Chicago.

Xu, J., Kerrigan, R.W., Callac, P. et al. (1997) Genetic structure of natural populations of *Agaricus bisporus*, the commercial mushroom. *J. Hered.* **88**: 482–488.

CHAPTER 20

Search for Horizontal Gene Transfer from Transgenic Crops to Microbes

Michael Syvanen

After US regulatory agencies approved Calgene's Flavr Savr tomato, those in the scientific and business worlds rejoiced that a new era in farming technology had begun. Crops could be rationally designed by genetic engineering and by new advances in cell and molecular biology. Many of us dismissed the early protests against this new technology as a predictable reaction of modern-day Luddites. Unfortunately, industry and the US Food and Drug Administration (FDA) handled some of these criticisms clumsily, which resulted in political problems that currently threaten an entire industry.

The technology for making transgenic plants in 1986 required using genes that conferred resistance to antibiotics; these genes were actually normal bacterial genes adapted for expression in plant cells. The fear was raised that these antibiotic-resistance genes being released into the environment in large amounts could be taken up by native bacteria and pose public health risks.

Indeed, some scientists had recommended early in the development of this technology that transgenic plants carrying antibiotic-resistance genes from bacteria should not be approved for that reason. And it appears that use of these genes for crops in development today is now being avoided. However, even if the use of drug-resistance genes ceases tomorrow, the problem raised by the genes in transgenic plants already approved remains. Many have already been approved for use in

the United States, at least 10 of them are now in production and many millions of acres of maize, cotton and soybeans have been planted. The investment of time and money is huge and these products will not be withdrawn from the market without some evidence that they are dangerous.

When transgenic plants were first approved, the FDA responded to fears of the spread of antibiotic-resistance genes with the assertion that the genes from the plants would not be incorporated into bacteria. Even as recently as 1995, the FDA wrote, "There is no known mechanism by which a gene can be transferred from a plant chromosome to a microbe. Thus, the possibility that such transfer would generate new resistant organisms is very small, especially when compared to the high rate of spread of resistance through known mechanisms of microbe to microbe transfer of antibiotic resistance genes" (US FDA, 1995). When these claims were made, I was in the process of attempting to convince my colleagues that horizontal gene transfer between life's kingdoms was a natural process and that the many different demonstrations of DNA uptake by bacteria, fungi, and animal cells reflected the operation of natural processes used to facilitate horizontal gene transfer. If these speculations were true, it would make more sense to defend the transgenic crop industry by arguing that gene transfer is a natural phenomena than by arguing that it does not occur. As is well documented throughout this book,

237

Copyright © 2002 by Academic Press.
All rights of reproduction in any form reserved.

we can infer numerous cases cross-species gene transfers from the whole genome sequencing studies. However, this does not necessarily mean that the frequency of transfer is high enough for gene transfer on the farm may to occur.

Stimulated by the challenging assertion from the FDA and some industry spokesmen that horizontal gene transfer from transgenic plants to microorganisms does not occur, a number of investigators have embarked on studies to show that this statement is false. At this time the challenge remains intact, but just barely. First, we know that mechanisms that facilitate the movement of DNA from plants to bacteria do exist, and now the only question is whether they do in fact operate in natural environments.

Two types of experiments have been done. In the first, transgenic plant materials have been fed to mice and the coliform bacteria isolated from their feces have been screened for the presence of antibiotic-resistance genes originating from the plants. Numerous trials in my laboratory and others recovered none. In the second approach, researchers in two labs have looked at bacteria that cause spoilage of vegetables and examined whether, after growth on transgenic vegetables, the bacteria had taken up transgenic DNA. In both cases, the bacterium *Erwinia chrysanthemum* was grown on a variety of vegetables but no transformants were detected (unpublished data) (Schluter, 1995). There has been a recent report that a fungus that inhabits the honey bee gut will incorporate transgenic plant DNA; this report has not yet appeared in the peer-reviewed scientific literature, so its validity remains untested.[*]

Other studies have used experiments in which the likelihood of gene transfer occurring was maximized in order to determine whether mechanisms were present to facilitate horizontal gene transfer or, possibly, to act as a barrier to it. Using this approach, researchers were able to show movement of a plant transgene into a naturally occurring *Acinetobacter* species selected for the study based on its inherently robust DNA uptake ability (de Vries and Wackernagel, 1998; Gebhard and Smalla, 1998). These authors used a high-frequency recombinational repair assay between a neomycin phosphotransferase gene (*npt*) with an internal deletion carried by the bacterium and an intact *npt* gene in transgenic plants. If the bacteria could take up DNA with the intact *npt* gene from the plants, neomycin resistance would be restored. Using this system with its high transformation and recombination rates, they were able to demonstrate horizontal gene transfer from transgenic plants to bacteria; in one trial with an optimized protocol, 5 mg of transgenic potato DNA yielded 100 transformants. It would not be difficult to imagine that this frequency could be observed in a container storing tons of spoiling transgenic vegetables. In the most recent experiments along these lines, Nielson et al. (2000) showed that these bacteria would take up pure DNA when grown in sterile soil. These results clearly show that the bacterium, *Acinetobacter* will take up plant DNA and when a high frequency recombination system is provided, it will incorporate that DNA to give an antibiotic resistant strain.

In another set of experiments, the fate of plasmid DNA, which is highly enriched in antibiotic-resistance genes compared with transgenic plant DNA, was monitored after it was fed to mice (Schubbert et al., 1997, 1998). Most DNA was degraded rapidly, but a fraction remained in a high molecular weight form for many hours and was even detected in the feces. Uptake of this DNA by gut bacteria was not detected. The biggest surprise, however, was finding intact plasmid DNA in both mouse white blood cells and in the fetuses of pregnant females.

In summary, no-one has been able to show that native bacteria or other microbes will take up antibiotic-resistance genes when exposed to transgenic plants under natural, non-sterile conditions. But the experiments with pure cultures of *Acinetobacter* growing in liquid and sterile soils and with purified DNA demonstrate the existence of mechanisms that make such an event possible. If and when

---

[*]Institute for Global Communications (IGC) is a web site for environmental activists. See IGC report for a description of the work on microbes isolated from bees (forum.igc.apc.org:20080/WebX?13@188.ZowCaMb7aqp^0@.ee6b6af). http://forum.igc.apc.org:20080/WebX?13@188.ZowCaMb7aqp^0@.ee6b6af

experiments prove that horizontal transfer occurs between genetically modified crops and microbes in non-laboratory conditions, the implications will be debatable. In such a scenario, some vectors might, indeed, create an unacceptable risk. The case of neomycin phosphotransferase (the most common antibiotic-resistance gene used in transgenic plants) is more ambiguous. This gene is already ubiquitous, so it is far from clear that its presence in genetically engineered plants will add substantially to the existing danger. Most of the neomycin produced in the world is fed to chickens, and enteric bacteria from chickens are universally neomycin resistant, with most resistance caused by *nptI* and *nptII*.

ACKNOWLEDGMENTS

This chapter is reprinted with modifications from a SCOPE forum (http://scope.educ.washington.edu/gmfood/member/index.php) that is supported by the National Science Foundation and carried out by the Universities of California and Washington.

REFERENCES

de Vries, J. and Wackernagel, W. (1998) Detection of nptII (kanamycin resistance) genes in genomes of transgenic plants by marker-rescue transformation. *Mol. Gen. Genet.* **257**: 606–613.

Gebhard, F., and Smalla, K. (1998) Transformation of *Acinetobacter* sp. strain BD413 by transgenic sugar beet DNA. *Appl. Environ. Microbiol.* **64**: 1550–1554.

Nielsen KM; van Elsas JD; Smalla K. (2000) Transformation of *Acinetobacter* sp. strain BD413(pFG4DeltanptII) with transgenic plant DNA in soil microcosms and effects of kanamycin on selection of transformants. *Appl. Environ. Microbiol.* **66**:1237–1242.

Schluter, K. (1995) Risk assessment. In *Gene Transfer to Plants* (eds, I. Potrykus and G. Spangenberg), pp. 350–361, Springer-Verlag, New York.

Schubbert, R., Renz, D., Schmitz, B. and Doerfler, W. (1997) Foreign (M13) DNA ingested by mice reaches peripheral leukocytes, spleen, and liver via the intestinal wall mucosa and can be covalently linked to mouse DNA. *Proc. Natl Acad. Sci. USA* **94**: 961–966.

Schubbert, R., Hohlweg, U., Renz, D. and Doerfler, W. (1998) On the fate of orally ingested foreign DNA in mice: chromosomal association and placental transmission to the fetus. *Mol. Gen. Genet.* **259**: 569–576.

US Food and Drug Administration (1995) FDA's Policy for Foods Developed by Biotechnology. CFSAN Handout (Center for Food Safety and Applied Nutrition, Washington, DC). CFSAN statement (vm.cfsan.fda.gov/~lrd/biopolcy.html).

CHAPTER 21

Gene Transfer in the Fungal Host–Parasite System *Absidia glauca–Parasitella parasitica* Depends on Infection

Johannes Wöstemeyer, Anke Burmester, Anke Wöstemeyer, Kornelia Schultze and Kerstin Voigt

Infection of the zygomycete *Absidia glauca* (Mucorales, Absidiaceae) by the facultative mycoparasite *Parasitella parasitica* (Mucorales, Mucoraceae) is accompanied by the formation of a limited cytoplasmic continuum between both partners. Nuclei of the parasite invade the host's mycelium. Simple genetic experiments show that many different auxotrophic mutants of *A. glauca* are complemented by acquiring *Parasitella*'s genetic material. Artificial plasmids coding for neomycin resistance are efficiently transferred from *P. parasitica* to *A. glauca*. In all cases genetic transfer depends on the formation of the typical infection structures.

Successful infection of *A. glauca* by *P. parasitica* requires that the partners belong to complementary mating types. This observation points towards a physiological relationship between parasitism and the sexual pathway. In all mucoralean fungi analyzed sexual differentiation depends on the synthesis of the sex pheromone trisporic acid. In *A. glauca* trisporic acid is synthesized via the complementary biosynthetic action of both mating types. Also complementary combinations between *A. glauca* and *P. parasitica* produce trisporic acid although this interspecific complementation does not induce the sexual pathway across genus borders. Therefore we hypothesize that trisporic acid is also involved in host–parasite recognition and presumably in mediating the first steps during formation of infection structures. Mutants of *A. glauca* with defects in sexual spore (zygospore) formation were analyzed for their ability to serve as hosts for *P. parasitica*. In general sexually defective mutants can also not be infected, pointing at common steps for both differentiation pathways. One of the genes for an enzyme from the trisporic acid biosynthetic pathway was cloned from host and parasite. Expression studies at the transcriptional level provide further clues for understanding relationships between sexuality and parasitism.

INTRODUCTION

In 1924, the German botanist Hans Burgeff described the interaction between the myco-parasite *Parasitella parasitica* (syn. *P. simplex*) and one of its many hosts, *Absidia glauca*. *P. parasitica* is a facultative parasite, which grows better parasitically and in addition depends completely on the parasitic lifestyle with respect to forming sexual zygospores, but it can be maintained easily on synthetic media, thus facilitating biochemical and genetic analysis. By careful microscopic observation he recognized that both partners of this interaction undergo fusion of

Copyright © 2002 by Academic Press.
All rights of reproduction in any form reserved.

their mycelia with each other. Nuclei of the parasite were seen invading the host's mycelium. He was not able to follow their fate or measure their activity. Today, *P. parasitica* is recognized as a fusion biotroph. Burgeff is probably the first to report an experimental system for studying gene transfer between organisms belonging to different species. Few people have read this paper, and those who did were interested in mycology, not in genetics or mechanisms of evolution. One of the few exceptions was Joshua Lederberg (1952), who cited Burgeff in a review article and assumed that the transfer of genetic material might be associated with this kind of parasitism.

Biotrophic fusion parasites are not very widespread in the fungal world (Jeffries and Young, 1994). Two additional fusion biotrophs among zygomycetes, *Absidia parricida* (Absidiaceae) and *Chaetocladium brefeldi* Lii (Chatocladiaceae) were described at the microscopic level, and *A. parricida* has been shown to transfer genetic material to its host organisms (A. Wöstemeyer, unpublished).

THE INFECTION PATHWAY OF *PARASITELLA PARASITICA*

When complementary mating types of *A. glauca* and *P. parasitica* are grown on the same agar medium in a Petri dish, typical infection structures are formed in the aerial mycelium after several days of growth. The morphology of this differentiation pathway is complex (Figure 21.1). Infection starts with directed growth of a *Parasitella* hypha towards its host. Host hyphae can be any part of *A. glauca*'s aerial mycelium, including the sporangiophore. As soon as the contact is made, *P. parasitica*'s hyphal tip swells and cytoplasm (including nuclei) is transported into this bulb. Distant to the bulb a septum is formed. All fungi belonging into the *Mucor* relationship grow as a syncytium. Normally only the sporangiophores for mitotic sporulation and the zygophores for the formation of sexual spores are delimited by septae. After septum formation *P. parasitica*'s infection bulb fuses with the host hypha. The enzymology of this process is not clear. Looking for cell wall

degrading enzymes in compatible co-cultures of host and parasite has revealed an increase in proteolytic activity (unpublished).

Although the physiology of the fusion process is not understood, fusion was shown to be the prerequisite for delivering the parasite's genetic material into the host. At this stage roughly 20 *P. parasitica* nuclei invade *A. glauca*'s mycelium. By an unknown process a complex morphogenetic program is induced in the contact zone. *A. glauca* develops a branched, gall-like structure that surrounds *P. parasitica*'s secondary bulb. Burgeff called it a secondary sikyotic cell; the Greek-derived word "sikyotic" means "like a cupping glass," which is an allusion to the assumed function of this cell – the resorption of nutrients. After 3–4 days, the secondary sikyotic cell differentiates into a spore-like structure, the sikyospore, which germinates under favorable conditions.

INFECTION LEADS TO EFFICIENT GENE TRANSFER

As an immediate consequence of the fusion event, nuclei of both partners come into close contact. The obvious question to address is: does this intimate intergeneric contact lead to a parasexual recombination event? This question has been addressed by infecting many different amino acid auxotrophic mutants of the potential recipient *A. glauca* with a compatible prototrophic donor strain of *P. parasitica* (Kellner et al., 1993; Wöstemeyer et al., 1995). The recipient survives the infection and forms normal sporangia containing hundreds of uninucleate, vegetative spores, even close to the infection sites. After one sporulation cycle, the spores can be removed from the agar medium and plated on minimal medium with glucose as carbon and ammonium as nitrogen source under conditions, where only prototrophic *Absidia* offspring grow. Germination of *Parasitella* spores can be suppressed by low levels of neomycin. Between 0.1% and 1.5% prototrophic *A. glauca* spores are monitored among the vegetative spore progeny of infection plates. In our laboratory a mean value for nine independent gene transfer experiments

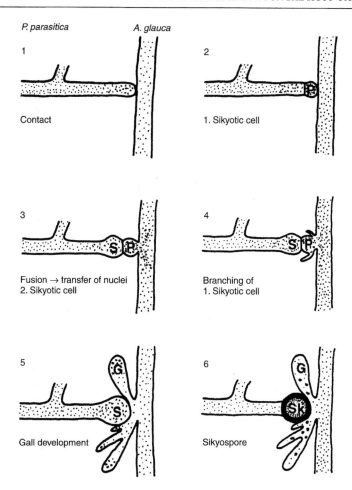

FIGURE 21.1 Sikyotic parasitism of *Parasitella parasitica* on its host *Absidia glauca*. (1) contact formation mediated by volatile derivatives of trisporic acid; (2) septum and bulb formation (primary sikyotic cell); (3) fusion between *A. glauca* and the primary sikyotic cell of *P. parasitica*, transfer of nuclei from the parasite to the host; (4) branching of the primary sikyotic cell and formation of the secondary sikyotic cell by *P. parasitica*; (5) gall development at the interphase between the partners; the secondary sikyotic cell acts as a sorption organ; (6) differentiation of a resting spore, the sikyospore, from the secondary sikyotic cell.

was 0.42%. This frequency is extraordinarily high for parasexual gene transfer systems: it exceeds the reversion rate of the recipient auxotrophs by more than 10^4 even in experiments with as little as 0.1% transfer. These results imply that intergeneric recombinants were formed as a consequence of infection. The term para-recombinants was coined for these hybrids, which alludes to both, parasexuality and parasitism. The appearance of para-recombinants depends strictly on the formation of infection structures. Incompatible pairs of donor and recipient, such as.

strains belonging to identical mating types, do not lead to prototrophic offspring (Kellner et al., 1993).

At the molecular level gene transfer has also been proven by transforming the donor, *P. parasitica*, with a plasmid which also replicates in the recipient (Figure 21.2). We have constructed an autonomously replicating plasmid that confers neomycin resistance to both fungi (Burmester, 1992; Kellner et al., 1993). All *Mucor*-like fungi tend to propagate plasmids following transformation extrachromosomally. The promoters, used for controlling the expression of the neomycin resistance gene, also function as ARS elements. In contrast to other fungi stable integration is a very rare event in zygomycetes. After infection, neomycin-resistant *Absidia* recipients have been found at high frequency. These para-recombinants contained the plasmid, as shown by Southern hybridization and retransformation of *Escherichia coli* (Kellner et al. 1993).

TRANSFERRED MARKERS TEND TO BE UNSTABLE

In most instances para-recombinants have been shown to be genetically somewhat unstable. Figure 21.3 documents the loss frequencies of the *met*$^+$ phenotype acquired after parasitic transfer. The initial culture directly on the infection plate had 0.2% of para-recombinants. For subsequent sporulation cycles, single uni-nucleated spores were used for inoculation of agar media under non-selective conditions. After one week approximately 10^9 progeny spores were formed. Such a vegetative sporulation cycle corresponds to roughly 30 subsequent mitotic divisions between the starting sporangiospore and the total number of

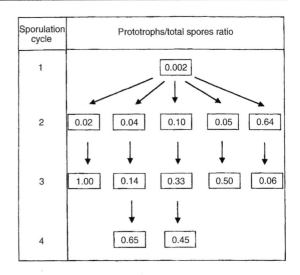

FIGURE 21.3 Loss frequency of the met$^+$-phenotype of a para-recombinant in subsequent mitotic sporulation cycles.

mature spores ($2^{30} = 10^9$). Thus the stability of para-recombinants in terms of loss per mitotic cycle is much higher than that of an ARS plasmid in yeast, but lower than expected for an integrative recombination event. The reason for this instability is unknown. This question is directly connected with the mechanism of gene transfer. Two basic possibilities are obvious (Figure 21.4).

(1) After entering the host mycelium. *Parasitella* nuclei fuse with *Absidia* recipients. During subsequent mitotic divisions the resulting intergeneric hybrid nucleus stabilizes via several aneuploid intermediates. The karyotype of the final stabilized nucleus could depend on the selection procedure. To test this hypothesis electrophoretic karyotype patterns of donor (Burmester and Wöstemeyer, 1994) and recipient (Kayser and Wöstemeyer, 1991) were compared with several para-recombinants. Although we frequently observed aberrations, we never obtained evidence for the introduction of a complete *Parasitella* chromosome into *Absidia*'s chromosomal complement.

(2) After invading the host's cytoplasm, *Parasitella* nuclei disintegrate. The DNA

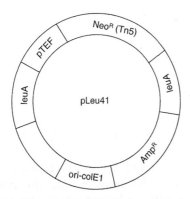

FIGURE 21.2 The autonomously replicating plasmid pLeu41 was used to show genetic transfer from *P. parasitica* to *A. glauca*. A Neo^R-transformant of *P. parasitica* was used for infection of *A. glauca*. After one sporulation cycle Neo^R-colonies of *A. glauca* were selected and shown to contain the plasmid by Southern blot hybridization analysis and by transformation in *E. coli*.

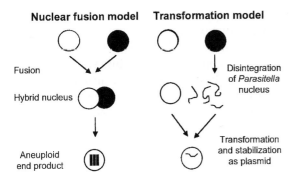

FIGURE 21.4 Hypothetical mechanisms for the formation of para-recombinants.

may enter *Absidia* nuclei via a pathway analogous to transformation. In this case the amount of *Parasitella* DNA in recombinant nuclei should be lower than that after nuclear fusion. For estimating the amount of DNA that establishes in the host genome during the parasexual event hybridization, studies with more than 100 different, highly repetitive species-specific interspersed *Parasitella* DNA elements were performed. The amount of transferred DNA is low and definitely below 1%.

Taken together, these observations are not compatible with a model based on establishing complete *Parasitella* chromosomes in an *Absidia* background. The hybridization experiments with many *Parasitella*-specific repetitive elements favor pathway of nuclear disintegration. However, it cannot be ruled out that following intergeneric nuclear fusion as the primary event, the donor chromosomes disintegrated to a final size below the magnitude of complete chromosomes.

The instability of transferred chromosomal markers can best be accounted for by the nuclear disintegration model if it is assumed that the donor DNA is established in the recipient by a mechanism other than integration. *Parasitella* DNA could circularize within *Absidia* nuclei and could be propagated as an extrachromosomal nuclear replicon. Several observations support this hypothesis. In contrast to ascomycetous fungi, *A. glauca* contains many small circular plasmids (Hänfler et al.,

1992). We have analyzed one of these plasmids down to the sequence level. We know that it codes for a single protein, and that it has no chromosomal counterpart. This shows that the replication and the partitioning machinery of *A. glauca* copes well with small circular plasmids. Perhaps DNA from *P. parasitica* also forms analogous small plasmids that are less stable than native plasmids.

SEXUAL AND PARASITIC INTERACTIONS SHARE A COMMON RECOGNITION PATHWAY

Successful infection depends on compatible host/parasite combinations with complementary mating types. Culture supernatants of mating type pairs of host and parasite contain trisporic acid. This observation led to the hypothesis that recognition of the partners in both differentiation pathways, formation of infection structures and formation of sexual zygospores, share a common recognition pathway. Trisporic acid or one of its derivatives is the sexual pheromone of all mucoraceous fungi (Figure 21.5). Complementarity of mating types in parasitic interactions as a prerequisite for infection can be explained if trisporic acid is also required for host/parasite recognition. Trisporic acid is synthesized as a cooperative action of complementary mating types (Werkman, 1976; van den Ende, 1978; Figure 21.5). Starting from β-carotene, mating type specific precursors are produced, which are passed on to the partner, where they are processed to the active pheromone.

To study dependencies between sexual and parasitic interactions at the molecular level, the (–) type specific enzyme dihydromethyltrisporic acid dehydrogenase was purified from *Mucor mucedo*. Based on amino acid sequences of subfragments generated by digestion with endopeptidase Lys-C, the corresponding gene (*tdh*) was cloned by a PCR-approach (Czempinski et al., 1996). The complete gene including several kilobases of upstream and downstream sequences were cloned and sequenced. Despite many efforts in several laboratories there

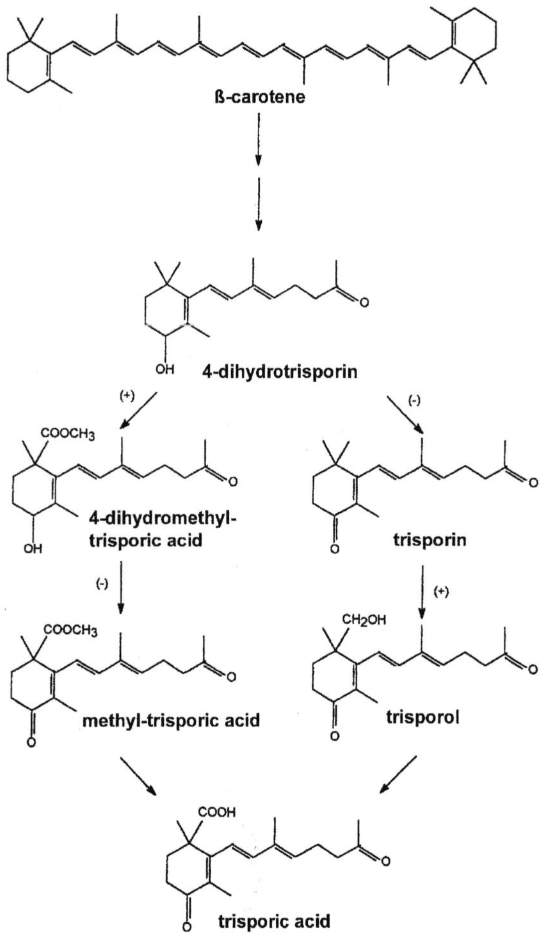

FIGURE 21.5 Cooperative biosynthesis of the general mucoralean sex pheromone, trisporic acid (according to van den Ende, 1978). Biochemical reactions specific for one of the mating types are labelled (+) or (–).

With respect to more general and evolutionary relationships between parasitism and sexuality in fungi the *Parasitella* system offers many possibilities for rewarding studies. Mechanisms of horizontal gene transfer can efficiently be studied in laboratory experiments. It is not known if gene transfer by fusion biotrophs has contributed to evolutionary biology of mucoraceous fungi. These aspects can only be investigated by sequencing appropriate genes within the host range of *P. parasitica* and looking for an unexpected phylogenetic bias in comparison with the same genes outside the host range. Towards this goal, the actin gene from 82 different species from all major genera of Mucorales were amplified, cloned and sequenced (Voigt et al., 1999; Voigt and Wöstemeyer, 2000, 2001). The actin gene sequences did not reflect phylogenetic consequences of the parasexual system at the genomic level; a similarly comprehensive study based on the *tdh* gene has been started.

ACKNOWLEDGMENTS

This study was partly supported by grants from "Deutsche Forschungsgemeinschaft" and "Fonds der Chemischen Industrie." The authors thank Christine Schimek and Annett Petzold for contributing by chemical analysis of trisporic acid derivatives. We thank especially Joshua Lederberg for encouraging *Parasitella* gene transfer research by "seeking hints of this for many years."

is to date no technique to construct site-specific mutants in zygomycetes by gene disruption or gene replacement. *Mucor mucedo* harbors a single copy of the *tdh* gene, whereas *A. glauca* and *P. parasitica* display two signals in Southern hybridization experiments. The regulation of *tdh* expression is complicated: *A. glauca* and *P. parasitica* have at least three transcripts; expression is much lower in the parasite than in the host. Two of the transcripts are expressed constitutively, whereas the third one is stimulated by mating and during infection.

REFERENCES

Burgeff, H. (1924) Untersuchungen über Sexualität und Parasitismus bei Mucorineen I. *Botanische Abhandlungen* 4: 1–135.

Burmester, A. (1992) Transformation of the mycoparasite *Parasitella simplex* to neomycin resistance. *Curr. Genet.* 21: 121–124.

Burmester, A. and Wöstemeyer, J. (1994) Variability in genome organization of the zygomycete *Parasitella parasitica. Curr. Genet.* 26: 456–460.

Czempinski, K., Kruft, V., Wöstemeyer, J. and Burmester, A. (1996) Purification of 4-dihydromethyl-trisporate dehydrogenase from *Mucor mucedo*, an enzyme of the sexual pheromone pathway, and cloning of the corresponding gene. *Microbiology* **141**: 2647–2654.

Hänfler, J., Teepe, H., Weigel, C. et al. (1992) Circular extrachromosomal DNA codes for a surface protein in the (+) mating type of the zygomycete *Absidia glauca*. *Curr. Genet.* **22**: 319–325.

Jeffries, P. and Young, T.W.K. (1994) *Interfungal Parasitic Relationships*, pp. 143–146, CAB International, Wallingford.

Kayser, T. and Wöstemeyer, J. (1991) Electrophoretic karyotype of the zygomycete *Absidia glauca*: evidence for differences between mating types. *Curr. Genet.* **19**: 279–284.

Kellner, M., Burmester, A., Wöstemeyer, A. and Wöstemeyer, J. (1993) Transfer of genetic information from the mycoparasite *Parasitella parasitica* to its host *Absidia glauca*. *Curr. Genet.* **23**: 334–337.

Lederberg, J. (1952) Cell genetics and hereditary symbiosis. *Physiological Rev.* **32**: 403–430.

Van den Ende, H. (1978) Sexual morphogenesis in the phycomycetes. In *The Filamentous Fungi* (eds, J.E. Smith and D.R. Berry), vol. III, pp. 257–274, Edward Arnold, London.

Voigt, K., Matthäi, A. and Wöstemeyer J. (1999) Phylogeny of zygomycetes: A molecular approach towards systematics of Mucorales. *Cour. Forsch.-Inst. Senckenberg* **215**: 207–213.

Voigt, K. and Wöstemeyer, J. (2000) Reliable amplification of actin genes facilitates deep-level phylogeny. *Microbiol. Res.* **155**: 179–195.

Voigt, K. and Wöstemeyer, J. (2001) Phylogeny and origin of 82 zygomycetes from all 54 genera of the Mucorales and Mortierellales based on combined analysis of actin and translation elongation factor EI-1α genes. *Gene* **270**: 113–120.

Werkman, B. (1976) Localization and partial characterization of a sex-specific enzyme in homothallic and heterothallic mucorales. *Arch. Microbiol.* **109**: 209–213.

Wöstemeyer, J., Wöstemeyer, A., Burmester, A. and Czempinski, K. (1995) Relationships between sexual processes and parasitic interactions in the host–pathogen system *Absidia glauca–Parasitella parasitica*. *Can. J. Bot.* **73**: S243-S250.

C H A P T E R *22*

Automatic Eukaryotic Artificial Chromosomes: Possible Creation of Bacterial Organelles in Yeast

George Chisholm, Lynne M. Giere, Carole I. Weaver, Chin Y. Loh, Bryant E. Fong, Meghan E. Bowser, Nathan C. Hitzeman and Ronald A. Hitzeman

These experiments demonstrate a new technology that could be used for transferring an entire prokaryotic genome or other large DNA molecules, in circular form, into a eukaryotic cell where the circular DNA is automatically converted into an artificial linear chromosome *in vivo*. This research requires the expression of an endonuclease, with an infrequent recognition sequence, which can be either introduced into or is already present in the eukaryotic organism. We have chosen yeast as our model system to test for the function of automatic yeast artificial chromosomes (AYACs). We have modified plasmid pYAC5 (used *in vitro* for YAC production) to contain very rare endonuclease recognition sequences at the *Bam*HI sites adjacent to the telomeres, flanking the yeast *HIS3* gene. Two infrequent recognition sites, *HO* (24 bp) and PI-*Sce*I (31 bp), were used in this study. Upon transformation of yeast containing the desired endonuclease, either constitutively expressed or induced, plasmids (pAYACs) containing these sites were converted into linear artificial chromosomes *in vivo*, while the control plasmids without the added sites remained circular. We postulate that this new technology could be used to make hybrid eukaryotes with new or recovered functions. For example, placing the pAYAC vector into a bacterial genome followed by fusion of this bacteria with yeast may allow both nuclear and "organellular" localization of the bacterial DNA. Such vectors could be used to make photosynthetic yeast, nitrogen fixing plants, study organelle biogenesis, or as a mechanism for human gene therapy.

INTRODUCTION

Evolutionary theory proposes that mitochondria, plastids, as well as chloroplasts originated by engulfment or cell fusion of prokaryotes by eukaryotes. As this endosymbiotic relationship evolved, the size of the bacterial DNA genome decreased and the functions of genes lost from the bacterial genome were assumed by the eukaryotic chromosome (Cavalier-Smith, 1987). Support for this theory is found in the fungus, *Geosiphon pyriforme*, which contains in its hyphal system cyanobacteria engulfed during growth and belonging to the genus *Nostoc* (Mollenhauer, 1992). The *Nostoc*, as "organelles," perform photosynthesis and nitrogen fixation pathways in the chimera. However, new fungal cells must engulf *Nostoc* species anew to obtain these capabilities again. Additional support for

Copyright © 2002 by Academic Press.
All rights of reproduction in any form reserved.

endosymbiosis is found in algae which have plastids containing DNA that has a significant level of homology and similar gene organization to cyanobacteria but the plastids have lost most of the cyanobacterial genes to the cell nucleus (Douglas, 1994). The reason and mechanism for the relocation of a large proportion (>90%) of the bacterial genes to the nucleus are unknown (Valentin et al., 1992) but such a mutational interaction of nucleus and plastid must make the chimeric organism more stable.

Bacteria called *Wolbachia* have also been shown to be present within the reproductive cells of many insects or more generally arthropods affecting reproduction, development, and nutrition (Werren et al., 1995). Furthermore, within the guts of some insects are unicellular anaerobic protists or ciliates which contain endosymbiotic archaea that produce methane from the hydrogen the eukaryotes produce (van Hoek et al., 2000). These bacterial engulfments have provided their host's adaptation to various ecological niches.

Another present day example of a eukaryotic and prokaryote interaction is the nodulation of legume plants by Rhizobia bacteria (Pueppke, 1996). The interactions between these two organisms are very strain specific, complex, and involve many genes from both. The final form of these nodules is the *Rhizobium* bacteria inside the root cells of the plant where they perform nitrogen fixation. As with *Geosiphon pyriforma* described above, the seeds of the legume do not carry the bacterial "organelle." Therefore new plants must go through the same process of interaction to fix nitrogen.

From 1974 to 1978, papers were published that reported the introduction of nitrogen fixing bacteria (e.g. *Azotobacter* and the cyanobacteria, *Anabaena variabilis* and *Gloeocapsa*) into a eukaryotic fungus and plant cells (maize and tobacco) by protoplast fusion. Low nitrogenase activity was observed for up to 212 days for the *Azotobacter*–fungal association while nitrogen fixation in other chimeras was observed for much shorter periods (Meeks et al., 1978).

It would therefore be useful to develop a system to introduce an entire prokaryotic genome or circular DNA into the nucleus of a eukaryotic organism. Preferably, such a system would permit both nuclear and extranuclear localization of the bacterial genome. This system may provide a model for the evolution of mitochondria, chloroplasts, and other plastids by continual selections for the chimeric organism (biostat selection for many generations) as opposed to the individual organisms and may be used to add new functions to eukaryotes. These new vectors and methods additionally provide a means efficiently to introduce very large segments of circular DNA into eukaryotic cells without extracellular manipulation.

DESCRIPTION OF EXPERIMENTS

Plasmid structure for interconversion from circular DNA to linear chromosome

Figure 22.1 depicts the structure of the yeast plasmid pYAC5 (Burke et al., 1987; Larin et al., 1996), currently used to clone large linear fragments of DNA by isolation of the *Not*I–*Bam*HI telomere containing fragments and ligating them *in vitro* to large genomic DNA fragments followed by transformation of yeast cells. We envision an *in vivo* method where all the elements needed to establish an artificial chromosome in yeast are inserted into the genome of a target organism, bacterial in this case. The recombined circular DNA is introduced, intact, into a recipient yeast cell by protoplast fusion (Curran and Bugeja, 1996) with concomitant conversion of the molecule into a linear artificial chromosome *in vivo*.

The conversion of the circular DNA to linear chromosome is mediated by a site-specific endonuclease with an infrequent target site expressed *in vivo* by the recipient yeast cell. We have chosen the *HO* endonuclease (Kostriken and Heffron, 1984; Russell et al., 1986; Herskowitz and Jensen, 1991; Herskowitz et al., 1992) as our first test enzyme. Its action can be observed directly by looking for a switch in yeast mating type.

Strain GY5328 (*MAT*α *ura3*–52 *trp1*–Δ63 *his3*–Δ200 *GAL ho⁻ leu2*::pGAL10-*HO/URA3*) was transformed (Ausubel et al., 1994) with plasmids

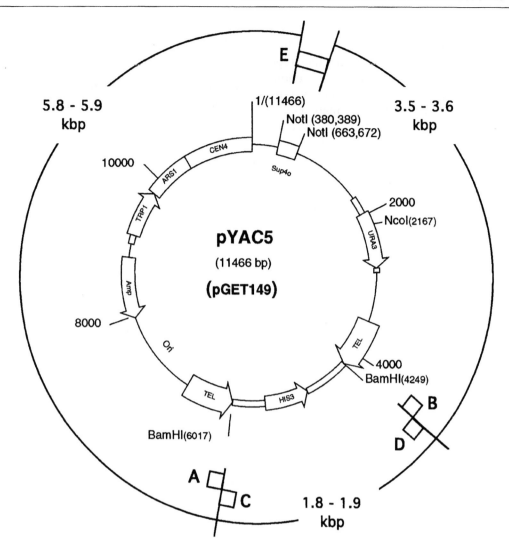

FIGURE 22.1 Structure of the AYAC plasmids. The structure of the basic plasmid, pYAC5 (pGET149) is drawn to scale above. The various structural elements are indicated by open boxes with the arrows indicating the direction of transcription or polarity on the elements. The thin lines represent pBR322 sequences while the thick regions indicate yeast chromosomal DNA. Sites where DNA was inserted into this plasmid are marked A–E. Endonuclease target sequences were introduced at the following sites: *HO* sites at A and B, and PI-*Sce*I sites at C and D. Site E was used to introduce a 8978 bp *Not*I/(*Bam*HI) fragment of yeast chromosomal DNA containing the *ADE2* gene. The plasmids used in this study are as follows: pGET149 – as shown above; pGET774 – *HO* sites added at A and B; pGET860 – *ADE2 Not*I fragment added at site E; pGET856 – *HO* sites added at A and B + *ADE2 Not*I fragment added at site E; and pGET1144 – PI-*Sce*I sites added at C and D + *ADE2 Not*I fragment added at site E.

pGET149 (no *HO* sites), pGET774 (+ *HO* sites), pGET860 (no *HO* sites + *ADE2* gene) and pGET856 (+ *HO* sites, + *ADE2* gene) (Figure 22.1; see Figure 22.2A for nucleotide sequence of *MAT*a and *MAT*α *HO* sites (Nickoloff et al., 1990)). Trp⁺, His⁺, Ura⁺, Leu⁻ transformants were

isolated and then grown overnight in liquid medium (0.67% YNB + 0.5% CAA) containing either glucose or galactose (inducer for *HO* expression) as sole carbon source. Following single colony selection on medium containing glucose the transformant phenotypes were analyzed.

When pre-grown on glucose, all colonies showed the parental phenotype (Table 22.1); His⁺, Trp⁺ and MATα, with no visible mating type switching. Pre-growth on galactose (HO endonuclease induced) caused marked mating type switching in all four original transformants (Table 22.1). Linearization of the pAYAC DNA by cleavage at both HO sites A and B (Figure 22.1) would release the HIS3 gene fragment and convert the strain to His⁻ Trp⁺ (H–, W+, Table 22.1). Strains of this phenotype were observed only from transformants pre-grown on galactose with plasmids that contain HO sites (Table 22.1; pGET774 and pGET856). The number of His⁻ colonies observed increased approximately 5.2 fold due to the presence of the large ADE2 containing chromosomal DNA fragment (8978 bp inserted at

site E, Figure 22.1) from an average of 0.63% for pGET774 to an average of 3.3% for pGET856.

Genetic evidence for linear DNA structure in His⁻ Trp⁺ transformants

Previous studies (Murray and Szostak, 1983) have indicated that larger linear YAC molecules are more stable. Linearized pGET856 DNA, containing the large ADE2 chromosomal DNA fragment, should be relatively more stable than linearized pGET774 DNA but less stable than a circular DNA molecule containing an ARS and CEN element. Five, His⁻ Trp⁺ clones (GY5328 + pGET856, putative linear molecules) and one His⁺ Trp⁺ clone (GY5328 + pGET860, circular plasmid) were tested for plasmid stability. The

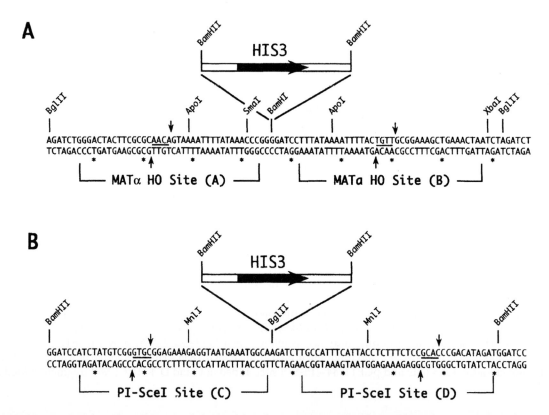

FIGURE 22.2 Nucleotide sequence of the endonuclease target sites. The nucleotide sequence of the endonuclease target sites used in this study are given above. All sites were constructed using synthetic DNA. (A) The sequence of the MATa and MATα HO recognition sites are given relative to the BamHI fragment containing the HIS3 gene (arrow above the sequence line). They were inserted between the telomere regions and the HIS3 gene fragment (Figure 22.1). (B) The sequence of the PI-SceI recognition sites are given relative to the BamHI fragment containing the HIS3 gene (arrow above the sequence line). They were inserted between the telomere regions and the HIS3 gene fragment (Figure 22.1).

TABLE 22.1 AYAC plasmid test

| Plasmid | HO sites | Carbon source | Total | Colony phenotype | | | | Mating type | | | % H−, W+ | Isolated DNA in strain GY5345 |
				H+, W+	H−, W−	H−, W+	L−	MATa	MATα	None		
GET149	−	Glu	109	109	0	0	109	0	89	20	0.00	
	−	Glu	124	124	0	0	124	0	99	25	0.00	
GET774	+	Glu	106	106	0	0	106	0	93	13	0.00	
	+	Glu	88	88	0	0	88	0	75	13	0.00	
GET860	−	Glu	173	173	0	0	173	0	169	4	0.00	
	−	Glu	128	128	0	0	128	0	119	9	0.00	
GET856	+	Glu	148	148	0	0	148	0	138	10	0.00	
	+	Glu	168	168	0	0	168	0	156	12	0.00	
GET149	−	Gal	120	116	4	0	120	75	0	45	0.00	
	−	Gal	85	85	0	0	85	63	2	20	0.00	
GET774	+	Gal	238	234	1	2	238	177	5	56	0.84	
	+	Gal	234	233	0	1	234	164	0	70	0.43	
GET860	−	Gal	270	270	0	0	270	168	43	59	0.00	GYT3678
	−	Gal	243	242	1	0	243	181	35	27	0.00	
GET856	+	Gal	191	185	0	6	191	109	28	54	3.14	GYT3677
	+	Gal	202	195	0	7	202	116	38	48	3.47	

Yeast strain GY5328; Glu = glucose, Gal = galactose, H = histidine, W = tryptophan, L = leucine, none = non-mater.

circular plasmid pGET860 was lost at an average rate of 3.7% while the five His⁻ Trp⁺ clones derived from pGET856 lost the *TRP1* marker at average rates of 59.9%, 55.5%, 75.1%, 54.7%, and 55.9% respectively. This is consistent with the His⁻ Trp⁺ phenotype resulting from a linear chromosomal form.

Physical demonstration of *in vivo* conversion of pAYAC to YAC

DNA isolated from the original transformants in Table 22.1 was transformed into yeast strain GY5345 (*MATa ura3–52 trp1-Δ63 ade2–101 his3–Δ200 leu2–Δ1 GAL ho⁻*) to remove *HO* endonuclease gene expression from consideration in further analysis. Physical confirmation of the nature of the DNA in transformants GYT3677 (GY5345/pGET856) and GYT3678 (GY5345/pGET860) (Table 22.1) was obtained by Southern Blot analysis. *Not*I digested DNA from GYT3678 (Figure 22.3A, lane 3) yielded a single band of approximately 11.3 kb consistent with a circular structure for plasmid pGET860 (Figure 22.1). In contrast, *Not*I digested DNA from GYT3677 (Figure 22.3A, lane 4) yielded two bands of >3.6 kb and

>5.9 kb which is consistent with transformed pGET856 existing as a linear artificial chromosome (Figure 22.1). DNA from GYT3678 digested with *Not*I + *Bam*HI (Figure 22.3A, lane 6) yielded two bands of 3.5 kb and 5.8 kb; the predicted pattern from digestion of circular pGET860 (Figure 22.1) DNA. The increase in size of the two fragments observed for GYT3677 DNA after identical digestion (Figure 22.3A, lane 7) is consistent with the formation of a linear molecule that has undergone telomere remodeling and extension to form functional telomeres (Zakian, 1996).

Effect of intact endonuclease site sequences on telomere function

To determine the effect of the *HO* recognition site sequences on telomere function *Bam*HI digested pGET860 (no *HO* sites + *ADE2* gene) and pGET856 (+ *HO* sites + *ADE2* gene) plasmid DNA were used to transform strain GY5345. All transformants were Trp⁺, Ade⁺ (white), Ura⁺, His⁻, and Leu⁻. Since both molecules were able efficiently to transform strain GY5345 to Trp⁺, there appears to be no gross effect on telomere function due to the presence of unrestricted *HO* sites on the ends of the telomere regions. Several transformants for each linear DNA were analyzed for DNA stability using a red/white Ade sectoring assay. Intact circular pGET860 and pGET856 plasmids (controls) showed high stability with loss between 0.79% to 2.46% (Table 22.2, rows 1–4). Five transformants from linearized pGET860 were tested and all showed similar DNA stability, loss between 19% and 35% (Table 22.2, rows 5, 6, 17–19). Analysis of the pGET856 linear transformants showed two classes; three with loss

identical to the linear pGET860 transformants (Table 22.2, rows 8, 13, 15), and five with little or no loss of DNA (Table 22.2, rows 7, 11, 12, 14, 16). Further characterization of the stable class by DNA sequencing and Southern Blot analysis (Figure 22.3B, lane 4) indicates that these circular molecules most likely arose by a recombination event between the ends of the molecule followed by deletion of the telomeric sequences (verified by sequencing). DNA from the unstable class generated the expected two band pattern consistent

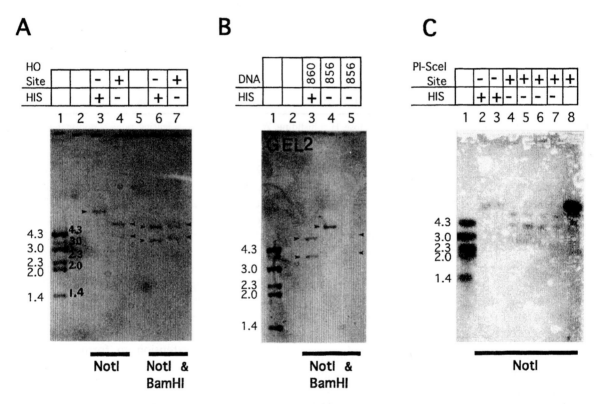

FIGURE 22.3 Analysis of AYAC DNA structure in yeast transformants. The results of Southern analysis on DNA preparations from various yeast transformants are depicted above. Biotin labeled pBR322 plasmid DNA was used as the probe for all experiments. (A) Lane 1 – molecular size standards (kb); lanes 3 and 6 – DNA from GYT3678 (GY5345/pGET860); lanes 4 and 7 – DNA from GYT3677 (GY5345/pGET856); lanes 2 and 5 – empty. The presence or absence of HO sites and the HIS3 phenotype of the transformant is presented above the blot. (B) Lane 1 – molecular size standards (kb); lane 3 – DNA from GYT3678; lane 4 – DNA from GYT3693 (stable transformant of GY5345 with linear pGET856 DNA); lane 5 – DNA from GYT3695 (unstable transformant of GY5345 with linear pGET856 DNA). The origin of the plasmid DNA and the HIS3 phenotype of the transformant is presented above the blot. (C) Lane 1 – molecular size standards (kb); lanes 2 and 3 – DNA from YPH499 transformed with pGET860 DNA (GYT3748 and GYT3749 respectively, see Table 22.2); lanes 4 to 7 – DNA from YPH499 transformed with pGET1144 DNA (lane 4 – GYT3750 and lane 5 – GYT3751, see Table 22.2); lane 8 – NotI digested pGET1144 plasmid DNA. The presence or absence of PI-SceI sites and the HIS3 phenotype of the transformant is presented above the blot. The molecular size standards were generated by mixing the following digests of pBR322 DNA; EcoRI, PvuII + EcoRI, and PvuII + SalI.

TABLE 22.2 Test of AYAC Linear DNAs

No.	Plasmid	DNA form	Endo. sites	Ade phenotype						Strain designation[b]
				Total	R	S	% R	% S	% R + S	
1	GET860(A)	P	−	203	1	4	0.49	1.97	2.46	
2	GET860(B)	P	−	254	1	1	0.39	0.39	0.79	
3	GET856(A)	P	+	214	1	1	0.47	0.47	0.93	
4	GET856(B)	P	+	210	2	0	0.95	0.00	0.95	
5	GET860(A)	L	−	183	27	24	14.75	13.11	27.87	
6	GET860(B)	L	−	216	22	54	10.19	25.00	35.19	
7	GET856(A)	L	+	175	0	2	0.00	1.14	1.14	
8	GET856(B)	L	+	154	27	34	17.53	22.08	39.61	
9	GET860(A)	L	−	805	129	110	16.02	13.66	29.68	
10	GET860(B)	L	−	1080	179	217	16.57	20.09	36.66	
11	GET856(C)	L	+	703	14	6	1.99	0.85	2.84	GYT3693
12	GET856(D)	L	+	808	0	0	0.00	0.00	0.00	
13	GET856(E)	L	+	676	26	109	3.85	16.12	19.97	
14	GET856(F)	L	+	547	0	0	0.00	0.00	0.00	
15	GET856(G)	L	+	889	147	154	16.53	17.32	33.85	
16	GET856(H)	L	+	667	2	0	0.30	0.00	0.30	
17	GET860(C)	L	−	1635	116	193	7.09	11.80	18.89	
18	GET860(D)	L	−	809	132	116	16.32	14.34	30.66	
19	GET860(G)	L	−	958	145	132	15.13	13.78	28.91	
20	GET856(G)	L	+	745	143	179	19.19	24.02	43.21	GYT3695
21	GET860(A[a])	P	−	180	1	3	0.56	1.67	2.23	GYT3748
22	GET860(B[a])	P	−	185	3	5	1.69	2.82	4.51	GYT3749
23	GET1144(A[a])	P	+	169	34	30	20.12	17.75	37.87	GYT3750
24	GET1144(B[a])	P	+	143	30	14	20.98	9.79	30.77	GYT3751

Yeast strain GY5345; A–H = individual transformants; R = Ade⁻ colony, S = sectored colony. Form: P = plasmid, L = linear; % R = (No. R/total)(100%); % S = (No. S/total)(100%). % R + S = % R + % S.
[a]Yeast strain GY5097 (YPH499).
[b]Strains used for DNA isolation and Southern analysis.

with a linear YAC structure (Figure 22.3B, lane 5). Both bands were slightly larger than those seen from pGET860 plasmid DNA (Figure 22.3B, lane 3) indicating normal telomere restructuring had occurred.

The results above indicate that the presence of complete *HO* site sequences on the ends of the telomere regions inhibits linear YAC formation by allowing a competing recircularization reaction to occur. No stable products were observed when starting with circular plasmid pAYAC DNA *in vivo* indicating that at least a complete *HO* site must be present for the recircularization to be favored.

Test of second endonuclease, PI-*Sce*I

We further tested the concept by making a new pAYAC plasmid, pGET1144 that contained sites for the PI-*Sce*I endonuclease (Gimble and Thorner, 1993) adjacent to the telomere regions (Figure 22.1, sites C and D) of plasmid pGET860. PI-*Sce*I is a constitutively expressed, naturally occurring, protein intein encoded endonuclease with a 31 bp recognition sequence (Figure 22.2B, (Gimble and Wang, 1996)). Upon transformation into yeast strain YPH499 (Silorski and Hieter, 1989) pGET1144 generated 16% Trp⁺ His⁻ transformants (seven out of 44 tested). The higher rate of His- generation observed may be due to the constitutive expression of the converting endonuclease in this yeast strain. The control plasmid, pGET860 (no PI-*Sce*I sites) generated only Trp⁺ His⁺ colonies.

DNA stability tests (Ade sectoring assay) were performed on two His⁺ pGET860 transformants and two His⁻ pGET1144 transformants (Table 22.2, bottom). Both His⁺ pGET860 transformants

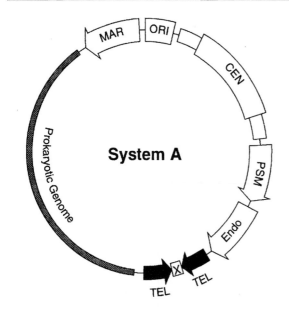

showed high DNA stability (loss of 2.2% and 4.5% respectively) consistent with circular DNA structure while both His⁻ transformants showed lower DNA stability (loss of 38% and 31% respectively) consistent with a linear DNA structure. Southern Blot analysis of the pGET1144 transformants demonstrated a linear structure for all four His⁻ transformants tested (Figure 22.3C, lanes 4–7) and a circular structure for the two His⁺ transformants containing pGET860 (Figure 22.3C, lanes 2 and 3). The above data indicate that the same linear YAC structure is generated regardless of the endonuclease or yeast strain used for *in vivo* conversion.

DISCUSSION

This publication shows a recombinant nucleic acid that functions as an automatic yeast artificial chromosome (AYAC) or more generally an automatic eukaryotic artificial chromosome (AEAC). pAYACs are circular plasmids composed of a yeast selectable marker(s), an *ARS* sequence for replication in yeast, a centromere as well as at least two inverted telomeres that function in yeast, one or more rare restriction endonuclease sites between the two inverted telomeres, a prokaryotic selectable marker(s), and optionally flanking prokaryotic sequences for integration into a bacterial chromosome by homologous recombination (and by Cre/loxP (Sauer, 1996), FLP/FRT (Huang et al., 1997), or other systems). If not encoded by the eukaryote, the AEAC should contain the chosen

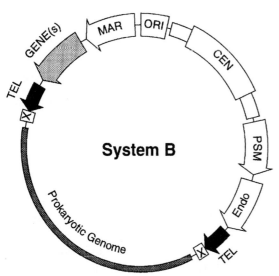

FIGURE 22.4 Automatic eukaryotic artificial chromosome (AEAC) systems. The large boxes define the functional elements with the arrows indicating direction of transcription or orientation. The telomere sequences are indicated by the black arrows. The medium, filled box represents the prokaryotic genome sequences while the thin lines represent pBR322 sequences. The open boxes containing the "X" depict the endonuclease cleavage sites of the AEAC system. System A depicts an AEAC system, not drawn to scale, for the introduction of a prokaryotic genome or large segment of genomic DNA into a eukaryotic cell. System B depicts an AEAC system, not drawn to scale, for the introduction of genetic material, a single gene or a group of genes (new pathway) into mammalian or plant cells without the incorporation of any prokaryotic genomic DNA. Abbreviations: TEL, telomeres that function in the host cell; MAR, eukaryotic selectable marker; ORI, eukaryotic origin(s) of replication, CEN, eukaryotic centromere and flanking DNA; PSM, prokaryotic selectable marker for prokaryotic maintenance; Endo, endonuclease gene under control of appropriate promoter and terminator elements; Gene(s), gene(s) to be expressed in eukaryote either single replacement gene (gene therapy) or a set of genes for a new pathway or function.

endonuclease gene expressed using the appropriate eukaryotic transcription/translation DNA signals. Figure 22.4 shows two different AEACs with endonuclease sites so upon conversion in the chosen eukaryote the bacterial sequence is included or not included when the EAC is formed. System A in Figure 22.4 may more likely contain a structure similar to Figure 22.1 with two inverted telomeres and a eukaryotic selectable marker between them to stabilize the inverted telomeric repeat in bacteria and to monitor the formation of the EAC from the AEAC.

Any DNA that is circular and contains only the added endonuclease recognition sites adjacent to telomeres may be transferred into the yeast or other eukaryotic organisms. The endonuclease site should not be present in the prokaryote or eukaryotic genomes; except for example, a restrictable *HO* endonuclease recognition site present in the *MAT* locus found on yeast chromosome III (Herskowitz and Jensen, 1991). Cutting at this site precedes mating type conversion and is not lethal to the cell.

It is surprising that there appears to be no significant telomeric silencing (Zakian, 1996) of the directly adjacent endonuclease sites in the AYAC vector by inhibiting the access of the endonuclease for both endonuclease sites in the eukaryotic nucleus.

One possible limitation for this work is that *E. coli* does not contain fortuitous replication sequences that function in yeast (*ARS*) (Stinchcomb et al., 1980); however, other bacteria have not been thoroughly checked for such functions. Without enough fortuitously functioning *ARS*s present in the prokaryotic genome, upon transfer to a yeast or other eukaryote, the bacterial genome may not replicate properly and thus be unstable. If this is true of a given bacterium, system A (Figure 22.4) may not generate one stable artificial chromosome but system B (Figure 22.4) would. However, in contrast to *E. coli* the genomic DNA of the bacterium, *Staphylococcus aureus*, contains many different fortuitous *ARS*s that function in yeast possibly due to the A+T richness of its DNA (Goze et al., 1985).

Further technical limitations associated with these hybrid bacterial/eukaryotic systems can be

anticipated. For example, restriction modification systems in various bacteria may degrade the yeast or eukaryotic DNA. Therefore, bacteria that are deficient in restriction systems are preferable.

A great advantage of this system over existing technologies is that an entire genome from one organism can be introduced into a second organism without *in vitro* manipulation. This is accomplished by first modifying a prokaryotic (e.g. bacterial) genome *in vivo* with inserted AEAC sequences and introducing the entire genome into a eukaryotic organism, such as a yeast, by protoplast fusion. Such fusions have been shown to transfer both plasmids and chromosomes. After establishment of the prokaryotic genome as a chromosome in the eukaryote, a second fusion with the same bacterium without the integrated AEAC sequences (but with a selectable system against its growth outside the chimera) may or may not be necessary for eventual organelle establishment. Establishment may require many generations of selections to obtain nuclear/"organellular" DNA interdependence. New functions can be selected for, such as new sugar use, mitochondrial function in *rho*⁻ or *rho*⁰ yeast (little or no mitochondrial DNA and loss of mitochrondrial function; (Fox et al., 1991)), photosynthesis, nitrogen fixation, growth at high temperatures following thermophilic bacterial fusions, etc.

Another advantage of this method is that large DNA segments are easily manipulated and introduced into eukaryotes. The length of YACs made by conventional methods are somewhat limited and average about 1 Mb (Larin et al., 1996). Although the technology has improved over the years, much larger linear pieces of DNA ($\geq$3.5 Mbp) have only been transferred by protoplast fusions (Allshire et al., 1987).

A photosynthetic, nitrogen-fixing yeast made by fusion of yeast and a cyanobacterium (with an integrated pAYAC) may be used for the environmentally friendly production of alcohol (for gasohol). This could be done using a *rho*⁰ yeast, which has no functional mitochondria for better production of alcohol aerobically. This would bypass the use of corn sugar to feed yeast which then produce the alcohol.

The production of AEAC-bacterial genomes for animals, human cells (AHACs), and plants preferably require different telomeres, centromeres, and selectable markers that function in these cells to make automatic chromosomes. All of these are of similar DNA size to yeast except for the centromere which for humans and plants is very much larger than yeast centromeres – up to several megabases compared with 125 bp for *Saccharomyces cerevisiae* (Harrington et al., 1997). The centromeres are composed of repeats of 171 bp α-satellite (alphoid) DNA which have been placed together in arrays of about 1 megabase in BACs (bacterial artificial chromosomes) or cloned as arrays of around 100 kb from chromosome parts in YACs to make artificial chromosomes in human cells (Grimes and Cooke, 1998). These systems suggest the ability to apply our automatic artificial chromosome system to nitrogen fixation in plants and gene therapy. Gene therapy has suffered badly from delivery systems and from the low frequency of cells expressing the needed gene for the necessary length of time (Prince, 1998). AHACs carrying genomic or cDNAs for the necessary genes to be transferred would be ideal due to their formation of automatic functional chromosomes upon reaching the nuclei of the human cells (see Figure 22.4, system B). The high percentage of genomic DNA being HAC DNA within the AHAC may facilitate the construction of vesicular delivery systems having restricted the AHAC to HAC *in vitro*. The use of attenuated bacteria, which promote their phagocytosis into human cells, could improve AHAC delivery (Pizarro-Cerda et al., 1997). Alternatively, the proteins that induce engulfment may be used to make artificial membrane vesicles (Cossart, 1998).

REFERENCES

Allshire, R.C., Cranston, G., Gosden, J.R. et al. (1987) A fission yeast chromosome can replicate autonomously in mouse cells. *Cell* **50**: 391–403.

Ausubel, F.M., Brent, R., Kingston, R.E. et al. (1994) *Current Protocols in Molecular Biology*, John Wiley., New York.

Burke, D.T., Carle, G.F. and Olson, M.V. (1987) Cloning of large DNA segments of exogenous DNA into yeast by means of artificial chromosome vectors. *Science* **236**: 806–812.

Cavalier-Smith, T. (1987) The simultaneous symbiotic origin of mitochondria, chloroplasts, and microbodies. *Ann. NY Acad. Sci.* **503**: 55–71.

Cossart, P. (1998) Interactions of the bacterial pathogen *Listeria monocytogenes* with mammalian cells: bacterial factors, cellular ligands, and signaling. *Folia Microbiol.* **43**: 291–303.

Curran, B.P.G. and Bugeja, V.C. (1996) Protoplast Fusion in *Saccharomyces cerevisiae*. In *Methods in Molecular Biology Vol53: Yeast Protocols* (ed., I. Evans), pp. 45–49, Humana Press, Totowa, NJ.

Douglas, S.E. (1994) Chloroplast origins and evolution. In *Molecular Biology of Cyanobacteria* (ed., D.A. Bryant), vol. 1, pp. 91–118, Kluwer, Boston.

Fox, T.D., Folley, L.S., Mulero, J.J. et al. (1991) Analysis and manipulation of yeast mitochondrial genes. *Methods Enzymol.* **194**: 149–155.

Gimble, F.S. and Thorner, J. (1993) Purification and characterization of VDE, a site-specific endonuclease from the yeast *Saccharomyces cerevisiae*. *J. Biol. Chem.* **263**: 21844–21853.

Gimble, F.S. and Wang, J. (1996) Substrate recognition and induced DNA distortion by the PI-SceI endonuclease, an enzyme generated by protein splicing. *J. Mol. Biol.* **263**: 163–180.

Goze, A., Dedieu, A., Goursot, R. and Ehrlich, S.D. (1985) *Saccharomyces cerevisiae* ARS on a plasmid from *Staphylococcus aureus*. *Plasmid* **14**(3): 255–260.

Grimes, B. and Cooke, H. (1998) Engineering mammalian chromosomes. *Human Mol. Genet.* **7**: 1635–1640.

Harrington, J.J., Bokkelen, G.V., Mays, R.W. et al. (1997) Formation of *de novo* centromeres and construction of first generation human artificial microchromosomes. *Nature Genetics* **15**: 345–355

Herskowitz, I. and Jensen, R.E. (1991) Putting the *HO* gene to work: practical uses for mating-type switching. *Methods Enzymol.* **194**: 132–146.

Herskowitz, I., Rhine, J. and Strathern, J. (1992) Mating-type determination and mating-type interconversion in *Saccharomyces cerevisiae*. In *The Molecular and Cellular Biology of the Yeast Saccharomyces* (eds., E.W. Jones, J.R. Pringle and J.R. Broach), vol. 2, pp. 583–656, Cold Spring Harbor Laboratory Press, Cold Spring Harbor, NY.

Huang, L-C., Wood, E.A. and Cox, M.M. (1997) Convenient and reversible site-specific targeting of exogenous DNA into a bacterial chromosome by use of the FLP recombinase: the FLIRT system. *J. Bacteriol.* **179**: 6076–6083.

Kostriken, R. and Heffron, F. (1984) The product of the *HO* gene is a nuclease: purification and characterization of the enzyme. *Cold Spring Harbor Symp. Quant. Biol.* **49**: 89–96.

Larin, Z., Monaco, A.P. and Lehrach, H. (1996) Generation of large insert YAC libraries. In *Methods in Molecular Biology, YAC Protocols* (ed., D. Markie), vol. 54, pp. 1–11, Humana Press, Totowa, NJ.

Meeks, J.C., Malmberg, R.L. and Wolk, C.P. (1978) Uptake of auxotrophic cells of a heterocyst-forming cyanobacterium by tobacco protoplasts, and the fate of their associations. *Planta* **139**: 55–60.

Mollenhauer, D. (1992) *Geosiphon pyriforme*. In *Algae and Symbiosis* (ed., W. Reisser), pp. 339–351, Biopress , Bristol.

Murray, A.W. and Szostak, J.W. (1983) Construction of artificial chromosomes in yeast. *Nature* 305: 189–193.

Nickoloff, J.A., Singer, J.D. and Heffron, F. (1990) *In vivo* analysis of the *Saccharomyces cerevisiae HO* nuclease recognition site by site-directed mutagenesis. *Mol. Cell. Biol.* **10**: 1174–1179.

Pizarro-Cerda, J., Moreno, E., Desjardins, M. and Gorve, J.P. (1997) When intracellular pathogens invade the frontiers of cell biology and immunology. *Histol. Histopathol.* **12**: 1027–1038.

Prince, H.M. (1998) Gene transfer: a review of methods and applications. *Pathology* **30**: 335–347.

Pueppke, S.G. (1996) The genetic and biochemical basis for nodulation of legumes by Rhizobia. *Critical Reviews in Biotechnology* **16**(1): 1–51.

Russell, D.W., Jensen, R., Zoller, M.J. et al. (1986) Structure of the *Saccharomyces cerevisiae HO* gene and analysis of its upstream regulatory region. *Mol. Cell. Biol.* **6**: 4281–4294.

Sauer, B. (1996) Multiplex Cre/lox recombination permits selective site-specific DNA targeting to both a natural and an engineered site in the yeast genome. *Nucl. Acids Res.* **24**: 4608–4613.

Sikorski, R.S. and Hieter, P. (1989) A system of shuttle vectors and yeast host strains designed for efficient manipulation of DNA in *S. cerevisiae*. *Genetics* **112**: 19–27.

Stinchcomb, D.T., Thomas, M., Kelly, J. et al. (1980) Eukaryotic DNA segments capable of autonomous replication in yeast. *Proc. Natl Acad. Sci. USA* **77**(8): 4559–4563.

Valentin, K., Cattolico, R.A. and Zetsche, K. (1992) Phylogenetic origin of the plastids. In *Origins of Plastids: Symbiogenesis, Prochlorophytes, and the origins of Chloroplasts* (ed., R.A. Lewin), pp. 193–221, Chapman and Hall, New York.

van Hoek, A.H., van Alen, T.A., Spakel, V.S. et al. (2000) Multiple acquisition of methanogenic archaeal symbionts by anaerobic ciliates. *Mol. Biol. Evol.* **17**: 251–258.

Werren, J.H., Zhang, W. and Guo, L.R. (1995) Evolution and phylogeny of *Wolbachia*: reproductive parasites of arthropods. *Proc. R. Soc. Lond. B. Biol. Sci.* **261**: 55–63.

Zakian, V.A. (1996) Structure, function, and replication of *Saccharomyces cerevisiae* telomeres. *Annu. Rev. Genetics* **30**: 141–172.

CHAPTER 23

Bacteria as Gene Delivery Vectors for Mammalian Cells

Catherine Grillot-Courvalin, Sylvie Goussard and Patrice Courvalin

Gene transfer can occur from bacteria to a very broad host range of recipient cells. It has been reported between distantly related bacterial genera, as demonstrated by Trieu-Cuot et al. (1987), between bacteria and yeast as demonstrated by Heinemann and Sprague (1989) and from bacteria to plants, as demonstrated by Buchanan-Wollaston et al. (1987). More recently, several laboratories have reported that bacteria can also transfer functional genetic information into mammalian cells. Transfer of replicons has been described after either *in vitro* co-incubation of *Shigella* or *Listeria* with phagocytic or non-phagocytic cells or *in vivo* administration of attenuated *Shigella* or *Salmonella*. We have developed a non-pathogenic bacterium, an invasive *Escherichia coli* deficient in cell wall synthesis, to study this type of transfer *in vitro*. This process is known as abortive or suicidal invasion since the bacteria have to lyse for DNA delivery to occur. The bacterial species which can transfer genes to professional and non-professional phagocytes have in common the ability to invade these cells. The plasmid DNA released by intracellular bacteria is transferred from the cytoplasm to the nucleus resulting in cellular expression of the transfected gene(s). Bacterial transfer of DNA can result in stimulation of humoral and cellular responses after *in vivo* administration. Gene delivery by abortive invasion of eukaryotic cells by bacteria may be valuable for *in vivo* and *ex vivo* gene therapy and for stimulation of mucosal immunity.

IN VITRO GENE TRANSFER

Mammalian cells can transiently express genes delivered intracellularly by strains of *Shigella flexneri* impaired in peptidoglycan synthesis, as demonstrated by Sizemore et al. (1995) and by Courvalin et al. (1995). A diaminopimelate auxotroph (*dap⁻*) mutant of *Shigella*, which undergoes lysis upon entry into mammalian cells because of impaired cell wall synthesis, was transformed with plasmid pCMVβ which directs the synthesis of *E. coli* β-galactosidase under the control of a eukaryotic promoter. Forty-eight hours after a 90 minute co-incubation of this suicidal bacterial vector with cultured BHK cells, 1% to 2% of cells expressed β-galactosidase. Similar results were obtained with the murine cell line P815.

The genes responsible for entry, intra- and intercellular mobility of *S. flexneri* are borne by the *ca.* 200 kb virulence plasmid pWR100. Transfer of the plasmid to *E. coli* confers to this otherwise extracellular bacterial species the ability to invade epithelial cells. Using this invasive strain of *E. coli*, designated BM 2710, we have shown that bacteria that undergo lysis upon entry into mammalian cells (because of impaired cell wall synthesis due to diaminopimelate (dap) auxotrophy) can deliver plasmid DNA to their hosts.

L. monocytogenes is able to invade a wide range of mammalian cell types. After internalization, these bacteria rapidly escape from the primary

Copyright © 2002 by Academic Press.
All rights of reproduction in any form reserved.

vacuole into the cytosol where they replicate. An attenuated strain, impaired in intra- and intercellular movements has been engineered to undergo self-destruction upon entry into mammalian cells by production of a phage lysin under the control of the promoter of the *actA* gene which is preferentially activated in the cytosol. This bacterial vector was able, as demonstrated by Dietrich et al. (1998), to deliver a plasmid carrying *gfp* (green fluorescent protein), *cat* (chloramphenicol acetyltransferase) or a portion of the *ova* (ovalbumine) gene under the control of the CMV promoter in the P388D macrophage cell line. Functional transfer was dramatically increased (from 0.001% to 0.2%) if there was lysis of the internalized *Listeria* in the cytoplasm, either by self-killing by lysin production or after antibiotic treatment. Efficient expression of chloramphenicol acetyltransferase or antigen presentation was observed in P388D macrophages. Plasmid DNA, from macrophage clones cultured in selective medium for more than 12 weeks, was found to be integrated into the genome of the new host at a frequency of approximately 10^{-7}.

In order to design a genetically more defined bacterial vector, we have cloned the 3.2 kb *inv* locus encoding the invasin of *Yersinia pseudotuberculosis* alone or combined with the 1.5 kb *hly* gene coding for the listeriolysin O from *L. monocytogenes* in the stable *dap⁻* auxotroph *E. coli* BM2710 (Grillot-Courvalin et al., 1998). We also introduced by transformation in that strain plasmid pEGFP-C1 which directs synthesis of the green fluorescent protein (GFP) in mammalian cells but not in bacteria. Between 5 and 20% of HeLa, CHO, and COS-1 cells producing GFP were observed 2 days after incubation with invasive bacteria. Co-expression in *E. coli* of the gene for listeriolysin enhanced transfer efficiency. Expression of the acquired genes occurred both in dividing and in quiescent cells, albeit at a lower efficiency in the latter. Transfer of functional plasmid DNA could be observed in J774 macrophages but not in a mouse dendritic cell line. Expression of acquired DNA was very stable, as tested after 2 months of culture. The number of viable intracellular bacteria decreased rapidly with no survivors after 24 to 72 h, and chromosomal DNA of the donor could not be detected after 23 to 48 days. Plasmid transfer was obtained, although at a lower frequency, from *E. coli* BM2711, the dap prototroph parental strain counterpart of BM2710, indicating that bacteria may transfer functional genetic information to non-professional phagocytes provided they can induce their own internalization.

Agrobacterium tumefaciens is a soil phytopathogen that elicits neoplastic growths on the host plant species. It has very recently been shown that *Agrobacterium* can also transfer DNA to mammalian cells, as demonstrated by Kunick et al. (2001). *Agrobacterium* infection requires two genetic components that are carried by the tumor-inducing plasmid: the transferred DNA (T-DNA) which is introduced into the plant genome and the virulence region which specifies the protein apparatus for T-DNA transfer. Experiments *in vitro* indicated that *A. tumefaciens* attaches to and transfers DNA to several types of human cells, by conjugation, the mechanism it uses to transform plant cells. Study of stable transfected HeLa cells indicate integration of T-DNA into their genome.

IN VIVO GENE TRANSFER IN ANIMAL MODELS

The ability of *Shigella* to enter intestinal epithelial cells and to evade from the endocytic vesicle was exploited to develop an *in vivo* gene delivery system as demonstrated by Sizemore et al. (1995). Mice were inoculated twice intranasally with 10^6 to 10^7 bacteria transformed with plasmid pCMVβ, 4 weeks apart; splenocytes from these mice proliferated after a 3 day culture *in vitro* in the presence of β-galactosidase and anti-β-galactosidase antibodies were detected in their sera as demonstrated by Sizemore et al. (1997). These results provided the first evidence that delivery of plasmid DNA by *Shigella in vivo* could elicit an immune response to the plasmid-encoded antigen. According to recent findings (Fennelly et al., 1999), after intranasal inoculation of mice, a highly attenuated strain of Δasd *Shigella* harboring a DNA measles vaccine plasmid induced a vigorous measles-specific immune-response of both Th1 and Th2 types.

Attenuated *Salmonella* expressing heterologous antigens have been used for oral immunization in mice, farm animals, and in humans resulting in efficient stimulation of mucosal and systemic immune responses. An attenuated strain of *S. typhimurium* was used successfully by Darji et al. (1997) as a vector for oral genetic immunization. In this work, oral administration of an *aroA⁻* auxotrophic mutant harboring plasmids encoding *E. coli* β-galactosidase or truncated ActA and listeriolysin (two virulence factors of *L. monocytogenes*), each under the control of an eukaryotic promoter, led to excellent humoral and cellular immune responses in mice. The immunized mice were protected from a lethal challenge with virulent *L. monocytogenes*, even after a single administration. β-galactosidase activity was detected in splenic macrophages 5 weeks after oral administration of *S. typhimurium* carrying plasmid pCMVβ. Plasmid DNA transfer was studied *in vitro* after 1 h infection of mouse primary peritoneal macrophages with attenuated *S. typhimurium* carrying pCMVβ. Evidence of β-galactosidase synthesis by the infected macrophages was provided but tetracycline had to be added during the entire culture period to inhibit endogenous residual synthesis of β-galactosidase by bacteria.

In a similar study by Paglia et al. (1998), oral administration to mice of an attenuated *Salmonella* harboring pCMVβ could generate humoral and cellular immune responses to β-galactosidase and a protective response against an aggressive murine fibrosarcoma transduced with the β-galactosidase gene. Expression of the transgene by antigen presenting cells after *per os* administration was documented with an eukaryotic vector expressing GFP. Among the 19% of splenocytes expressing GFP 28 days after oral administration, 50% were dendritic cells and 30% expressed macrophage markers. This provides evidence of direct *in vivo* gene transfer by orally administered bacteria to dendritic cells.

According to more recent findings by Paglia et al. (2000), oral administration of an attenuated *Salmonella*, carrier for an eukaryotic expression vector encoding the murine INFγ gene, resulted in the production of this cytokine in INFγ-deficient mice. This provides evidences that attenuated *Salmonella* can be used *in vivo* as a DNA delivery system for the correction of a genetic defect.

DISCUSSION

In very few instances, direct introduction of DNA from bacteria to mammalian cells has been reported, as demonstrated by Schaffner (1980), that plasmids carrying tandem copies of the SV40 virus genome could be transferred from *E. coli* to mammalian cells by exposing the cell culture to a bacterial suspension. However, transfer was found to occur at a very low frequency. Similar observations were made with *E. coli* harboring plasmids carrying poliovirus 1 cDNA and transfer took place in the presence of high concentrations of DNAse in the culture medium, as demonstrated by Heitmann and Lópes-Pila (1993). In these experiments, the vectors were human pathogens that could not be employed outside an experimental setting.

Attenuated intracellular bacteria, such as *Shigella*, *Salmonella*, *Listeria*, and invasive *E. coli* have been found to act as efficient gene delivery vectors in both phagocytic and non-phagocytic mammalian cells (Table 23.1). Observations from *in vitro* and *in vivo* experiments are consistent with the hypothesis that, after internalization into a primary vacuole, bacteria or their plasmid content have to escape into the cytosol to gain access to the nucleus.

Studies on microbial pathogenesis have expanded our understanding of the mechanisms designed by bacteria to achieve entry into host cells and to gain access to intracellular compartments. There are two major strategies for bacteria to gain access to host cells, as reviewed by Marra and Isberg (1996). For certain genera like *Salmonella* or *Shigella*, contact between the bacteria and the host cells results in the secretion, by the bacteria, of a set of invasion proteins that triggers signalling events into the cells, leading to cytoskeletal rearrangement, membrane ruffling and bacterial uptake by micropinocytose. For other genera like *Yersinia* or *Listeria*, binding of a single bacterial protein to a particular ligand on the host cell surface is necessary and sufficient to trigger entry into phagocytic and non-phagocytic cells by a zipper-like mechanism.

TABLE 23.1 Main studies on bacteria to eukaryote gene transfer

| Donor bacterial species | Transferred plasmid(s) | Recipient | | Reference |
		In vitro: cell lines transfected	*In vivo*: route of administration	
Shigella flexneri dapA⁻	pCMVβ	BHK	Intranasal	Sizemore et al.
		P815	Intracorneal	(1995, 1997)
	Measles-vaccine plasmid		Oral	Fennelly et al. (1999)
Invasive *E. coli dapA⁻*	β-gal replicative and integrative eukaryotic vectors pEGFP-C1	HeLa COS-1 CHO A549 J774	Not determined	Courvalin et al. (1995), Grillot-Courvalin et al. (1998)
Salmonella typhimurium aroA⁻	pCMVβ pCMV *actA-hly* INF-γ gene	Primary mouse macrophage	Oral Oral	Darji et al. (1997), Paglia et al. (1998) Paglia et al. (2000)
Listeria monocytogenes Δ *mpl actA plcB*	pCMV*gfp* pCMV*cat* pCMV*ova*	J774 P388D	Intraperitoneally	Dietrich et al. (1998), Spreng et al. (2000)

Binding of *Yersinia* invasin to β1 integrins results in entry into cells expressing this integrin at the surface; binding of internalin with E-cadherin mediates entry of *Listeria* into certain cell types. Similarly, introduction in *E. coli* of the virulence plasmid (pWR100) of *S. flexneri* or of the *inv* gene of *Y. pseudotuberculosis* confers to this otherwise extracellular bacteria the ability to invade non-phagocytic cells.

Access to the cell cytosol depends on the mechanisms by which bacteria survive inside the cells. *Listeria* and *Shigella* escape rapidly from the vacuole of entry after lysis of its membrane, by production of listeriolysin in the case of *Listeria*, and replicate in the cytoplasm of the cell. Other bacteria, like *Salmonella*, remain in the phagosomal vacuole and replicate within this compartment. The successful delivery of genes by *Salmonella* is therefore surprising since *Salmonella*-containing vacuoles have a unique trafficking pathway, uncoupled from the normal endocytic degradation pathway. Interestingly, infection by *Salmonella* resulted in functional DNA delivery into primary macrophages but not in macrophage cell lines, as demonstrated by Sizemore et al. (1995) and Darji et al. (1997).

DNA entering host cell cytoplasm by phagocytosis is less efficiently routed to the nucleus than when introduced directly into the cytoplasm (e.g. by gene gun). Consistent with this finding are the observations that more efficient gene transfer was observed when *Listeria* was destroyed in the cytoplasm of the cell (Dietrich et al., 1987) or if the invasive *E. coli* produced listeriolysin that triggers pore formation in the vacuolar membrane (Grillot-Courvalin et al., 1998). However, it cannot be excluded that DNA may gain access to the mammalian cell cytoplasm via leakage from host cell phagosomes, as has been proposed for transfer of certain antigens by Kovacsovics-Bankowski and Rock (1995).

The property of bacteria to act as a gene delivery system has been mainly exploited for DNA vaccination. Attenuated auxotroph mutants of *Salmonella* are already in use as live vaccines in man and in animals, and effective gene transfer to dendritic cells after oral administration of mice with these bacteria has been documented by Paglia et al. (1998). In these *in vivo* studies, the number of transfected cells varies greatly depending on the type of bacterial vector and on the cell type. However, dendritic cells are highly efficient antigen presenting cells and a small number of transfected dendritic cells has been shown to be sufficient to stimulate both primary and secondary T and B cell responses and to process and present antigens efficiently, as demonstrated by Banchereau and Steinman (1998). Moreover, *per os* administration of bacterial vectors could lead to transfection of gut-associated lymphoid cells and stimulation of mucosal immunity, as demonstrated by Darji et al. (1997), Paglia et al. (1998), and Fennelly et al. (1999).

The potential use of bacterial vectors for *in vivo* or *ex vivo* gene therapy has only been tested once (Paglia et al., 2000) for a genetic defect associated with monocyte/macrophage cell type. Despite the high numbers of bacteria and plasmids internalized by the host cells, transgene expression remains low and comparable to the levels obtained with other non-viral delivery systems such as polycation-DNA complexes. Following bacterial internalization, plasmid DNA has to be released into the cytosol before its nuclear entry can occur. The turnover of plasmid DNA delivered by microinjection in the cytosol has been shown by Lechardeur et al. (1999) to be rapid, with an apparent half-life of 50 to 90 min in HeLa and COS cells. Direct delivery into the cytosol of native plasmids by intracellular bacteria may constitute a means of protecting this DNA from degradation by cytosolic nucleases.

REFERENCES

Banchereau, J. and Steinman, R.M. (1998) Dendritic cells and the control of immunity. *Nature* **392**: 245–252.

Buchanan-Wollaston, V., Passiatore, J.E. and Cannon, F. (1987) The *mob* and *oriT* mobilization functions of a bacterial plasmid promote its transfer to plants. *Nature* **328**: 172–175.

Courvalin, P., Goussard, S. and Grillot-Courvalin, C. (1995) Gene transfer from bacteria to mammalian cells. *CR Acad. Sci.* **318**: 1207–1212.

Darji, A., Guzman, C.A., Gerstel, B. et al. (1997) Oral somatic transgene vaccination using attenuated *S. typhimurium*. *Cell* **91**: 765–775.

Dietrich, G., Bubert, A., Gentschev, I. et al. (1998) Delivery of antigen-encoding plasmid DNA into the cytosol of macrophages by attenuated suicide *Listeria monocytogenes*. *Nat. Biotechnol.* **16**: 181–185.

Fennelly, G.J., Khan, S.A., Abadi, M.A. et al. (1999) Mucosal DNA vaccine immunization against measles with a highly attenuated *Shigella flexneri* vector. *J. Immuno.* **162**: 1603–1610.

Grillot-Courvalin, C., Goussard, S., Huetz, F. et al. (1998) Functional gene transfer from intracellular bacteria to mammalian cells. *Nat. Biotechnol.* **16**: 862–866.

Heinemann, J.A. and Sprague, G.F. Jr. (1989) Bacterial conjugative plasmids mobilize DNA transfer between bacteria and yeast. *Nature* **340**: 205–209.

Heitmann, D. and Lópes-Pila, J.M. (1993) Frequency and conditions of spontaneous plasmid transfer from *E. coli* to cultured mammalian cells. *Biosystems* **29**: 37–48.

Kovacsovics-Bankowski, M. and Rock, K.L. (1995) A phagosome-to-cytosol pathway for exogenous antigens presented on MHC class I molecules. *Science* **267**: 243–246.

Kunik, T., Tzfira, T., Kapulnik, Y. et al. (2001) Genetic transformation of HeLa cells by *Agrobacterium*. *Proc. Natl Acad. Sci. USA* **98**: 1871–1876.

Lechardeur, D., Sohn, K.J., Haardt, M. et al. (1999) Metabolic instability of plasmid DNA in the cytosol: a potential barrier to gene transfer. *Gene Ther.* **6**: 482–497.

Marra, A. and Isberg, R.R. (1996) Common entry mechanisms. Bacterial pathogenesis. *Curr. Biol.* **6**: 1084–1086.

Paglia, P., Medina, E., Arioli, I. et al. (1998) Gene transfer in dendritic cells, induced by oral DNA vaccination with *Salmonella typhimurium*, results in protective immunity against a murine fibrosarcoma. *Blood* **92**: 3172–3176.

Paglia, P., Terrazzini, N., Schulze, K. et al. (2000) In vivo correction of genetic defects of monocyte/macrophages using attenuated *Salmonella* as oral vectors for targeted gene delivery. *Gene. Ther.* **7**: 1725–1730.

Schaffner, W. (1980) Direct transfer of cloned genes from bacteria to mammalian cells. *Proc. Natl Acad. Sci. USA* **77**: 2163–2167.

Sizemore, D.R., Branstrom, A.A, Sadoff and J.C. (1995) Attenuated *Shigella* as a DNA delivery vehicle for DNA-mediated immunization. *Science* **270**: 299–302.

Sizemore, D.R., Branstrom, A.A and Sadoff, J.C. (1997) Attenuated bacteria as a DNA delivery vehicle for DNA-mediated immunization. *Vaccine* **15**: 804–807.

Spreng, S., Dietrich, G., Niewiesk, S. et al. (2000) Novel bacterial systems for the delivery of recombinant protein or DNA. *FEMS Immuno. Med. Microbiol.* **27**: 299–304.

Trieu-Cuot, P., Carlier, C., Martin, P. and Courvalin, P. (1987) Plasmid transfer by conjugation from *Escherichia coli* to Gram-positive bacteria. *FEMS Microbiol. Lett.* **48**: 289–294.

Whole Genome Comparisons: The Emergence of the Eukaryotic Cell

The chapters in this section describe numerous examples of protein phylogenetic trees that are incongruent with the underlying species trees. Also, additional cases are described where a protein sequence found in a genome has no homologue among its closest relatives but where homologues can only be found in remotely related domains. With the recent completion of more than 30 entire genome sequences with representatives from all three of life's domains, the number of examples of putative horizontal gene transfer events has increased dramatically. Chapter 24, by Doolittle, begins with a cautionary tale on deducing gene transfer events when examining protein sequences that are highly diverged. Doolittle weighs the rate argument very heavily in concluding that horizontal gene transfer has occurred. It turns out that most of the purported examples of horizontal gene transfer among the three kingdoms – Archaea, Bacteria and Eukaryota – are evidenced by proteins that display only 25–35% similarity upon interkingdom comparisons; this is the case for most of the examples shown in Chapters 25, 26 and 27, by Koonin et al., Brown et al. and Henze et al., respectively. These chapters are the result of whole genome sequence analyses and represent major computational efforts. Most of these results were obtained using highly automated procedures that "mine" the information from numerous whole genomes. Using a variety of procedures, potential horizontal gene transfers can be identified. In most of these cases, the criteria for identifying such events are the phylogenetic congruency test (Syvanen, 1994) but, as Doolittle points out, they are not supported by rate arguments. This means that the formal possibility remains that paralogues, and not orthologues, are being compared. However, the argument can be made that horizontal gene transfer explanations are more parsimonious than selective loss of paralogues, as was described in the introduction to Section III. The argument from parsimony goes as follows: If horizontal gene transfer is *not* an acceptable explanation, and the incongruencies are the result of selective inheritance of paralogous copies, then we must further postulate that the last common ancestor of all life had a considerably larger genome than any known archaeal or bacterial genome – larger by many hundreds of paralogous genes. That is the case because the last common ancestor would have had to carry duplications of numerous ancestral genes. Since the number of incongruencies in the gene trees is in the many hundreds – if not thousands – we would have to postulate that the last common ancestor would have had to carry a duplicate of each of those examples.

The above argument from parsimony is also relevant to the recent dispute that has arisen over claims of bacterial genes in the human genome (see notes added in proof in Chapters 25 and 26). This claim is based on the observation that these genes were absent from the other sequenced eukaryotes – the yeasts, *Caenorhabditis elegans*, *Arabidopsis thaliana* and *Drosophila melanogaster* – but present in humans and many bacteria.

Subsequently, homologues for some of these genes have appeared in *Dictyostelium discoideum*, as is described in the note added in proof in Chapter 25. This fact certainly changes the nature of the evidence for horizontal gene transfer of bacterial genes into the human genome, but does not rule out that possibility.

If the hypothesis of horizontal gene transfer is supported by parsimony, it must further be postulated that most of the gene transfer events described here occurred more than 1 billion years ago – probably during the emergence of the modern eukaryotic cell. This follows because the proteins which are being compared between the different domains are so highly diverged.

Chapter 28 by Adkins and Li is relevant here because these authors show that many proteins shared by the three kingdoms appear to be much younger, on the basis of molecular clock arguments, than the underlying lineages. This raises the possibility that these genes were introduced into multiple lineages well after the last common ancestor. Indeed, once this reasoning is followed to its logical conclusion, there is no longer any need to postulate a last common ancestor – we can extrapolate multiple lineages back to the origin of life and account for shared genes through the mechanism of horizontal gene transfer.

Finally, whole genome sequencing projects have resulted in the demise of at least one claim of horizontal gene transfer. Syvanen published in the first edition of this book a chapter entitled "Cytochrome-c from *Stellaria longipes* and *Arabidopsis thaliana* was likely transferred from fungi since the radiation of terrestrial plants" (Syvanen and Kado, 1998). Upon completion of the *Arabidopsis* genome sequence, however, the cytochrome *c* sequence that was originally claimed to be from *Arabidopsis* could not be found; rather, other cytochrome *c*'s were detected that were homologous to those from other plants.

REFERENCES

Syvanen, M. (1994) Horizontal gene transfer: evidence and possible consequences. *Annu. Rev. Genet.* **28**: 237–261.

Syvanen, M and Kado, C. (eds) (1998) *Horizontal Gene Transfer*, 1st edn, Chapman and Hall, New York.

CHAPTER 24

Gene Transfers Between Distantly Related Organisms

Russell F. Doolittle

With the completion of numerous microbial genome sequences, reports of individual gene transfers between distantly related prokaryotes have become commonplace. On the other hand, transfers between prokaryotes and eukaryotes still excite the imagination. Many of these claims may be premature, but some are certainly valid. In this chapter I once again consider the kinds of supporting data needed to propose transfers between distantly related organisms and cite some interesting examples.

INTRODUCTION

In the interval since the first edition of this book was assembled, the complete genome sequences of several dozen Bacteria and Archaea have been reported. With those reports has come a great increase in the number of alleged horizontal gene transfers between these two kinds of prokaryotes. Indeed, the phrase "rampant exchange" has crept into many reviews of the subject, leading some to suggest that the amount of horizontal exchange has been so overwhelming that it may not be possible to trace a true phylogeny of the major groupings of life (Doolittle, 1999).

The apparent profusion of horizontal transfers observed among prokaryotes has also revived the idea, first put forth by Kandler (1994), that the Darwinian dogma of a single common ancestor for all living things on Earth is no longer tenable (Woese, 1998). Rather, this school of thought contends that, in the period before the differentiation of organisms into the three principal realms, lateral gene transfer was all pervasive, exceeding the vertical transmission of genes. As such, it is reasoned, each of the three major lineages must have descended from a *community* of heterosperse organisms. It is significant that converts to this way of thinking were driven by reports of transfers which must have occurred long *after* the divergences leading to the three groups, whether from the conventional common ancestor or the proposed community of undifferentiated organisms.

Until recently, suspicions of horizontal gene transfers typically began with the observation that a protein sequence from some organism was found to have a stronger resemblance to a sequence from a distantly related organism than to nearer kin; a phylogenetic tree would be inconsistent with a classical biological relationship. With the rise of genomics, however, first reports often arise during the initial gene identification process. If computer matching reports the highest scoring candidate sequence to be from a distantly related organism, horizontal gene transfer is an almost automatic consideration, whether or not genes from closer relatives are known to have been characterized. The presumption is that current databanks have a sufficient sampling of organisms to justify such an assignment.

Horizontal Gene Transfer
ISBN: 0-12-680126-6

Copyright © 2002 by Academic Press.
All rights of reproduction in any form reserved.

This is a dangerous gambit. Even when all the extant sequences in the world are reported, the problem of gene loss and the occasional gene conversion or displacement will retain the potential to confound. The loss of a gene from nearer relatives is not at all uncommon, especially when paralogues are involved, and not rare even for orthologues. When the first dozen completely sequenced bacterial genomes were compared, for example, only 34 genes were found to be common to all of them (Huynen and Bork, 1998). And when a large portion of the genome of a fission yeast was compared with that of a budding yeast, it was found that at least 300 genes had been lost along one or the other lineages in the time since their divergence (Aravind et al., 2000). Clearly, gene loss has been an important factor in evolution. The consequence of these absent genes for simple match expectations must be taken into account.

In the past, cases of possible horizontal gene transfer were scrutinized carefully on an individual basis. Is comparable care impractical in this era of sequence abundance? Can the automatic detection of horizontal transfers be improved? In my view, there are a few simple measures that can be taken. In particular, I once again urge consideration of an attribute that too

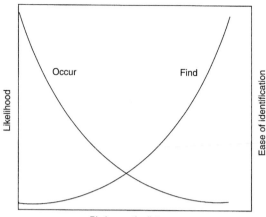

FIGURE 24.2 Likelihood of horizontal gene transfers as a function of evolutionary difference between the donor and recipient organisms (solid line). Ease of detectability as realized by sequence comparison (dashed line) (from Doolittle, 1998).

many sequence comparers ignore: the percent identity of the sequences in question.

Thus, in the first edition of this book I wrote:

> As a rule of thumb, if an enzyme (or other protein) sequence is found to be more than 60% identical between the two groups (viz. prokaryotes and eukaryotes), then horizontal gene transfer looms as a distinct possibility.

The usefulness of this criterion was based on a survey of 531 amino acid sequences for 57 enzymes in which analysis showed the average resemblance between Bacteria and Eukarya sequences to be 37% identity (Doolittle et al., 1996; Figure 24.1). Indeed, the highest percent identity observed was only 56%, and it was that specific observation that led to the rule of thumb. The rule still holds.

Naturally, in the immediate wake of a horizontal transfer, the genes in the two organisms involved must be identical, but the relentless rain of base substitutions promptly begins to alter the situation. The more time that has elapsed after the transfer, the more dissimilar the sequences. Presuming that transfers are still occurring (Jain et al., 1999), we should occasionally be finding extremely similar sequences reflecting very recent events. As an example, a very recent transfer between two quite

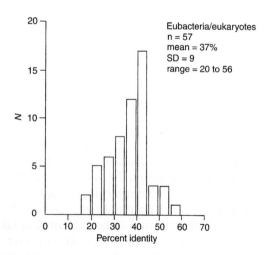

FIGURE 24.1 Average resemblances (percent identity) between multiple sequences from 57 enzymes of prokaryotes and eukaryotes as measured in blocks of five percentage points (from Doolittle et al., 1996).

TABLE 24.1 Some anomalously high resemblances between sequences from Eubacteria and Eukarya

Protein	Eubacterium	Eukaryote	Percent ID	Direction[a]
GAP dehydrogenase A	*Escherichia coli*	*Trypanosoma brucei*	88	K to B
Nitrite reductase	*Pseudomonas* sp.	*Fusarium oxysporum*	85	B to K
Glucose phosphate isomerase	*Escherichia coli*	*Drosophila melanogaster*	71	K to B
Nucleoside-diphosphokinase	*Synechococcus* sp.	*Dictyostelium discoideum*	69	K to B
Phosphofructokinase	*Treponema pallidum*	*Entamoeba histolytica*	68	B to K
Superoxide dismutase (Fe)	*Escherichia coli*	*Entamoeba histolytica*	58	B to K
Isopenicillin synthetase	*Streptomyces* sp.	*Acremonium crysogenum*	57	B to K
Serine protease	*Streptomyces griseus*	*Metarhyzium anisopliae*	55	B to K

[a]Presumes a horizontal transfer between ancestors of the species listed. B = Eubacteria; K = Eukarya.

TABLE 24.2 Some anomalously high resemblances between sequences from Eubacteria and Archaea

Protein	Eubacterium	Archaea	Percent ID	Direction[a]
Nitrogenase	*Clostridium pasteurianum*	*Methanosarcina barkeri*	74	A to B
Heat shock 70	*Clostr. acetabutyliricum*	*Methanosarcina mazei*	58	A to B
Enolase	*Zymomonas mobilis*	*Methanococcus jannaschii*	60	B to A
HMG-CoA reductase	*Pseudomonas* sp.	*Archaeoglobus fulgidus*	57	?
Nucleoside-diphosphokinase	*Staphylococcus aureus*	*Methanococcus jannaschii*	56	B to A
GAP dehydrogenase B	*Bacillus subtilis*	*Haloarcula vallismortisa*	52	B to A
Serine aminoacyl synthetase	*Bacillus subtilis*	*Haloarcula marismortui*	51	B to A

[a]Presumes a horizontal transfer between ancestors of the species listed. B = Eubacteria; A = Archaea.

different Archaea, *Pyrococcus furiosus* and *Thermococcus litoralis*, has been observed in which only 173 base changes occur in a 16 kb stretch (Di Ruggiero et al., 2000). The long DNA insert contained genes for maltose and trehalose transport, the protein sequences of which are virtually identical in the two organisms. There was evidence that IS elements were involved in the transfer. This is an unarguable case of recent gene transfer, even though it is between rather closely related members of the same superkingdom.

It has been argued that transfers between closely related organisms are more likely to occur than between distantly related ones (Figure 24.2). In line with this reasoning, we can ask how often such events occur between the Archaea and Bacteria or between either of these groups and the Eukarya.

We can begin our quest for recent transfers between distantly related organisms by asking how many proteins from Bacteria have counter-

parts among the Eukarya that are more than 60% identical? Not so many, it turns out (Table 24.1). The same is true of sequences from Archaea compared with Bacteria (Table 24.2). Indeed, there is little evidence here for "rampant exchange."

PRUNING AND GRAFTING THE TREE OF LIFE

The question arises as to how we can determine when and where alleged horizontal transfers occurred, if, as is the case, the community is still arguing about the fundamental nature of the Tree of Life (Doolittle, 2000). Granted that all cellular organisms can still be grouped on the basis of their ribosomal RNA sequences (Woese, 1998), is it hopeless to think that the vast majority of genes will reveal a concordant tripartite world? In fact, some adjustments are in order to accommodate special circumstances.

Thus, although the average sequence resemblance of the 57 bacterial and eukaryotic enzymes was 37% (Figure 24.1), the enzymes actually broke into three groups with different phylogenies (Feng et al., 1997). One of these had the Bacteria and Eukarya as sister groups, the average resemblance being 42%. This group, called the "import group," was probably acquired by early eukaryotes from endosymbionts, including but not limited to those that went on to become mitochondria. They may also have been acquired from the ingested bacteria that constituted their food. Enzymes in the other two groups were, on the average, only 34% identical.

Similarly, there were corresponding phylogeny-based classes of sequence relationship when the Archaea and Eukarya were compared. For those genes with phylogenetic trees that had the Archaea and Eukarya as nearer relatives, the average percent identity was 43%. In all other pairings the average resemblance dropped to 33%. The implication here is that eukaryotes acquired a large cohort of archaeal genes as a result of endosymbiotic or dietary means (Doolittle, 2000).

TRANSFERS AMONG THE PROKARYOTES

It has long been realized that horizontal gene transfers occur between bacteria. Indeed, the observation of transfer of virulence factors between strains of *Pneumococcus* heralded a turning point in biology. More recently, it has been reported that as much as 17% of the *Escherichia coli* and *Salmonella enterica* genomes have been exchanged since their divergence about 100 million years ago (Lawrence and Ochman, 1997). The transfer of almost identical or very similar genes from closely related organisms is obviously difficult to detect at the sequence level. At the same time, as sequences become increasingly different through normal divergence, the less likely it will be that opportunities for gene transfer will exist (Figure 24.2). Thus, we expect the proteins of prokaryotes and eukaryotes to be distinctly different, but, apart from endosymbiotic or parasitic situations, the opportunities for transfer seem slight.

In this regard, validation for horizontal gene transfers is often sought by considering the matter of opportunity as exemplified by symbiosis or parasitism or common habitat. Another consideration has to do with the natural advantage of acquiring some new feature, positive selection being so much stronger a force in the lottery of survival than neutral chance.

Thus, when considering horizontal gene transfers and the likelihood not only of their occurrence but also of their fixation, it is important to distinguish between simple displacements – a kind of neutral event in which both the donor and acceptor organisms have equivalent genes – from those where a new gene is being introduced that might confer some benefit on the acceptor. Examples of the former might be ubiquitous enzymes like gap dehydrogenases or aminoacyl-tRNA synthetases, the newly acquired gene simply displacing one that was already there. Genes that can confer a new advantage are obviously more narrowly restricted and not present in the potential recipient. Good examples are the antibiotic resistance genes that so often spread among bacteria on plasmids.

Another example is nitrogenase which, if accompanied by suitable accessory genes, can confer the ability to fix nitrogen on a previously less able organism. When some time ago it was discovered that the nitrogenase of *Clostridium pasteuranium* was more like that from a methanogen, workers were undecided about whether horizontal transfer was involved or whether it was merely descent from an ancient ancestral type (Chien and Zinder, 1994). At 74% identity (Table 24.2), the match is the highest sequence resemblance between an archaeon and a bacterium yet reported and seems certainly to be due to horizontal gene transfer.

A rule of thumb is never foolproof. There may be a few proteins that are so slow changing that they will have resemblances of more than 60% between the different superkingdoms even when horizontal transfers are not involved. Contrarily, there will be many more cases where horizontal transfers have likely occurred and the residual resemblances are significantly lower. As a case in point, consider the matter of sulfate reduction by the archaeon *Archaeoglobus fulgidus*. When the full genome sequence was determined (Klenk et al., 1997), a gene cluster for sulfate reduction was found that is absent in many other Archaea and which is arranged exactly the same as occurs in the bacterium *Desulfovibrio desulfuricans*. The percent identity for the key enzyme adenyl sulfate reductase (subunit A) between the two is only 51%, but the identical cluster arrangement and its absence from other Archaea seemed to cement the case for transfer of the cluster. Subsequently, an adenyl sulfate homologue was found in another archaeon, *Pyrobaculum islandicum*, and it was found to be less similar to the *Archaeoglobus* sequence than was the *Desulfovibrio* enzyme, consistent with the proposed horizontal transfer (Molitor et al., 1998).

There are other cases where clusters of genes are transferred and confer some obvious new benefit on an organism. Pathogenicity islands, for example, are large gene clusters that are transferred *en masse* among bacteria, transforming avirulent strains into virulent ones (Lee, 1996). There is often accompanying evidence for the involvement of phages or other transposable elements. The possibility of transferring clusters of genes gave rise to the concept of "selfish operons" (Lawrence and Roth, 1996). According to this hypothesis, the very existence of operons in Bacteria and Archaea is the consequence of the group transfer of genes from organism to organism. It follows that operons, and transferred clusters, should mostly involve potentially advantageous but non-essential genes.

Implicit in the discussion of some of the above examples is the matter of restricted occurrence. If a gene is found only among a very restricted set of Bacteria, for example, and then again only in a very small group of Archaea, the case for horizontal transfer may be greatly strengthened.

TRANSFERS BETWEEN BACTERIA AND EUKARYOTES

Given the limited range of similarities commonly observed for bacterial and eukaryotic enzyme sequences (Figure 24.1), it ought to be possible to spot good candidates for horizontal transfer on the basis of percent identity alone. In this regard, any individual horizontal gene transfers of enzymes between Bacteria and Eukarya that have occurred during the last billion years – well after the introduction of mitochondria – should have resemblances that are on average greater than 60% (Table 24.1).

Of these, the 88% identity between the GAP dehydrogenases of kinetoplastids like *Trypanosoma brucei* and γ proteobacteria like *E. coli* remains the most remarkable. How and when did this presumed transfer occur? Straightforward phylogenetic analysis suggests that it was the eukaryotic sequence that was introduced among the proteobacteria and, given the high resemblance, it must have been a quite recent event, perhaps within the last few hundred million years. On the other hand, the fact that the orthologous enzyme of *H. influenzae* is equally similar to the eukaryote type implies the event occurred before *E. coli* and *H. influenzae* went their separate ways.

There have been other reported cases of proteobacteria acquiring eukaryote genes, including the gene for glucose phosphate isomerase (Katz, 1996). In this case the amino acid sequences of the enzymes are more than 70% identical in animals and proteobacteria. A persuasive case has also been developed for the transfer of glutamine-tRNA synthetase from some presumably primitive eukaryote (Lamour et al., 1994; Handy and Doolittle, 2000). In this case the resemblance has diverged to approximately 50% identity. The mechanism for eukaryote-to-prokaryote transfer in these three cases can only be guessed at, but direct transformation is a possibility.

As for the transfer of genes from bacteria to eukaryotes, Takaya et al. (1998) have made a compelling case for there being a transfer of the nitrite reductase gene from a bacterium to a fungus, the enzyme being 85% identical in the two groups (Table 24.2). Other cases of transfers from bacteria to fungi include isopenicillin synthetase (Penalva et al., 1990; Smith et al., 1990; Buades and Moya, 1996) and a serine protease (Screen and St. Leger, 2000). All three of these cases are buttressed by the enzymes in question having a very restricted distribution within the fungi.

Good cases have also been made for the transfer of genes from bacteria to the bacteria-engulfing protist *Entamoeba histolytica* (Smith et al., 1992; Rosenthal et al., 1997), as well as the gap dehydrogenase B gene from bacteria *to Trichomonas vaginalis* (Markos et al., 1993). Other authors were cautious about the transfer of a phosphofructokinase gene from bacteria to protists (Mertens et al., 1998), but the fact that the percent identity is greater than 68% strongly suggests horizontal transfer (Table 24.1).

A CONCLUDING COMMENT

Anomalous phylogenetic trees are not always sufficient grounds to make a case for horizontal gene transfer; paralogy and gene loss often provide better explanations. Other considerations, including unusually high sequence resemblance and restricted occurrence, can strengthen the case for horizontal gene transfer considerably. Experience has taught us to be cautious in accepting first reports of horizontal transfers. At this point, however, it seems that horizontal transfers between distantly related organisms are not so frequent that they seriously jeopardize phylogenetic classifications.

POSTSCRIPT

Just as this chapter was being submitted, the first full reports of the sequence of the human genome appeared in *Science* and *Nature*. Apart from the unexpectedly low number of ORFs – somewhere between 30 000 and 35 000 – the biggest news is reportedly the large number of apparent horizontal transfers. Indeed, newspapers have been carrying lurid reports about the large number of genes that humans have acquired from bacteria. What are all these genes? So far as I can tell, they are ORFs which during the automatic searching scored higher hits against entries from Bacteria and, perhaps, Archaea, than were registered against the fully sequenced eukaryotic genomes of *Saccharomyces cerevisiae*, *Caenorhabditis elegans* or *Drosophila melanogaster*.

How strong are the resemblances that led to these reports? Not all the data are easily accessible as yet, but in one case singled out by Natalie Angier of the *New York Times* (February 12, 2001), that of a monoamine oxidase, I tried to track down all the relevant observations. What I found was that in no case did the resemblance between the human gene – or any other vertebrate animal – and any bacterial entry in the database exceed 35% identity. Whatever else may have transpired, this is not a recent acquisition by animals, the gene products in question having a lower resemblance than the average enzyme vertically descended from Bacteria and Eukarya (Figure 24.1). For the moment, it may be prudent to exercise more than the usual caution before accepting these reports as authentic horizontal gene transfers. In the absence of supporting data, the best cases for horizontal gene transfer between Bacteria and eukaryotes remain those with very high sequence identity.

REFERENCES

Aravind, L., Watanabe, H., Lipman, D.J. and Koonin, E.V. (2000) Lineage-specific loss and divergence of functionally linked genes in eukaryotes. *Proc. Natl Acad. Sci. USA* **97**: 11319–11324.

Buades, C. and Moya, A. (1996) Phylogenetic analysis of the isopenicillin-N-synthetase horizontal gene transfer. *J. Mol. Evol.* **42**: 537–542.

Chien, Y.-T. and Zinder, S.H. (1994) Cloning, DNA sequencing and characterization of nifD-homologous gene from the archaeon *Methanosarcina barkeri* 227 which resembles nifD1 from the eubacterium *Clostridium pasteurianum*. *J. Bact.* **176**: 6590–6598.

Di Ruggiero, J., Dunn, D., Maeder, D.L. et al. (2000) Evidence of recent lateral gene transfer among hyperthermophilic Archaea. *Mol. Microbiol.* **381**: 684–693.

Doolittle, R.F. (1998) The case for gene transfers between very distantly related organisms. In *Horizontal Gene Transfer* (eds, M. Syvanen and C. Kado), pp. 311–320, Chapman and Hall, London.

Doolittle, R.F. (2000) Searching for the common ancestor. *Res. Microbiol.* **151**: 85–89.

Doolittle, R.F., Feng, D.F., Tsang, S. et al. (1996) Determining divergence times of the major kingdoms of living organisms with a protein clock. *Science* **271**: 470–477.

Doolittle, W.F. (1999) Phylogenetic classification and the universal tree. *Science* **284**: 2124–2128.

Feng, D.-F., Cho, G. and Doolittle, R.F. (1997) Determining divergence times with a protein clock: update and reevaluation. *Proc. Natl Acad. Sci. USA* **94**: 13028–13033.

Handy, J. and Doolittle, R.F. (2000) An attempt to pinpoint the phylogenetic introduction of glutaminyl-tRNA synthetase among bacteria. *J. Mol. Evol.* **49**: 709–715.

Huynen, M. and Bork, P. (1998) Measuring genome evolution. *Proc. Natl Acad. Sci. USA* **95**: 5849–56.

Jain, R., Rivera, M.C. and Lake, J. (1999) Horizontal gene transfer among genomes: the complexity hypothesis. *Proc. Natl Acad. Sci. USA* **96**: 3801–3806.

Kandler, O. (1994) The early diversification of life and the origin of the three domains: a proposal. In *Thermophiles: The keys to molecular evolution and the origin of life* (eds, J. Wiegel and M.W.W. Adams), pp. 19–31, Taylor and Francis, London.

Katz, L. (1996) Transkingdom transfer of the phosphoglucose isomerase gene. *J. Mol. Evol.* **43**: 453–459.

Klenk, H.-P., Clayton, R.A., Tomb, J.F. et al. (1997) The complete genome sequence of the hyperthermophilic, sulfate reducing archaeon *Archaeoglobus fulgidus*. *Nature* **390**: 364–370.

Lamour, V., Quevillon, S., Diriong, S. et al. (1994) Evolution of the Glx-tRNA synthetase family. The glutaminyl enzyme as a case of horizontal gene transfer. *Proc. Natl Acad. Sci. USA* **91**: 8670–8674.

Lawrence, J.G. and Ochman, H. (1997) Amelioration of bacterial genomes: rates of change and exchange. *J. Mol. Evol.* **44**: 363–397.

Lawrence, J.G. and Roth, J.R. (1996) Selfish operons: horizontal transfer may drive the evolution of gene clusters. *Genetics* **143**: 1843–1860.

Lee, C.A. (1996) Pathogenicity islands and the evolution of bacterial pathogens. *Infectious Agents Dis.* **5**: 1–7.

Markos, A., Miretsky, A. and Muller, M. (1993) A glyceraldehyde-3-phosphate dehydrogenase with eubacterial features in the amitochondriate eukaryote, *Trichomonas vaginales*. *J. Mol. Evol.* **37**: 631–643.

Mertens, E., Lador, U.S., Lee, J.A. et al. (1998) The pyrophosphate-dependent phosphofructokinase of the protist, *Trichomonas vaginalis*, and the evolutionary relationships of protist phosphofructokinases. *J. Mol. Evol.* **47**: 739–750.

Molitor, M., Dahl, C., Molitor, I. et al. (1998) A dissimilatory sirohaem-sulfite reductase-type protein from the hyperthermophilic archaeon *Pyrobaculum islandium*. *Microbiology* **144**: 529–541.

Peñalva, M.A., Moya, A., Dopazo, J. and Ramón, D. (1990) Sequences of isopenicillin N synthetase genes suggest horizontal gene transfer from prokaryotes to eukaryotes. *Proc. Royal Soc. London [Biol.]* **241**: 161–169.

Rosenthal, B., Mai, Z. Caplivski, D. et al. (1997) Evidence for the bacterial origin of genes encoding fermentation enzymes of the amitochondriate protozoan parasite *Entamoeba histolytica*. *J. Bact.* **179**: 3736–3745.

Screen, S.E. and St. Leger, R.J. (2000) Cloning, expression and substrate specificity of a fungal chymotrypsin. Evidence for a horizontal transfer from an actinomycete bacterium. *J. Biol. Chem.* **275**: 6689–6694.

Smith, D.J., Burnham, M.K., Bull, J.H. et al. (1990) β-Lactam antibiotic biosynthetic genes have been conserved in clusters in prokaryotes and eukaryotes. *EMBO J.* **9**: 741–747.

Smith, M.W., Feng, D.F. and Doolittle, R.F. (1992) Evolution by acquisition: the case for horizontal gene transfers. *TIBS* **17**: 489–493.

Takaya, N., Kobayashi, M. and Shoun, H. (1998) Fungal denitrification, a respiratory system possibly acquired by horizontal gene tranfer from prokaryotes. In *Horizontal Gene Transfer* (eds, M. Syvanen and C. Kado), pp. 321–327, Chapman and Hall, London.

Woese, C. (1998) The universal ancestor. *Proc. Natl Acad. Sci. USA* **95**: 6854–6859.

Woese, C. (2000) Interpreting the universal phylogenetic tree. *Proc. Natl Acad. Sci. USA* **97**: 8392–8396.

Horizontal Gene Transfer and its Role in the Evolution of Prokaryotes

Eugene V. Koonin, Kira S. Makarova, Yuri I. Wolf and L. Aravind

In the pre-genomic era, horizontal gene transfer has been typically considered a marginal phenomenon without much general evolutionary significance. However, comparative analysis of bacterial, archaeal and eukaryotic genomes shows that a significant proportion of the genes, at least in prokaryotes, have been subject to horizontal transfer. In some cases, the amount and source of apparent gene acquisition by horizontal transfer could be linked to an organism's phenotypes. For example, bacterial hyperthermophiles seem to have scavenged many more archaeal genes than other bacteria, whereas transfer of certain classes of eukaryotic genes is most common in parasitic and symbiotic bacteria. In eukaryotes, in addition to the horizontal transfer of bacterial genes from the progenitors of the endosymbiotic organelles and probable other ancient horizontal transfers, considerable lineage-specific acquisition of bacterial genes seems to have taken place. Horizontal transfer events fall into distinct categories of acquisition of genes that are new to the given lineage, acquisition of paralogues of pre-existing genes, and xenologous gene displacement whereby a gene is displaced by a horizontally transferred orthologue from another lineage. The fixation and long-term retention of horizontally transferred genes indicates that they confer a selective advantage upon the recipient organism. In most cases, the nature of this advantage is not understood, but for several cases of acquisition of eukaryotic genes by bacteria, the biological significance seems clear. Examples include isoleucyl-tRNA synthetase whose acquisition from eukaryotes by several bacteria is linked to antibiotic resistance and ATP/ADP translocases acquired by intracellular parasitic bacteria, *Chlamydia* and *Rickettsia*, apparently from plants.

BACKGROUND: HORIZONTAL GENE TRANSFER AND THE NEW GENOMICS

Horizontal (lateral) gene transfer of genes between different species is an evolutionary phenomenon whose extent and significance have been the subject of an intense debate, particularly with regard to apparent cases that involve eukaryotes (Sprague, 1991; Syvanen, 1994). Indeed, should one accept that horizontal gene transfer is a major evolutionary phenomenon, rather than a collection of inconsequential anecdotes, certain central theoretical tenets of modern biology might be challenged. First, if a substantial fraction of the genes in each genome has been acquired via horizontal gene transfer, the traditional, tree-based view of the evolution of life must be considered incomplete or even dubious (Pennisi, 1998, 1999; Doolittle, 1999a,b, 2000; Martin, 1999; Koonin et al., 2000;). Secondly, horizontal gene transfer, at least in eukaryotes, seems to fly in the face of the core

Copyright © 2002 by Academic Press.
All rights of reproduction in any form reserved.

Neo-Darwinist belief in the central role of reproductive isolation between species in evolution. Prior to the genome sequencing era, although striking anecdotal examples of horizontal gene transfer have been described (Smith et al., 1992) and prescient speculation on the potential major evolutionary impact of such events has been published (Syvanen, 1985), the prevailing view seemed to hold that these events were rare enough not to affect our general understanding of evolution. The only instance where the impact of horizontal gene transfer had been clearly recognized was the apparent massive flow of genes from the genomes of endosymbiotic organelles, mitochondria in all eukaryotes and chloroplasts in plants, to the eukaryotic nuclear genome (Gray, 1992, 1999; Martin and Herrmann, 1998; Lang et al., 1999).

However, comparative analysis of complete genome sequences has quickly shown that apparent horizontal gene transfers were too common to be dismissed as inconsequential (Doolittle, 1999b; Ochman et al., 2000). The first strong indication of the possible extent of this phenomenon came from the multifactorial analysis of codon frequencies in portions of the *Escherichia coli* genome that revealed significant deviations from the general pattern of codon usage in approximately 15% of this bacterium's genes (Medigue et al., 1991). Because some of these genes showed a clear affinity with bacteriophage genes, the hypothesis has been proposed that all of the deviant genes were "alien" to *E. coli* and have been acquired horizontally from various sources. This type of observation seemed strongly to support substantial and relatively recent horizontal gene flow. The possibility of numerous ancient horizontal transfers has been suggested by the lack of congruence between phylogenetic trees for different sets of orthologous genes from a wide range of organisms. For example, some archaeal genes showed a clear affinity to their eukaryotic counterparts, in agreement with the rRNA tree topology, whereas others equally strongly clustered with bacterial homologues (Golding and Gupta, 1995; Gupta and Golding, 1996; Brown and Doolittle, 1997).

Comparison of multiple, complete prokaryotic genomes brought about the new age of "lateral genomics" (Doolittle, 1999a). Major differences in gene repertoires even among bacteria that belong to the same evolutionary lineage, such as, for example, *E. coli* and *Haemophilus influenzae* (Tatusov et al., 1996), indicated that genome evolution could not be reasonably described in terms of vertical descent alone. Much of the difference is attributable to lineage-specific gene loss, particularly in parasites, but there is little doubt that horizontal gene transfer is the other major evolutionary factor that should be taken into account to explain the complex relationships between prokaryotic genomes. The archaeal genomes presented a particularly striking "genomescape," which is strongly suggestive of massive horizontal gene exchange with bacteria. In agreement with the earlier indications from phylogenetic studies, but now on the whole-genome scale, archaeal proteins could be classified into those that were most similar to bacterial homologues and those that looked "eukaryotic" (Koonin et al., 1997; Doolittle and Logsdon, 1998; Makarova et al., 1999). With some exceptions, the "bacterial" and "eukaryotic" proteins in archaea were divided along functional lines, with those involved in information processing (translation, transcription and replication) showing the eukaryotic affinity, and metabolic enzymes, structural components and a variety of uncharacterized proteins that appeared "bacterial." Because the informational components generally appear to be less subject to horizontal gene transfer (however, some important exceptions are discussed below) and in accord with the "standard model" of early evolution whereby eukaryotes share a common ancestor with archaea (Woese et al., 1990; Doolittle and Handy, 1998), these observations have been tentatively explained by massive gene exchange between archaea and bacteria (Koonin et al., 1997). This hypothesis was further supported by the result of genome analysis of two hyperthermophilic bacteria, *Aquifex aeolicus* and *Thermotoga maritima*. Each of these genomes contained a significantly greater proportion of "archaeal" genes than any of the other bacterial genomes, in an obvious correlation between the similarity in the lifestyles of evolutionarily very distant organisms (bacterial and archaeal hyperthermophiles) and the apparent rate of horizontal gene exchange between them (Aravind et al., 1998; Nelson et al., 1999). These findings also emphasized the dilemma of adaptive versus opportunistic nature of horizontal gene transfer – do the genes that probably have been acquired from archaea enable *Aquifex* and *Thermotoga* to thrive under hyperthermal conditions, or these bacteria have acquired more archaeal genes than others simply because they have been more exposed to contacts with archaea because of their thermophily? Another case of apparent non-randomness in horizontal gene transfer has been observed for the cyanobacterium *Synechocystis* sp. which encodes a variety of proteins that seem to be associated with different forms of signaling and have been thought of as "eukaryotic" (Kaneko and Tabata, 1997; Ponting et al., 1999).

The finding that the contributions of horizontal gene transfer and lineage-specific gene loss to the gene composition of prokaryotic genomes was comparable to that of vertical descent amounted to a major shift in our understanding of evolution. Indeed, it became apparent that, in many cases, phylogenetic trees for different genes were incongruent not because of artifacts of tree-construction methods, but rather because of genuine differences in the evolutionary histories of genes caused by horizontal transfer. Thus, a true tree of life, a species tree, could not be constructed not because of the complexity of the problem and erosion of the phylogenetic signal associated with ancient divergence events, but perhaps in principle (Doolittle, 1999b, 2000). The best one could hope for was a consensus tree that might depict the evolution of a gene core that is conserved in all or the majority of species and is not subject to horizontal gene transfer. But the very existence of such a stable core, and more so its actual delineation remain questionable.

In retrospect, the apparent high incidence of horizontal gene transfer in prokaryotes perhaps should not have come as a complete surprise. The ability of microbes to absorb DNA from the environment and integrate it into the genome has been dramatically demonstrated by the classical Avery-McLeod-McCarthy experiment of 1943 that proved the role of DNA as the substrate of heredity (Avery et al., 1944). Subsequently, high transformability has been

demonstrated for many microbial species (Lorenz and Wackernagel, 1994). Bacteriophages and plasmids, some of which readily cross species barriers, provide additional, potentially highly effective vehicles for horizontal gene transfer (Hartl et al., 1984; Sundstrom, 1998; Moreira, 2000). Given that microbes typically coexist in tightly knit communities such as microbial mats and the microflora of animal guts (Risatti et al., 1994; Moyer et al., 1995; Ward et al., 1998; Jeanthon, 2000; Tannock, 2000), opportunities should abound for DNA transfer by various means between diverse prokaryotes and potentially even between eukaryotes and prokaryotes although, in the latter case, the additional problem of getting rid of introns resident in eukaryotic genes is involved.

Despite its growth in stature with the progress of sequence-based comparative genomics, the issue of horizontal gene transfer being a major evolutionary force remains highly contested (Aravind et al., 1999a; Kyrpides and Olsen, 1999; Logsdon and Faguy, 1999). One reason seems to be that the paradigm shift in evolutionary biology that seems inevitable if the notion of "lateral genomics" is vindicated elicits healthy skepticism in many quarters. The other problem is that, whereas the general significance of horizontal transfer seems to ensue from genome comparisons, proving individual cases beyond reasonable doubt often proves to be difficult.

In this chapter, we discuss the criteria used to ascertain horizontal gene transfer, present conservative quantitative estimates of the number of probable horizontal gene transfer between prokaryote and eukaryote genomes and discuss examples of such transfers in some of which the nature of the selective advantage conferred to the recipient by the acquired gene seems to be clear.

DETECTING HORIZONTAL GENE TRANSFER EVENTS

The criteria for identifying probable horizontal gene transfer events, or, more precisely, cases of acquisition of "alien" genes by a particular genome, inevitably rely upon some unusual feature(s) of subsets of genes that distinguishes them from the bulk of the genes in the given genome. Traditional tests for horizontal gene transfer involve phylogenetic tree analysis and inherit both the power and the inherent pitfalls of these methods. Accumulation of multiple genome sequences provides for new, sometimes simpler criteria. "Direct proofs" of horizontal gene transfer may be impossible to obtain simply because there is no record of such events, other than what could be deciphered by comparison of extant genomes. Therefore, all indications for horizontal transfer are inherently probabilistic, the goal of all methods used to analyze this phenomenon being to maximize the likelihood of correct identification of these events.

Unusual phyletic patterns

The notion of a phyletic (phylogenetic) pattern has been introduced with the systematic delineation of clusters of orthologues, direct evolutionary counterparts related by vertical descent (Tatusov et al., 1997; Gaasterland and Ragan, 1998; Galperin and Koonin, 2000). In the most straightforward formulation, a phyletic pattern is simply the pattern of species that are present or missing in the given orthologous cluster. Analysis of the collection of Clusters of Orthologous Groups of proteins (COGs) revealed an extreme diversity of phyletic patterns (Tatusov et al., 1997, 2001), most of which

include only a small number of genomes (Figure 25.1). This distribution of COGs by the number of represented species immediately suggests that lineage-specific gene loss and horizontal gene transfer have contributed extensively to the observed gene repertoires of most, if not all, prokaryotic genomes. Certain types of phyletic patterns may provide more direct evidence of horizontal transfer (Table 25.1). For example, when a set of orthologues shows the presence of a typical "archaeal-eukaryotic" protein in a single bacterial lineage, the odds for horizontal gene transfer underlying this pattern seem to be high. The B-family DNA polymerase (*E. coli* DNA polymerase II) is a clear-cut case of such obvious horizontal gene acquisition. The γ proteobacterial Pol II sequences are highly similar to those of their archaeal and eukaryotic orthologues, which renders irrelevant, for all practical purposes, the typical objections that other bacteria might in fact encode orthologues that have diverged beyond recognition. The presence of the Pol II gene in several sequenced γ proteobacterial genomes rules out the possibility of artifacts such as contamination of a bacterial genome with eukaryotic sequences. In this case, there even seems to be an indication as to the likely vehicle of gene transfer because Family B polymerases are encoded by numerous bacteriophages and animal viruses (Braithwaite and Ito, 1993; Knopf, 1998).

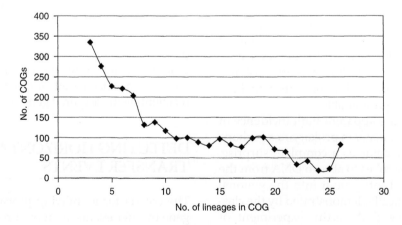

FIGURE 25.1 Distribution of the clusters of orthologous groups of proteins (COGs) by the number of species they include. Each COG consists of predicted orthologues from at least three genomes that belong to 26 distinct lineages (Tatusov et al., 2001).

TABLE 25.1 Probable horizontal gene transfers detected using phyletic patterns[a]

COG No.	Function	Occurrence in complete bacterial genomes	Occurrence in complete archaeal and eukaryotic genomes
From archaea or eukaryotes to bacteria			
0417	DNA polymerase, B family	Ec, Pa, Vc	All (also many viruses and bacteriophages)
0430	RNA phosphate cyclase	Ec, Pa, Aa	All
0467	KaiC-like ATPase of RecA-superfamily (implicated in signal transduction)	Ssp, Aa, Tm	All
0615	Predicted cytidylyltransferase	Aa, Bs	All
1257	Hydroxymethylglutaryl-CoA reductase	Vc, Bb	All
1577	Mevalonate kinase	Bb	All
2519	Predicted SAM-dependent methyltransferase involved in tRNA^{Met} maturation	Aa, Mtu	All
From bacteria to archaea and eukaryotes			
0847	DNA polymerase III, epsilon subunit/domain(3′ -5′ exonuclease)	All except Bb, mycoplasmas	Af, Sc, Ce, Dm
0566	RRNA methylase	All	Af, Sc, Ce, Dm
0188	DNA gyrase (topoisomerase II) A subunit	All	Af, Sc, Ce, Dm
bl0187	DNA gyrase (topoisomerase II) B subunit	All	Af, Sc, Ce, Dm
0138	AICAR transformylase/IMP cyclohydrolase (PurH)	All except Hp, Rp, spirochetes, chlamydia, mycoplasmas	Af, Sc,Ce, Dm
0807	GTP cyclohydrolase II (riboflavin bisynthesis)	All except Rp, spirochaetes, mycoplasmas	Af, Sc

[a]Species name abbreviations in this and subsequent tables: Aa, *Aquifex aeolicus*, Af, *Archaeoglobus fulgidus*, Ap, *Aeropyrum pernix*, Bb, *Borrelia burgdorferi*, Bh, *Bacillus halodurans*, Bs, *Bacillus subtilis*, Ce, *Caenorhabiditis elegans*, Cj, *Campylobacter jejuni*, Cp, *Chlamydia pneumoniae*, Ct, *Chlamydia trachomatis*, Dm, *Drosophila melanogaster*, Dr, *Deinococcus radiodurans*, Ec, *Escherichia coli*, Hi, *Haemophilus influenzae*, Hp, *Helicobacter pylori*, Mth, *Methanobacterium thermoautotrophicum*, Mtu, *Mycobacterium tuberculosis*, Nm, *Neisseria meningitidis*, Pa, *Pseudomonas aeruginosa*, Rp, *Rickettsia prowazekii*, Sc, *Saccharomyces cerevisiae*, Ssp, *Synechocystis sp*, Tm, *Thermotoga maritima*, Tp, *Treponema pallidum*, Vc, *Vibrio cholerae*, Uu, *Ureaplasma urealyticum*, Xf, *Xylella fastidiosa*.

Unexpected distribution of sequence similarity among orthologues

The suspicion that horizontal gene transfer might have been part of the evolutionary history of a particular gene usually emerges when the gene sequence (or rather the encoded protein sequence because database searches are typically performed at the protein level) from a particular organism shows the strongest similarity to a homologue from a distant taxon. For example, when all protein sequences encoded in a bacterial genome are compared with the entire protein database and the detected (probable) homologues (or hits, in the computational biology parlance) are classified according to the taxonomic position of the corresponding species, a certain fraction of proteins will show the greatest similarity to eukaryotic homologues, rather than to those from other bacteria. The size of this fraction depends on the genome and also on the cut-off (usually expressed in terms of alignment score or expect value) that is used to define "more similar" (the taxonomic distribution of the best hits for all proteins encoded in completely sequenced prokaryotic genomes is available through the Genome Division of the Entrez retrieval system (Tatusova et al., 1999) at http://www.ncbi.nlm.nih.gov:80/PMGifs/Genomes/org.html; see the "Distribution of BLAST protein homologues by taxa" for individual genomes). These genes make a list of candidates for horizontal gene exchange between the given bacterium (or, more precisely, the evolutionary lineage it represents) and eukaryotes. The strength of the claim depends on the cut-off used, but generally, although the evidence from sequence comparisons is indispensable for obtaining a genome-wide picture of probable horizontal transfers, validation of each individual case requires phylogenetic analysis in addition.

Unexpected phylogenetic tree topology

Analysis of phylogenetic tree topologies is traditionally the principal means to decipher evolutionary scenarios, including horizontal gene transfer events (Syvanen, 1994). Indeed, if, for example, in a well-supported tree, a bacterial protein groups with its eukaryotic homologues, to the exclusion of homologues from other bacteria and, best of all, shows a reliable affinity with a particular eukaryotic lineage, the evidence of horizontal gene transfer seems to be strong. In a convincing case like this, even the most likely direction of transfer, from eukaryotes to bacteria, seems clear. Unfortunately, however, phylogenetic analysis does not offer such clear-cut solutions in all suspected cases of horizontal gene transfer, not necessarily even in a majority of them. The common pitfalls of phylogenetic methods, in particular long-branch attraction (Moreira and Philippe, 2000), are particularly relevant for the analysis of probable horizontal gene transfer because these events may be accompanied by accelerated evolution, resulting in long branches in phylogenetic trees. Tree topology is a good indicator of the probable course of evolution only in cases when the critical nodes are strongly supported statistically, by bootstrap analysis or, preferably, by likelihood estimates for different topologies

(Hasegawa et al., 1991; Brown, 1994; Efron et al., 1996). However, many gene families seem to have undergone "star evolution," with very short internal branches. In such cases, the actual tree topology remains uncertain and phylogenetic analysis is useless for verifying the candidate horizontal transfer events. More practically, phylogenetic analysis is time- and labor-consuming, critically depends on correct sequence alignments and is hard to automate without compromising the robustness of the results. All these problems notwithstanding, an attempt has been recently undertaken automatically to construct the complete sets of phylogenetic trees for seven prokaryotic genomes and systematically to compare the topologies in search of horizontal transfer events (Sicheritz-Ponten and Andersson, 2001). The results seem to be coherent with those produced by analysis of unexpected distribution of sequence similarity revealing significant amounts of horizontal transfer (see also below), which indicates that, statistically, even such a crude system for tree analysis is suitable for revealing major trends in gene flow.

Conservation of gene order between distant taxa

Evolution of bacterial and archaeal genomes involved extensive genome rearrangements, and there is little conservation of gene order between distantly related genomes (Mushegian and Koonin, 1996a; Dandekar et al., 1998; Itoh et al., 1999). The presence of three or more genes in the same order in distant genomes is extremely unlikely unless these genes form an operon (Wolf et al., 2001). The same analysis implies that each operon typically emerges only once in the course of evolution and is subsequently maintained by selection (Lawrence, 1997, 1999). Therefore, when a (predicted) operon is found only in a few distantly related genomes, horizontal gene transfer seems to be the most likely scenario. If phylogenetic analysis confirms that genes comprising the operon form monophyletic groups in the corresponding trees, such observations figure among the strongest available indications of horizontal transfer.

Horizontal mobility of operons that encode restriction-modification systems is probably the most compelling example of horizontal transfer of operons (Naito et al., 1995; Kobayashi et al., 1999), but for many "normal" operons, dissemination by horizontal transfer also appears highly probable. The archaeal-type H^+-ATPase operon is a well-characterized example of such apparent horizontal dissemination of an operon among bacteria, with displacement of the classical bacterial ATPase operon (Hilario and Gogarten, 1993; Olendzenski et al., 2000).

Anomalous nucleotide composition

This criterion is perhaps best known, but is applicable only to recent horizontal transfers. The approach is based on the "genome hypothesis," according to which codon usage and GC content are signatures of each genome (Grantham et al., 1980a,b). Thus, genes whose nucleotide or codon composition significantly deviate from the mean for a given genome are considered to be probable horizontal acquisitions (Medigue et al., 1991; Lawrence and Ochman, 1997; Mrazek and Karlin, 1999; Garcia-Vallve et al., 2000; Ochman et al., 2000). A significant fraction of prokaryotic genomes, up to 15–20% of the genes, appears to belong to this class of recent horizontal acquisitions (Medigue et al., 1991; Lawrence and Ochman, 1997; Garcia-Vallve et al., 2000). Many of the horizontally transferred genes revealed by this approach are prophages, transposons and other mobile elements, for which this mode of evolution is not unexpected.

Directionality of horizontal gene transfer

Even harder than proving that horizontal gene transfer has occurred during the evolution of a particular gene family is to determine unequivocally which organism is the donor and which one is the recipient in each case. The available collection of genome sequences is but a tiny sampling of the genome universe (Pace, 1997), because of which it is impossible to identify the actual source of any gene present in a given genome. At best, it might be possible to identify the probable donor lineage. The logic used to formulate such hypotheses is based on the assumption that, if horizontal transfer has indeed occurred, the taxon with the broadest representation of the given family is the most likely source. The examples shown in Table 25.1 illustrate this approach; these gene families are either widely (usually, universally) represented in archaea and eukaryotes, but are found in only one or a few bacterial species, or vice versa, are present in most bacteria, but only one or two archaeal species, which strongly suggest the transfer directionality. However, in the rather frequent cases of limited representation of the given family in two distant taxa, the fact of horizontal gene transfer may be (almost) certain, but the direction of transfer cannot be established with any confidence.

DIFFERENT TYPES OF HORIZONTAL GENE TRANSFER EVENTS: A ROUGH QUANTITATIVE ASSESSMENT

With respect to the relationships between the horizontally transferred gene and homologous genes (if any) pre-existing in the recipient lineage, at least three distinct types of horizontal transfer events are identifiable:

(1) acquisition of a new gene that is otherwise missing in the given lineage (to the extent sequence data is available);
(2) acquisition of a paralogue of a pre-existing gene;
(3) xenologous gene displacement, or acquisition of a phylogenetically distant orthologue followed by the elimination of the resident gene (xenology has been defined as homology of genes that is not congruent with the species tree and so implies horizontal gene transfer (Patterson, 1988; Gogarten, 1994)).

The first two categories of events in many cases may reflect *non-orthologous gene displacement*, that is, acquisition of an unrelated (or distantly related) gene with the same function as an essential gene typical of the given clade, with subsequent elimination of the latter (Koonin et al., 1996; Mushegian and Koonin, 1996b).

We produced rough quantitative estimates of the amount of horizontal gene transfer in bacterial and archaeal genomes and attempted to classify the transfer events into the above categories. All its caveats notwithstanding, taxonomic classification of database hits is the only practicable method to identify candidate horizontal gene transfer events on genome scale for all sequenced genomes. We applied this approach with conservative cut-offs and detected proteins that are significantly more similar to homologues from other taxa than to those from the taxon to which the given species belongs. Protein sets from 31 complete prokaryotic genomes (nine archaeal and 22 bacterial) were obtained from the Genome division of the Entrez retrieval system (Tatusova et al., 1999) and used as queries to search the non-redundant (NR) protein sequence database at the National Center for Biotechnology Information (NIH, Bethesda) with the gapped BLASTP program (Altschul et al., 1997).

From the results of these searches, three sets of "paradoxical" best hits were identified using the Tax_Collector program of the SEALS package (Walker and Koonin, 1997). The first set includes candidate horizontal transfers between phylogenetically most distant species. Thus, for nine archaeal species, all proteins were selected whose best hits to bacterial or eukaryotic homologues had expectation (E) values significantly lower than the E-value of the best hit to an archaeal homologue or that did not have detectable archaeal homologues at all (see legend to Table 25.2). For the 22 bacterial proteomes, the "paradoxical" best hits to archaeal and eukaryotic proteins were collected.

The second group includes genes that might have been exchanged between major bacterial lineages. This type of analysis currently is best applicable to small genomes of parasitic bacteria when at least one larger genome sequence of a related species is available because this provides reasonable confidence that the selected proteins indeed do not have highly conserved homologues within their own bacterial lineage. For example, for *Haemophilus influenzae* and *Rickettsia prowazekii*, hits to Proteobacteria (two large Proteobacterial genomes, those of *E. coli* and *Pseudomonas aeruginosa*, have been sequenced)

were compared with hits to all other bacteria and "paradoxical" hits were collected. To isolate probable horizontal transfers between distantly related bacteria, the transfers to and from Archaea and Eukaryotes included in the first set were subtracted.

The third set included probable recent horizontal transfers. To reveal such events, two pairs of closely related species, namely *Chlamydophyla pneumoniae–Chlamydia trachomatis* and *Mycoplasma genitalium–Mycoplasma pneumoniae* were compared.

A tally of candidate horizontal transfers that involve different primary domains of life is shown in Table 25.2. In most free-living bacteria, these "long-distance" transfers seem to involve ~3% of the genes. This fraction was much lower in parasitic bacteria, with the exception of *Chlamydia* and *Rickettsia*. Archaea had a greater fraction of candidate horizontal transfers of this type, typically between 4% and 8% of the genes. One should keep in mind that the protocol employed to obtain these numbers detects primarily relatively recent horizontal transfer events because ancient ones, for example, those that could have occurred prior to the divergence of the analyzed archaeal species, are obscured by inter-archaeal hits. Given these limitations and the conservative cut-off values used, the level of apparent gene exchange between different domains of life seems to be quite substantial.

Several species stand out with respect to the number of genes that probably have been horizontally acquired from a different domain of life. In two of these, *Synechocystis* sp. (a cyanobacterium, representative of the bacterial lineage from which chloroplasts have been derived (Gray, 1992, 1999)) and *Rickettsia prowazekii* (an α proteobacterium, the group of bacteria to which the progenitor of the mitochondria is thought to belong (Andersson et al., 1998)), the apparent high level of interdomain horizontal gene transfer probably reflects the well-recognized gene flow between chloroplasts and mitochondria, and eukaryotic nuclear genomes. As noticed previously, hyperthermophilic bacteria, *Aquifex aeolicus* and especially *Thermotoga maritima*, are markedly enriched in genes that apparently have been horizontally transferred from archaea, compared with mesophilic bacteria

TABLE 25.2 Probable interdomain horizontal transfers: a conservative estimate[a]

Species	Reference lineage	Paralogue acquisition or xenologous displacement (number and % of the genes)	Acquisition of new genes (number and % of the genes)
Aeropyrum pernix	Archaea	34 (1.8)	47 (2.5)
Archaeoglobus fulgidus	Archaea	103 (4.3)	100 (4.2)
Methanobacterium thermoautotrophicum	Archaea	100 (5.3)	61 (3.3)
Methanococcus jannaschii	Archaea	43 (2.5)	39 (2.3)
Pyrococcus horikoshii	Archaea	55 (2.7)	39 (1.9)
Pyrococcus abyssi	Archaea	72 (4.1)	39 (2.2)
Thermoplasma acidophilum	Archaea	112 (7.8)	54 (3.7)
Halobacterium sp.	Archaea	204 (8.4)	174 (7.2)
Aquifex aeolicus	Bacteria	87 (5.7)	45 (3.0)
Thermotoga maritima	Bacteria	207 (11.1)	53 (2.9)
Deinococcus radiodurans	Bacteria	47 (1.5)	45 (1.5)
Bacillus subtilis[b]	Bacteria	71 (1.7)	28 (0.7)
Bacillus halodurans[b]	Bacteria	79 (1.9)	40 (1.0)
Mycobacterium tuberculosis	Bacteria	50 (1.3)	62 (1.7)
Escherichia coli	Bacteria	26 (0.6)	13 (0.3)
Haemophilus influenzae	Bacteria	3 (0.2)	3 (0.2)
Rickettsia prowazekii	Bacteria	23 (2.8)	7 (0.8)
Pseudomonas aeruginosa	Bacteria	66 (1.2)	39 (0.7)
Neisseria meningitidis	Bacteria	6 (0.3)	5 (0.2)
Vibrio cholerae	Bacteria	12 (0.3)	16 (0.4)
Xylella fastidiosa	Bacteria	22 (0.8)	8 (0.3)
Buchnera sp.	Bacteria	0 (0.0)	0 (0.0)
Treponema pallidum	Bacteria	10 (1.0)	4 (0.4)
Borrelia burgdorferi	Bacteria	3 (0.4)	6 (0.7)
Synechocystis PCC6803	Bacteria	219 (6.9)	115 (3.6)
Chlamydia pneumoniae[c]	Bacteria	23 (2.2)	9 (0.9)
Mycoplasma pneumoniae	Bacteria	0 (0.0)	1 (0.1)
Ureaplasma urealyticum	Bacteria	1 (0.2)	1 (0.2)
Helicobacter pylori	Bacteria	5 (0.3)	3 (0.2)
Campylobacter jejuni	Bacteria	5 (0.3)	4 (0.2)

[a]All protein sequences from each genome were compared with the NR database using the BLASTP program (expect (E) value cut-off 0.001, no filtering for low complexity) and the results were searched for paradoxical best hits, i.e. those that either had a hit to a homologue from a non-reference lineage with an E-value 10 orders of magnitude lower (more significant) than that of the best hit to a homologue from the reference taxon, or had statistically significant hits to homologues from non-reference taxa only. All automatically detected paradoxical best hits were manually checked to eliminate possible false positives.

(Aravind et al., 1998; Nelson et al., 1999; Worning et al., 2000). Conversely, the archaea *Thermoplasma acidophilum* and especially *Halobacterium* sp. seem to have acquired many more bacterial genes than other archaeal species, perhaps because these organisms are moderate thermophiles and share their habitats with numerous bacterial species. Along the same lines, *Bacillus halodurans*, a halotolerant bacterium, seems to have a much greater number of "archaeal" genes than the closely related *B. subtilis*. In this case, the difference is plausibly explained by co-habitation of *Bacillus halodurans* with halophilic archaea. Notably, a recent genome-wide phylogenetic tree analysis suggested an even higher level of gene exchange with archaea and eukaryotes for several bacterial species (Sicheritz-Ponten and Andersson, 2001).

In this type of genome-wide analysis, apparent acquisition of new genes, that is cases when a given protein simply has no detectable homologues in its own lineage, could be automatically distinguished from paralogue acquisition/xenologous displacement, but to differentiate between the latter two types of events, additional analysis was required. For most genomes, the amount of paralogue acquisition/xenologous displacement was comparable to that of new gene acquisition (Table 25.2).

The data on the probable horizontal gene transfers between major bacterial lineages is summarized in Table 25.3. The estimates of the number horizontal transfer events produced by this approach differ substantially, from the modest 1.6% of the genes for *M. genitalium* to the striking 32.6% in *T. pallidum*. The spirochete data could be an overestimate caused by differential gene loss in the two parasitic spirochetes with very different lifestyles (Subramanian et al., 2000), in the absence of a sequence of a larger genome from the same lineage. Nevertheless, in general, the data seem to support the notion of a major horizontal gene flow between different branches of bacteria.

ACQUISITION OF EUKARYOTIC GENES BY BACTERIA AND ARCHAEA

Acquisition of eukaryotic genes by bacteria may be of particular interest because of the possible role of such horizontally transferred genes in bacterial pathogenicity (Groisman and Ochman, 1997; Ochman et al., 2000). We produced conservative estimates of the number of acquired eukaryotic genes in each of the completely sequenced prokaryotic genomes by using the "paradoxical best hit" approach. For this particular purpose, all prokaryotic proteins that showed significantly greater similarity to eukaryotic homologues than to bacterial ones (with a possible exception for closely related bacterial species) were detected using the Tax_Collector program. The number of apparent eukaryotic acquisitions for most of the prokaryotic genomes seems to be relatively small, typically in the order of 1% of the genes (Table 25.4; the high number of "plant" genes in *Synechocystis* is an artifact caused by the direct evolutionary relationship between Cyanobacteria and chloroplasts).

TABLE 25.3 Probable horizontal gene transfers between major bacterial lineages: conservative estimate[a]

Species	Reference lineage	Paralogue acquisition or xenologous displacement (number and % of the genes)	Acquisition of a new gene (number and % of the genes)
Mycoplasma genitalium	Firmicutes	6 (1.2)	2 (0.4)
Mycoplasma pneumoniae	Firmicutes	9 (0.9)	8 (1.2)
Bacillus subtilis	Firmicutes	685 (16.7)	383 (9.3)
Bacillus halodurans	Firmicutes	772 (19.0)	400 (9.8)
Treponema pallidum	Spirochetales	132 (12.8)	204 (19.8)
Borrelia burgdorferi	Spirochetales	109 (12.8)	141 (16.6)
Haemophilus influenzae	Proteobacteria	32 (1.9)	21 (1.2)
Rickettsia prowazekii	Proteobacteria	49 (5.9)	32 (3.8)
Escherichia coli	Proteobacteria	223 (5.2)	102 (2.4)
Pseudomonas aeruginosa	Proteobacteria	448 (8.1)	275 (5.0)
Neisseria meningitidis	Proteobacteria	55 (2.7)	34 (1.7)
Vibrio cholerae	Proteobacteria	130 (3.4)	85 (2.2)
Xylella fastidiosa	Proteobacteria	88 (3.2)	83 (3.0)
Buchnera sp.	Proteobacteria	0 (0.0)	0 (0.0)
Mycoplasma genitalium	Mycoplasma	0 (0.0)	0 (0.0)
Chlamydia pneumoniae	Chlamydiales	4 (0.4)	25 (2.4)

[a]The schema for detection of candidate horizontal transfers was the same as in Table 25.2.

TABLE 25.4 Acquisition of eukaryotic aaRS genes by different bacterial lineages[a]

Bacterial group	Eukaryotic aaRS	Comment
Spirochetes	Pro (*Borrelia* only), Ile, Met, Arg, His, Asn	
Spirochetes	Ser, Glu	Apparent acquisition of the mitochondrial gene
Chlamydia	Ile, Met, Arg, Asn??	
	Glu	Apparent acquisition of the mitochondrial gene
Bacillus	Asn	
Mycobacteria	Ile, Asn??, Pro (*M. leprae* only)	
Mycoplasma	Pro, Asn	
γ Proteobacteria	Asn, Gln	
Helicobacter	His	
Deinococcus	Gln, Asn	
Cyanobacteria	Arg, Asn	
Pyrococcus (archaeon)	Trp	

[a]An update of the results presented in (Wolf et al., 1999).

There seems to be a modest but consistent excess of acquired eukaryotic genes in at least some parasites, such as *M. tuberculosis*, *Pseudomonas aeruginosa*, *Xylella fastidiosa* and *C. pneumoniae* (data not shown). However, for other parasitic bacteria, such as spirochetes, only a very small number of probable acquired eukaryotic genes was detected with this approach. Apart from the obvious Cyanobacterium–chloroplast relationship notwithstanding, *Synechocystis* has an unusual excess of "eukaryotic" genes, including some that are otherwise animal-specific, such as several proteins that share conserved domains with cadherins and other extracellular receptors (Ponting et al., 1999). A possible explanation could be that, similarly to the extant cyanobacterial symbionts of sponges, the ancestors of *Synechocystis* have passed through a phase of animal symbiosis in their evolution (Paerl, 1996).

Unexpectedly, apparent acquisition of some eukaryotic genes is seen in each of the archaeal genomes, with the greatest number detected in *Halobacterium* sp, the archaeon that also appears to have acquired the greatest number of bacterial genes (Table 25.3 and discussion above). A more detailed phyletic breakdown of the eukaryotic acquisitions shows some limited correlation with parasite–host affinities. For example, *P. aeruginosa* has an excess of "animal"

genes, whereas the plant pathogen *Xylella fastidiosa* seems to have acquired a greater number of "plant" genes (data not shown). *Chlamydia* is an unusual case in that this animal pathogen seems to have acquired a greater number of genes from plants than from animals. It seems plausible that Chlamydiae and their close relatives had a long history of parasitic or symbiotic relationships with eukaryotes and, at some stages of their evolution, could have been parasites of plants or their relatives.

It should be emphasized that the approach employed here probably underestimates, perhaps considerably, the true number of eukaryotic genes transferred to prokaryotes, because some of the proteins encoded by transferred genes may not show highly significant similarity to their eukaryotic ancestors and are therefore missed. For example, this is the case for most of the signaling proteins discussed below; detection of such subtle cases requires a more detailed and not easily automatable analysis. "Eukaryotic" genes are particularly abundant in certain functional classes of bacterial genes.

Aminoacyl-tRNA synthetases

Generally, genes that encode components of the translation machinery appear to belong to the conserved core of the genome that is less prone to horizontal transfer than other groups of genes (Jain et al., 1999). A notable exception is ribosomal protein S14, for which several probable horizontal transfer events have been revealed by phylogenetic analysis (Brochier et al., 2000). However, the evolution of one group of essential components of the translation machinery, aminoacyl-tRNA synthetases (aaRS), appears to have involved numerous horizontal transfers, which could reflect the relative functional autonomy of these enzymes compared, for example, with ribosomal proteins that function as subunits of a tight complex (Doolittle and Handy, 1998; Wolf et al., 1999; Woese et al., 2000). Phylogenetic analysis shows that varying amounts of horizontal transfer probably have contributed to the evolution of aaRS of all 20 specificities (Wolf et al., 1999). On many independent occasions, eukaryotic aaRS appear to have displaced the resident bacterial ones (Table 25.4). Acquisition of eukaryotic aaRS genes is most typical of two groups of parasitic bacteria, *Chlamydia*, and particularly spirochetes. It may not be particularly surprising that parasites have acquired more aaRS genes than free-living bacteria, but why are spirochetes so unusual in this respect, remains unclear.

A highly unexpected case of probable horizontal transfer from eukaryotes is the TrpRS from the genus *Pyrococcus* that includes free-living, hyperthermophilic archaea (Wolf et al., 1999). Given this lifestyle, the presence of an aaRS in the *Pyrococci* appears extremely surprising, but the result of phylogenetic analysis in this case has been unequivocal. Thus, *Pyrococcus* probably has acquired a eukaryotic TrpRS gene from one of the few eukaryotic species that have been found in hyperthermophilic habitats, for example, a polychaete annelid (Sicot et al., 2000).

In most cases, the apparent horizontal transfer of eukaryotic aaRS genes into prokaryotes involves xenologous gene displacement, i.e. the corresponding ancestral prokaryotic aaRS is not present along with the eukaryotic one. A special case is GlnRS that apparently first emerged in eukaryotes through a duplication of the GluRS gene and subsequently has been horizontally acquired by γ proteobacteria. In most of the other bacteria and

archaea, glutamine incorporation into protein is mediated by a completely different mechanism, namely transamidation, whereby glutamine is formed from Glu-tRNA^{Gln} in a reaction catalyzed by the specific transamidation complex, GatABC (Curnow et al., 1997; Ibba et al., 1997). The gatABC are missing in γ proteobacteria indicating that, in this case, horizontal gene transfer, followed by non-orthologous gene displacement, has resulted in a switch to a completely different pathway for an essential biochemical process. In several other proteobacteria and in *Deinococcus radiodurans*, GlnRS and the GatABC complex co-exist (Handy and Doolittle, 1999), suggesting that, in γ proteobacteria, the elimination of the transamidation system has been a relatively late event compared with the acquisition of the eukaryotic GlnRS gene. The finding that *Mycobacterium leprae* encodes a eukaryotic-type ProRR, in contrast to the closely related *M. tuberculosis* that has a typical bacterial form (Cole et al., 2001), supports the notion that xenologous displacement of aaRS is an active, ongoing process. This case is particularly interesting as a "smoking gun" for xenologous gene displacement of an individual gene that has occurred without disrupting the conserved order of surrounding genes.

It has been noticed that the topology of some of the aaRS trees, particularly that for IleRS, could be accounted for by a single horizontal gene transfer from eukaryotes, with subsequent lateral dissemination among bacteria (Brown et al., 1998). This hypothesis is compatible with the reliable clustering of all bacterial species, that are suspected to have acquired the respective eukaryotic gene, in the IleRS and HisRS trees. However, the topologies of the trees for MetRS, ArgRS and Asp-AsnRS do not seem to agree with the single-transfer scenario, suggesting instead multiple cases of acquisition of the respective eukaryotic genes by different bacterial lineages (Wolf et al., 1999).

The gene for the eukaryotic-type IleRS disseminates through bacterial populations on plasmids, conferring resistance to the antibiotic mupirocin (Brown et al., 1998). This is one of the rare cases when not only the vehicle of horizontal gene transfer, but also the nature of the selective pressure that results in the fixation of the transferred gene in the bacterial population seems obvious.

Signal transduction systems

Eukaryotes have vastly more complex signal transduction systems than most bacteria and archaea (Davie and Spencer, 2000; Hunter, 2000; Lengeler et al., 2000), although in some bacteria, such as Cyanobacteria, Myxobacteria and Actinomycetes, these systems also show remarkable versatility (Hopwood et al., 1995; Dunny and Leonard, 1997). A detailed survey of the phyletic distribution of "eukaryotic" protein domains involved in various forms of signaling has revealed, rather unexpectedly, their frequent presence, sometimes in highly divergent forms, in prokaryotes (Ponting et al., 1999). Based on their provenance among bacteria and archaea, these domains have been classified into those that probably have been inherited from the Last Universal Common Ancestor and those that have evolved in eukaryotes and have been subsequently horizontally transferred to bacteria, and less commonly, archaea (Table 25.5).

When a domain is present in all eukaryotes, but, in contrast, is found in only one or two bacterial lineages, the case for horizontal transfer appears compelling. For example, the SWIB domain, which is present in subunits of the SWI/SNF chromatin-associated complex in all eukaryotes (Cairns et al., 1996), was detected in only one bacterial lineage, the *Chlamydia* (Stephens et al., 1998). The chlamydial SWIB domain, one copy of which is fused to topoisomerase I, might participate in chromatin condensation, a distinguishing feature of this group of intracellular bacterial parasites (Barry et al., 1993). Similarly, the SET domain, a characteristic eukaryotic histone methylase (Rea et al., 2000), has been detected in *Chlamydia* (Stephens et al., 1998) and *Xylella fastidiosa* (Schultz et al., 2000). The functions of the SET methylases in these bacteria remain unclear, and their elucidation might uncover regulatory mechanisms so far unsuspected in prokaryotes.

Some of the "eukaryotic" signaling domains acquired by bacteria probably perform functions that, at least mechanistically, are very similar to their functions in eukaryotic systems. An example of such probable functional conservation is the phosphoserine-peptide-binding FHA domain (Li et al., 2000) whose partners in signal transduction, protein kinases and phosphatases, are present in both eukaryotes and prokaryotes, suggesting that this domain functions in similar phosphorylation-based signaling pathways (Leonard et al., 1998).

Other signaling domains of eukaryotic origin probably have been exapted (Gould, 1997) for distinct functions in bacteria. Such apparent exaptation is exemplified by predicted cysteine proteases of two distinct families that have been detected in *Chlamydia* (Stephens et al., 1998; Makarova et al., 2000) (Table 25.5). In eukaryotes, one of these protease families, the adenovirus-type proteases, is known (Li and Hochstrasser, 1999, 2000) and the other one has been predicted (Makarova et al., 2000) to participate in ubiquitin-dependent protein degradation. The ubiquitin system does not exist in bacteria, which rules out functional conservation for these proteins. It seems most likely that, in *Chlamydia*, these proteases contribute to the pathogen–host-cell interaction as indicated, in particular, by the predicted membrane localization of the adenovirus-family proteases in *C. trachomatis* (Stephens et al., 1998).

Miscellaneous eukaryotic genes acquired by prokaryotes

In addition to the apparent common horizontal gene transfer in certain functional categories, many genes of very diverse functions seem to have been horizontally transferred from eukaryotes to bacteria or archaea (Table 25.6). On most occasions, the selective advantage that could be conferred on the prokaryote by the acquired eukaryotic gene remains unclear. However, the few exceptions when this is feasible allow biologically interesting inferences. Perhaps the most clear-cut example is the chloroplast-type ATP/ADP translocase that is present in the intracellular parasites *Chlamydia* and *Rickettsia* and the plant pathogen *X. fastidiosa*. For *Chlamydia* and *Rickettsia*, the advantage of having this enzyme is obvious because it allows them to scavenge ATP from the host, thus becoming, at

Table 25.5 Eukaryotic-bacterial transfer of genes coding for proteins and domains involved in signal transduction[a]

Domain	Occurrence in eukaryotes	Occurrence in prokaryotes	Prevalent protein context		Functions in eukaryotes
			Eukaryotes	Prokaryotes	
WD40	All	Ssp, Hi, Dr, Tm, *Streptomyces*, *Cenarchium symbiosum* (archaeon)	Multiple repeats in various regulatory proteins including G-protein subunits	Stand-alone, multiple repeats, Ser/Thr protein kinases	Regulatory protein–protein interactions
Ankyrin	All	Ec, Ssp, Bb, Tp, Nm, Rp, Dr, *Streptomyces*	Multiple repeats in various regulatory proteins	Stand-alone, multiple repeats	Signal transduction including transcription regulation, cell cycle control and PCD
TIR	Animals, plants	Bs, Ssp, *Streptomyces*, *Rhizobium*	Toll/interleukin receptors, apoptosis adaptors	Stand-alone, AP-ATPases, WD40	Apoptosis, interleukin signaling
EF-hand	All	Ssp, *Streptomyces*	Calmodulins, calcineurin phosphatase regulatory subunits, other Ca-binding proteins	Predicted Ca-binding proteins; transaldolase	Various forms of Ca-dependent regulation
FHA (forkhead-associated domain)	All	Ct, Cp, Mtu, Ssp, Xf, *Streptomyces*, *Myxococcus*	Protein Ser/Thr kinases and phosphatases various nuclear regulatory proteins	Stand-alone, adenylate cyclase, histidine kinase	Phosphoserine binding, protein kinase-mediated nuclear signaling
SET	All	Ct, Cp, Xf	Various multi-domain chromatin proteins	Stand-alone	Histone Methylase, chromatin remodeling
SWIB	All	Ct, Cp	SWI/SNF complex subunits	Stand-alone, Topoisome-rase type I	Chromatin remodeling, transcription regulation
Sec7	All	Rp	Guanine-nucleotide exchange factors; protein transport system components	Stand-alone (protein transport?)	Protein transport, GTPase regulation
Kelch	All	Ec, Hi, Af	Multiple repeats, fusions with POZ and other signaling domains; actin-binding proteins, transcription regulators	Stand-alone, multiple repeats	Transcription regulation; cytoskeleton assembly

Table 25.5 Contd

| Domain | Occurrence in eukaryotes | Occurrence in prokaryotes | Prevalent protein context | | Functions in eukaryotes |
			Eukaryotes	Prokaryotes	
Adenovirus-type protease	All; DNA-viruses	Ct, Ec	Stand-alone; ubiquitin-like protein hydrolases	Stand-alone; membrane-associated (Ct)	Regulation of ubiquitin-like protein-dependent protein degradation
OTU-family protease	All; RNA- and DNA-viruses	Cp	Stand-alone or fused to ubiquitin hydrolase	Large protein with non-globular domains	Unknown; possible role in ubiquitin-mediated protein degradation
START domain	Animals and plants	Pa	Stand-alone or fused to homeodomains and GTPase regulatory domains	Stand-alone	Lipid binding

[a]An updated excerpt of the results presented in (Ponting et al., 1999).

least in part, "energy parasites" (Winkler and Neuhaus, 1999). In contrast, the presence of the ATP/ADP translocase in *X. fastidiosa* (M.Y. Galperin, V. Anantharaman, L. Aravind and E.V. Koonin, unpublished observations) is unexpected and might indicate that such use of the energy-producing facilities of the host is not limited to intracellular parasites.

Another case when the adaptive value of the horizontally transferred gene seems clear is the sodium/phosphate co-transporter that was detected in *Vibrio cholerae*, but so far not in any other bacterium; this transporter probably facilitates the survival of the bacterium in the host intestines and could be involved in pathogenicity.

Many enzymes that apparently have been acquired by bacteria from eukaryotes seem to have been exapted for various functions, including interactions with their eukaryotic hosts. One probable example is a hemoglobinase-like protease that is present in *Pseudomonas* (PA4016) and might function as a virulence factor that degrades host proteins. The fukutin-like enzymes found in certain pathogenic bacteria, including *Haemophilus* and *Streptococcus*, are another case of horizontally transferred eukaryotic proteins that probably have been adapted by these bacteria for the modification of their own surface molecules (Aravind and Koonin, 1999).

Acquisition and utilization of eukaryotic genes by prokaryotes reveals the fundamental functional plasticity of many cellular systems, which is manifest in the compatibility of components evolved in phylogenetically very distant organisms. A striking example is topoisomerase IB, an enzyme that is ubiquitous in eukaryotes, but has not been detected in prokaryotes until the genome of the extreme radioresistant bacterium *D. radiodurans* has been sequenced (White et al., 1999). The major differences in the repertoires of the involved proteins and molecular mechanisms between the bacterial and eukaryotic repair systems (Aravind et al., 1999b; Eisen and Hanawalt, 1999) notwithstanding, this typical eukaryotic topoisomerase contributes to the UV-resistance of *D. radiodurans* and hence apparently does have a function in repair (Makarova et al., 2001).

TABLE 25.6 Examples of eukaryotic-prokaryotic transfer of functionally diverse genes

Gene function	Representative (gene name_species)	Occurrence in prokaryotes	Occurrence in eukaryotes	Probable eukaryotic source	Category of horizontal transfer event
ATP/ADP translocase	XF1738_Xf	Xf, *Chlamydia, Rickettsia*	Plants	Plant	Acquisition of a new gene
Hydrolase, possibly RNase	AF2335_Af	Af (archaeon)	Animals, *Leishmania*	Animal?	Acquisition of paralogue
Heme-binding protein	MTH115_Mth	Mth	Plants, animals	Plant?	Acquisition of a new gene
Glutamate-cysteine ligase	XF1428_Xf	Xf, *Zymomonas, Bradyrhizobium*	Plants	Plant	Xenologous gene displacement
Fructose-bisphosphate aldolase	XF0826_Xf	Xf, Cyanobacteria	Plants, animals	Plant	Non-orthologous gene displacement (of the typical bacterial FBA)
Sulfotransferase	BH3370_Bh	Bh	Animals	Animal	Acquisition of a new gene
Gamma-D-glutamyl-L-diamino acid endopeptidase I	ENP1_Bs	Bacillus	Animals	Animal	Acquisition of a new gene
General stress protein	GsiB_Bs	Bs only	Plants	Plant	Acquisition of a new gene
Superfamily I helicase	Cj0945c_Cj	Cj only	Fungi, animals, plants	Eukaryotic	Acquisition of paralogue
Guanylate cyclase	Rv1625c_Mtu	Mtu only	Animals, slime mold	Animal	Acquisition of paralogue
Purple acid phosphatase	Rv2577_Mtu	Mtu only	Plants, fungi, animals	Plant?	Acquisition of paralogue
α/β hydrolase (possible cutinase or related esterase)	Rv1984_Mtu, Rv3451_Mtu, Rv2301_Mtu, Rv1758_Mtu	Mycobacteria; multiple paralogues	Fungi	Fungal	Acquisition of a new gene
C-5 sterol desaturase	Slr0224_Ssp	*Synechocystis,* Mtu, *Vibrio*	Fungi, animals, plants	Eukaryotic	Acquisition of a new gene
Carnitine O-palmitoyl-transferase	MPN114_Mp	Mp only	Fungi, animals	Eukaryotic	Acquisition of a new gene
Arylsulfatase	B1498_Ec	Ec only	Animals	Animal	Acquisition of paralogue
Cation transport system component	ChaC_Ec	Ec only	Fungi, animals, plants	Eukaryotic	Acquisition of a new gene
Thiamine pyrophosphokinase	NMB2041_Nm	Nm only	Fungi, animals	Fungal?	Non-orthologous gene displacement?

TABLE 25.6 Contd

Gene function	Representative (gene name_species)	Occurrence in prokaryotes	Occurrence in eukaryotes	Probable eukaryotic source	Category of horizontal transfer event
Phospholipase A2	VC0178_Vc	Vc only	Plants, animals	Plant	Acquisition of paralogue
Sodium/phosphate cotransporter	VC0676_Vc	Vc only	Animals	Animal	Acquisition of a new gene
Topoisomerase IB	DR0690_Dr	Dr only	All eukaryotes	Eukaryotic	Acquisition of a new gene
RNA binding protein Ro	DR1262_Dz	Dr, *Streptomyces*	Animals	Animal	Acquisition of a new gene

THE OPPOSITE DIRECTION OF THE GENE FLOW: ACQUISITION OF PROKARYOTIC GENES BY EUKARYOTES

Gene transfer from bacteria to eukaryotes is the only direction of horizontal gene flow that is universally recognized as a major evolutionary phenomenon, in the context of relocation of genes from the genomes proto-organellar symbionts to the nuclear genome of the eukaryotic host (Gray, 1992, 1999; Martin and Herrmann, 1998; Rujan and Martin, 2001). Somewhat ironically, we are unaware of any definitive studies delineating the entire complement of nuclear genes that originate from proto-organelles. Many critical questions still await a comprehensive analysis, in particular, what fraction of the genes from proto-organellar genomes has been transferred to the nucleus, and whether the sets of the transferred genes are the same in different eukaryotes. That transfer of mitochondrial genes to the nuclear genome has not been a one-time affair, is evidenced by the structure of the mitochondrial genome from the early-branching eukaryotes *Reclinomonas americana*, which contains a substantially larger and more diverse repertoire of genes than the mitochondrial genomes of crown-group eukaryotes (Lang et al., 1997). Some recent estimates suggest that only a small fraction, perhaps as few as 40 genes, of the original gene complement of the α-proteobacterial ancestor of the mitochondria had been transferred to the nuclear

genome, whereas the rest perished during evolution of the endosymbiotic relationship (Karlberg et al., 2000). These estimates, however, are based on the unrealistic assumption that all genes acquired from the pro-mitochondrial endosymbiont actually function in the mitochondrion, and therefore are unlikely to give us the full picture of gene transfer. Similarly, a recent attempt on quantitative evaluation of the number of genes of chloroplast origin in the *Arabidopsis* genome produced only very crude estimates, from 1. 6% to 9.2% of the plant nuclear genes (Rujan and Martin, 2001).

Anecdotally, many eukaryotic genes show specific sequence similarity to genes from bacterial lineages other than Proteobacteria or Cyanobacteria. These observations, although not as systematic as might be desirable, have led to the ingenious proposal that, at least at the early stages of their evolution, eukaryotes have been entering numerous, sometimes transient, symbiotic or commensal relationships with various prokaryotes and, in the process, probably have captured many prokaryotic genes. This trend has been memorably encapsulated in the "You are what you eat" maxim of Ford Doolittle (Doolittle, 1998). As with the proto-organellar gene acquisition, this probable ancient influx of prokaryotic genes into the eukaryotic genomes have not been systematically investigated and there is currently no estimate of their extent, not even a crude one. Furthermore, major questions about this class of horizontal gene transfers remained unanswered, in particular whether there had been preferred prokaryotic sources that supplied genes to eukaryotes, other than the progenitors of organelles.

A distinct type of apparent horizontal acquisition of prokaryotic genes by eukaryotes has recently come into focus, namely the recent or, more accurately, lineage-specific gene transfer. A breakdown of the taxonomic distribution of sequence conservation for all proteins encoded in the nematode *Caenorhabditis elegans* genome revealed 185 proteins (~1% of the entire predicted proteome) that showed a significantly greater similarity to bacterial homologues (using the same criteria as described above for detection of probable horizontal transfer between domains and between bacterial lineages) than to homologues from other eukaryotes (except, possibly, for other nematodes). Most of these proteins are widespread among bacteria and, accordingly, it has been proposed that the majority, if not all, of the respective genes have been acquired by the nematodes after their divergence from the common ancestor with other animal lineages (at least fungi, flowering plants, insects and vertebrates, for which representative complete genome sequences are available). Notably, these apparently recently acquired genes had the same density of introns as the genes that are conserved among eukaryotes (and may be shared with archaea and bacteria) indicating that the transferred genes are rapidly saturated with introns (Wolf et al., 2000).

A similar analysis that has been recently performed as part of the annotation of the draft human genome sequence showed that the products of 223 of the ~30 000 human genes were significantly more similar to bacterial homologues than to homologues from other eukaryotes (International Human Genome Sequence Consortium, 2001). A detailed case-by-case examination of these genes suggested that 113 of them (~0.3% of the predicted proteome) were common among bacteria and represented probable cases of vertebrate-specific horizontal transfer (Table 25.7). These cases included all categories of horizontal transfer delineated above, namely acquisition of genes that appear to be completely new to eukaryotes, acquisition of paralogues of conserved eukaryotic genes, and xenologous gene displacement (Table 25.7).

There was no apparent strongly preferred bacterial source of these horizontally transferred genes, which favors the view that these are indeed relatively recent acquisitions, rather than part of the proto-organellar heritage that has been selectively retained by vertebrates. It may be more difficult to rule out that these are ancient acquired genes that have been lost by other eukaryotic lineages or have rapidly diverged in them. However, this explanation is decidedly less parsimonious than vertebrate-specific horizontal transfer. Functionally, the recently acquired genes are a mixed group, but the considerable number of enzymes implicated in the metabolism of different kinds of xenobiotics, particularly oxidoreductases, is notable (Table 25.7).

Interestingly, some of the genes that made the list of apparent vertebrate-specific horizontal transfers are involved in vertebrate-specific physiological functions. The most dramatic example of apparent recruitment of a bacterial gene for such a function is monoamine oxidase (MAOs), an enzyme involved in metabolism of an essential neuromediator and a drug target in psychiatric disorders (Abell and Kwan, 2000). Figure 25.2 (see color plates) shows a phylogenetic tree of monoamine oxidases and related L-amino acid oxidases. The clustering of vertebrate MAOs with those from several diverse bacteria, namely *Mycobacterium*, *Synechocystis* and *Micrococcus*, is evident and is strongly supported by bootstrap analysis (Figure 25.3). This tree essentially proves that evolution of MAO involved multiple horizontal gene transfer events. The direction of transfer, however, is a more complicated issue. Given the presence of a distinct MAO subfamily in vertebrates and a set of bacterial species that represent different bacterial lineages, xenologous displacement of the ancestral animal gene in vertebrates by a horizontally transferred bacterial genes seems to be the most parsimonious explanation. However, the alternative scenarios, namely multiple horizontal transfers from vertebrates to bacteria or a single transfer with subsequent horizontal dissemination among bacteria, cannot be ruled out. An extensive sampling of bacteria and eukaryotes for the presence of the given subfamily of MAOs will allow a more accurate

TABLE 25.7 Probable vertebrate-specific horizontal acquisitions of bacterial genes[a]

Protein function	Representation in vertebrates	Representation in bacteria	Likely type of horizontal transfer event
Formiminotransferase cyclodeaminase	Human, pig, rat, chicken	Scattered, best hit in *Thermotoga*	Acquisition of a new gene
Predicted methyltransferase	Human, mouse	Many bacteria, best hit in Streptomyces	Acquisition of a paralogue
Fatty acid synthase component, thioesterase	Human, rat	Most bacteria, best hits in Gram-positive	Acquisition of a paralogue
Membrane protein of cholinergic synaptic vesicles, quinone oxidoreductase	Many mammals, electric ray	Many bacteria, best hit in Mycobacterium	Acquisition of a paralogue
Na/glucose (and other solutes) cotransporter	Many mammals	Many bacteria, best hit in *Vibrio*	Acquisition of a paralogue
neuraminidase	Many mammals	Gram-positive bacteria, best hit in Streptomyces	Acquisition of a paralogue
UDP-*N*-acetylglucosamine-2-epimerase/*N*-acyl-mannosamine kinase	Human, rat, mouse	Many bacteria, best hit in *Legionella*	Acquisition of a paralogue
Methionine sulfoxide reductase	Human, cow	Most bacteria, best hit in Synechocystis	Acquisition of a new gene
Hyaluronan synthase	Many mammals, chicken, *Xenopus*, *Danio*	Many bacteria, best hit in *Rhizobium*	Xenologous displacement?
Predicted arylsulfatase	Human, mouse	Many bacteria and archaea, best hit in *Synechocystis*	Acquisition of a paralogue
Predicted oxidoreductase fused to multitrans-membrane domain (bacterial homologues have oxidoreducatse domain only)	Human, mouse, pig	Actinomycetes and several other bacteria, best hit in *Streptomyces*	Acquisition of a paralogue
Betaine-homocysteine methyltransferase	Human, rat, pig	Gram-positive bacteria, *Thermotoga*	Acquisition of a new gene
Acetyl-CoA synthetase	Human, mouse, rat	Most bacteria, best hit in *Bacillus*	Acquisition of a paralogue
Acetyl-CoA synthetase	Human, mouse, rat	Most bacteria, best hit in *Methanothermobacter*	Acquisition of a paralogue
Soluble adenylate cyclase	Human, rat	Many bacteria, best hit in *Anabaena*	Acquisition of a paralogue
Glycine amidotransferase	Human, rat, pig, *Xenopus*	Actinomycetes, *Pseudomonas* (best hit)	Acquisition of a new gene
Predicted methyltransferase	Human, mouse	Many bacteria, best hit in *Streptomyces*	Acquisition of a paralogue

TABLE 25.7 Contd

Protein function	Representation in vertebrates	Representation in bacteria	Likely type of horizontal transfer event
CMP-*N*-acetylneuraminic acid synthase	Human, mouse	Many bacteria, best hit in *Aquifex*	Acquisition of a new gene
Aspartoacylase	Human, mouse, rat, cow	Cyanobacteria, γ proteobacteria, best hit in *Synechocystis*	Acquisition of a new gene
Unknown	Human, zebrafish	γ proteobacteria	Acquisition of a new gene
Thiopurine *S*-methyl-transferase	Human, mouse, rat	γ proteobacteria	Acquisition of a new gene
Monoamine oxidase	Many mammals, rainbow trout	Many bacteria, best hits in *Mycobacterium, Micrococcus, Synechocystis*	Xenologous displacement?
L-Amino acid oxidase	Human, mouse, rattlesnake, scomber	Gram-positive bacteria, Cyanobacteria, best hit in *Bacillus*	Xenologous displacement?
Quinone reductase	Human, rat, mouse, *Oryzias* (fish)	γ proteobacteria, Gram-positive bacteria, best hit in *Pseudomonas*	Acquisition of a paralogue
Aldo-keto reductase	Human, rat	Many bacteria, best hit in *Thermotoga*	Acquisition of a paralogue
Predicted epoxide hydrolase	Human, mouse, zebrafish, fugu	Many bacteria, best hit in *Pseudomonas*	Acquisition of a paralogue
Phosphoglycerate mutase 1	Human, rat, mouse	Many bacteria, best hit in *E. coli*	Acquisition of a paralogue
Acyl-CoA-thioesterase	Human, rat	Most bacteria, best hit in *Bacillus*	Acquisition of a paralogue
Glucose-6-phosphate transporter	Human, mouse, rat	Many bacteria, best hit in *Chlamydia*	Acquisition of a paralogue
Interphotoreceptor retinol-binding protein (tail-specific protease in bacteria)	Many mammals, chicken, *Xenopus*, zebrafish etc	Many bacteria, best hit in *Vibrio*	Acquisition of a new gene
Arylamine *N*-acetyl-transferase	Many mammals, chicken	Many bacteria, best hit in *Mycobacterium*	Acquisition of a paralogue
Methylated-DNA-protein-cysteine methyltransferase	Human, mouse, hamster	Many bacteria, best hit in *Bacillus*	Xenologous displacement
N-Methylpurine DNA glycosylase	Human, mouse, rat	Many bacteria, best hit in *Streptomyces*	Acquisition of a new gene
Hydroxyindole *O*-methyltransferase	Human, cow, chicken	Actinomycetes	Acquisition of a paralogue
Galactokinase	Human, mouse	Most bacteria, best hit in *E. coli*	Acquisition of a paralogue

TABLE 25.7 Contd

Protein function	Representation in vertebrates	Representation in bacteria	Likely type of horizontal transfer event
Glucokinase regulator, predicted sugar phosphate isomerase	Human, rat, *Xenopus*	Many bacteria, best hit in *Vibrio*	Acquisition of a new gene
8-Oxo-dGTPase	Human, mouse, rat	Most bacteria, best hit in *E. coli*	Acquisition of a paralogue
Mitochondrial DNA polymerase, regulatory subunit (glycyl-tRNA synthetase in bacteria)	Human, mouse, chicken	Many bacteria, best hit in *Thermus*	Acquisition of a paralogue
Unknown	Human, mouse	All bacteria, best hit in *Thermotoga*	Acquisition of a new gene
3-Ketoacyl-acyl-carrier protein reductase	Human	Many bacteria, best hit in *Thermotoga*	Acquisition of a paralogue
Ribosomal protein L33 homologue	Human	All bacteria, best hit in *Deinococcus*	Acquisition of a new gene
Acyl-CoA dehydrogenase	Human	Most bacteria, best hit in *Pseudomonas*	Acquisition of a paralogue
Unknown	Human	γ proteobacteria	Acquisition of a new gene
Glutamine synthetase	Human	Most bacteria, best hit in *Pseudomonas*	Acquisition of a new gene
Surfactant protein B, adenylosuccinate lyase	Human	Most bacteria, best hits in Gram-positive bacteria	Acquisition of a paralogue
Neutral sphyngomyelinase	Human	Most bacteria, best hit in *Bacillus*	Acquisition of a new gene
Oxygen-independent coproporphyrinogen oxidase	Human	Most bacteria, best hit in *Rickettsia*	Acquisition of a new gene
Predicted oxidoreductase	Human	*Synechocystis, Pseudomonas*	Acquisition of a paralogue
Predicted a/b superfamily hydrolase	Human	Many bacteria, best hit in *Rickettsia*	Acquisition of a paralogue
ADP-ribosylglycohydrolase	Human	Many bacteria, best hit in *Aquifex*	Acquisition of a new gene
In bacteria, protein involved in conjugal DNA transfer and secretion (VirB10)	Human	Many bacteria, best hit in *Agrobacterium*	Acquisition of a new gene
Ribosomal protein S6 – glutamic acid ligase	Human	Most bacteria, best hit in *Haemophilus*	Acquisition of a paralogue
Thiol-disulfide isomerase	Human	γ proteobacteria	Acquisition of a new gene

TABLE 25.7 Contd

Protein function	Representation in vertebrates	Representation in bacteria	Likely type of horizontal transfer event
In bacteria, protein involved in conjugal DNA transfer (VirB4)	Human	Many bacteria, best hit in *Rickettsia*	Acquisition of a new gene
Homoserine dehydrogenase	Human	Most bacteria, bect hit in *Corynebacterium*	Acquisition of a paralogue
4-Oxalomesaconate hydratase	Human	*Cyanobacteria, Actinomycetes, Sphingomonas* (best hit in *Synechocystis*)	Acquisition of a new gene
Predicted helicase	Human	γ proteobacteria	Acquisition of a paralogue
Predicted phosphatase (homologue of histone H2A macro domain)	Human	Many bacteria, best hit in *Thermotoga*	Acquisition of a paralogue
Cyclopropane-fatty-acid-phospholipid synthase	Human	Most bacteria, best hit in *E. coli*	Acquisition of a paralogue
Predicted membrane transporter	Human	Many bacteria, best hit in *Synechocystis*	Acquisition of a paralogue
Acyl-CoA thioester hydrolase	Human	Most bacteria, best hit in *Chlamydophila*	Acquisition of a paralogue
dCTP deaminase	Human	Many bacteria and archaea, best hit in *Chlamydia*	Acquisition of a new gene
Ribonucleoside-diphosphate reductase	Human	Most bacteria, best hit in *Salmonella*	Acquisition of a paralogue
N-Acetylneuraminate lyase	Human	Many bacteria, best hit in *Vibrio*	Acquisition of a paralogue
Triacylglycerol lipase	Human	Many bacteria, best hit in *Pseudomonas*	Acquisition of a paralogue
Predicted membrane symporter	Human	Many bacteria, best hit in *Synechocystis*	Acquisition of a paralogue
Anaerobic ribonucleide-diphosphate reductase	Human	γ proteobacteria	Acquisition of a paralogue
Predicted metal-dependent hydrolase	Human	All bacteria, best hit in *Borrelia*	Acquisition of a new gene
Orotate phosphoribosyl transferase	Human	Most bacteria, best hit in *Bacillus*	Acquisition of a paralogue
Quinone oxidoreductase	Human	Gram-positive bacteria, best hit in *Bacillus*	Acquisition of a paralogue
Selenophosphate synthase	Human	Many bacteria, best hit in *E. coli*	Acquisition of a new gene
GTPase involved in ferrous iron transport	Human	Many bacteria, best hit in *Xylella*	Acquisition of a paralogue

TABLE 25.7 Contd

Protein function	Representation in vertebrates	Representation in bacteria	Likely type of horizontal transfer event
ATP-dependent nuclease subunit, helicase	Human	Gram-positive bacteria	Acquisition of a paralogue
Acetylornithine aminotransferase	Human	Most bacteria, best hit in *Bacillus*	Acquisition of a paralogue
Polyphosphate kinase	Human	Many bacteria, best hit in *Bacillus*	Acquisition of a new gene
Mg-chelatase ATPase subunit	Human	Most bacteria, best hit in *Synechocystis*	Acquisition of a paralogue
Predicted membrane transporter	Human	γ proteobacteria	Acquisition of a new gene
Non-ribosomal peptide synthetase	Human	Many bacteria, best hit in *Bacillus*	Acquisition of a new gene
Beta-xylosidase	Human	Proteobacteria, Gram-positive bacteria, best hit in *E. coli*	Acquisition of a new gene
Di-tripeptide ABC transporter, transmembrane subunit	Human	Most bacteria, best hits in γ proteobacteria	Acquisition of a paralogue
tRNA (guanine N1)-methyltransferase	Human	Most bacteria, best hit in *Neisseria*	Acquisition of a new gene
Thymidine phosphorylase	Human	Many bacteria, best hit in *Deinococcus*	Acquisition of a new gene
Melanoma-associated antigen; DNA repair ATPase RecN in bacteria	Human	Most bacteria, best hit in *Neisseria*	Acquisition of a paralogue
NAD-dependent DNA ligase	Human	All bacteria, best hit in *Bacillus*	Acquisition of a new gene
Predicted phosphatase; homologues of histone H2A macro domain	Human	Many bacteria, best hit in *Thermotoga*	Acquisition of a paralogue
Myelin transcription factor; plasmid replication protein in bacteria	Human	Many bacterial plasmids; best hit in *Leuconostoc*	Acquisition of a new gene

[a]Adopted, with modifications and additions, from International Human Genome Sequencing Consortium (2001); identifications and sequences of all listed proteins are available as a Supplement to International Human Genome Sequencing Consortium (2001).

evaluation of the relative likelihoods of each evolutionary alternative.

CONCLUSIONS

Estimates based on the analysis of taxon-specific best hits suggest a significant amount of horizontal gene transfer between the three primary domains of life, and particularly between major bacterial lineages. For selected genomes that have been analyzed in detail, this conclusion is largely supported by phylogenetic analysis. Probable horizontal transfer events could be classified into three distinct categories:

(1) acquisition of a new (for the recipient lineage) gene;
(2) acquisition of a paralogue of a gene(s) pre-existing in the recipient lineage, which may followed by non-orthologous gene;
(3) acquisition of phylogenetically distant orthologues followed by xenologous gene displacement.

Acquisition of eukaryotic genes by bacteria, particularly parasites and symbionts and, to a lesser extent, by archaea is one of the important directions of horizontal gene flow. Apparent horizontal gene transfer events of this class was detected in different functional categories of genes, although it is particularly characteristic of certain categories such as, for example, aminoacyl-tRNA synthetases and signal transduction systems.

The opposite direction of horizontal gene flow, from bacteria to eukaryotes, seems to include three distinct classes of events:

(1) relocation of genes from proto-organellar genomes to the nuclear genome;
(2) ancient horizontal transfer events that probably occurred according to the "You are what you eat" principle;
(3) relatively recent, lineage-specific acquisitions of bacterial genes.

The first two classes of acquisition of bacterial genes by eukaryotes have been studied largely anecdotally and, as a consequence, there are no robust estimates of the number of genes acquired via these routes. A comprehensive quantitative and qualitative examination of these genes is an important challenge for the near future. Lineage-specific acquisition of bacterial genes by vertebrates has been recently investigated in some detail A conservative estimate suggests that at least 100 human genes (~0.3% of all genes) have been acquired after the divergence of vertebrates from the common ancestor with other major eukaryotic lineages; similar data were obtained for the nematode *C. elegans*.

From an evolutionary–theoretical viewpoint, horizontal gene transfer, particularly between eukaryotes and bacteria, emphasizes the remarkable unity of molecular-biological mechanisms in all life forms which is manifest in the compatibility of eukaryotic and bacterial proteins that have evolved in their distinct milieus for hundreds of millions of years prior to the gene transfer. Co-adaptation of proteins might impede (but apparently not preclude altogether) horizontal transfer of certain types of genes, for example, those coding for ribosomal proteins of RNA polymerase subunits, components of many functional systems appear to be fully compatible. To be fixed in the bacterial population and retained in the long term, an acquired eukaryotic gene must confer selective advantage on the recipient bacteria. In particular, in cases of xenologous gene displacement, the acquired version of a gene should immediately become superior, from the standpoint of selection, to the resident version of the recipient species. In one case, that of eukaryotic isoleucyl-tRNA synthetase displacing the original gene in some bacteria, this has been clearly explained by antibiotic resistance that is conferred by the transferred gene. This observation may have general implications for xenologous gene displacement. Acquisition of a new gene, for example, the ATP/ADP translocases in the intracellular parasitic bacteria, *Chlamydia* and *Rickettsia*, sometimes also leads to an obvious selective advantage. Most often, however, comparative genomics can only identify the genes that probably have entered a particular genome by horizontal transfer; understanding the biological significance of horizontal gene transfer will require direct experimental analysis of these genes.

NOTE ADDED IN PROOF

After the publication of the paper on the draft human genome sequence, in which the possibility of horizontal transfer of bacterial genes into the vertebrate lineage was proposed (International Human Genome Sequencing Consortium, 2001), three independent groups reported that orthologues of the candidate horizontally acquired genes were detected in *Dictyostelium discoideum* through searches of dedicated slime mold genome databases (Andersson et al., 2001; Ponting, 2001; Roelofs and Van Haastert, 2001; Salzberg et al., 2001; Stanhope et al., 2001). Moreover, the slime mold orthologues typically clustered with the vertebrate proteins in phylogenetic trees. From these observations, the authors concluded that the original findings had to be explained primarily, if not exclusively, by multiple events of lineage-specific gene loss, whereas acquisition of bacterial genes by vertebrates (and, by implication, other multicellular eukaryotes) should have been extremely rare. These results emphasize caution that is needed in the interpretation of indications for horizontal gene transfer in the absence of a representative set of genomes from the major lineages of eukaryotes. At the same time, we believe that the data can be explained through several alternative scenarios, including multiple horizontal transfers and a combination of a relatively early horizontal transfer (e.g. to the common ancestor of *Dictyostelium* and animals, if a phylogeny including such an ancestor is accepted) with subsequent multiple gene losses. Undoubtedly, the whole issue needs to be revisited periodically, once a more representative collection of sequenced genomes becomes available.

REFERENCES

Abell, C.W. and Kwan, S.W. (2000) Molecular characterization of monoamine oxidases A and B. *Prog. Nucleic Acid Res. Mol. Biol.* **65**: 129–156.

Altschul, S.F., Madden, T.L., Schaffer, A.A. et al. (1997) Gapped BLAST and PSI-BLAST: a new generation of protein database search programs. *Nucleic Acids Res.* **25**: 3389–402.

Andersson, J.O., Doolittle, W.F. and Nesbo, C.L. (2001) Genomics. Are there bugs in our genome? *Science* **292**: 1848–1850.

Andersson, S.G., Zomorodipour, A., Andersson, J.O. et al. (1998) The genome sequence of *Rickettsia prowazekii* and the origin of mitochondria. *Nature* **396**: 133–140.

Aravind, L. and Koonin, E.V. (1999) The fukutin protein family--predicted enzymes modifying cell-surface molecules. *Curr. Biol.* **9**: R836–R837.

Aravind, L., Tatusov, R.L., Wolf, Y.I. et al. (1998) Evidence for massive gene exchange between archaeal and bacterial hyperthermophiles. *Trends Genet.* **14**: 442–444.

Aravind, L., Tatusov, R.L., Wolf, Y.I. et al. (1999a) Reply. Archaeal and bacterial hyperthermophiles: horizontal gene exchange or common ancestry? *Trends Genet.* **15**: 299–300.

Aravind, L., Walker, D.R. and Koonin, E.V. (1999b) Conserved domains in DNA repair proteins and evolution of repair systems. *Nucleic Acids Res.* **27**: 1223–1242.

Avery, O.T., MacLeod, C.M. and McCarty, M. (1944) Studies on the chemical nature of the substance inducing transformation of pneumococcal types. Inductions of transformation by a desoxyribonucleic acid fraction isolated from pneumococcus type III. *J. Exp. Med.* **149**: 297–326.

Barry, C.E.D., Brickman, T.J. and Hackstadt, T. (1993) Hc1-mediated effects on DNA structure: a potential regulator of chlamydial development. *Mol. Microbiol.* **9**: 273–283.

Braithwaite, D.K. and Ito, J. (1993) Compilation, alignment, and phylogenetic relationships of DNA polymerases. *Nucleic Acids Res.* **21**: 787–802.

Brochier, C., Philippe, H. and Moreira, D. (2000) The evolutionary history of ribosomal protein RpS14: horizontal gene transfer at the heart of the ribosome. *Trends Genet.* **16**: 529–533.

Brown, J.K. (1994) Bootstrap hypothesis tests for evolutionary trees and other dendrograms. *Proc. Natl Acad. Sci. USA* **91**: 12293–12297.

Brown, J.R. and Doolittle, W.F. (1997) Archaea and the prokaryote-to-eukaryote transition. *Microbiol. Mol. Biol. Rev.* **61**: 456–502.

Brown, J.R., Zhang, J. and Hodgson, J.E. (1998) A bacterial antibiotic resistance gene with eukaryotic origins. *Curr. Biol.* **8**: R365–367.

Cairns, B.R., Lorch, Y., Li, Y. et al. (1996) RSC, an essential, abundant chromatin-remodeling complex. *Cell* **87**: 1249–1260.

Cole, S.T., Eiglmeier, K., Parkhill, J. et al. (2001) Massive gene decay in the leprosy bacillus. *Nature* **409**(6823): 1007–1011.

Curnow, A.W., Hong, K., Yuan, R. et al. (1997) Glu-tRNAGln amidotransferase: a novel heterotrimeric enzyme required for correct decoding of glutamine codons during translation. *Proc. Natl Acad. Sci. USA* **94**: 11819–11826.

Dandekar, T., Snel, B., Huynen, M. and Bork, P. (1998) Conservation of gene order: a fingerprint of proteins that physically interact. *Trends Biochem. Sci.* **23**: 324–328.

Davie, J.R. and Spencer, V.A. (2000) Signal transduction pathways and the modification of chromatin structure. *Prog. Nucleic Acid Res. Mol. Biol.* **65**: 299–340.

Doolittle, R.F. and Handy, J. (1998) Evolutionary anomalies among the aminoacyl-tRNA synthetases. *Curr. Opin. Genet. Dev.* **8**: 630–636.

Doolittle, W.F. (1998) You are what you eat: a gene transfer ratchet could account for bacterial genes in eukaryotic nuclear genomes. *Trends Genet.* **14**: 307–311.

Doolittle, W.F. (1999a) Lateral genomics. *Trends Cell Biol.* **9**: M5–8.

Doolittle, W.F. (1999b) Phylogenetic classification and the universal tree. *Science* **284**: 2124–2129.

Doolittle, W.F. (2000) Uprooting the tree of life. *Sci. Am.* **282**: 90–95.

Doolittle, W.F. and Logsdon, J.M., Jr. (1998) Archaeal genomics: do archaea have a mixed heritage? *Curr. Biol.* **8**: R209–211.

Dunny, G.M. and Leonard, B.A. (1997) Cell-cell communication in Gram-positive bacteria. *Annu. Rev. Microbiol.* **51**: 527–564.

Efron, B., Halloran, E. and Holmes, S. (1996) Bootstrap confidence levels for phylogenetic trees [corrected and republished article originally printed in *Proc. Natl Acad. Sci. USA* 1996 Jul 9; 93(14): 7085–90]. *Proc. Natl Acad. Sci. USA* **93**: 13429–13434.

Eisen, J.A. and Hanawalt, P.C. (1999) A phylogenomic study of DNA repair genes, proteins, and processes. *Mutat. Res.* **435**: 171–213.

Felsenstein, J. (1996) Inferring phylogenies from protein sequences by parsimony, distance, and likelihood methods. *Methods Enzymol.* **266**: 418–427.

Fitch, W.M. and Margoliash, E. (1967) Construction of phylogenetic trees. *Science* **155**: 279–284.

Gaasterland, T. and Ragan, M.A. (1998) Microbial genescapes: phyletic and functional patterns of ORF distribution among prokaryotes. *Microb. Comp. Genomics* **3**: 199–217.

Galperin, M.Y. and Koonin, E.V. (2000) Who's your neighbor? New computational approaches for functional genomics. *Nat. Biotechnol.* **18**: 609–613.

Garcia-Vallve, S., Romeu, A. and Palau, J. (2000) Horizontal gene transfer in bacterial and archaeal complete genomes. *Genome Res.* **10**: 1719–1725.

Gogarten, J.P. (1994) Which is the most conserved group of proteins? Homology-orthology, paralogy, xenology, and the fusion of independent lineages. *J. Mol. Evol.* **39**: 541–543.

Golding, G.B. and Gupta, R.S. (1995) Protein-based phylogenies support a chimeric origin for the eukaryotic genome. *Mol. Biol. Evol.* **12**: 1–6.

Gould, S.J. (1997) The exaptive excellence of spandrels as a term and prototype. *Proc. Natl Acad. Sci. USA* **94**: 10750–10755.

Grantham, R., Gautier, C. and Gouy, M. (1980a) Codon frequencies in 119 individual genes confirm consistent choices of degenerate bases according to genome type. *Nucleic Acids Res.* **8**: 1893–912.

Grantham, R., Gautier, C., Gouy, M. et al. (1980b) Codon catalog usage and the genome hypothesis. *Nucleic Acids Res.* **8**: r49–r62.

Gray, M.W. (1992) The endosymbiont hypothesis revisited. *Int. Rev. Cytol.* **141**: 233–357.

Gray, M.W. (1999) Evolution of organellar genomes. *Curr. Opin. Genet. Dev.* **9**: 678–687.

Groisman, E.A. and Ochman, H. (1997) How Salmonella became a pathogen. *Trends Microbiol.* **5**: 343–349.

Gupta, R.S. and Golding, G.B. (1996) The origin of the eukaryotic cell [see comments]. *Trends Biochem. Sci.* **21**: 166–171.

Handy, J. and Doolittle, R.F. (1999) An attempt to pinpoint the phylogenetic introduction of glutaminyl-tRNA synthetase among bacteria. *J. Mol. Evol.* **49**: 709–715.

Hartl, D.L., Dykhuizen, D.E. and Berg, D.E. (1984) Accessory DNAs in the bacterial gene pool: playground for coevolution. *Ciba Found. Symp.* **102**: 233–245.

Hasegawa, M., Kishino, H. and Saitou, N. (1991) On the maximum likelihood method in molecular phylogenetics. *J. Mol. Evol.* **32**: 443–445.

Hilario, E. and Gogarten, J.P. (1993) Horizontal transfer of ATPase genes – the tree of life becomes a net of life. *Biosystems* **31**: 111–119.

Hopwood, D.A., Chater, K.F. and Bibb, M.J. (1995) Genetics of antibiotic production in Streptomyces coelicolor A3(2), a model streptomycete. *Biotechnology* **28**: 65–102.

Hunter, T. (2000) Signaling – 2000 and beyond. *Cell* **100**: 113–127.

Ibba, M., Curnow, A.W. and Soll, D. (1997) Aminoacyl-tRNA synthesis: divergent routes to a common goal. *Trends Biochem. Sci.* **22**: 39–42.

International Human Genome Sequencing Consortium (2001) Initial sequencing and analysis of the human genome. *Nature* **409**: 860–921.

Itoh, T., Takemoto, K., Mori, H. and Gojobori, T. (1999) Evolutionary instability of operon structures disclosed by sequence comparisons of complete microbial genomes. *Mol. Biol. Evol.* **16**: 332–346.

Jain, R., Rivera, M.C. and Lake, J.A. (1999) Horizontal gene transfer among genomes: the complexity hypothesis. *Proc. Natl Acad. Sci. USA* **96**: 3801–3806.

Jeanthon, C. (2000) Molecular ecology of hydrothermal vent microbial communities. *Antonie van Leeuwenhoek* **77**: 117–133.

Kaneko, T. and Tabata, S. (1997) Complete genome structure of the unicellular cyanobacterium Synechocystis sp. PCC6803. *Plant Cell Physiol.* **38**: 1171–1176.

Karlberg, O., Canback, B., Kurland, C.G. and Andersson, S.G. (2000) The dual origin of the yeast mitochondrial proteome. *Yeast* **17**: 170–187.

Knopf, C.W. (1998) Evolution of viral DNA-dependent DNA polymerases. *Virus Genes* **16**: 47–58.

Kobayashi, I., Nobusato, A., Kobayashi-Takahashi, N. and Uchiyama, I. (1999) Shaping the genome – restriction-modification systems as mobile genetic elements. *Curr. Opin. Genet. Dev.* **9**: 649–656.

Koonin, E.V., Mushegian, A.R. and Bork, P. (1996) Non-orthologous gene displacement. *Trends Genet.* **12**: 334–336.

Koonin, E.V., Mushegian, A.R., Galperin, M.Y. and Walker, D.R. (1997) Comparison of archaeal and bacterial genomes: computer analysis of protein sequences predicts novel functions and suggests a chimeric origin for the archaea. *Mol. Microbiol.* **25**: 619–637.

Koonin, E.V., Aravind, L. and Kondrashov, A.S. (2000) The impact of comparative genomics on our understanding of evolution. *Cell* **101**: 573–576.

Kyrpides, N.C. and Olsen, G.J. (1999) Archaeal and bacterial hyperthermophiles: horizontal gene exchange or common ancestry? *Trends Genet.* **15**: 298–299.

Lang, B.F., Burger, G., O'Kelly, C.J. et al. (1997) An ancestral mitochondrial DNA resembling a eubacterial genome in miniature. *Nature* **387**: 493–497.

Lang, B.F., Gray, M.W. and Burger, G. (1999) Mitochondrial genome evolution and the origin of eukaryotes. *Annu. Rev. Genet.* **33**: 351–397.

Lawrence, J. (1999) Selfish operons: the evolutionary impact of gene clustering in prokaryotes and eukaryotes. *Curr. Opin. Genet. Dev.* **9**: 642–648.

Lawrence, J.G. (1997) Selfish operons and speciation by gene transfer. *Trends Microbiol.* **5**: 355–359.

Lawrence, J.G. and Ochman, H. (1997) Amelioration of bacterial genomes: rates of change and exchange. *J. Mol. Evol.* **44**: 383–397.

Lengeler, K.B., Davidson, R.C., D'Souza, C. et al. (2000) Signal transduction cascades regulating fungal development and virulence. *Microbiol. Mol. Biol. Rev.* **64**: 746–785.

Leonard, C.J., Aravind, L. and Koonin, E.V. (1998) Novel families of putative protein kinases in bacteria and archaea: evolution of the "eukaryotic" protein kinase superfamily. *Genome Res.* **8**: 1038–1047.

Li, J., Lee, G., Van Doren, S.R. and Walker, J.C. (2000) The FHA domain mediates phosphoprotein interactions. *J. Cell Sci.* **113**: 4143–4149.

Li, S.J. and Hochstrasser, M. (1999) A new protease required for cell-cycle progression in yeast. *Nature* **398**: 246–251.

Li, S.J. and Hochstrasser, M. (2000) The yeast ULP2 (SMT4) gene encodes a novel protease specific for the ubiquitin-like Smt3 protein. *Mol. Cell Biol.* **20**: 2367–2377.

Logsdon, J.M. and Faguy, D.M. (1999) Thermotoga heats up lateral gene transfer. *Curr. Biol.* **9**: R747–751.

Lorenz, M.G. and Wackernagel, W. (1994) Bacterial gene transfer by natural genetic transformation in the environment. *Microbiol. Rev.* **58**: 563–602.

Makarova, K.S., Aravind, L., Galperin, M.Y. et al. (1999) Comparative genomics of the Archaea (Euryarchaeota): evolution of conserved protein families, the stable core, and the variable shell. *Genome Res.* **9**: 608–628.

Makarova, K.S., Aravind, L. and Koonin, E.V. (2000) A novel superfamily of predicted cysteine proteases from eukaryotes, viruses and *Chlamydia pneumoniae*. *Trends Biochem. Sci.* **25**: 50–52.

Makarova, K.S., Aravind, L., Wolf, Y.I. et al. (2001) Genome of the extremely radiation-resistant bacterium *Deinococcus radiodurans* viewed from the perspective of comparative genomics. *Mol. Biol. Microbiol. Rev.* **65**: 44–79.

Martin, W. (1999) Mosaic bacterial chromosomes: a challenge en route to a tree of genomes. *Bioessays* **21**: 99–104.

Martin, W. and Herrmann, R.G. (1998) Gene transfer from organelles to the nucleus: how much, what happens, and Why? *Plant Physiol.* **118**: 9–17.

Medigue, C., Rouxel, T., Vigier, P. et al. (1991) Evidence for horizontal gene transfer in *Escherichia coli* speciation. *J. Mol. Biol.* **222**: 851–856.

Moreira, D. (2000) Multiple independent horizontal transfers of informational genes from bacteria to plasmids and phages: implications for the origin of bacterial replication machinery. *Mol. Microbiol.* **35**: 1–5.

Moreira, D. and Philippe, H. (2000) Molecular phylogeny: pitfalls and progress. *Int. Microbiol.* **3**: 9–16.

Moyer, C.L., Dobbs, F.C. and Karl, D.M. (1995) Phylogenetic diversity of the bacterial community from a microbial mat at an active, hydrothermal vent system, Loihi Seamount, Hawaii. *Appl. Environ. Microbiol.* **61**: 1555–1562.

Mrazek, J. and Karlin, S. (1999) Detecting alien genes in bacterial genomes. *Ann. NY. Acad. Sci.* **870**: 314–29.

Mushegian, A.R. and Koonin, E.V. (1996a) Gene order is not conserved in bacterial evolution. *Trends Genet.* **12**: 289–290.

Mushegian, A.R. and Koonin, E.V. (1996b) A minimal gene set for cellular life derived by comparison of complete bacterial genomes [see comments]. *Proc. Natl Acad. Sci. USA* **93**: 10268–10273.

Naito, T., Kusano, K. and Kobayashi, I. (1995) Selfish behavior of restriction-modification systems. *Science* **267**: 897–899.

Nelson, K.E., Clayton, R.A., Gill, S.R. et al. (1999) Evidence for lateral gene transfer between Archaea and bacteria from genome sequence of *Thermotoga maritima*. *Nature* **399**: 323–329.

Ochman, H., Lawrence, J.G. and Groisman, E.A. (2000) Lateral gene transfer and the nature of bacterial innovation. *Nature* **405**: 299–304.

Olendzenski, L., Liu, L., Zhaxybayeva, O. et al. (2000) Horizontal transfer of archaeal genes into the Deinococcaceae: detection by molecular and computer-based approaches. *J. Mol. Evol.* **51**: 587–599.

Pace, N.R. (1997) A molecular view of microbial diversity and the biosphere. *Science* **276**: 734–740.

Paerl, H.W. (1996) Microscale physiological and ecological studies of aquatic cyanobacteria: macroscale implications. *Microsc. Res. Tech.* **33**: 47–72.

Patterson, C. (1988) Homology in classical and molecular biology. *Mol. Biol. Evol.* **5**: 603–625.

Pennisi, E. (1998) Genome data shake tree of life. *Science* **280**: 672–674.

Pennisi, E. (1999) Is it time to uproot the tree of life? *Science* **284**: 1305–1307.

Ponting, C.P. (2001) Plagiarized bacterial genes in the human book of life. *Trends Genet.* **17**: 235–237

Ponting, C.P., Aravind, L., Schultz, J. et al. (1999) Eukaryotic signalling domain homologues in archaea and bacteria. Ancient ancestry and horizontal gene transfer. *J. Mol. Biol.* **289**: 729–745.

Rea, S., Eisenhaber, F., O'Carroll, D. et al. (2000) Regulation of chromatin structure by site-specific histone H3 methyltransferases [see comments]. *Nature* **406**: 593–599.

Risatti, J.B., Capman, W.C. and Stahl, D.A. (1994) Community structure of a microbial mat: the phylogenetic dimension. *Proc. Natl Acad. Sci. USA* **91**: 10173–10177.

Roelofs, J. and Van Haastert, P.J. (2001) Genes lost during evolution. *Nature* **411**: 1013–1014.

Rujan, T. and Martin, W. (2001) How many genes in *Arabidopsis* come from cyanobacteria? An estimate from 386 protein phylogenies. *Trends Genet.* **17**: 113–120.

Salzberg, S.L., White, O., Peterson, J. and Eisen J.A. (2001) Microbial genes in the human genome: lateral transfer or gene loss? *Science* **292**: 1903–1906.

Schultz, J., Copley, R.R., Doerks, T. et al. (2000) SMART: a web-based tool for the study of genetically mobile domains. *Nucleic Acids Res.* **28**: 231–234.

Sicheritz-Ponten, T. and Andersson, S.G. (2001) A phylogenomic approach to microbial evolution. *Nucleic Acids Res.* **29**: 545–552.

Sicot, F.X., Mesnage, M., Masselot, M. et al. (2000) Molecular adaptation to an extreme environment: origin of the thermal stability of the pompeii worm collagen. *J. Mol. Biol.* **302**(4): 811–820.

Smith, M.W., Feng, D.F. and Doolittle, R.F. (1992) Evolution by acquisition: the case for horizontal gene transfers. *Trends Biochem. Sci.* **17**: 489–493.

Sprague, G.F., Jr. (1991) Genetic exchange between kingdoms. *Curr. Opin. Genet. Dev.* **1**: 530–533.

Stanhope, M.J., Lupas, A., Italia, M.J. et al. (2001) Phylogenetic analyses do not support horizontal gene transfers from bacteria to vertebrates. *Nature* **411**: 940–944.

Stephens, R.S., Kalman, S., Lammel, C. et al. (1998) Genome sequence of an obligate intracellular pathogen of humans: *Chlamydia trachomatis*. *Science* **282**: 754–759.

Subramanian, G., Koonin, E.V. and Aravind, L. (2000) Comparative genome analysis of the pathogenic spirochetes *Borrelia burgdorferi* and *Treponema pallidum*. *Infect. Immun.* **68**: 1633–1648.

Sundstrom, L. (1998) The potential of integrons and connected programmed rearrangements for mediating horizontal gene transfer. *APMIS Suppl.* **84**: 37–42.

Syvanen, M. (1985) Cross-species gene transfer; implications for a new theory of evolution. *J. Theor. Biol.* **112**: 333–343.

Syvanen, M. (1994) Horizontal gene transfer: evidence and possible consequences. *Annu. Rev. Genet.* **28**: 237–261.

Tannock, G.W. (2000) The intestinal microflora: potentially fertile ground for microbial physiologists. *Adv. Microb. Physiol.* **42**: 25–46.

Tatusov, R.L., Mushegian, A.R., Bork, P. et al. (1996) Metabolism and evolution of *Haemophilus influenzae* deduced from a whole- genome comparison with *Escherichia coli*. *Curr. Biol.* **6**: 279–291.

Tatusov, R.L., Koonin, E.V. and Lipman, D.J. (1997) A genomic perspective on protein families. *Science* **278**: 631–637.

Tatusov, R.L., Natale, D.A., Garkavtsev, I.V. et al. (2001) The COG database: new developments in phylogenetic classification of proteins from complete genomes. *Nucleic Acids Res.* **29**: 22–28.

Tatusova, T.A., Karsch-Mizrachi, I. and Ostell, J.A. (1999) Complete genomes in WWW Entrez: data representation and analysis. *Bioinformatics* **15**: 536–543.

Thompson, J.D., Higgins, D.G. and Gibson, T.J. (1994) CLUSTAL W: improving the sensitivity of progressive multiple sequence alignment through sequence weighting, position-specific gap penalties and weight matrix choice. *Nucleic Acids Res.* **22**: 4673–4680.

Walker, D.R. and Koonin, E.V. (1997) SEALS: a system for easy analysis of lots of sequences. *Ismb* **5**: 333–339.

Ward, D.M., Ferris, M.J., Nold, S.C. and Bateson, M.M. (1998) A natural view of microbial biodiversity within hot spring cyanobacterial mat communities. *Microbiol. Mol. Biol. Rev.* **62**: 1353–1370.

White, O., Eisen, J.A., Heidelberg, J.F. et al. (1999) Genome sequence of the radioresistant bacterium *Deinococcus radiodurans* R1. *Science* **286**: 1571–1577.

Winkler, H.H. and Neuhaus, H.E. (1999) Non-mitochondrial ATP transport. *Trends Biochem. Sci.* **24**: 64–68.

Woese, C.R., Kandler, O. and Wheelis, M.L. (1990) Towards a natural system of organisms: proposal for the domains Archaea, Bacteria, and Eucarya. *Proc. Natl Acad. Sci. USA* **87**: 4576–4579.

Woese, C.R., Olsen, G.J., Ibba, M. and Soll, D. (2000) Aminoacyl-tRNA synthetases, the genetic code, and the evolutionary process. *Microbiol. Mol. Biol. Rev.* **64**: 202–236.

Wolf, Y.I., Aravind, L., Grishin, N.V. and Koonin, E.V. (1999) Evolution of aminoacyl-tRNA synthetases – analysis of unique domain architectures and phylogenetic trees reveals a complex history of horizontal gene transfer events. *Genome Res.* **9**: 689–710.

Wolf, Y.I., Kondrashov, F.A. and Koonin, E.V. (2000) No footprints of primordial introns in a eukaryotic genome. *Trends Genet.* **16**: 333–334.

Wolf, Y.I., Rogozin, I.B., Kondrashov, A.S. and Koonin, E.V. (2001) Genome alignment, evolution of prokaryotic genome organization and prediction of gene function using genomic context. *Genome Res.* **11**: 356–372

Worning, P., Jensen, L.J., Nelson, K.E., Brunak, S. and Ussery, D.W. (2000) Structural analysis of DNA sequence: evidence for lateral gene transfer in *Thermotoga maritima*. *Nucleic Acids Res.* **28**: 706–709.

CHAPTER 26

Horizontal Gene Transfer and the Universal Tree of Life

James R. Brown, Michael J. Italia, Christophe Douady and Michael J. Stanhope

In the late 1980s to mid 1990s, universal trees were derived which convincingly showed that (1) Archaea and eukaryotes were sister groups with the rooting in the branch leading to Bacteria and (2) all three domains were unique and exclusive or mono/holophyletic. The sequencing of entire genomes from a multitude of organisms held the promise of refining the details of the universal tree. Instead, whole genome sequences seem to "uproot" the universal tree. New genomic sequence data from species belonging to all three domains but in particular single cell organisms, strongly suggest the past occurrence of extensive, and sometimes complex, horizontal gene transfers (HGT). Herein we review the development of the concept of the universal tree and highlight current anomalies and controversies.

THE DIVERSITY OF LIFE

The field of molecular evolution has revolutionized our view of the living world. Many diverse evolutionary questions have been convincingly resolved through the phylogenetic analysis of nucleotide and protein sequence data. However, the definite resolution of the universal tree, which, as its name suggests, is the evolutionary relationships among the major groups of all living organisms, remains one of the greatest challenges to molecular evolutionary biology.

The tacit supposition of the universal tree is that all living things are related genetically, however distant. This basic assumption is supported by the universal existence of the genetic code as well as the occurrence of fundamentally similar modes of DNA replication, transcription and translation in all cellular organisms and viruses. Although the concept of the universal tree has long been recognized, the lack of morphological traits to link distantly related groups of organisms, such as animals, plants, fungi and bacteria, made rational phylogenetic reconstruction impossible.

Molecular approaches are particularly important in the determination of the evolution of microbes (both prokaryotes and single cell eukaryotes) where distinguishable morphological traits are rare. In the late 1970s, Woese, Fox and co-workers tackled prokaryotic systematics by digesting *in vivo* labeled 16S ribosomal RNA (rRNA) using T1 ribonuclease to produce oligonucleotide "words" which could be analyzed using dendograms. The results were startling. For decades, biologists had been comfortable with the simple, fundamental subdivision of living organisms into prokaryotes and eukaryotes. Since evolution probably progresses from simple to more complex entities, it was assumed that eukaryotes evolved from prokaryotes. However, dendograms that included rRNA sequences of some unusual methanogenic "bacteria" revealed a significant offshoot (Fox et al., 1977). So deep was the split

Copyright © 2002 by Academic Press.
All rights of reproduction in any form reserved.

in the prokaryotes that Woese and Fox (1977) named the methanogens and their relatives "archaebacteria," which relayed their distinctness from the true bacteria or "eubacteria" as well as met contemporary preconceptions that these organisms might have thrived in the environmental conditions of a younger Earth.

Subsequently Woese et al. (1990) formally proposed the replacement of the bipartite prokaryote–eukaryote division with a new tripartite scheme based on three urkingdoms or domains: the Bacteria (formally eubacteria), Archaea (formally archaebacteria) and Eucarya (eukaryotes, still the more often used name). The rationale behind this revision came from a growing body of biochemical, genomic and phylogenetic evidence which, when viewed collectively, suggested that the Archaea were unique from eukaryotes and the Bacteria. The discovery of the Archaea was a significant event, which added a new dimension to the construction of the universal tree since the evolutionary relationships between three rather than two major subdivisions had to be considered (Figure 26.1).

ROOTING THE UNIVERSAL TREE

A central issue surrounding the universal tree is the location of its root. In other words, which of the three domains evolved first from the cenancestor (Fitch and Upper, 1987)? Assuming that each domain emerged from a single

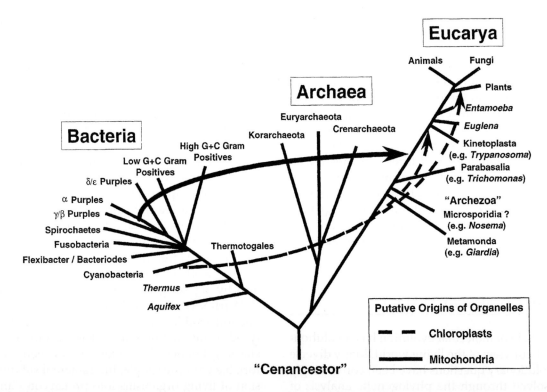

FIGURE 26.1 Schematic drawing of a universal rRNA tree showing the relative positions of evolutionary pivotal groups in the domains Bacteria, Archaea, and Eucarya. The location of the root (the cenancestor) corresponds with that proposed by reciprocally rooted gene phylogenies (Gogarten et al., 1989; Iwabe et al., 1989). The question mark beside Microsporidia denotes recent suggestions that it might branch higher in the eukaryotic portion of the tree (Keeling and MacFadden, 1998). (Branch lengths have no meaning in this tree.) The putative bacterial origins of eukaryotic organelles, mitochondria and chloroplast, are indicated. The timing of the introduction of mitochondria into eukaryotes, prior or post emergence of amitochondrial protists also known as the "Archezoa," is still an open question.

ancestor, there are three possible answers (depicted respectively in Figure 26.2A, B and C): (1) Bacteria diverged first from a lineage producing Archaea and eukaryotes (called here the AK tree) or (2) Eukaryotes diverged from a fully prokaryotic clade, consisting of Bacteria and Archaea (the AB tree) or (3) the Archaea diverged first such that eukaryotes and Bacteria are sister groups (the BK tree). Is there any convincing genetic or biochemical evidence which would give clues about the relatedness of the domains?

The distinction between Bacteria and eukaryotes had long been recognized, but the Archaea are novel territory. As a group, the Archaea are remarkable organisms, having successfully adapted to life in the harshest of environments thereby earning the title "extremophiles." The Archaea also occur in "mesophilic" or less extreme, widely dispersed environments such as oceans, lakes, soil, and even animal guts (Stein and Simon, 1996). Therefore, in terms of species diversity and carbon biomass, the Archaea are far from insignificant. Prior to whole genomic sequence data, considerable knowledge had accumulated on the comparative biochemistry, and cellular and molecular biology of the Archaea (Danson, 1993; Kates et al., 1993; Keeling et al., 1994; Brown and Doolittle, 1997). Briefly, the Archaea have a few unique biochemical and genetic traits, such as isopranyl ether-linked lipids, the absence of acyl ester lipids and fatty acid synthetase, modified tRNA molecules, and a split gene coding one of the RNA polymerase subunits. Also, Archaea have a variety of metabolic regimes, which deviate from known metabolic pathways of Bacteria and eukaryotes, and are not simply particular environmental adaptations. A recent genome comparison found 351 archaea-specific "phylogenetic footprints" or combinations of genes uniquely shared by two or more archaeal species but not found in either bacteria or eukaryotes (Graham et al., 2000). However, functional catalytic orthologues might actually be lower since both hyperthermophilic Archaea and Bacteria tend to have more split genes than mesophilic species (Snel et al., 1999).

Archaea and Bacteria are united in the "realm of prokaryotes" by generally similar cell sizes, the

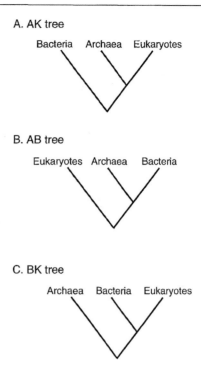

FIGURE 26.2 Three possibilities for the rooting of the universal tree. (A) Bacteria diverged first from a lineage producing Archaea and eukaryotes (called here the AK tree); (B) Eukaryotes diverged from a fully prokaryotic clade, consisting of Bacteria and Archaea (the AB tree) or; (C) the Archaea diverged first such that eukaryotes and Bacteria are sister groups (the BK tree).

absence of nuclear membrane and organelles, operon organization of many genes, and the presence of a large circular chromosome, occasionally accompanied by one or more smaller circular DNA plasmids. Archaea and eukaryotes share significant components of DNA replication, transcription, and translation, which are either not found in Bacteria or replaced by an evolutionary unrelated (analogous) enzyme. Many DNA replication and repair proteins are homologous between Archaea and eukaryotes but completely absent in Bacteria (Edgell and Doolittle, 1997). While the archaebacterium, *Pyrococcus abyssi*, was recently shown to have a bacteria-like origin of DNA replication, most of its replication enzymes are eukaryote-like (Kelman, 2000; Myllykallio et al., 2000). Archaeal DNA scaffolding proteins are remarkable similar to eukaryotic histones (Reeve et al., 1997).

Eukaryotes and the Archaea have similar transcriptional proteins, such as multi-subunit DNA-dependent RNA polymerases (Langer et al., 1995), as well as sharing translation initiation factors not found in the Bacteria (Olsen and Woese, 1997; Kyrpides and Woese, 1998). On the basis of genetic components, the Archaea seem to occupy a middle ground between the Bacteria and eukaryotes. Thus rooting the universal tree on the basis of cellular synapomorphies would be somewhat ambiguous.

Phylogenetic approaches to determining the rooting of any group of species or sequences requires the inclusion of an outgroup in the analysis. As an example, a tree of all animal species can only be rooted using homologous sequences from non-animal groups such as plants, fungi or bacteria. Obviously, suitable outgroups do not exist for the universal tree of all living things (minerals are not useful, in this regard!). However, in 1989, two separate research teams cleverly solved the rooting of the universal tree. Both groups used a similar approach but with different molecular sequence data. They reasoned that although no organism can be an outgroup, it is possible to root the universal tree utilizing ancient duplicated genes (Figure 26.3). Iwabe and co-workers (1989) aligned amino acids from five conserved regions shared by the elongation factors (EF) Tu/1α and EF-G/2 genes of the archaebacterium, *Methanococcus vannielii*, and several species of Bacteria and eukaryotes. According to protein sequence similarity and neighbor-joining trees, both EF-1α and EF-2 genes of Archaea were more similar to their respective eukaryotic, rather than bacterial, homologues. Gogarten and co-workers (1989) developed composite trees based on duplicated ATPase genes where the V-type A and V-type B occurs in Archaea and eukaryotes and the F_0F_1-type β and F_0F_1-type α occurs in Bacteria. In agreement with the elongation factor rooting, reciprocally rooted ATPase subunits trees also showed that the Archaea, represented by a sole species *Sulfolobus acidocaldarius*, were closer to eukaryotes than to Bacteria.

Subsequent paralogous protein rootings based on aminoacyl-tRNA synthetases (Brown and Doolittle, 1995; Brown et al., 1997) and carbamoylphosphate synthetase (Lawson et al.,

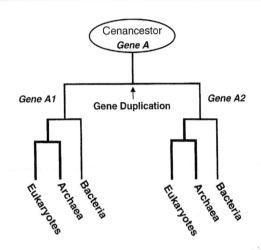

FIGURE 26.3 Conceptual rooting of the universal tree using paralogous genes. Gene A was duplicated in the cenancestor such that all extant organisms have paralogous copies, gene A1 and gene A2. The two genes have sufficiently similar for phylogenetic signal, allowing for the construction of reciprocally rooted trees. The topology depicted here, Archaea and eukaryotes as sister groups with the root in Bacteria, has been consistently supported by paralogous trees.

1996) confirmed the rooting in the Bacteria and linking Archaea and eukaryotes as sister groups. If one argues that enzymes involved in DNA replication, transcription and translation, so-called "information" genes, are core to living things then the evolutionary scenario suggested by paralogous gene trees seems particularly reasonable. Thus emerged the "canonical" universal tree with the Archaea and eukaryotes being sister groups, the rooting in the Bacteria, and all three domains as monophyletic groups.

IS THE ROOTING CORRECT?

Despite the convincing results from paralogous gene trees, the rooting of the universal tree has not been without controversy. Phylogenetic analyses using alternative methods and expanded datasets raised questions about the rooting of the universal tree and the monophyly of the Archaea (Lake, 1988; Rivera and Lake, 1992; Baldauf et al., 1996). Philippe and co-workers (Forterre and Phillipe, 1999; Lopez et al., 1999; Philippe and Forterre, 1999) have maintained that phylogenies of distantly related species are strongly affected by saturation for multiple mutations at nearly every amino acid position in a protein. Unequal mutation rates between different species can lead to long branch attraction effects. However, a greater issue is the degree to which horizontal gene transfers (HGT) between the domains of life have affected the actual viability of constructing a definitive universal tree.

The increasing size of sequence databases adds to the species richness of universal trees. Perhaps not surprisingly, nature provides plenty of exceptions to the canonical universal tree paradigm. Among the first reported involved ATPase subunits. Archaeal V-type ATPases were reported for two bacterial species, *Thermus thermophilus* (Tsutsumi et al., 1991) and *Enterococcus hiraea* (Kakinuma et al., 1991), and an F1- ATPase β subunit gene was found in the Archaea, *Methanosacrina barkeri* (Sumi et al., 1992). Forterre et al. (1993) suggested that the ATPase subunit gene family had not been fully determined, and that other paralogous family members might be discovered. Hilario and Gogarten (1993) believed that the observed distribution of ATPase subunits was the result of a few, rare lateral gene transfers. In support of their view, broader surveys have failed to detect archaeal V-type ATPases in other bacterial species (Gogarten et al., 1996).

Furthermore, there are many examples of single gene trees, although not uniquely rooted, which have irreconcilable topologies with the canonical universal tree (Smith et al., 1992). Gupta and Golding (1995) examined the phylogenetic trees for 24 universally conserved proteins and found only nine had the AK tree topology. Eight trees showed the Bacteria to be paraphyletic with Gram-positive bacteria clustering with the Archaea and "Gram-negative" bacteria grouping with eukaryotes. Subsequent works by Golding and Gupta (1996) and Roger and Brown (1996) slightly modified the number of proteins in each of the categories but, nonetheless, several gene trees clearly conflicted the universal tree.

Feng et al. (1997) attempted to estimate the timing of ancient divergence events by calibrating the rates of amino acid substitution in 64 different proteins to the vertebrate fossil record (a necessity since the unicellular fossil record is poor for species identification). In their analysis, 38 proteins represented species from all three domains hence suited for universal tree construction. (They were also interested in the timing of multicellular organism evolution, thus not all of their data included bacterial or archaeal species.) AK, AB and BK topologies occurred in the phylogenies for 8, 11, and 15 proteins, respectively. Relying on the AK tree, time estimates were made of 3200–3800 Myr ago since the split of Archaea and Bacteria and 2300 Myr ago since the emergence of eukaryotes.

A broader survey involving phylogenetic analysis of 66 proteins was completed concurrent with the release of whole archaeal genome sequences (Brown and Doolittle, 1997). In that study, 34 protein trees had the AK topology, 21 protein trees depicted the AB topology and 11 protein trees showed the BK topology with the remaining trees being indeterminate. Recent genome sequence data have eroded the AK list with several new examples of horizontal gene transfer between eukaryotes and bacteria, such as isoleucyl-tRNA synthetase (Brown et al., 1998).

GENOMES AND GENE TRANSFERS

Phylogeny and the detection of HGT

Genomes from over 60 different organisms have now been completely sequenced and the progress of nearly twice that number can be followed at NCBI Genome Database (http://www.ncbi.nlm.nih.gov/PMGifs/Genomes/bact.html) or TIGR Microbial Database (http://www.tigr.org/tdb/mdb/mdb.html). The abundance of sequence data has resulted in a more, not less, confusing picture of the universal tree. Comparative analysis of archaea, bacterial and eukaryotic genomes suggest that relatively few genes are entirely conserved across all genomes. Important biochemical pathways appear to be incomplete in some organisms. In some instances, a protein has been discovered to take over the catalytic role of an unrelated protein, so-called non-orthologous gene replacement (Koonin et al., 1996).

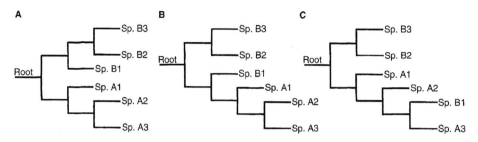

FIGURE 26.4 Detection of horizontal gene transfer (HGT) from phylogeny. Hypothetical protein trees for three bacterial species (B1–B3) and three archaeal species (A1–A3). (A) The true rooting of the tree postulates a split between the Archaea and Bacteria, which results in two monophyletic clusters. (B) The lowest branching bacterial species, B1, has a more rapid rate of amino acid substitution than other bacterial species which results in phylogenetic as well as homology searching software implicating the Archaea as the closest relatives. At first glance, the tree would suggest HGT between B1 and Archaea. However, the clustering of species is actually the result of the new position of the root, which was shifted by the "attraction" of the B1 branch to the outgroup, the Archaea. (C) Strong phylogenetic evidence for HGT is the "imbedding" of a distantly related ingroup species within the outgroup and away from the root. In this example, bacterial species B1 clusters with a more derived archaeal species, A3, which strongly suggests HGT occurred from the Archaea to Bacteria.

Comparative studies suggest that HGT has extensively occurred between Archaea and Bacteria. Koonin et al. (1997) found that 44% of the gene products of the archaebacterium, *Methanococcus jannaschii* were more similar to bacterial over eukaryotic proteins while only 13% were more like eukaryotic proteins. Nelson et al. (1999) reported that 24% of proteins from *Thermotoga maritima*, a thermophilic bacterium with a deep rRNA tree lineage, were most similar to archaeal proteins. On the basis of such analyses, HGT has been suggested to be a formidable influence in genome evolution and a perplexing factor in universal tree reconstruction (Hilario and Gogarten, 1993; Martin and Müller, 1998; Doolittle, 1999a; Ochman et al., 2000).

However, reports of homology without supporting phylogeny need to be carefully scrutinized. In particular, deep branching species are susceptible to arbitrary clustering with species belonging to the outgroup, as might be the case of *T. maritima* with respect to the Archaea and eukaryotes (Logsdon and Faguy, 1999). Differences in evolutionary rate can lead to an incorrect rooting which will result in mistaken occurrences of HGT between the deep branching species and the outgroup (Figure 26.4A and 4B). Conversely, protein trees where an ingroup species is solidly imbedded within an outgroup clade provide strong evidence for HGT (Figure 26.4C).

Consequently, phylogenetic analysis suggests that *T. maritima* received far fewer genes from the Archaea than first estimated by homology searches (Logsdon and Faguy, 1999; Nelson et al., 1999). Phylogenetic analyses of putative archaeal-like proteins from *Deinococcus radiodurans*, a bacterium which branches nearly as deeply as *Thermotoga*, suggests that HGT involving either Archaea or eukaryotes occurred for fewer than 1% of its total genome complement (Olendzenski et al., 2000).

Construction of universal trees based on the distribution of genes is a logical use of genomic sequence data in evolutionary biology. Although different approaches have been used, most studies arrive at tree topologies remarkably similar to the rRNA tree (Huyanen et al., 1999; Snel et al., 1999; Lin and Gerstein, 2000). However, W.F. Doolittle (1999b) argued that while genome inventories might tell us about the similarities in the contents of genomes from different species, the nuisances of HGT involving universally conserved genes are lost. In this respect, phylogenetic methods are the key to documenting HGT events.

Gene transfer and cellular function

While inventories of genome homologues have critical caveats with respect to interpretation, phylogenetic evidence for HGT remains particularly compelling and supported. Gene trees suggest that HGT can affect many types of genes including those coding proteins essential to cell viability. The aminoacyl-tRNA synthetases are an example. In protein synthesis, aminoacyl-tRNA synthetases are responsible for the attachment of a tRNA to its cognate amino acid. As such, a specific aminoacyl-tRNA synthetase exists for each amino acid. On the basis of structure, biochemical activity, and sequence similarity, aminoacyl-tRNA synthetases can be divided nearly equally into two evolutionary distinct protein families known as class I and II. Despite their critical role in protein synthesis and ancient origins (without them interpretation of the genetic code would be impossible), aminoacyl-tRNA synthetases have been extensively shuttled between genomes (Brown, 1998; Wolf et al., 1999; Woese et al., 2000). Phylogenetic trees suggest that class I isoleucyl-tRNA synthetases may have been transferred from an early eukaryote to bacteria as a specific adaptation to resist a natural antibiotic compound (Brown et al., 1998). Orthologous genes to eukaryotic glutaminyl-tRNA synthetase occur in many proteobacteria and *D. radiodurans* but not in other Bacteria or the Archaea (Brown and Doolittle, 1999).

Even more surprising are the lysyl-tRNA synthetases, which not only cross species boundaries but also exist in both class I and II families. Until recently, all examples of lysyl-tRNA synthetase were class II type enzymes. However, novel class I type lysyl-tRNA synthetases were discovered in the Archaea and subsequently in bacterial spirochetes, both of which lack the more typical class II isoforms (Ibba et al., 1997). While class I and II lysyl-tRNA synthetases do not share any sequence or structural similarity, class I lysyl-tRNA synthetases in spirochetes and the Archaea are clearly related. A less dramatic, but nonetheless, clear example of HGT involves phenylalanyl-tRNA synthetase where phylogenies of both the α and β subunits clearly show that genes were transferred from the Archaea to spirochetes (Teichmann and Mitchison, 1999; Woese et al., 2000; Brown, in press). The mechanism or timing of this HGT event remains unclear. Spirochetes are human

parasites, while the Archaea are not known to be pathogenic thus ruling out any recent genetic exchange.

Lake and colleagues suggest that, based on their propensity for HGT, genes can be divided into two categories, informational and operational genes (Rivera et al., 1998). Informational genes, which include the central components of DNA replication, transcription and translation, are less likely to be transferred between genomes than operational genes involved with cell metabolism. The fact that informational gene products, at least qualitatively, have more complex interactions might restrict their opportunities for genetic exchange and fixation (Jain et al., 1999). Additional support for this view is the conservation of genomic context for translation-associated genes in bacteria (Lathe et al., 2000).

Conversely, metabolic genes can have surprising species distributions. An example is the genes involved in the mevalonate pathway for isoprenoid biosynthesis. The mevalonate pathway has been well studied in humans because 3-hydroxy-3-methylglutaryl coenzyme A (HMG-CoA) reductase is the target for the statin class of cholesterol-lowering drugs. The mevalonate pathway was long believed to be specific to eukaryotes since bacteria utilize an evolutionary unrelated metabolic route for isoprenoid biosynthesis, the pyruvate/GAP pathway. Recent genome surveys and phylogenetic analyses have found not only HMGCoA reductase but also four other enzymes in the mevalonate pathway in Gram-positive coccal bacteria (Doolittle and Logsdon, 1998; Boucher and Doolittle, 2000; Wilding et al., 2000). The genes are also found in the Archaea and the bacterial spirochete, *Borrelia burgoderi*. However, the mevalonate pathway is absent from the completely sequenced genome of a closely related spirochete, *Treponema pallidum* and the Archaea have likely substituted an analogous protein for one of the enzymes in the pathway (Smit and Mushegian, 2000). In the Bacteria with the mevalonate pathway, the genes encoding component enzymes are tightly linked suggesting that all genes might have been transferred simultaneously. Entire pathways might be more highly selected for gene transfer over individual genes, which, in turn, would select for the organization of genes encoding common pathway into tightly linked operons (Lawrence and Roth, 1996; Lawrence, 1997).

REVISITING EVOLUTIONARY TRENDS IN CONSERVED PROTEINS

By definition, a universally conserved protein is one that occurs in every organism. The increasing number of completely sequenced genomes invariably leads to the shrinking of this inventory since exceptions are found. For example, the 70 kilodalton heat shock protein (HSP70), once thought to be highly conserved from the perspective of both amino acid substitutions and species distribution, is absent from several species of Archaea (Gribaldo et al., 1999). In many cases, the biochemical function is still required but an evolutionary unrelated enzyme serves as the catalyst. Arguably, only those proteins found in all completely sequenced genomes are conserved enough to provide a continuous picture of all lineages back to the last universal common ancestor.

Previously, one of us (JRB) reviewed the universal tree from the perspective of 66 different protein phylogenetic trees (Brown and Doolittle, 1997). Since then, the majority of those proteins have proven not to be universally conserved. Proteins were either selectively lost or never acquired by particular groups. As an example, animals, unlike fungi, Archaea or Bacteria, lack the enzymes needed to synthesize aromatic amino acids. Taking a very conservative and preliminary approach, an updated view of the universal tree is provided below.

Methods and caveats

Sequences were collected from 45 species, for which complete or nearly complete genomes were available as of December 1, 2000 (Table 26.1). A previously described relational bacterial genomic database was used to identify orthologous proteins (Brown and Warren, 1998). The core of this database is an array of protein-by-protein sequence similarity scores (smallest sum probabilities) across multiple

genomes as calculated by BLASTP v.2.0 (Altschul et al., 1997). Using the smallest genome, *Mycoplasma genitalium*, as the "driver query," individual homologous protein datasets were assembled based on two criteria. First, there was significant amino acid sequence homology to the driver protein (P[N] ≤ 1.0e-05). Secondly, the protein was found in all sampled genomes. A total of 65 proteins were initially collected. Proteins from any genome, which occurred in more than one protein family, were detected using an automated procedure then assigned to the correct family after visually evaluating annotation, multiple sequence alignments and phylogenies. Redundant protein families were discarded giving a final set of 23 orthologous proteins found across 45 species.

Phylogenetic analyses followed standard methods. Individual homologous protein datasets were initially aligned using the program CLUSTALW v.1.7 (Thompson et al., 1994) with default settings. Multiple sequence alignments were further refined manually using the program SEQLAB of the GCG v10.0 software package (Genetics Computer Group, Madison WI, USA). Regions with residues that could not be unambiguously aligned or that contained insertions or deletions were removed from the alignments. Phylogenetic trees were constructed by distance neighbor-joining (NJ) and maximum parsimony (MP) methods. NJ trees were based on pairwise distances between amino acid sequences using the programs NEIGHBOR and PROTDIST (Dayhoff option) of the PHYLIP 3.57c package (Felsenstein, 1993). The programs SEQBOOT and CONSENSE were used to estimate the confidence limits of branching points from 100 bootstrap replications. MP analyses were done using the software package PAUP* (Swofford, 1999) and 100 bootstrap replicates were also performed.

Results and discussion: phylogenetic signal in conserved proteins

The determined number of conserved proteins, 23 families, is fewer than previous genomic studies (Table 26.2). For example, the Clusters of Orthologous Groups of proteins (COGs)

TABLE 26.1 Genomes searched for conserved proteins

Domain	Kingdom	Species
Bacteria	γ Proteobacteria (purple bacteria)	*Actinobacillus actinomycetemcomitans*
		Escherichia coli
		Haemophilus influenzae
		Pseudomonas aeruginosa
		Vibrio cholerae
		Xyella fastidiosa
	β Proteobacteria	*Neisseria meningitidis*
	α Proteobacteria	*Rickettsia prowazekii*
	ε Proteobacteria	*Helicobacter pylori* J99
		Helicobacter pylori 26695
		Campylobacter jejuni
	Bacillus/Clostridium (low G + C Gram positives)	*Bacillus subtilis*
		Clostridium acetobutylicum
		Enterococcus faecalis
		Staphylococcus aureus
		Streptococcus pneumoniae
		Streptococcus pyogenes
		Mycoplasma genitalium
		Mycoplasma pneumoniae
		Ureaplasma urealyticum
	Actinobacteria (high G + C Gram positives)	*Mycobacterium leprae*
		Mycobacterium tuberculosis
		Streptomyces coelicolor
	Chlamydiales	*Chlamydia pneumoniae*
		Chlamydia trachomatis
	Spirochetes	*Borrelia burgdorferi*
		Treponema pallidum
	Cyanobacteria	*Synechocystis* sp. PCC6803
	Green sulfur	*Chlorobium tepidum*
	CFB[a]	*Porphyromonas gingivalis*
	Thermus/Deinococcus	*Deinococcus radiodurans*
	Thermotogales	*Thermotoga maritima*
	Aquificaceae	*Aquifex aeolicus*
Archaea	Euryarchaeota	*Archaeoglobus fulgidus*
		Methanobacterium thermoautotrophicus
		Methanococcus jannaschii
		Pyrococcus abyssi
		Pyrococcus furiosus
		Pyrococcus horikoshii
		Thermoplasma acidophilum
	Crenarchaeota	*Aeropyrum pernix*
Eukaryotes	Metazoa	*Caenorhabditis elegans*
		Drosophila melanogaster
		Homo sapiens
	Fungi	*Saccharomyces cerevisiae*

[a]Cytophaga–Flexibacter–Bacteroides

TABLE 26.2 Universal conserved proteins and support for domain of life monophyly

	Cell role	Protein	Alignment length[a]	Domain monophyly[b]		
				Archaea	Bacteria	Eucarya
1	Translation	Alanyl-tRNA synthetase	502	100	–	100
2		Aspartyl-tRNA synthetase	249	–	100	100
3		Glutamyl-tRNA synthetase	199	50 (–)	100	100
4		Histidyl-tRNA synthetase	166	–	–	100 (93)
5		Isoleucyl-tRNA synthetase	552	–	–	–
6		Leucyl-tRNA synthetase	358	–	100	100
7		Methionyl-tRNA synthetase	306	–	–	99
8		Phenylalanyl-tRNA synthetase β subunit	177	–	–	100
9		Threonyl-tRNA synthetase	305	–	– (34)	100
10		Valyl-tRNA synthetase	538	–	–	100
11		Initiation factor 2	337	–	100	100
12		Elongation factor G	536	64 (87)	100	100
13		Elongation factor Tu	340	– (42)	100	100
14		Ribosomal protein L2	192	46 (–)	100	100
15		Ribosomal protein S5	131	46 (19)	100	100 (99)
16		Ribosomal protein S8	118	–	100	100
17		Ribosomal protein S11	110	–	100	100
18		Aminopeptidase P	95	–	–	–
19	Transcription	DNA-directed RNA polymerase β chain	537	99 (78)	100	100
20	DNA	DNA topoisomerase I	236	–	100	100
21	Replication	DNA polymerase III subunit	194	46 (49)	100	100 (95)
22	Metabolism	Signal recognition particle protein	298	71 (39)	100	100
23		rRNA dimethylase	126	–	–	100 (98)

[a]Number of amino acids.
[b]Percent occurrence of monophyletic nodes in neighbor-joining and maximum parsimony (in parentheses where values differ from neighbor-joining consense tree) analysis of 100 bootstrap replicated datasets. Dash indicates that the nodes were not monophyletic.

database (http://www.ncbi.nlm.nih.gov/COG/xindex.html) reports for 34 complete genomes, a total of 78 completely conserved proteins (Tatusov et al., 1997). However, the present survey includes more genomes (45 species), some of which were incomplete at the time of analysis. In addition, if the collection of organisms is diverse, then the likelihood increases that particular lineages, by chance, have lost a particular pathway or replaced components with analogous proteins. Arguably, the list presented in Table 26.2 represents the most highly conserved or widely found proteins known to date.

The majority of the 23 proteins are involved in translation, which reflects the general conservation of the process of protein synthesis across the domains of life. Ten proteins are aminoacyl-tRNA synthetases, which is half the potential complement of 20 synthetases. However, several aminoacyl-tRNA synthetases are known to be sporadically missing from archaeal and bacterial genomes while others are so highly divergent as to fall below the preset threshold of detection of homology (Brown, 1998; Woese et al., 2000). Other conserved proteins in translation include four ribosomal proteins, two elongation factors, an initiation factor and aminopeptidase P. For transcription, only RNA polymerase β chain was found to be conserved in all genomes while two proteins involved in DNA replication, DNA topoisomerase I and DNA polymerase III subunit, were recovered. Only two "non-information" pathway enzymes were found to be highly conserved, signal recognition particle protein component and rRNA dimethylase.

After editing to remove gaps and ambiguously aligned regions, alignment lengths ranged from 95 to 522 amino acids (Table 26.2). Phylogenetic trees for the conserved proteins are shown in Figures 26.5 to 26.27. A total of five proteins depicted all three domains as monophyletic according to both maximum parsimony and neighbor-joining methods (Table 26.2). Only two protein trees, elongation factor G and DNA-directed RNA polymerse β chain, supported the monophyly of all three domains by both phylogenetic methods with bootstrap values over 604. However, 14 protein trees support the monophyly of at least two domains.

Inspection of individual protein trees reveals that Bacteria and eukaryotes are usually monophyletic while the Archaea are often paraphyletic. In such cases, eukaryotes and Archaea together form a monophyletic clade. Paraphyly of the Archaea occurs when certain archaeal species branch at the base of eukaryotes. Such instances are not conclusive evidence for HGT since unequal rates of evolution among the Archaea could also cause some species to be attracted to the outgroup, eukaryotes (Figure 26.4B). Archaeal species branching at the base of eukaryote clade were from either kingdom, Crenarchaeota or Euryarchaeota. Thus, the occurrence of paraphyly among the Archaea does not provide evidence for some specific association between eukaryotes and the Crenarchaeota (represented by the species, *Aeropyrum pernix*) as postulated by the eocyte hypothesis for the origin of eukaryotes (Rivera and Lake, 1992). Similar conclusions were reached in earlier surveys of universal protein gene trees, which included archaeal species of both major kingdoms (Brown and Doolittle, 1997). However, instances of HGT involving eukaryotes might not be apparent given that the four eukaryotic species used here belong to the "crown group." Alanyl-tRNA synthetase (Chidade et al., 2000) and methionyl-tRNA synthetase (see below) are examples of early and late evolved eukaryotes having different evolutionary relationships with prokaryotes.

In contrast, paraphyly among the Bacteria does provide a strong phylogenetic footprint of past HGT events. As an example, in the phenylalanyl-tRNA synthetase β subunit, the bacteria *Borrelia burgdorferi* and *Treponema pallidum* cluster with the Archaea thus indicating a past HGT from an archaeal species to spirochetes. Multiple transfer events from bacteria to eukaryotes and from bacteria to two archaeal species are suggested by the threonyl-tRNA synthetase tree topology. Bacterial valyl-tRNA synthetases were transferred to euaryotes (Hashimoto et al., 1998) while an archaeal version found its way into the proteobacterium, *Rickettsia prowazekii*. While this phylogenetic analysis of highly conserved proteins finds general support for the topology of the universal tree, it also highlights the frequent occurrence of HGT even among essential cellular proteins.

HORIZONTAL GENE TRANSFER AMONG DOMAINS: PATTERN AND PROCESS

Searching for a signal

While HGT between the domains of life appears to have frequently occurred, few specifics are clear on the magnitude, directionality or timing of the phenomena. Are there discernible patterns in species and genes relationships? This question is very difficult to answer definitely. As the diversity of life is further exposed at the genomic level, the only certainty will be that exceptions will emerge to every generality. However, there are a few trends in HGT, which are noteworthy.

Archaea with eukaryotes: the canonical universal tree?

The close evolutionary relationship between Archaea and eukaryotes (AK tree) is evident from the phylogenetic analyses of many proteins involved in DNA replication, transcription and translation (Brown and Doolittle, 1997). In addition, Archaea utilize a wider range of eukaryote-type proteins for these processes than Bacteria. Paralogous gene trees also position Archaea and eukaryotes as sister groups although it has been suggested that such results are idiosyncratic due to more rapid rates of evolutionary change in Bacteria (Brinkmann and Philippe, 1999).

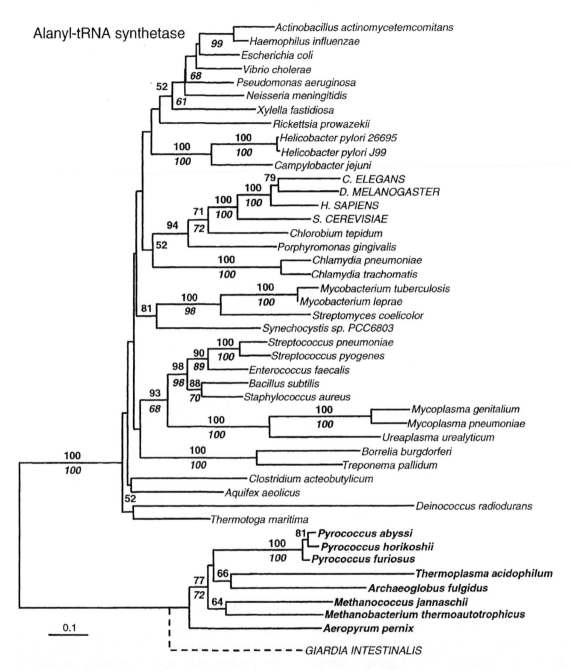

FIGURE 26.5 Phylogenetic tree for alanyl-tRNA synthetase. In this phylogenetic figure and all others, the different typefaces indicate whether the species belong to the domain Eukaryote (all uppercase), Archaea (bold lowercase), or Bacteria (normal, lowercase). Full eukaryotic species names are *Caenorhabditis elegans, Drosophila melanogaster, Homo sapiens* and *Saccharomyces cerevisiae*. Protein trees were constructed using the neighbor-joining method based on pairwise distance estimates of the expected number of amino acid replacements per site (0.1 in the scale bars). Numbers above and below (also in italics) branches show the percent occurrence of nodes in 100 bootstrap replications of neighbor-joining and maximum parsimony analysis, respectively. Only values greater than 50% are shown. Sequence alignment and phylogenetic methods are described in the text. Dashed line indicates the approximate branch location of *Giardia intestinalis* alanyl-tRNA synthetase (Chihade et al., 2000).

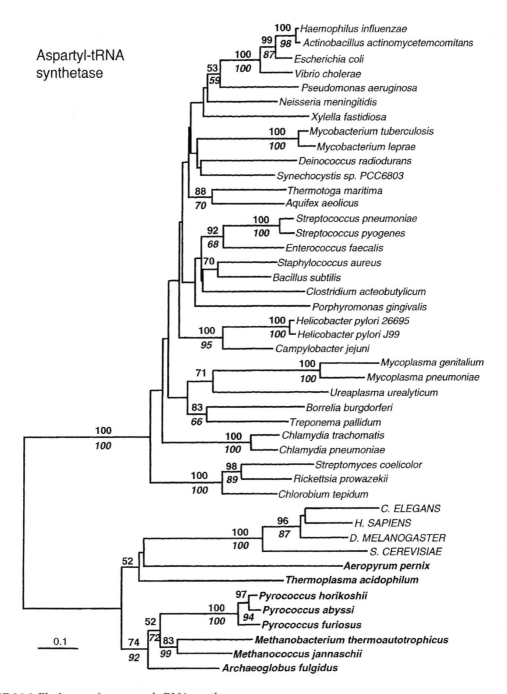

FIGURE 26.6 Phylogeny for aspartyl-tRNA synthetase.

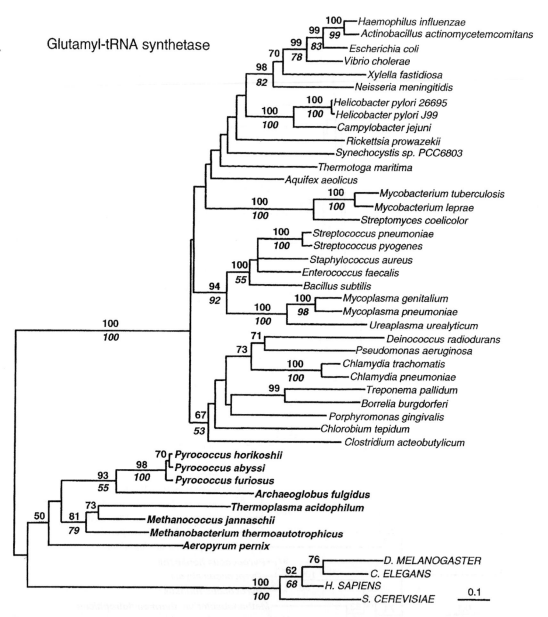

FIGURE 26.7 Phylogeny for glutamyl-tRNA synthetase.

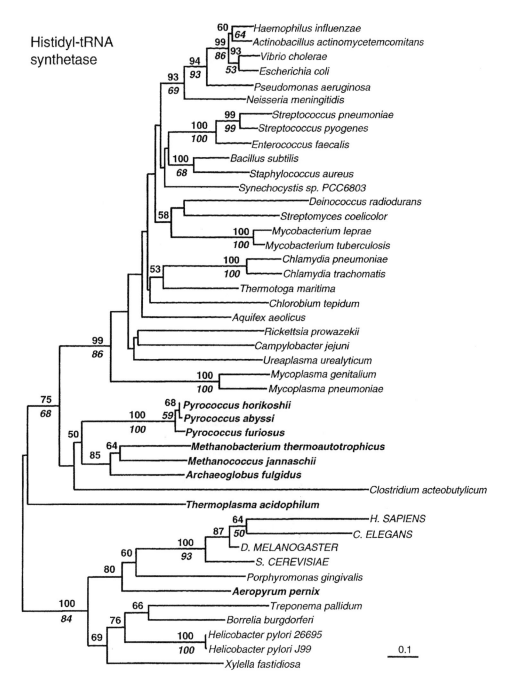

FIGURE 26.8 Phylogeny for histidyl-tRNA synthetase.

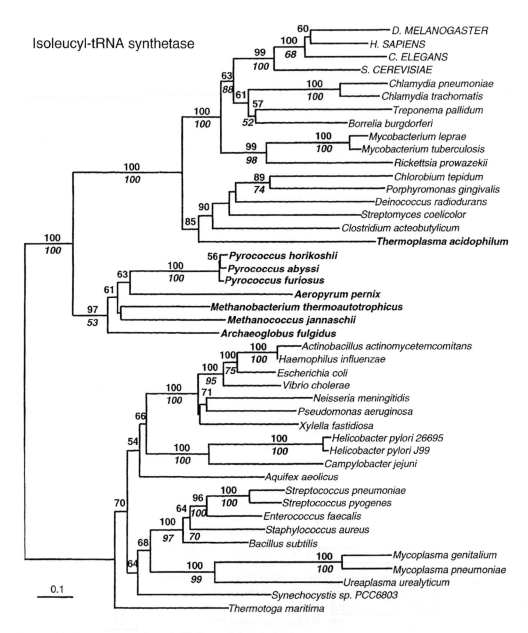

FIGURE 26.9 Phylogeny for isoleucyl-tRNA synthetase.

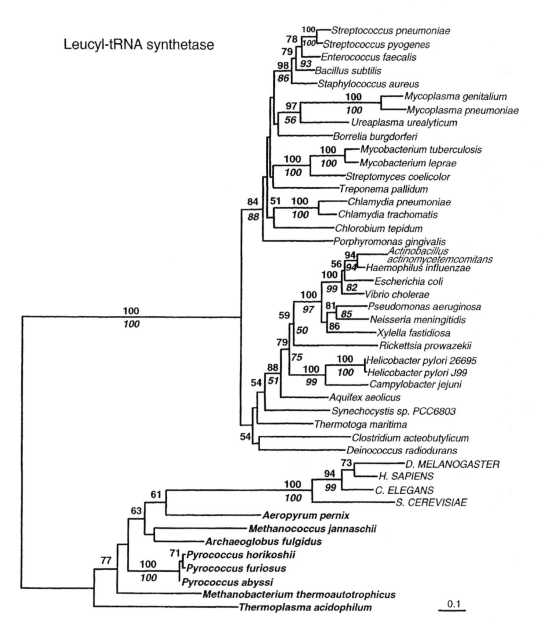

FIGURE 26.10 Phylogeny for leucyl-tRNA synthetase.

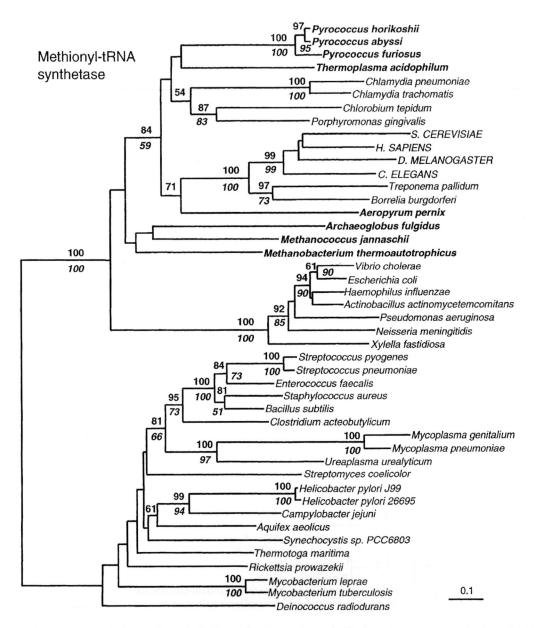

FIGURE 26.11 Phylogeny for methionyl-tRNA synthetase.

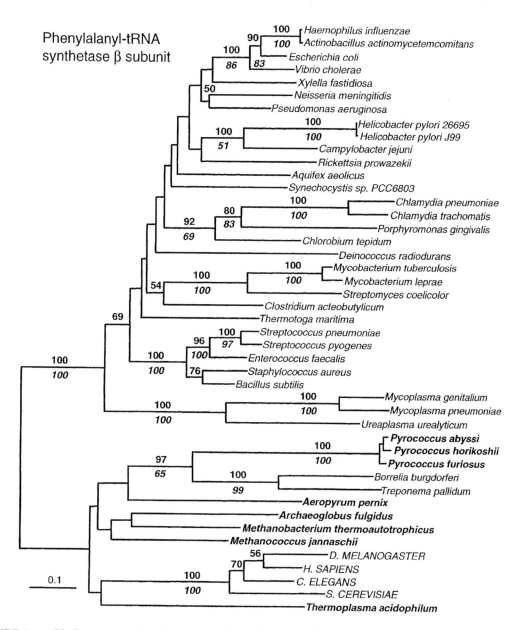

FIGURE 26.12 Phylogeny for phenylalanyl-tRNA synthetase β subunit.

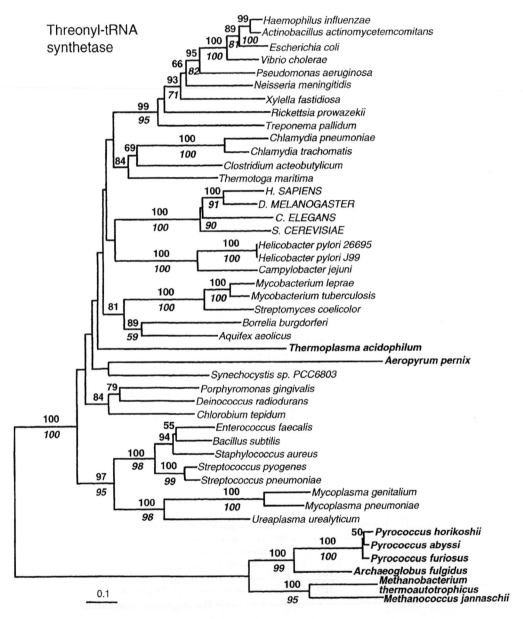

FIGURE 26.13 Phylogeny for threonyl-tRNA synthetase.

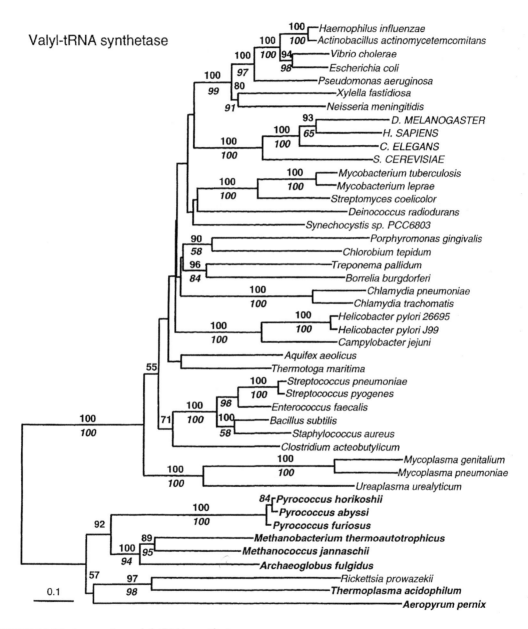

FIGURE 26.14 Phylogeny for valyl-tRNA synthetase.

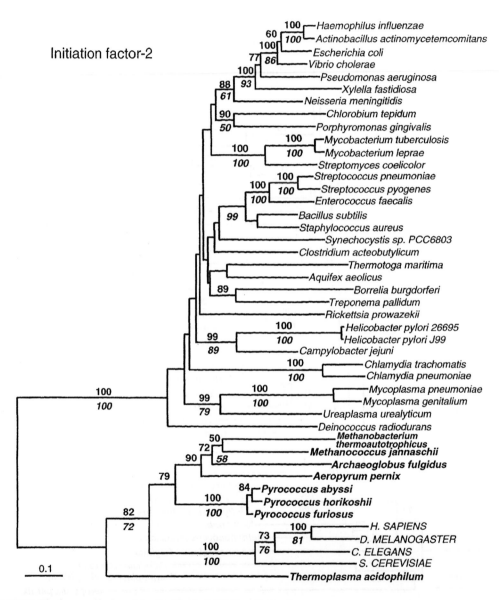

FIGURE 26.15 Phylogeny for protein synthesis initiation factor-2.

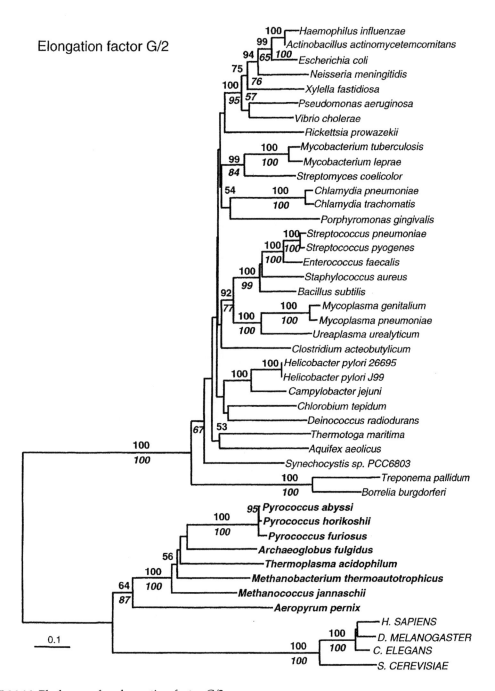

FIGURE 26.16 Phylogeny for elongation factor G/2.

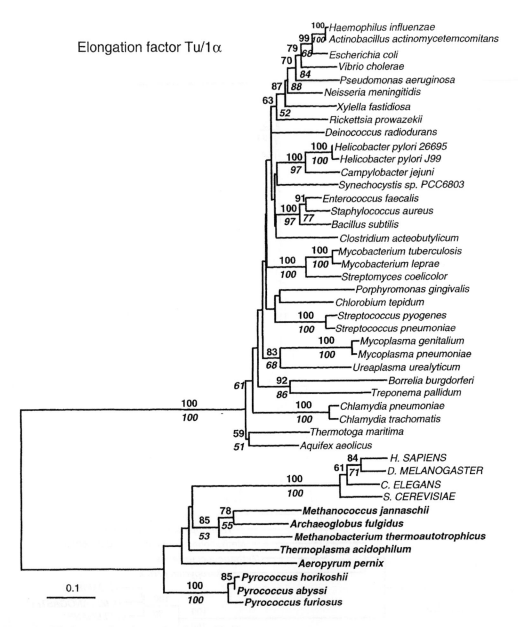

FIGURE 26.17 Phylogeny for elongation factor Tu/1α.

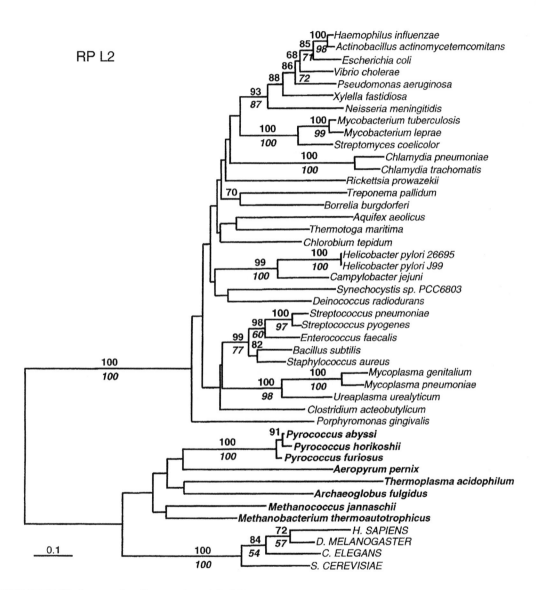

FIGURE 26.18 Phylogeny for ribosomal protein L2.

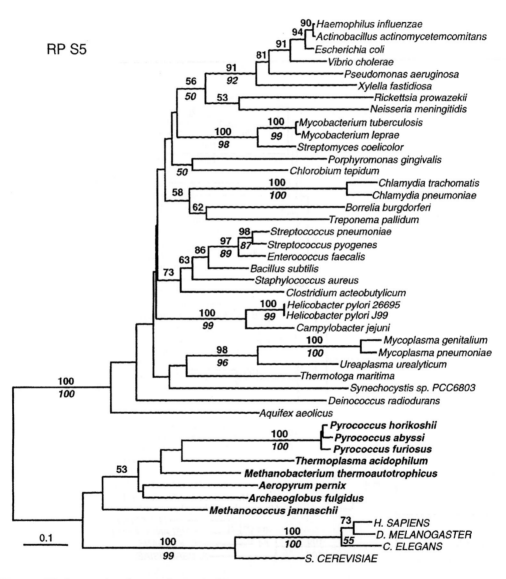

FIGURE 26.19 Phylogeny for ribosomal protein S5.

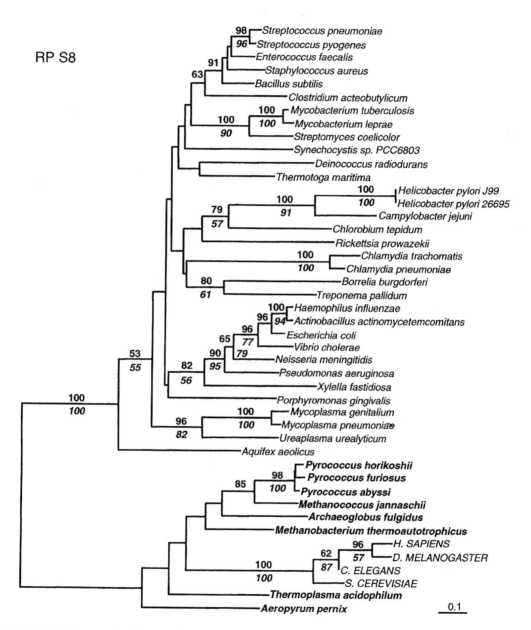

FIGURE 26.20 Phylogeny for ribosomal protein S8.

RP S11

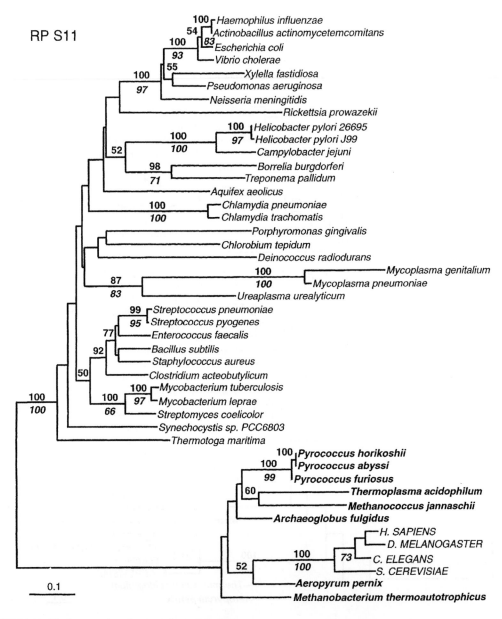

FIGURE 26.21 Phylogeny for ribosomal protein S11.

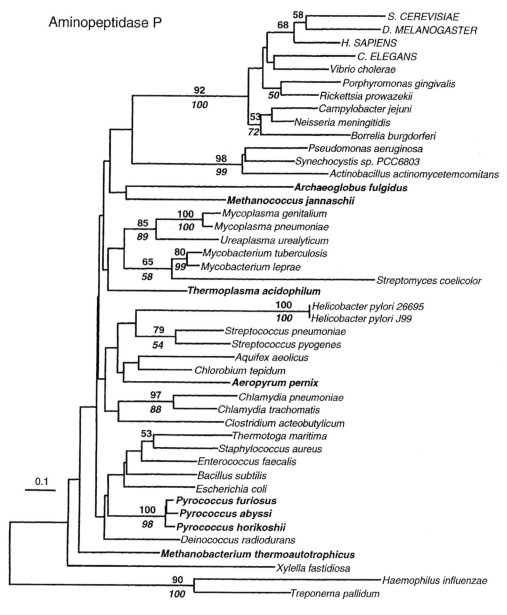

FIGURE 26.22 Phylogeny for aminopeptidase P.

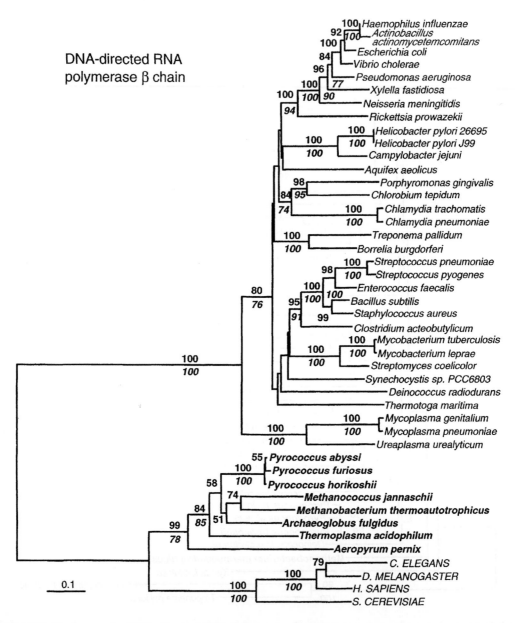

FIGURE 26.23 Phylogeny for DNA-directed RNA polymerase β chain.

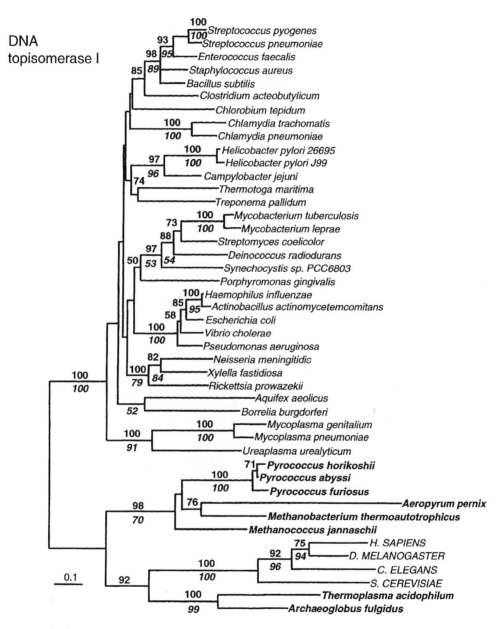

FIGURE 26.24 Phylogeny for DNA topoisomerase I.

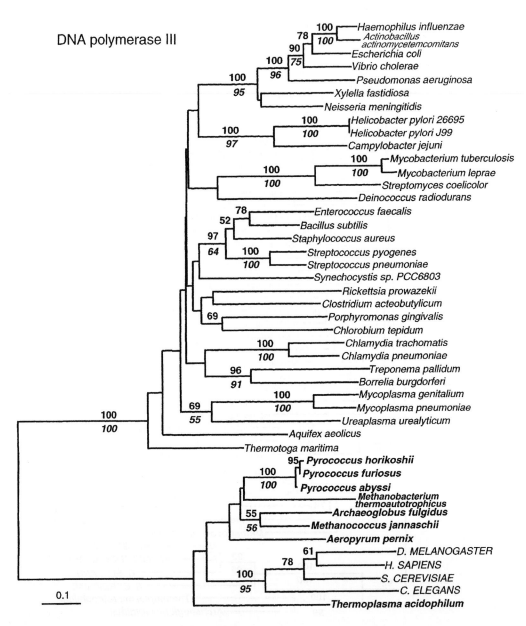

FIGURE 26.25 Phylogeny for DNA polymerase III.

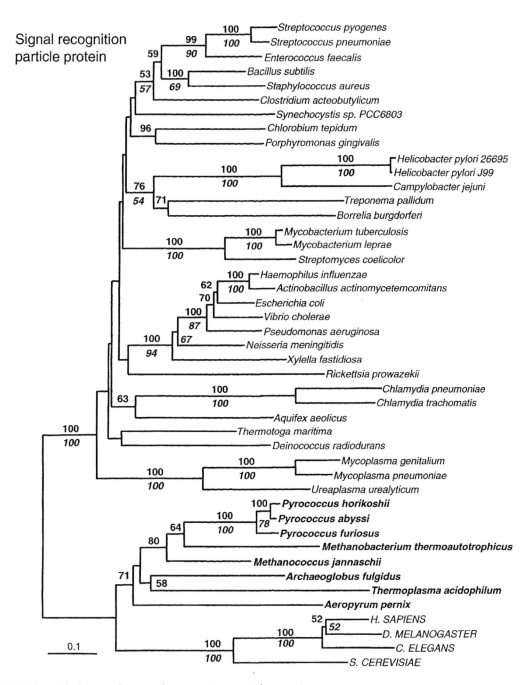

FIGURE 26.26 Phylogeny for signal recognition particle protein component.

rRNA methylase

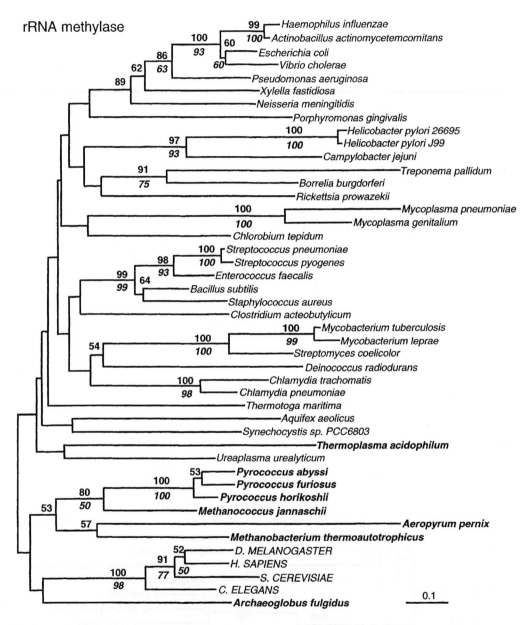

FIGURE 26.27 Phylogeny for rRNA methylase.

Interestingly, among the three possible universal tree scenarios, only trees depicting the AK clustering ever show, even if occasionally, all three domains to be monophyletic simultaneously (Brown and Doolittle, 1997). If extensive polyphyly (species from different domains in the same clade) is evidence for HGT then, by default, monophyly indicates evolution in the absence of HGT. Given the universe of genes, domain monophyly appears to be a rare occurrence. However, even the existence of a few monophyletic gene trees should give some hope that their topology reflects the underlying evolutionary trajectory of the species involved without the complication of HGT. If true, then the overall scenario of cellular evolution, heavily diluted by HGT events, remains the canonical universal tree with its rooting in the Bacteria with Archaea and eukaryotes as sister groups.

However, the persistence of monophyly in universal trees is highly dependent upon the number and type of species. As discussed above, the sequencing of new species of Bacteria and Archaea created revisions of gene-species distributions as well as polyphyletic trees for many conserved proteins. Similar situations are emerging for eukaryotes, where until recently most sequence data were available from the "crown" species, fungi, plants, and animals. There are many species of protists, which as anaerobes share metabolic genes with bacterial and archaeal species but not higher eukaryotes. Furthermore, several aminoacyl-tRNA synthetases from the protist *Giardia intestinalis* (see below) show differential origins from those of other eukaryotes (Chihade et al., 2000). As the diversity of eukaryotes becomes appreciated, there will no doubt emerge further evidence for HGT with particular groups of Archaea and Bacteria.

Archaea with Bacteria: genetic exchanges between specific groups

Unlike the trees showing Archaea and eukaryotes as sister groups, protein trees clustering Archaea and Bacteria together (AB tree) always portray one or both domains as para/polyphyletic groups. Such tree topologies are

evidence for HGT between Archaea and Bacteria, the patterns for which can be often complex (Figure 26.4C). The alternative hypothesis, the rooting of the universal tree in eukaryotes, would require both the Archaea and Bacteria to have undergone extreme genome simplification or "streamlining" (Forterre and Phillipe, 1999). While possible, this scenario requires accommodation for some very complex issues surrounding the requisite genome size reduction, such as the loss of introns, the reorganization of genes into operons, and the evolution of a single origin of replication. Furthermore, models need to account for the retention of eukaryotic DNA replication, transcription and translation proteins by the Archaea, on the one hand, and the commonality of bacterial and eukaryotic membrane structures, on the other. Therefore, close evolutionary relationships between Archaea and Bacteria depicted in many gene trees are probably evidence for past HGT events.

The genes and species implicated in Archaea-Bacteria HGT are highly varied. Glutamine synthetases (Brown et al., 1994), glutamate dehydrogenase (Benachenhou-Lahfa et al., 1993) and HSP70 (Gupta and Golding, 1993) of Archaea are closely related with orthologues from Gram-positive bacteria. Hyperthermophilic archaeal and bacterial species share a reverse gyrase as a unique adaptation to life at extremely high temperatures (Forterre et al., 2000). Catalase-peroxidase genes appear to have exchanged between Archaea and pathogenic proteobacteria (Faguy and Doolittle, 2000). Two component signal transduction systems in the Archaea as well as fungi and slime molds were probably acquired from bacteria (Koretke et al., 2000).

However, not all similarities between Bacteria and Archaea in terms of either phylogeny or gene occurrence should be strictly interpreted as the result of HGT. As discussed above and illustrated in Figure 26.4B, species forming low branches in the two domains can be attracted or cluster together because of rooting artifacts. In addition, gene distributions shared by Bacteria and Archaea but not eukaryotes might be caused by gene loss or replacement in eukaryotes rather than HGT between Archaea and Bacteria. (However, some "lost" eukaryotic genes might be present, but undiscovered, in lower branching protists.) Regardless, there are many striking examples of shared gene transfers between Archaea and Bacteria, which require careful re-examination of genome evolution and species relationships.

Bacteria with eukaryotes: endosymbiosis hypothesis extended

Obviously, gene transfers between domains can be bi-directional. Clues to the directionality of gene transfers can sometimes, but not always, be inferred from the distribution of a particular gene family. If a particular protein is widespread in one domain but restricted to a few species in the other, then it is likely that a species from the more populated group was the origin for the transferred gene. Phylogeny must be used to provide additional verification of the relative branching order among different gene orthologues, thereby better localizing the species or lineages that probably participated in the gene transfer. An example of eukaryote to bacteria transfer is glutaminyl-tRNA synthetase. This protein occurs in all eukaryotes, even lower protists, and appears to have evolved from an early, specific gene duplication in eukaryotes which resulted in glutamyl-tRNA synthetase and glutaminyl-tRNA synthetase (Lamour et al., 1994; Brown and Doolittle, 1999). Glutaminyl-tRNA synthetases occur in a number of bacteria, including species of the subdivisions Thermus/Deinococcus, γ and β subdivision Proteobacteria and the Cytophaga–Flexibacter–Bacteroides (CFB), but are entirely absent from others. All bacterial glutaminyl-tRNA synthetases cluster with those of eukaryotes thus the overall conclusion is that an early eukaryote contributed the gene for this enzyme to bacteria one or more times.

Gene transfers from eukaryotes to bacteria can provide interesting insights into the capacity of bacteria to adapt to new environmental conditions by foreign gene acquisition. However, the reciprocal exchange, bacteria to early eukaryotes, is being increasingly viewed as a seminal event(s) in eukaryotic evolution. The origin of mitochondria and plastids from different bacterial endosymbionts has been a widely accepted hypothesis for several decades (Margulis, 1970). However, the extent of additional gene transfer from bacteria to the eukaryotic genome and the relative timing of such events are still being revealed. Phylogenies for many universal proteins, such as the glycolytic pathway enzymes, show Archaea, rather than Bacteria, as the outgroup, which suggests a close relationship between the Bacteria and eukaryotes (reviewed in Brown and Doolittle, 1997). Further evidence for the early integration for bacterial genes into the eukaryotic genome comes from studies of proteins from simple protists such as G. intestinalis (previously G. lamblia; Diplomonadida) and Trichomonas vaginalis (Parabaslia), which commonly lack mitochondria (amitochondria) and appear to be the earliest evolved eukaryotic lineages from rRNA phylogenies.

T. vaginalis has both mitochondria-specific HSP60 (Roger et al., 1996) and mitochondria targeted HSP70 genes (Germot et al., 1996). Clark and Roger (1995) describe HSP60 and a second mitochondrion specific gene, pyridine nucleotide transhydrogenase from a more highly evolved amitochondrial protist Entamoeba histolytica. Plausibly, amitochondria protists either secondarily lost their organelles or underwent some kind of endosymbosis, which resulted in the successful fixation of several bacterial genes in the nuclear genome but not an intracellular organelle. In phylogenetic trees, triosephosphate isomerase (TPI) of G. intestinalis clusters with other eukaryotic versions, which, in turn, are related to an α-proteobacteria TPI gene (Keeling and Doolittle, 1997). In addition, the glyceraldehyde 3-phosphate dehydrogenase (GAPDH) genes of E. histolytica and G. intestinalis fall within the same cluster of eukaryotic homologues suggested to have emanated from a proteobacterium (Henze et al., 1995).

Considering the fate of a single orthologous gene found in both the host (either an archaebacterium or proto-eukaryote) and the bacterial endosymbiont, there are four possible outcomes (Figure 26.28). First, the gene was retained in the genomes of both the organelle (formerly endosymbiont) and the host. Mitochondria genomes, even the highly derived ones

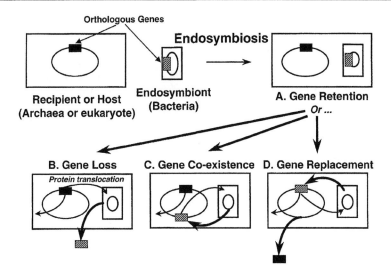

FIGURE 26.28 Possible fates of orthologous genes resulting from endosymbosis. Considering the fate of a single orthologous gene found in both the host (either an archaebacterium or proto-eukaryote) and bacterial endosymbiont, there are four possible outcomes: (A) The gene was retained in the genomes of both the organelle (formerly endosymbiont) and the host; (B) the organelle gene was lost from the organism, and its encoded metabolic function assumed by the host gene; (C) the organelle gene was transferred to the host genome where it co-exists with the host copy or (D) the organelle gene is transferred into the eukaryotic host genome and either contributes a new function or replaces the host orthologues. Martin (1996) called the latter phenomena endosymbiotic gene replacement.

of animals, encoded for a number of proteins, rRNAs and tRNAs which are critical for organelle function. However, the overwhelming evolutionary trend is for a reduction in mitochondria genome size. The largest sequenced mitochondria genome recorded in GenBank, that of the sugar beet, *Beta vulgaris*, is 368 799 bp (Kubo et al., 2000) while the smallest known bacterial genome, that of *Mycoplasma genitalium*, is 580 074 bp, nearly twice as large. A second possible fate is that the organelle gene was lost from the organism and its encoded metabolic function was assumed by genes in the host genome. Thirdly, the organelle gene was transferred to the host genome where it co-exists with the host copy. Eukaryotes have duplicate copies of several aminoacyl-tRNA synthetases, one isoform which functions in the cytoplasm and is more closely related to archaeal homologues while the second copy is targeted to the mitochondria compartment and is most similar to bacterial homologues. Finally, the organelle gene was transferred into the eukaryotic host genome and retained as a single copy. The organelle gene either

contributes a new function or it replaces the original orthologous gene of the host. Martin (1996) called the phenomena of substituting a gene from the bacterial endosymbiont for the original nuclear copy endosymbiotic gene replacement. The more specialized instances where amitochondrial protists have the organelle gene but not the organelle are termed cryptic endosymbiosis, which invokes the notion of a temporal state of endosymbiosis, followed by loss of the bacterial endosymbiont (Henze et al., 1995).

Aminoacyl-tRNA synthetase evolution – alternative views of endosymbiosis?

However, analyses of new genomic sequence data suggest that extending the endosymbosis hypothesis to the wider issue of the origin of eukaryotes requires further thought. In particular, phylogenetic analyses of aminoacyl-tRNA synthetases, which incorporate data from lower eukaryotes, suggest that endosymbosis was not

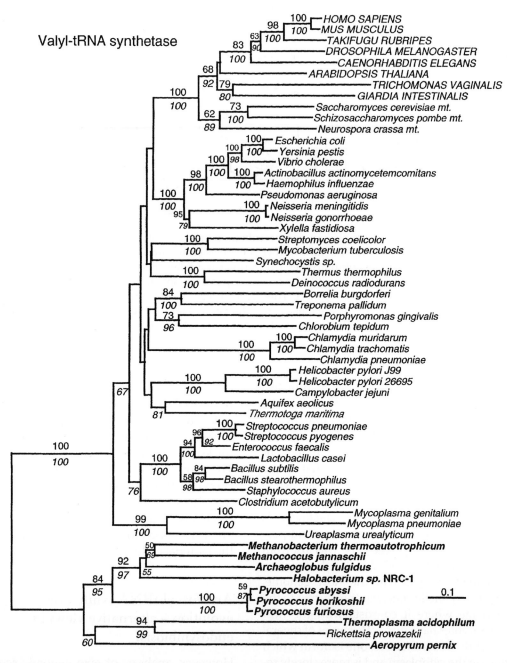

FIGURE 26.29 Phylogeny of valyl-tRNA synthetase, which includes all public available sequences including lower protists (*Trichomonas vaginalis* and *Giardia intestinalis*; Hashimoto et al., 1998). Note that eukaryote cytoplasmic and mitochondrial synthetases do not share a recent common ancestor with the α-Proteobacteria, *Rickettsia prowazekii*, which is the nearest contemporary lineage of the putative prokaryotic endosymbiont.

a simple event. Initial phylogenetic analysis showed that valyl-tRNA synthetase of eukaryotes, including that of *T. vaginalis*, and bacteria were similar (Brown and Doolittle, 1995). In fungi, the same gene encodes valyl-tRNA synthetase targeted for the mitochondria and cytoplasm (Jordana et al., 1987; Martindale et al., 1991), which further hints that endosymbiotic gene replacement is at play. Subsequent phylogenetic trees, which included archaeal species as well as additional protists, seemed to confirm this scenario (Hashimoto et al., 1998). However, missing from both these analyses were representative species from the α-Proteobacteria, the widely held bacterial endosymbiont progenitor of the mitochondria. Recently, the entire genome of a member species, *Rickasettia prowazekii*, was completed (Andersson et al., 1998.). In phylogenetic analyses, the majority of *R. prowazekii* proteins show strong similarities to mitochondria-targeted eukaryotic orthologues thus supporting the view that α-Proteobacteria were the progenitors of the mitochondria (Kurland and Andersson, 2000). Surprisingly, *R. prowazekii* valyl-tRNA synthetase clusters the Archaea rather than with mitochondria derived eukaryotes or bacteria (Figure 26.29). Either α-Proteobacteria and Archaea exchanged valyl-tRNA synthetase after mitochondria biogenesis or, more important, eukaryotic mitochondria valyl-tRNA synthetase did not arise from an α-Proteobacteria endosymbiont.

Alanyl-tRNA synthetase is a second example of a protein phylogeny conflicting with current views on the endosymbosis hypothesis (Chihade et al., 2001). Nuclear-encoded cytoplasmic alanyl-tRNA synthetases from mitochodriate eukaryotes are rooted by mitochondria targeted isoforms with both isoforms firmly nested within the bacterial alanyl-tRNA synthetases. This tree topology suggests that endosymbiotic gene-replacement occurred for cytoplasmic alanyl-tRNA synthetase. However, alanyl-tRNA synthetase from the amitochondriate protist *G. intestinalis* is more closely related to archaeal alanyl-tRNA synthetase than to either bacterial or other eukaryotic homologues. (In Figure 26.5, *G. intestinalis* alanyl-tRNA synthetase would branch at the base of the Archaea.) Thus, the gene for alanyl-tRNA synthetase shows different patterns of acquisition and retention between an amitochondriate and mitochondriate eukaryotes.

One scenario would be an archaebacterium having transferred its alanyl-tRNA synthetase to *Giardia* after it lost its mitochondria. However, this possibility is unlikely because at least two rounds of gene transfers and replacements are necessary to have occurred. Furthermore, the branching of *Giardia* before the deepest archaeal split that separates the kingdoms Crenarchaeota and Euryarchaeota also suggests that any postulated gene transfer event would have occurred before the evolution of contemporary Archaea. Alternatively, *Giardia* did not fully participate in the genetic transfers between mitochondria and nuclei to the same extent as other eukaryotes. If the latter scenario is true, then *Giardia* and its diplomonad kin represent a truly ancient group of eukaryotes, which diverged before the full integration of the mitochondria genome.

A third intriguing evolutionary pattern involves methionyl-tRNA synthetase (Figure 26.30). The overall tree topology shows two distinct clades, which are arbitrarily labeled here as Group A and Group B. Eukaryotes, the Archaea and several bacterial species are found in Group A. Eukaryotes and Archaea are closely related but interspersed with bacterial species belonging to various subdivisions including Chlamydia and spirochaetes. Proteobacteria, with the exception of *R. prowazekii* and the ε-Proteobacteria *Helicobacter pylori* and *Campylobacter jejuni*, belong to Group A but as a divergent subgroup. Methionyl-tRNA synthetase Group B sequences include all remaining bacteria, eukaryote mitochondria-targeted enzymes and isoforms, and *Giardia* (data from McArthur et al., 2000). The clade of *R. prowazekii*, mitochondria-targeted and *Giardia* methionyl-tRNA synthetase suggests a common evolutionary ancestor, which is compatible with the endosymbosis hypothesis. However, unlike alanyl-tRNA synthetase, it is diplomonads that underwent endosymbiotic gene replacement and mitochondrial eukaryotes retained the archaeal (ancestral) version of methionyl-tRNA synthetase. The evolution of Group A methionyl-tRNA synthetases from mitochondrial eukaryotes have further complexities with other HGT events between different bacterial groups.

Methionyl-tRNA Synthetase

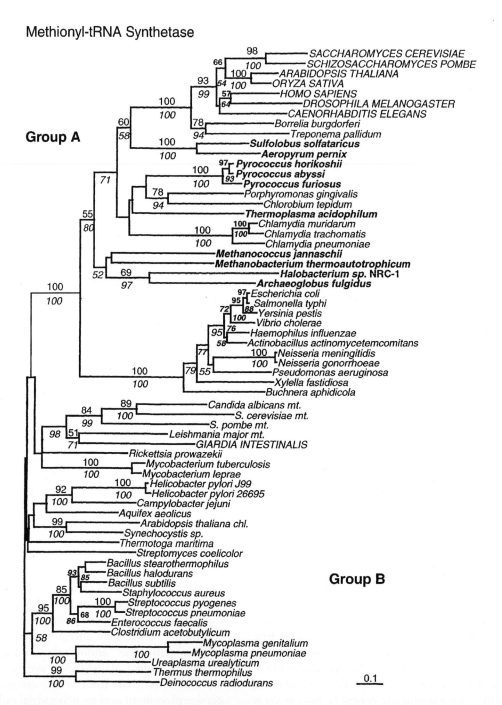

FIGURE 26.30 Phylogeny of methionyl-tRNA synthetase, which includes all public available sequences including that of the amitochondrial protist, *Giardia intestinalis*. In contrast to valyl-tRNA synthetase, methionyl-tRNA synthetases in eukaryotes targeted to the mitochondria, as well as synthetases from *G. intestinalis*, and *Rickettsia prowazekii* all share a recent common ancestor. However, synthetases targeted to the cytoplasm and those of other Proteobacteria are highly divergent.

Collectively, these data suggest that the integration of the bacterial mitochondrial progenitor overlapped with the formation of the eukaryote nucleus. If the diplomonads represent an intermediate stage in mitochondrial genesis, and considering the nucleated nature of *Giardia*, then the appearance of mitochondria may not have preceded the genesis of modern nuclei. These aminoacyl-tRNA synthetase trees hold the promise of finding more examples of intermediate phases of endosymbosis in extant eukaryotes, which could help illuminate the pattern and process of mitochondria biogenesis.

CONCLUDING REMARKS

The apparent occurrence of extensive of HGT across the domains of life has prompted much speculation on its significance to early cellular evolution. Networks of genetic interactions at the base of the universal tree have been suggested to be so intense as to render useless the concept of a single cellular ancestor for contemporary lineages (Hilario and Gogarten, 1993; Woese 1998). Other radical positions discuss the emergence of eukaryotes from the complete fusion of genomes from an archaebacterium and bacterium (reviewed in Brown and Doolittle, 1997). Martin and Müller (1998) proposed a more stepwise progression to eukaryotes beginning with a hydrogen-dependent host, probably an archaebacterium, and a respiring bacterial symbiont. Doolittle (1998) suggests a ratchet-like addition of bacterial content to the eukaryotic genomes from either a prokaryotic food source or gene transfers as a consequence of multiple but brief endosymbiotic associations.

Such controversies will either be resolved or amplified as genomes from more taxa are sequenced. Limitations to present analyses are database biases toward particular species. Bacterial databases are skewed towards pathogenic bacteria, some of which are highly DNA competent, such as the respiratory tract infection pathogens, *Haemophilus influenzae* and *Streptococcus pneumoniae*. Similarly, archaeal genomes are biased towards species living in extreme environments, in particular high temperatures. The uptake and fixation of foreign genes might be positive adaptations towards survival under stressful conditions. Eukaryotic genomes sequenced thus far are from model genetic organisms, which are closely related crown species. Even the malaria agent *Plasmodium falciparum* is highly diverged from amitochondriate groups, such as the diplomonads (Cavalier-Smith, 1993). Thus the genetic diversity of eukaryotes has yet to be fully explored.

While HGT has certainly unsettled the universal tree of life, it is premature to say that the tree has been permanently uprooted. The evidence linking the evolutionary histories of Archaea and eukaryotes is too overwhelming to be adequately explained by any alternative scenario. The excitement in the field of molecular evolution now lies ahead in trying to unravel the spectacular process of genome evolution and the role of HGT.

NOTE ADDED IN PROOF

Recently, Brown et al. (2001) published universal trees based on the combined alignments of those 23 orthologous proteins conserved across 45 species. While individual protein trees were variable in their support of domain integrity, trees based on combined protein datasets strongly supported separate monophyletic domains. Interestingly, trees based on the entire set of 23 proteins placed spirochaetes rather the thermophiles *Aquifex* and *Thermotoga* as the earliest derived bacterial group. Brown et al. (2001) hypothesized that this result might be due to HGT between Bacteria and Archaea/eukaryotes. As a test, they removed from the combined dataset the alignments of nine proteins where individual trees constructed by either the MP or NJ method (or both) did not depict Bacteria as monophyletic (see Table 26.2). They found that universal trees based on the combined alignment of the remaining 14 proteins also showed three monophyletic domains as well as placed thermophiles as the earliest evolved bacterial lineages which was highly congruent with rRNA tree topologies.

Evidence collected thus far suggests that HGT events between eukaryotes and bacteria occurred very early in eukaryotic evolution, and findings to the contrary are unsubstantiated. In their publication of the human genome sequence, the International Human Genome Sequencing Consortium (2001) made an extraordinary claim that as many as 113 vertebrate genes, some only found in humans, were the result of direct horizontal transfers from bacteria. However, two independent studies, using more thorough searches to detect non-vertebrate homologues (i.e. the National Center for Biotechnology Information 'EST others' database) as well as more rigorous phylogenetic analyses, concluded that there was no compelling support for direct bacteria to vertebrate gene transfers (Salzburg et al., 2001; Stanhope et al., 2001).

ACKNOWLEDGMENTS

Preliminary sequence data were obtained from the following sources. *Chlorobium tepidum*, *Enterococcus faecalis*, *Porphyromonas gingivalis*, and *Streptococcus pneumoniae* data were obtained from The Institute for Genomic Research (TIGR) through the Web site at http://www.tigr.org, which is supported by grants from the US Department of Energy (DOE), the National Institute of Allergy and Infectious Disease (NIAID) of NIH, TIGR, and the Merck Genome Research Institute, NIAID, and TIGR. *Actinobacillus actinomycetemcomitans* data were obtained from the Actinobacillus Genome Sequencing Project, University of Oklahoma ACGT and B.A. Roe, F.Z. Najar, S. Clifton, Tom Ducey, Lisa Lewis and D.W. Dyer through the Web site http://www.genome.ou.edu/act.html which is supported by a USPHS/NIH grant from the National Institute of Dental Research. *Pyrococcus furiosus* data were obtained from the Utah Genome Center, Department of Human Genetics, University of Utah through the Web site http://www.genome.utah.edu/sequence. html, which is supported by the DOE. *Streptomyces coelicolor* data were obtained from The Sanger Center through the Web site at http://www.sanger.ac.uk/Projects/S_coelicolor/. *Giardia intestinalis* (*lamblia*) Genome Project data were obtained via the Web site http://www.mbl.edu/Giardia.

REFERENCES

Altschul, S.F., Madden, T.L., Schäffer, A.A. et al. (1997) Gapped BLAST and PSI-BLAST: a new generation of protein database search programs. *Nucleic Acids Res.* **25**: 3389–3402.

Andersson, S.G., Zomorodipour, A., Andersson, J.O. et al. (1998) The genome sequence of *Rickettsia prowzekii* and the origin of mitochondria. *Nature* **396**: 133–140.

Baldauf, S.L., Palmer, J.D. and Doolittle, W.F. (1996) The root of the universal tree and the origin of eukaryotes based on elongation factor phylogeny. *Proc. Natl Acad. Sci. USA* **93**: 7749–7754.

Benachenhou-Lahfa, N., Forterre, P. and Labedan, B. (1993) Evolution of glutamate dehydrogenase genes: Evidence for paralogous protein families and unusual branching patterns of the archaebacteria in the universal tree of life. *J. Mol. Evol.* **36**: 335–346.

Boucher, Y. and Doolittle, W.F. (2000) The role of lateral gene transfer in the evolution of isoprenoid biosynthesis pathways. *Molec. Microbiol.* **37**: 703–716.

Brinkman, H. and Philippe, H. (1999) Archaea sister group of bacteria? Indications from tree reconstruction artifacts in ancient phylogenies. *Mol. Biol. Evol.* **16**: 817–825.

Brown, J.R. (2001) Genomic and phylogenetic perspectives on the evolution of prokaryotes. *Systematic Biol.* **50**: 497–512.

Brown, J.R. (1998) Aminoacyl-tRNA synthetases: evolution of a troubled family. In *Thermophiles – the Keys to Molecular Evolution and the Origin of Life?* (eds, J. Wiegel and M. Adams), pp. 217–230, Taylor and Francis, London.

Brown, J.R. and Doolittle, W.F. (1995) Root of the universal tree of life based on ancient aminoacyl-tRNA synthetase gene duplications. *Proc. Natl Acad. Sci. USA* **92**: 2441–2445.

Brown, J.R. and Doolittle, W.F. (1997) Archaea and the prokaryote to eukaryotes transition. *Microbiol. Molec. Biol. Rev.* **61**: 456–502.

Brown, J.R. and Doolittle, W.F. (1999) Gene descent, duplication, and horizontal transfer in the evolution of glutamyl- and glutaminyl-tRNA synthetases. *J. Mol. Evol.* **49**: 485–95.

Brown, J.R. and Warren, P.V. (1998) Antibiotic discovery: is it in the genes? *Drug Discovery Today* 3: 564–566.

Brown, J.R., Masuchi, Y., Robb, F.T. et al. (1994) Evolutionary relationships of bacterial and archaeal glutamine synthetase genes. *J. Mol. Evol.* **38**: 566–576.

Brown, J.R., Robb, F.T., Weiss, R. and Doolittle, W.F. (1997) Evidence for the early divergence of tryptophanyl- and tyrosyl-tRNA synthetases. *J. Mol. Evol.* **45**: 9–16.

Brown, J.R., Zhang, J. and Hodgson, J.E. (1998) A bacterial antibiotic resistance gene with eukaryotic origins. *Current Biology* **8**: R365–R367.

Brown, J.R., Douady, C.J., Italia, M.J. et al. (2001) Universal trees based on large combined protein sequence datasets. *Nature Genet.* **28**: 281–285.

Cavalier-Smith, T. (1993) Kingdom protozoa and its 18 phyla. *Microbiol. Rev.* **57**: 953–994.

Chihade, J., Brown, J.R., Schimmel, P. et al. (2000) Detection of an intermediate stage of mitochondria genesis. *Proc. Natl Acad. Sci. USA* **97**: 12153–12157.

Clark, C.G. and Roger, A.J. (1995) Direct evidence for secondary loss of mitochondria in *Entamoeba histolytica*. *Proc. Natl Acad. Sci. USA* **92**: 6518–6521.

Danson, M.J. (1993) Central metabolism of the Archaea. In *The Biochemistry of Archaea (Archaebacteria)* (eds, M. Kates, D.J. Kushner and A.T. Matheson), pp. 1–24, Elsevier, Amsterdam.

Doolittle, W.F. (1998) You are what you eat: a gene transfer ratchet could account for bacterial genes in eukaryotic nuclear genomes. *Trends Genet.* **14**: 307–311.

Doolittle, W.F. (1999a) Phylogentic classification and the universal tree. *Science* **284**: 2124–2128.

Doolittle, W.F. (1999b) Lateral gene transfer, genome surveys and the phylogeny of prokaryotes. Technical comments. *Science* **284**: 2124–2128

Doolittle, W.F. and Logsdon, J.M. Jr (1998) Archaeal genomics: do archaea have a mixed heritage? *Current Biology* **8**: R209–R211.

Edgell, D.R. and Doolittle, W.F. (1997) Archaea and the origin(s) of DNA replication proteins. *Cell* **89**: 995–998.

Faguy, D.M. and Doolittle, W.F. (2000) Horizontal transfer of catalase-peroxidase genes between Archaea and pathogenic bacteria. *Trends Genet.* **16**: 196–197.

Felsenstein, J. (1993) PHYLIP (Phylogeny Inference Package) version 3.57c. Distributed by the author: http://evolution.genetics.washington.edu/phylip.html, Department of Genetics, University of Washington, Seattle.

Feng, D.-F., Cho, G. and Doolittle, W.F. (1997) Determining divergence times with a protein clock: update and reevaluation. *Proc. Natl Acad. Sci. USA* **94**: 13028–13033.

Fitch, W.M., and Upper, K. (1987) The phylogeny of tRNA sequences provides evidence for ambiguity reduction in the origin of the genetic code. *Cold Spring Harbor Symp. Quantative Biol.* **52**: 759–767.

Forterre, P. and Philippe, H. (1999) Where is the root of the universal tree of life? *BioEssays* **21**: 871–879.

Forterre, P., Benachenhou-Lahfa, N., Confalonieri, F. et al. (1993) The nature of the last universal ancestor and the root of the tree of life, still open questions. *Biosystems* **28**: 15–32.

Forterre, P., Bouthier de la Tour, C, Philippe, H. and Duguet, M. (2000) Reverse gyrase from thermophiles: probable transfer of a thermoadaptation trait from Archaea to Bacteria. *Trends Genet.* **16**: 152–154.

Fox, G.E., Magrum, L.J., Balch, W.E. et al. (1977) Classification of methanogenic bacteria by 16S ribosomal RNA characterization. *Proc. Natl Acad. Sci. USA* **74**: 4537–4541.

Germot, A., Philippe, H. and Le Guyader, H. (1996) Presence of a mitochondrial-type 70-kDa heat shock protein in *Trichomonas vaginalis* suggests a very early mitochondrial endosymbiosis in eukaryotes. *Proc. Natl Acad. Sci. USA* **93**: 14614–14617.

Gogarten, J.P., Kibak, H., Dittrich, P. et al. (1989) Evolution of the vacuolar H^+-ATPase: Implications for the origin of eukaryotes. *Proc. Natl Acad. Sci. USA* **86**: 6661–6665.

Gogarten, J.P., Hilario, E. and Olendzenski. L. (1996) Gene duplications and horizontal transfer during early evolution. In *Evolution of Microbial Life* (eds, D.M. Roberts, G. Alderson, P. Sharp and M. Collins), pp. 267–292, Society for General Microbiology Symposia 54, Cambridge University Press, Cambridge.

Golding, G.B. and Gupta, R.S. (1995) Protein-based phylogenies support a chimeric origin for the eukaryotic genome. *Mol. Biol. Evol.* **12**: 1–6.

Graham, D.E., Overbeek, R., Olsen, G.J. and Woese, C.R. (2000) An archaeal genomic signature. *Proc. Natl Acad. Sci. USA* **97**: 3304–3308.

Gribaldo, S., Lumia, V., Creti, R. et al. (1999) Discontinous occurrence of the *hsp70* (*dnaK*) gene among *Archaea* and sequence features of HSP70 suggest a novel outlook on phylogenies inferred from this protein. *J. Bacteriol.* **181**: 434–443.

Gupta, R.S. and Golding, G.B. (1993) Evolution of HSP70 gene and its implications regarding relationships between archaebacteria, eubacteria and eukaryotes. *J. Mol. Evol.* **37**: 573–582.

Gupta, R.S. and Golding, G.B. (1996) The origin of the eukaryotic cell. *Trends Biochem. Sci.* **21**: 166–171.

Hashimoto, T., Sánchez, L.B., Shirakura, T. et al. (1998) Secondary absence of mitochondria in *Giardia lamblia* and *Trichomonas vaginalis* revealed by valyl-tRNA synthetase phylogeny. *Proc. Natl Acad. Sci. USA* **95**: 6860–6865.

Henze, K.A., Badr, A., Wettern, M. et al. (1996) A nuclear gene of eubacterial origin in *Euglena gracilis* reflects cryptic endosymbioses during protist evolution. *Proc. Natl Acad. Sci. USA* **92**: 9122–9126.

Hilario, E. and Gogarten, J.P. (1993) Horizontal transfer of ATPase genes – the tree of life becomes the net of life. *BioSystems* **31**: 111–119.

Huyanen, M., Snel, B. and Bork, P. (1999) Lateral gene transfer, genome surveys and the phylogeny of prokaryotes. Technical comments. *Science* **284**: 2124–2128

Ibba, M., Morgan, S., Curnow, A.W. et al. (1997) A euryarchaeal lysyl-tRNA synthetase: resemblance to class I synthetases. *Science* **278**: 1119–1122

International Human Genome Sequencing Consortium. (2001) Initial sequencing and analysis of the human genome. *Nature* **409**: 860–921.

Iwabe, N., Kuma, K.-I., Hasegawa, M. et al. (1989) Evolutionary relationship of Archaea, Bacteria, and eukaryotes inferred from phylogenetic trees of duplicated genes. *Proc. Natl Acad. Sci. USA* **86**: 9355–9359.

Jain, R., Rivera, M.C. and Lake, J.A. (1999) Horizontal gene transfer among genomes: the complexity hypothesis. *Proc. Natl Acad. Sci. USA* **96**: 3801–3806.

Jordana, X., Chatton, B., Paz-Weisshaar et al. (1987) Structure of the yeast valyl-tRNA synthetase gene (*VAS1*) and the homology of its translated amino acid sequence with *Escherichia coli* isoleucyl-tRNA synthetase. *J. Biol. Chem.* **262**: 7189–7194.

Kakinuma, Y., Igarishi, K., Konishi, K. and Yamato, I. (1991) Primary structure of the alpha-subunit of vacuolar-type Na^+-ATPase in *Enterococcus hirae*, amplification of a 1000 bp fragment by polymerase chain reaction. *FEBS Lett.* **292**: 64–68.

Kates, M., Kushner, D.J. and Matheson, A.T. (1993) *The Biochemistry of Archaea (Archaebacteria)*, Elsevier, Amsterdam.

Keeling, P.J., Charlebois, R.L. and Doolittle, W.F. (1994) Archaebacterial genomes: eubacterial form and eukaryotic content. *Curr. Opin. Genet. Dev.* **4**: 816–822.

Keeling, P.J. and Doolittle, W.F. (1997) Evidence that eukaryotic triosephosphate isomerase is of alpha-proteobacterial origin. *Proc. Natl Acad. Sci. USA* **94**: 1270–1275.

Keeling, P.J. and McFadden, G.I. (1998) Origins of microsporidia. *Trends Microbiol.* **6**: 19–23.

Kelman, Z. (2000) The replication origin of archaea is finally revealed. *Trends Biochem. Sci.* **25**: 521–523.

Koonin E.V., Mushegian, A.R. and Bork, P. (1996) Non-orthologous gene displacement. *Trends Genet.* **12**: 334–336.

Koonin, E.V., Mushegian, A.R., Galperin, M.Y. and Walker, D.R. (1997) Comparison of archaeal and bacterial genomes: computer analysis of protein sequences predicts novel functions and suggests a chimeric origin for the Archaea. *Molec. Microbiol.* **25**: 619–637.

Koretke, K.K., Lupas, A.N., Warren, P.V. et al. (2000) Evolution of two-component signal transduction. *Mol. Biol. Evol.* **17**: 1956–1970.

Kubo, T., Nishizawa, S., Sugawara, A. et al. (2000) The complete nucleotide sequence of the mitochondrial genome of sugar beet (*Beta vulgaris* L.) reveals a novel gene for tRNA(Cys)(GCA). *Nucleic Acids Res.* **28**: 2571–2576.

Kurland, C. and Andersson S.G.E. (2000) Origin and evolution of the mitochondrial proteome. *Microbiol. Mol. Biol. Rev.* **64**: 786–820.

Kyrpides, N.C. and Woese, C.R. (1998) Universally conserved translation initiation factors. *Proc. Natl Acad. Sci. USA* **95**: 224–228.

Lake, J.A. (1988) Origin of the eukaryotic nucleus determined by rate-invariant analysis of rRNA sequences. *Nature* **331**: 184–186.

Lamour, V., Quevillon, S., Diriong, S. et al. (1994) Evolution of the Glx-tRNA synthetase family: the glutaminyl enzyme as a case for horizontal gene transfer. *Proc. Natl Acad. Sci. USA* **91**: 8670–8674.

Langer, D., Hain, J., Thuriaux, P. and Zillig, W. (1995) Transcription in Archaea: similarity to that in Eucarya. *Proc. Natl Acad. Sci. USA* **92**: 5768–5772.

Lathe, W.C., Snel, B. and Bork, P. (2000) Gene context conservation of a higher order than operons. *Trends Biochem. Sci.* **25**: 474–479.

Lawrence, J.G. (1997) Selfish operons and speciation by gene transfer. *Trends Microbiol.* **5**: 355–359.

Lawrence, J.G. and Roth, J.R. (1996) Selfish operons – horizontal transfer may drive the evolution of gene clusters. *Genetics* **143**: 1843–1860.

Lawson, F.S., Charlebois, R.L. and Dillon, J.-A.R. (1996) Phylogenetic analysis of carbamoylphosphate synthetase genes: evolution involving multiple gene duplications, gene fusions, and insertions and deletions of surrounding sequences. *Mol. Biol. Evol.* **13**: 970–977.

Lin, J. and Gerstein, M. (2000) Whole-genome trees based on the occurrence of folds and orthologs: implications for comparing genomes on different levels. *Genome Res.* **10**: 808–818.

Logsdon, J.M.Jr. and Faguy, D.M. (1999) Evolutionary genomics: *Thermotoga* heats up lateral gene transfer. *Current Biol.* **9**: R747-R751.

Lopez, P., Forterre, P. and Philippe, H. (1999) The root of the tree of life in the light of the covarion model. *J. Mol. Evol.* **49**: 496–508.

Margulis, L. (1970) *Origin of Eukaryotic Cells*, Yale University Press, New Haven, CT.

Martin, W. (1996) Is something wrong with the tree of life? *Bioessays* **18**: 523–527.

Martin, W. and Müller, M. (1998) The hydrogen hypothesis for the first eukaryote. *Nature* **392**: 37–41.

Martindale, D.W., Gu, Z.M. and Csank C. (1989) Isolation and complete sequence of the yeast isoleucyl-tRNA synthetase gene (ILS1). *Curr. Genet.* **15**: 99–106.

McArthur, A.G., Morrison, H.G., Nixon, J.E. et al. (2000) The *Giardia* genome project database. *FEMS Microbiol Lett.* **189**: 271–273.

Myllykallio, H., Lopez, P., López-Garcia, P. et al. (2000) Bacterial mode of replication with eukaryotic-like machinery in a hyperthermophilic archaeon. *Science* **288**: 2212–2215.

Nelson, K.E., Clayton, R.A., Gill, S.R. et al. (1999) Evidence for lateral gene transfer between Archaea and bacteria from genome sequence of *Thermotoga maritima*. *Nature* **399**: 323–329

Ochman, H., Lawrence, J.G. and Groisman, E.A. (2000) Lateral gene transfer and the nature of bacterial innovation. *Nature* **405**: 299–304.

Olendzenski, L., Liu, L., Zhaxybayeva, O. et al. (2000) Horizontal transfer of archaeal genes into the deinococcaceae: detection by molecular and computer-based approaches. *J. Mol. Evol.* **51**: 587–599.

Olsen, G.J. and Woese, C.R. (1997) Archaeal genomics – an overview. *Cell* **89**: 991–994.

Philippe, H. and Forterre, P. (1999) The rooting of the universal tree of life is not reliable. *J. Mol. Evol.* **49**: 509–523.

Reeve, J.N., Sandman, K. and Daniels, C.J. (1997) Archaeal histones, nucleosomes and transcription initiation. *Cell* **89**: 999–1002.

Rivera, M.C. and Lake, J.A. (1992) Evidence that eukaryotes and eocyte prokaryotes are immediate relatives. *Science* **257**: 74–76.

Rivera, M.C., Jain, R., Moore, J.E. and Lake, J.A. (1998) Genomic evidence for two functionally distinct gene classes. *Proc. Natl Acad. Sci. USA* **95**: 6239–6244.

Roger, A.J. and Brown, J.R. (1996) A chimeric origin for eukaryotes re-examined. *Trends Biochem. Sci.* **21**: 370–371.

Roger, A.J., Clark, C.G. and Doolittle, W.F. (1996) A possible mitochondrial gene in the early-branching amitochondriate protist *Trichomonas vaginalis*. *Proc. Natl Acad. Sci. USA* **93**: 14618–14622.

Salzberg, S.L., White, O., Peterson, J. et al. (2001) Microbial genes in the human genome: lateral transfer or gene loss? *Science* **292**: 1903–1906.

Smit, A. and Mushegian, A. (2000) Biosynthesis of isoprenoids via mevalonate in Archaea: the lost pathway. *Genome Res.* **10**: 1468–1484.

Smith, M.W., Feng, D.-F. and Doolittle, R.F. (1992) Evolution by acquisition: the case for horizontal gene transfers. *Trends Biochem. Sci.* **17**: 489–493.

Snel, B., Bork, P. and Huynen, M.A. (1999) Genome phylogeny based on gene content. *Nature Genet.* **21**: 108–110.

Stanhope, M.J., Lupas, A.N., Italia, M.J. et al. (2001) Phylogenetic analyses of genomic and EST sequences do not support horizontal gene transfers between bacteria and vertebrates. *Nature* 411: 940–944.

Stein, J.L. and Simon, M.I. (1996) Archaeal ubiquity. *Proc. Natl Acad. Sci. USA* **93**: 6228–6230.

Sumi, M., Sato, M.H., Denda, K. et al. (1992) A DNA fragment homologous to F1-ATPase β subunit amplified from genomic DNA of *Methanosarcina barkeri*: Indication of an archaebacterial F-type ATPase. *FEBS Lett.* **314**: 207–210.

Swofford, D.L. (1999) PAUP*. Phylogenetic Analysis Using Parsimony (*and Other Methods). Version 4, Sinauer Associates, Sunderland, MA.

Tatusov, R.L., Koonin, E.V. and Lipman, D.J. (1997) A genomic perspective on protein families. *Science* **278**: 631–637.

Teichmann, S.A. and Mitchison, G. (1999) Is there a phylogenetic signal in prokaryote proteins? *J. Mol. Evol.* **49**: 98–107.

Thompson, J.D., Higgins, D.G. and Gibson, T.J. (1994) CLUSTAL W: improving the sensitivity of progressive multiple sequence alignment through sequence weighting, positions-specific gap penalties and weight matrix choice. *Nucleic Acids Res.* **22**: 4673–4680.

Tsutsumi, S., Denda, K., Yokoyama, K. et al. (1991) Molecular cloning of genes encoding major subunits of a eubacterial V-type ATPase from *Thermus thermophilus*. *Biochim. Biophys. Acta* **1098**: 13–20.

Wilding E.I., Brown, J.R., Bryant, A. et al. (2000) Identification, evolution and essentiality of the mevalo- nate pathway for isopentenyl diphosphate biosynthesis in Gram-positive cocci. *J. Bacteriol.* **182**: 4319–4327.

Woese, C.R. (1998) The universal ancestor. *Proc. Natl Acad. Sci. USA* **51**: 221–271.

Woese, C.R. and Fox, G.E. (1977) Phylogenetic structure of the prokaryotic domain: the primary kingdoms. *Proc. Natl Acad. Sci. USA* **51**: 221–271.

Woese, C.R., Kandler, O. and Wheelis, M.L. (1990) Towards a natural system of organisms: Proposal for the domains Archaea, Bacteria and Eucarya. *Proc. Natl Acad. Sci. USA* **87**: 4576–4579.

Woese, C.R., Olsen, G.J., Ibba, M. and Söll, D. (2000) Aminoacyl-tRNA synthetases, the genetic code, and the evolutionary process. *Microbiol. Molec. Biol. Rev.* **64**: 202–236.

Wolf, Y.I., Aravind, L., Grishin, N.V. and Koonin, E.V. (1999) Evolution of aminoacyl-tRNA synthetases – analysis of unique domain architectures and phylogenetic trees reveals a complex history of horizontal gene transfer events. *Genome Res.* **9**: 689–710.

Endosymbiotic Gene Transfer: A Special Case of Horizontal Gene Transfer Germane to Endosymbiosis, the Origins of Organelles and the Origins of Eukaryotes

Katrin Henze, Claus Schnarrenberger and William Martin

Endosymbiotic gene transfer describes the process through which chloroplasts and mitochondria relinquished the majority of their genes to the nucleus while not having surrendered the majority of proteins integral to the eubacterial nature of their metabolism. It is a special case of lateral gene transfer that was very important for the establishment of biochemical compartmentation during the evolution of eukaryotic cells. Many examples of endosymbiotic gene transfer in the history of higher plant genes for chloroplast-cytosolic isoenzymes of the Calvin cycle and glycolysis have been considered in the past few years. The data indicate that nuclear genes for almost all glycolytic enzymes of the eukaryotic cytosol were acquired from eubacteria early in eukaryotic evolution, probably in the course of the endosymbiotic event that gave rise to the origin of mitochondria. The genes for enzymes that are possessed both by the symbiont and host during a given endosymbiosis are redundant. In such cases, it is observed that the intruding gene contributed by the symbiont had a very high likelihood of successful gene transfer. But the protein products of genes that were transferred from symbiont to host chromosomes have been surprisingly often rerouted to compartments other than that from which the genes were donated. Because of this, the localization of a protein within the eukaryotic cell is not *per se* an indicator of the evolutionary origin of its gene. The evolutionary history of the enzymes of central carbon metabolism in eukaryotes and the role of lateral gene transfer in that process is summarized. These results are considered in light of the ability of competing alternative theories for the origins of eukaryotes to account for the observations.

INTRODUCTION

Genes encoded in organellar genomes attest beyond all reasonable doubt to the eubacterial ancestry of chloroplasts (Martin et al., 1998) and mitochondria (Gray et al., 1999). But how does endosymbiotic theory account for the origin of organellar proteins that are not encoded in organellar DNA? Their fate is commonly explained with the help of a scenario elegantly argued by Weeden (1981) that is known as the "product specificity corollary" (also known as the "gene transfer corollary") to endosymbiotic theory. It posits that during the course of eukaryotic history, the majority of genes for proteins integral to organellar metabolism were

Copyright © 2002 by Academic Press.
All rights of reproduction in any form reserved.

transferred to the nucleus where they became integrated into the regulatory hierarchy of the nucleus and acquired a transit peptide, so that the functional gene products could be reimported into the organelle of their genetic origin on a daily basis since. This is a reasonable and logical scenario that satisfyingly explains why organelles have retained so much of their biochemically eubacterial heritage while having relinquished to the nucleus the majority of genes necessary to have done so. Though ultimately incorrect in the majority of specific cases in which it has been tested (Martin and Schnarrenberger, 1997; Martin, 1998), primarily due to the unfulfilled predictions that it generates about the origin of cytosolic enzymes, the gene transfer corollary does manage adequately to account for several organelle-to-nucleus gene transfer events (Martin and Cerff, 1986; Baldauf and Palmer, 1990; Brennicke et al., 1993; Martin et al., 1993).

However, the simple phrase "organelle-to-nucleus gene transfer" raises a different, much more difficult and in our view much more pressing question, namely: what is the origin of the nuclear genome under endosymbiotic theory? When Weeden (1981) formulated the gene transfer corollary, it was not known whether chloroplasts and mitochondria were descendants of endosymbionts or not, and his considerations generated testable predictions that could help to muster evidence in favor (or in disfavor) of that view (Gray and Doolittle, 1982). A salient element of that reasoning was that nuclear-encoded organellar enzymes should reflect the evolutionary history of the symbiont while cytosolic enzymes in eukaryotes should reflect the evolutionary history of the host (Weeden, 1981; Gray and Doolittle, 1982). Although we now know with certainty that plastids and mitochondria are descendants of free-living eubacteria, and although we are reasonably (but not completely) confident that the origin of mitochondria preceeded the origin of chloroplasts in evolution, it is perhaps surprising that endosymbiotic theory is just as much in the dark about the origin of the host that acquired mitochondria today as it was 20 years ago. In other words, in order to have a null hypothesis for the patterns of similarity to

be expected for a given eukaryotic nuclear gene, one has to have a biological model for the origins of eukaryotes that generates such predictions.

What was the host? Biologists have a number of different suggestions, but no generally accepted answer to that question, but there are a few important clues that can serve as a guideline. Discovery of the archaebacterial nature of components of the nuclear genetic apparatus (Pühler et al., 1989; Ouzounis and Sander, 1992; Rowlands et al., 1994; Langer et al., 1995) and the results of extensive molecular phylogenetic work have slowly unveiled the host that acquired the mitochondrion as a descendant (in part or in whole) of an archaebacterium (Zillig et al., 1989; Ouzounis and Sander, 1992; Brown and Doolittle, 1995, 1997; Langer et al., 1995; Doolittle, 1996; Ribiero and Golding, 1998; Rivera et al., 1998). Under the simplest set of premises possible, this means that the host cell that acquired the mitochondrion was an archaebacterium (Doolittle, 1996; Martin and Müller, 1998). However, many other more complicated models involving more than two symbiotic partners to give rise to a mitochondrion-bearing eukaryote have been elaborated. It is worthwhile briefly to recapitulate a few of these models to see what sorts of predictions they generate, in particular about the evolution of molecules that occur in the cytosol of contemporary eukaryotes.

Probably the most popular model up until a few years ago is what is best described as the achezoa model (Cavalier-Smith 1987a,b), also sometimes called the "classical hypothesis" (Doolittle, 1980, 1998). Under this view, a primitively amitochondriate, nucleus-bearing cell (an archezoon) is suggested to have arisen from an archaebacterial ancestor in a process that involved the origin of typical eukaryotic features (cytoskeleton, endocytosis, nucleus and mitosis). A member of that ancient group acquired the mitochondrion through endosymbiosis, while others (the suspectedly most primitive eukaryotes) remained amitochondriate up to the present day. This model has not withstood the scrutiny of molecular phylogeneticists, who have found that all amitochondriate eukaryotes studied to date possessed a mitochondrion in their evolutionary past, but lost the organelle subsequently (Clark and Roger, 1995; Germot et al., 1997; Embley and Hirt, 1998; Martin and Müller, 1998; Gray et al., 1999; Roger, 1999; Tovar et al., 1999; Müller, 2000; Philippe et al., 2000), the evidence for which is founded in nuclear genes in these amitochondriate organisms which reflect a mitochondrial ancestry (that is, they were transferred from the organelle to the nucleus via endosymbiotic gene transfer). Furthermore, the achezoa model clearly predicts all eukaryotes in general, and amitochondriate eukaryotes in particular, to possess archaebacterial enzymes of cytosolic metabolism, because under this model, amitochondriate protists are, on the bottom line, to be interpreted as direct descendants of archaebacteria (Cavalier-Smith, 1987b). A version of the achezoa model has been published under the name of the "ox-tox" hypothesis (Andersson and Kurland, 1999), but it is not fundamentally different from the achezoa model, rather it is very similar to 1975 formulations of the endosymbiont hypothesis (John and Whatley, 1975). For a further discussion of this matter, see Rotte et al. (2000) and Martin (2000).

A second model is that of Lynn Margulis, which entails a symbiosis between a spirochete and a *Thermoplasma*-like archaebacterium (Sagan, 1967; Margulis, 1970, 1996) to give rise to a primitive, flagellated, mitochondrion-lacking (amitochondriate) eukaryote that subsequently acquired the mitochondrion through endosymbiosis. In the most recent formulation, that model suggests that the spirochete endosymbiont also gave rise to the nuclear compartment (Margulis et al., 2000). This model predicts that eukaryotic enzymes should, in general, be more similar to *Thermoplasma* enzymes than to homologues from other prokaryotes, but the sequence of the *Thermoplasma* genome did not bear out that prediction (Cowan, 2000).

A third model that has been discussed at length in the literature is the idea that the nucleus was an endosymbiont. This model is exactly as old as the notion that chloroplasts descend from cyanobacteria, because the first paper thoroughly to argue the latter case (almost 100 years ago) also argued the former, albeit more briefly (Mereschkowsky, 1905). The

idea that the nuclear compartment is the remnant of an ancient endosymbiosis suffered extinction for many decades (Mereschkowsky, 1905), but it was resuscitated in 1994 (Lake and Rivera, 1994), was rejuvenated in 1996 (Gupta and Golding, 1996), came more or less fully back to life in 1998 (Lopez-Garcia and Moreira, 1999) and has even been claimed to receive direct support from data (Horiike et al., 2001). In these formulations, the nucleus is viewed as an archaebacterial symbiont that took up residence in a eubacterial host. In Margulis's formulation, a spirochete symbiont (that became the flagellum and the nucleus) took up residence in an archaebacterial host. It should be noted that there are several severe cell biological problems (discussed in Martin, 1999b; Rotte and Martin, 2001) with this model that are usually overlooked by its advocates, most notably (1) that the nucleus is not bounded by a double membrane, like chloroplasts and mitochondria, but rather by a folded single membrane, (2) that the nuclear membrane disintegrates at mitosis (viable chloroplasts and mitochondria never lose their surrounding membranes), and (3) the way that the cytosol is separated from the nucleus has no similarity whatsoever to the way that free-living prokaryotes are separated from their environment. Perhaps most importantly, "nucleosymbiotic" models derive an archezoon (a nucleus-bearing but amitochondriate eukaryote), whereby – as mentioned above – available data indicate that all eukaryotes, including those that lack mitochondria, once possessed mitochondria in their evolutionary past, but lost them secondarily. Therefore, these models predict eukaryotes that never possessed mitochondria (achaezoa) to be found, descendants of the stem that acquired mitochondria. Furthermore, with regard to their statements about the cytosol, these models would ultimately predict eukaryotes to possess DNA both in the nucleus and in the cytosol (in addition to genomes in chloroplasts and mitochondria).

The fourth, and arguably simplest, current model to derive a mitochondrion-bearing eukaryote is called the "hydrogen hypothesis" (Martin and Müller, 1998). It posits that the host cell of the mitochondrial symbiosis was an archaebacterium, specifically an autotrophic and H_2-dependent archaebacterium, with an energy metabolism possibly similar to that found in modern methanogens. It accounts for the common ancestry of mitochondria and hydrogenosomes – the H_2-producing organelles of ATP-synthesis in anaerobic eukaryotes (Müller, 1988, 1993; Embley and Hirt, 1998). It posits that eukaryotes acquired their heterotrophic lifestyle, including the genes and enzymes necessary for that lifestyle, from the mitochondrial symbiont via endosymbiotic gene transfer from the heterotrophic symbiont's (eubacterial) to the autotrophic host's (archaebacterial) chromosomes (Martin and Müller, 1998). Under this model, the origins of mitochondria/hydrogenosomes and the origins of eukaryotes are identical and the nucleus is a subsequent eukaryotic invention, the origin of which probably relates mechanistically to the origin of eubacterial lipid synthesis in eukaryotes (Martin, 1999b). This model explicitly predicts eukaryotes to have a eubacterial glycolytic pathway in the cytosol as the result of specifically selected endosymbiotic gene transfer from symbiont to host prior to the evolution of a mitochondrial protein import machinery.

RESULTS AND DISCUSSION

Evolution of two further enzymes of central sugar phosphate metabolism in higher plants

In previous work, we have studied the evolution of several enzymes involved in sugar phosphate metabolism in the chloroplast and cytosol of higher plants (Henze et al., 1994; Martin and Schnarrenberger, 1997; Martin and Herrmann, 1998) and of the cytosolic homologues in several protists (Henze et al., 1995, 1998, 2001; Wu et al., 2001). Two enzymes whose gene origins have been unclear are ribose-5-phosphate isomerase (RPI), that catalyzes the freely reversible isomerization of ribose-5-phosphate and ribulose-5-phosphate, and phosphoglucomutase (PGluM), that catalyzes the freely reversible interconversion of glucose-1-phosphate and glucose-6-phosphate. Homologues for both genes

from a variety of sources were collected from GenBank and subjected to phylogenetic analysis (Figure 27.1).

The tree for RPI, which is a relatively short enzyme of about 220 amino acids, does not provide very sharp resolution, but it does reveal that the enzymes from plants, animals, and fungi are clearly more similar to eubacterial homologues than they are to archaebacterial homologues (Figure 27.1A). In spinach, there is only one isoenzyme of RPI known; cell fractionation has shown it to be located in the plastid (Schnarrenberger et al., 1995). The subcellular localization of the two *Arabidopsis* enzymes inferred from the genome sequence is currently unknown. The plant enzyme does not branch specifically with the cyanobacterial homologue, nor does the enzyme from animals and fungi branch specifically with proteobacterial homologues, as one might expect if these enzymes were donated to eukaryotes from the ancestors of plastids and mitochondria, respectively. This can be due to a number of factors, including the simple phylogenetic resolution problems involved with a protein that is only 220 amino acids long or, alternatively, to horizontal gene transfer between free-living eubacteria subsequent to the origins of organelles (see below). It is noteworthy that the plant enzyme does not branch with its homologues from animals and yeast, suggesting that the eubacterial donor of the plant gene was a different bacterium from that which donated the gene to the heterotrophic eukaryotes sampled.

The tree for PGluM (Figure 27.1B) reveals that the eukaryotic genes are derived from a single acquisition from a eubacterial donor, and that the higher plant chloroplast-cytosol isoenzymes are related by a gene duplication in the plant lineage, as has been observed for the majority of chloroplast-cytosol isoenzymes studied to date (Martin and Schnarrenberger, 1997), and in contrast to the predictions from the product specificity corollary (Weeden, 1981). Archaebacterial homologues of PGluM were not identified, but further searching revealed that eubacterial PGluM is related to phosphomannomutase (PMM) which does have homologues in archaebacterial genomes (Figure 27.1B). The plant PMM genes surveyed are descendants of

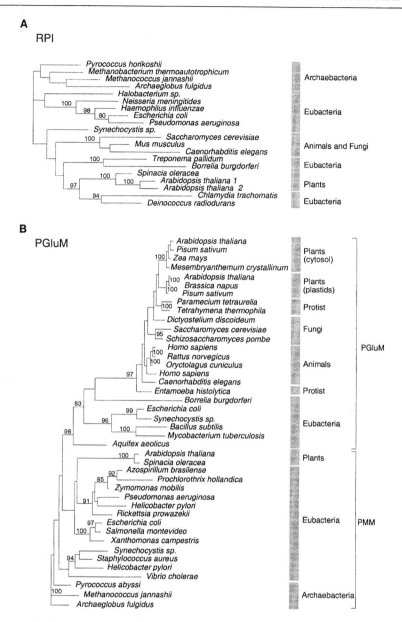

FIGURE 27.1 Protein phylogenies for (A) ribose-5-phosphate isomerase and (B) phosphoglucomutase. The RPI tree was inferred using ProtML (Adachi and Hasegawa, 1996) with local rearrangements and the JTT-F matrix starting from the neighbor-joining topology. The PGluM tree was inferred with the neighbor-joining method (Saitou and Nei, 1987) from the Dayhoff distance matrix using Phylip (Felsenstein, 1993). Bootstrap proportions >80 are indicated at branches.

eubacterial PMM genes, but again, not specifically related to the cyanobacterial homologue (see below). Notably, Figure 27.1B shows that the PGluM sequence from the amitochondriate protist *Entamoeba histolytica* (Ortner et al., 1997) is also an acquisition from eubacteria. In this

respect it is important to recall that although *Entamoeba histolytica* does not have either mitochondria or hydrogenosomes, it was recently discovered to possess a small, double membrane-bounded organelle in the cytosol – termed the mitosome – that is most easily interpreted as an intermediate stage in the evolutionary process of mitochonrion and hydrogenosome reduction (Tovar et al., 1999).

The evolution of compartmentalized sugar phosphate metabolism in higher plants

Figure 27.2 (see color plates) summarizes the observations from a large number of individual gene phylogenies published over the past years concerning the enzymes of compartmentalized sugar phosphate metabolism in higher plants. Importantly, all of the enzymes represented in the figure, with the exception of the large subunit of Rubisco, are encoded in the nucleus. Color coding indicates whether the nuclear encoded enzyme is (1) more similar to cyanobacterial homologues (green), (2) more similar to eubacterial (but not specifically cyanobacterial) homologues (blue), or (3) more similar to archaebacterial homologues (red). For the enzymes shaded in gray, no statement is currently possible. The results summarized in the figure have been discussed in further detail elsewhere (Martin and Schnarrenberger, 1997; Henze et al., 1998; Martin and Herrmann, 1998; Nowitzki et al., 1998; Wu et al., 2001) In general, what we observe is the following.

In the chloroplast, many of the enzymes of carbohydrate metabolism are indeed encoded by nuclear genes that were acquired from the cyanobacterial antecedent of plastids, as the product specificity corollary predicts. Examples are glyceraldehyde-3-phosphate dehydrogenase (GAPDH) (Martin et al., 1993), phosphoglycerate kinase (PGK) (Brinkmann and Martin, 1996; Martin and Schnarrenberger, 1997), transketolase (TKL) (Martin, 1998) and glucose-6-phosphate isomerase (GPI) (Nowitzki et al., 1998). But there are many exceptions to this rule. Notably, many of the enzymes localized in the plastid compartment are encoded by genes that were acquired from eubacteria, but – on the basis of available data – not from cyanobacteria. Examples are fructose-1,6-bisphosphatase (FBP) (Martin et al., 1996), triosephosphate isomerase (TPI) (Henze et al., 1994; Schmidt et al., 1995; Keeling

and Doolittle, 1997), glucose-6-phosphate dehydrogenase (G6PDH) (Wendt et al., 1999) and 6-phosphogluconate dehydrogenase (6PGDH) (Krepinsky et al., 2001). One enzyme in the chloroplast, enolase (ENO) (Hannaert et al., 2000) is more similar to archaebacterial homologues than to eubacterial homologues. In those cases where the chloroplast enzyme does not reflect a cyanobacterial origin, the nuclear gene for the enzyme arose through duplication of a pre-existing nuclear gene in the higher plant lineage. But as mentioned above, in all cases except enolase those pre-existing nuclear genes were themselves acquisitions from eubacteria.

In the cytosol, almost without exception, all of the enzymes of central carbon metabolism in higher plants are acquisitions from eubacteria. The exception is again enolase, which reflects an archaebacterial origin (Hannaert et al., 2000). Another is phosphoglycerate kinase, where a cyanobacterial enzyme has taken up residence in the cytosol. Very importantly, if one were to prepare Figure 27.2 for yeast or humans, the cytosolic portion of the figure would be exactly the same, except that PP_i-dependent PFK would be missing, both PGK and 6GPDH would be blue (instead of green), the oxidative pentose phosphate cycle would be in the cytosol, and transketolase (TKL) would be blue in fungi and red in animals.

Yet even more importantly, for those glycolytic enzymes that have been studied from amitochondriate protists (Markos et al., 1993; Henze et al., 1995, 1998; Keeling and Doolittle, 1997) a similar picture emerges, namely that amitochondriate protists also possess eubacterial enzymes of the glycolytic pathway (Martin and Müller, 1998). Specific examples include glucokinase from *Trichomonas vaginalis* (Wu et al., 2001), PP_i-PFK from *Trichomonas vaginalis* and *Giardia lamblia* (Mertens et al., 1998), aldolase from *Giardia lamblia* (Henze et al., 1998) and from *Trichomonas vaginalis* (Henze and Müller, unpublished), TPI from *Giardia lamblia* (Keeling and Doolittle, 1997; Martin, 1998), GAPDH from *Trichomonas vaginalis* (Markos et al., 1993), *Entamoeba histolytica* and *Giardia lamblia* (Markos et al., 1993; Henze et al., 1995), and PGluM from *Entamoeba histolytica* (Figure 27.1B).

The endosymbiont hypothesis in its various formulations and the archezoa hypothesis is at a loss to account for the observation that eukaryotes have a eubacterial glycolytic pathway in their cytosol. By contrast, the hydrogen hypothesis specifically predicts eukaryotes to have a eubacterial glycolytic pathway because the mechanism that it posits to have associated the mitochondrial (hydrogenosomal) symbiont with its H_2-dependent host entails the acquisition by the autotrophic host of the symbiont's heterotrophic lifestyle (glycolysis in particular) through endosymbiotic gene transfer, a special case of horizontal gene transfer (Martin and Müller, 1998).

It is noteworthy that the only enzyme of the glycolytic pathway in eukaryotes that seems to be a direct inheritance from the archaebacterial host lineage, enolase, is more similar to homologues from methanogens than to homologues from other eukaryotes (Hannaert et al., 2000), in line with the prediction of the hydrogen hypothesis that the host lineage should ultimately reflect a methanogenic ancestry (Martin and Müller, 1998). This view is also supported by that finding that methanogens are the only prokaryotes that possess true histones (Sandman and Reeve, 1998). Also in line with the hydrogen hypothesis is the finding that some protists have even preserved glycolytic enzymes in their mitochondria (Liaud et al., 2000). Furthermore in line with the predictions of the hydrogen hypothesis is the finding that a highly characteristic enzyme of hydrogenosomes, pyruvate:ferredoxin oxidoreductase (Müller, 1993), is found in the mitochondria of *Euglena gracilis* and that the mitochondrial and hydrogenosomal enzymes share a common eubacterial origin (Rotte et al., 2001).

Horizontal gene transfer between free-living eubacteria adds an additional level of complexity

Since the first edition of this book was published, it has become apparent that horizontal gene transfer between free-living eubacteria is a very widespread process, at least in evolutionarily recent times (Lawrence and Ochman, 1998; Doolittle, 1999; Eisen 2000; Ochman et al., 2000). Since horizontal gene transfer is well known

to occur at appreciable rates among free-living prokaryotes today, we should assume it also to have occurred in the distant past. This simple logic is essential to keep in mind when considering the origin of eukaryotic genes that were acquired from eubacterial symbionts through endosymbiotic gene transfer (Martin and Schnarrenberger, 1997; Martin, 1999a). The reason is because very many of the eukaryotic genes studied to date that clearly come from eubacteria do not branch specifically with their homologues from cyanobacteria and α-proteobacteria, the antecedents of chloroplasts and mitochondria. The interpretation of such findings requires a bit of thought.

For example, in a recent survey of several thousand *Arabidopsis* genes, all of 20 different sequenced prokaryotic genomes sampled would have appeared to have donated genes to the *Arabidopsis* lineage, if the gene phylogenies are interpreted at face value (Rujan and Martin, 2001). In other words, for all 20 prokaryotic genomes in that study (16 eubacteria and four archaebacteria), at least one gene tree was observed where the *Arabidopsis* nuclear gene branched with a homologue from the given genome. At face value, that would suggest that all 20 lineages of prokaryotes donated genes to *Arabidopsis*. Had 40 different prokaryotic genomes been studied, then 40 different donors to the *Arabidopsis* lineage would have been implied, if the trees are taken at face value.

But if we think things through in full and consider the process of horizontal gene transfer between free-living prokaryotes as a continuum back through time, it becomes immediately evident that it is unrealistic to expect all plant nuclear genes that were donated by the antecedent of plastids to branch with homologues from contemporary cyanobacteria, and it is unrealistic to expect all genes that were donated to the eukaryotic lineage by the antecedent of mitochondria to branch with α-proteobacterial homologues. This is shown in Figure 27.3 (see color plates), which was modified from Martin (1999a). All available evidence indicated a single origin of plastids (Martin et al., 1998; Moreira et al., 2000) and mitochondria (Gray et al., 1999). When the cyanobacterium that gave rise to plastids about 1.5 billion years ago (Doolittle, 1997) entered into its symbiosis, it contained exactly one genome's worth of genes (plus a plasmid or two, probably). Its free-living relative among the cyanobacteria that existed then contained exactly the same set of genes. But the cyanobacterium that became the plastid was cut off from gene exchange with other free-living prokaryotes, because it was genetically isolated within the eukaryotic cell, a condition that persisted for another 1.5 billion years to the present. By contrast, its free-living relatives were free to exchange their genes with other cyanobacteria and with other non-cyanobacterial prokaryotes for another 1.5 billion years. Exactly the same reasoning applies to the α-proteobacterium that gave rise to mitochondria, except that the free-living cousins of the mitochondrial (hydrogenosomal) symbiont had even more time, *about 2 billion years* (Doolittle, 1997), to exchange genes with other prokaryotes, whereas the mitochondrial symbiont did not.

Given that, let us consider the blazing speed (in terms of geological time) with which *E. coli*, as a well-studied example, has exchanged its genes with other eubacteria (Lawrence and Ochman, 1998). Lawrence and Ochman (1998) estimated that *E. coli* has acquired foreign DNA at a rate of about 16 kb per million years over the last 100 million years, and by inference, that about the same amount of DNA has been lost from its genome in return. The values for other prokaryotes may be similar (Doolittle, 1999; Eisen, 2000; Ochman et al., 2000). If that rate were to be projected back in time into the depths of history where mitochondria and plastids arose, then the current *E. coli* genome would not share any genes with the genome that existed in ancestors of *E. coli* that lived 1 billion years ago (Martin, 1999a).

Thus, it is extremely *unlikely* that any free-living prokaryote contains exactly the same set of genes as either the plastid symbiont and or the mitochondrial symbiont (Martin and Schnarrenberger, 1997; Martin 1999a; Rujan and Martin, 2001). For exactly the same reasons, it is extremely likely that many of the genes that are currently found in eukaryotic genomes and that are clearly acquisitions from eubacteria, probably come from the ancestors of mitochondria (hydrogenosomes) and the ancestors of plastids, even though they do not appear to stem from α-proteobacteria and cyanobacteria respectively. Conversely, one can expect some (or many) genes that were donated to the nucleus from mitochondria and chloroplasts not to branch with α-proteobacterial or cyanobacterial homologues. Rather, they will branch with any number of eubacterial groups, depending upon where these ancestrally donated genes have ended up in contemporary eubacteria. This is the essence of what is shown in Figure 27.3 (Martin, 1999a; Rujan and Martin, 2001).

CONCLUSION

Horizontal gene transfer is a very powerful process in evolution when it comes to shaping the gene content of genomes, both prokaryotic and eukaryotic. Plastids and mitochondria donated many genes to eukaryotic genomes. Some of the products of those genes are reimported as proteins into the organelle, many others are not – rather they are imported into other organelles or localized in the cytosol or the nucleus, which is not an organelle. Contrary to what was believed 20 years ago (and is sometimes still believed today), the compartmentation of a protein is not a good indicator of its evolutionary origin. Molecular phylogenetics generally has a hard time coming to grips with the problem of horizontal gene transfer because it makes it more difficult to draw direct inferences about the evolutionary past from the phylogenies of contemporary genes. But the situation may not be so dire. If we do not allow ourselves to digress into the realm of descriptive hypothesis-free science, that is if we have good biological theories with which to interpret molecular phylogenies and which we can test with molecular phylogenies, then horizontal gene transfer will probably enrich, rather than hamper, our understanding of life's history, in particular the origin of eukaryotic genes.

ACKNOWLEDGMENTS

We thank Miklos Müller for numerous stimulating discussions and Marianne Limpert for help in preparing the manuscript. Generous financial support from the Deutsche Forschungsgemeinschaft is gratefully acknowledged.

REFERENCES

Adachi, J. and Hasegawa, M. (1996) *Computer Science Monographs, No. 28. MOLPHY Version 2.3: Programs for Molecular Phylogenetics Based on Maximum Likelihood,* Institute of Statistical Mathematics, Tokyo.

Andersson, S.G.E. and Kurland, C.G. (1999) Origins of mitochondria and hydrogenosomes. *Curr. Opin. Microbiol.* **2**: 535–541.

Baldauf, S. and Palmer, J.D. (1990) Evolutionary transfer of the chloroplast *tufA* gene to the nucleus. *Nature* **344**: 262–265.

Brennicke, A., Grohmann, L., Hiesel, R. et al. (1993) The mitochondrial genome on its way to the nucleus: different stages of gene transfer in higher plants. *FEBS Letters* **325**: 140–145.

Brinkmann, H. and Martin, W. (1996) Higher plant chloroplast and cytosolic 3-phosphoglycerate kinases: a case of endosymbiotic gene replacement. *Plant Mol. Biol.* **30**: 65–75.

Brown, J.R. and Doolittle, W.F. (1995) Root of the universal tree of life based on ancient aminoacyl-tRNA synthetase gene duplications. *Proc. Natl Acad. Sci. USA* **92**: 2441–2445.

Brown, J.R. and Doolittle, W.F. (1997) Archaea and the prokaryote-to-eukaryote transition. *Microbiol. Mol. Biol. Rev.* **61**: 456–502.

Cavalier-Smith, T. (1987a) Eukaryotes with no mitochondria. *Nature* **326**: 332–333.

Cavalier-Smith, T. (1987b) The origin of eukaryote and archaebacterial cells. *Ann. NY Acad. Sci.* **503**: 17–54.

Clark, C.G. and Roger, A.J. (1995) Direct evidence for secondary loss of mitochondria in *Entamoeba histolytica. Proc. Natl Acad. Sci. USA* **92**: 6518–6521.

Cowan, D. (2000) Use your neighbor's genes. *Nature* **407**: 466–467.

Doolittle, W.F. (1980) Revolutionary concepts in evolutionary biology. *Trends Biochem. Sci.* **5**: 146–149.

Doolittle, W.F. (1996) Some aspects of the biology of cells and their possible evolutionary significance. In *Evolution of Microbial Life* (eds, D. Roberts, P. Sharp, G. Alserson and M. Collins), pp. 1–21, 54th Symp. Soc. Gen. Microbiol., Cambridge University Press, Cambridge.

Doolittle, W.F. (1997) Fun with genealogy. *Proc. Natl Acad. Sci. USA* **94**: 12751–12753.

Doolittle, W.F. (1998) A paradigm gets shifty. *Nature* **392**: 15–16.

Doolittle, W.F. (1999) Phylogenetic classification and the universal tree. *Science* **284**: 2124–2128.

Eisen, J. (2000) Horizontal gene transfer among microbial genomes: new insights from complete genome analysis. *Curr. Opin. Genet. Dev.* **10**: 606–611.

Embley, T.M. and Hirt, R.P. (1998) Early branching eukaryotes? *Curr. Opin. Genet. Dev.* **8**: 655–661.

Embley, T.M. and Martin, W. (1998) A hydrogen-producing mitochondrion. *Nature* **396**: 517–519.

Felsenstein, J. (1993) *Phylip (Phylogeny Inference Package) Manual,* version 3.5c. Distributed by the author. University of Washington, Seattle, Department of Genetics.

Germot, A., Philippe, H. and Le Guyader, H. (1997) Evidence for loss of mitochondria in microsporidia from a mitochondrial-type HSP70 in *Nosema locustae. Mol. Biochem. Parasitol.* **87**: 159–168.

Gray, M.W. and Doolittle, W.F. (1982) Has the endosymbiont hypothesis been proven? *Microbiol. Rev.* **46**: 1–42.

Gray, M.W., Burger, G. and Lang, B.F. (1999) Mitochondrial evolution. *Science* **283**: 1476–1481.

Gupta, R.S. and Golding, G.B. (1996) The origin of the eukaryotic cell. *Trends Biochem. Sci.* **21**: 166–171.

Hannaert, V., Brinkmann, H., Nowitzki, U. et al. (2000) Enolase from *Trypanosoma brucei,* from the amitochondriate protist *Mastigamoeba balamuthi,* and from the chloroplast and cytosol of *Euglena gracilis:* pieces in the evolutionary puzzle of the eukaryotic glycolytic pathway. *Mol. Biol. Evol.* **17**: 989–1000.

Hashimoto, T., Sanchez, L.B., Shirakura, T. et al. (1998) Secondary absence of mitochondria in *Giardia lamblia* and *Trichomonas vaginalis* revealed by valyl-tRNA synthetase phylogeny. *Proc. Natl Acad. Sci. USA* **95**: 6860–6865.

Henze, K., Schnarrenberger, C., Kellermann, J. and Martin, W. (1994) Chloroplast and cytosolic triosephosphate isomerases from spinach: purification, microsequencing and cDNA cloning of the chloroplast enzyme. *Plant Mol. Biol.* **26**: 1961–1973.

Henze, K., Badr, A., Wettern, M. et al. (1995) A nuclear gene of eubacterial origin in *Euglena gracilis* reflects cryptic endosymbioses during protist evolution. *Proc. Natl Acad. Sci. USA* **92**: 9122–9126.

Henze, K., Morrison, H.G., Sogin, M.L. and Müller, M. (1998) Sequence and phylogenetic position of a class II aldolase gene in the amitochondriate protist, *Giardia lamblia. Gene* **222**: 163–168.

Henze, K., Suguri, S., Moore, D.V. et al. (2001) Cyanobacterial relationships of the glycolytic enzymes of Giardia intestinalis, glucokinase and glucose-6-phosphate isomerase. *Gene,* in press.

Horiike, T., Hamada, K., Kanaya, S. and Shinozawa, T. (2001) Origin of eukaryotic cell nuclei by symbiosis of Archaea in Bacteria revealed is revealed by homology hit analysis. *Nature Cell Biol.* **3**: 210–214.

John, P. and Whatley, F.R. (1975) *Paracoccus denitrificans* and the evolutionary origin of the mitochondrion. *Nature* **254**: 495–498.

Keeling, P.J. and Doolittle, W.F. (1997) Evidence that eukaryotic triosephosphate isomerase is of alpha-proteobacterial origin. *Proc. Natl Acad. Sci. USA* **94**: 1270–1275.

Keeling, P.J. and Palmer, J.D. (2000) Parabasalian flagellates are ancient eukaryotes. *Nature* **405**: 635–637.

Krepinsky, K., Plaumann, M., Martin, W. and Schnarrenberger, C. (2001) Purification and cloning of chloroplast 6-phosphogluconate dehydrogenase from spinach: cyanobacterial genes for chloroplast and cytosolis isoenzymes encoded in eukaryotic chromosomes. *Eur. J. Biochem.* **268**: 2678–2686.

Lake, J.A. and Rivera, M.C. (1994) Was the nucleus the first endosymbiont? *Proc. Natl Acad. Sci. USA* **91**: 2880–2881.

Langer, D., Hain, J., Thuriaux, P. and Zillig, W. (1995) Transcription in Archaea: similarity to that in Eukarya. *Proc. Natl Acad. Sci. USA* **92**: 5768–5772.

Lawrence, J.G. and Ochman, H. (1998) Molecular archaeology of the *Escherichia coli* genome. *Proc. Natl Acad. Sci. USA* **95**: 9413–9417.

Liaud, M.-F., Lichtlé, C., Apt, K. et al. (2000) Compartment-specific isoforms of TPI and GAPDH are imported into diatom mitochondria as a fusion protein: evidence in favor of a mitochondrial origin of the eukaryotic glycolytic pathway. *Mol. Biol. Evol.* **17**: 213–223.

Lopez-Garcia, P. and Moreira, D (1999) Metabolic symbiosis at the origin of eukaryotes. *Trends Biochem. Sci.* **24**: 88–93.

Margulis, L. (1970) *Origin of Eukaryotic Cells*, Yale University Press, New Haven, CT.

Margulis, L. (1996) Archaeal–eubacterial mergers in the origin of Eukarya: phylogenetic classification of life. *Proc. Natl Acad. Sci. USA* **93**:1071–1076.

Margulis, L., Dolan, M.F. and Guerrero, R. (2000) The chimeric eukaryote: origin of the nucleus from the karyomastigont in amitochondriate protists. *Proc. Natl Acad. Sci. USA* **97**: 6954–6959.

Markos, A., Miretsky, A. and Müller, M. (1993) A glyceraldehyde-3-phosphate dehydrogenase with eubacterial features in the amitochondriate eukaryote *Trichomonas vaginalis. J. Mol. Evol.* **37**: 631–643.

Martin, W. (1998) Endosymbiosis and the origins of chloroplast-cytosol isoenzymes: a revision of the gene transfer corollary. In *Horizontal Gene Transfer: Implications and Consequences* (eds, M. Syvanen and C. Kado), pp. 363–379, Chapman and Hall, London.

Martin, W. (1999a) Mosaic bacterial chromosomes – a challenge en route to a tree of genomes. *BioEssays* **21**: 99–104.

Martin, W. (1999b) A briefly argued case that mitochondria and plastids are descendants of endosymbionts, but that the nuclear compartment is not. *Proc. R. Soc. Lond. B* **266**: 1387–1395.

Martin, W. (2000) Primitive anaerobic protozoa: the wrong host for mitochondria and hydrogenosomes? *Microbiology* **146**: 1021–1022.

Martin, W. and Cerff, R. (1986) Prokaryotic features of a nucleus encoded enzyme: cDNA sequences for chloroplast and cytosolyic glyceraldehyde-3-phosphate dehydrogenases from mustard (*Sinapis alba*). *Eur. J. Biochem.* **159**: 323–331.

Martin, W. and Herrmann, R.G. (1998) Gene transfer from organelles to the nucleus: How much, what happens and why? *Plant Physiol.* **118**: 9–17.

Martin, W. and Müller, M. (1998) The hydrogen hypothesis for the first eukaryote. *Nature* **392**: 37–41.

Martin, W. and Schnarrenberger, C. (1997) The evolution of the Calvin cycle from prokaryotic to eukaryotic chromosomes: A case study of functional redundancy in ancient pathways through endosymbiosis. *Curr. Genet.* **32**: 1–18.

Martin, W., Brinkmann, H., Savona, C. and Cerff, R. (1993) Evidence for a chimaeric nature of nuclear genomes: eubacterial origin of eukaryotic glyceraldehyde-3-phosphate dehydrogenase genes. *Proc. Natl Acad. Sci. USA* **90**: 8692–8696.

Martin, W., Mustafa, A.-Z., Henze, K. and Schnarrenberger, C. (1996) Higher plant chloroplast and cytosolic fructose-1,6-bisphophosphatase isoenzymes: origins via duplication rather than prokaryote-eukaryote divergence. *Plant Mol. Biol.* **32**: 485–491.

Martin, W., Stoebe, B., Goremykin, V. et al. (1998) Gene transfer to the nucleus and the evolution of chloroplasts. *Nature* **393**: 162–165.

Mereschkowsky, C. (1905) Über Natur und Ursprung der Chromatophoren im Pflanzenreiche. *Biol. Centralbl.* **25**: 593–604 (English translation in Martin, W. and Kowallik, K.V. (1999) *Eur. J. Phycol.* **34**: 287–295).

Mertens, E., Ladror, U.S., Lee, J.A. et al. (1998) The pyrophosphate-dependent phosphofructokinase of the protist, *Trichomonas vaginalis*, and the evolutionary relationships of protist phosphofructokinases. *J. Mol. Evol.* **47**: 739–750.

Moreira, D., Le Guyader, H. and Philippe, H. (2000) The origin of red algae and the evolution of chloroplasts. *Nature* **405**: 69–72.

Müller, M. (1988) Energy metabolism of protozoa without mitochondria. *Annu. Rev. Microbiol.* **42**: 465–488.

Müller, M. (1993) The hydrogenosome. *J. Gen. Microbiol.* **139**: 2879–2889.

Müller, M. (2000) A mitochondrion in *Entamoeba histolytica*? *Parasitol. Today* **16**: 368–369.

Nowitzki, U., Flechner, A., Kellermann, J. et al. (1998) Eubacterial origin of eukaryotic nuclear genes for chloroplast and cytosolic glucose-6-phosphate isomerase: sampling eubacterial gene diversity in eukaryotic chromosomes through symbiosis. *Gene* **214**: 205–213.

Ochman, H., Lawrence, J.G. and Groisman, E.S. (2000) Lateral gene transfer and the nature of bacterial innovation. *Nature* **405**: 299–304.

Ortner, S., Binder, M., Schiener, O. et al. (1997) Molecular and biochemical characterization of phosphoglucomutases from *Entamoeba histolytica* and *Entamoeba dispar. Mol. Biochem. Parasitol.* **90**: 121–129.

Ouzounis, C. and Sander, C. (1992) TFIIB, an evolutionary link between the transcription machineries of archaebacteria and eukaryotes. *Cell* **71**: 189–190.

Philippe, H., Germot, A. and Moreira, D. (2000) The new phylogeny of eukaryotes. *Curr. Opin. Gen. Dev.* **10**: 596–601.

Plaxton, W.C. (1996) The organization and regulation of plant glycolysis. *Annu. Rev. Plant Physiol. Plant Mol. Biol.* **47**: 185–214.

Pühler, G., Leffers, H., Gropp, F. et al. (1989) Archaebacterial DNA dependent RNA polymerases testify to the evolution of the eukaryotic nuclear genome. *Proc. Natl Acad. Sci. USA* **86**: 4569–4573.

Ribiero, S. and Golding, G.B. (1998) The mosaic nature of the eukaryotic nucleus. *Mol. Biol. Evol.* **15**: 779–788.

Rivera, M.C., Jain, R., Moore, J.E. and Lake, J.A. (1998) Genomic evidence for two functionally distinct gene classes. *Proc. Natl Acad. Sci. USA* **95**: 6239–6244.

Roger, A.J. (1999) Reconstructing early events in eukaryotic evolution. *Am. Nat.* **154**: 146–163.

Rotte, C. and Martin, W. (2001) Endosymbiosis does not explain the origin of the nucleus. *Nature Cell. Biol.* **8**: E173.

Rotte, C., Henze, K., Müller, M. and Martin, W. (2000) Origins of hydrogenosomes and mitochondria. *Curr. Opin. Microbiol.* **3**: 481–486.

Rotte, C., Stejskal, F., Zhu, G. et al. (2001) Pyruvate:NADP$^+$ oxidoreductase from the mitochondrion of *Euglena gracilis* and from the apicomplexan *Cryptosporidium parvum*: a fusion of pyruvate:ferredoxin oxidoreductase and NADPH-cytochrome P450 reductase. *Mol. Biol. Evol.* **18**: 710–720.

Rowlands, T., Baumann, P. and Jackson, S.P. (1994) The TATA-binding protein: a general transcription factor in eukaryotes and archaebacteria. *Science* **264**: 1326–1329.

Rujan, T. and Martin, W. (2001) How many genes in Arabidopsis come from cyanobacteria? An estimate from 386 protein phylogenies. *Trends Genet.*, in press

Sagan, L. (1967) On the origin of mitosing cells. *J. Theoret. Biol.* **14**: 225–274.

Saitou, N. and Nei, M. (1987) The neighbor-joining method: a new method for reconstructing phylogenetic trees. *Mol. Biol. Evol.* **4**: 406–425.

Sandman, K. and Reeve, J. (1998) Origin of the eukaryotic nucleus. *Science* **280**: 501–503.

Schmidt, M., Svendsen, I. and Feierabend, J. (1995) Analysis of the primary structure of the chloroplast isozyme of triosephosphate isomerase from rye leaves by protein and cDNA sequencing indicates a eukaryotic origin of its gene. *Biophys. Acta* **1261**: 257–264.

Schnarrenberger, C., Flechner, A. and Martin, W. (1995) Enzymatic evidence for a complete oxidative pentose phosphate pathway in chloroplasts and an incomplete pathway in the cytosol of spinach leaves. *Plant Physiol.* **108**: 609–614.

Tovar, J., Fischer, A. and Clark, C.G. (1999) The mitosome, a novel organelle related to mitochondria in the amito-chondrial parasite *Entamoeba histolytica*. *Mol. Microbiol.* **32**: 1013–1021.

Van de Peer, Y., Rensing, S., Maier, U.G. and De Wachter, R. (1996) Substitution rate calibration of small subunit RNA identifies chlorarachniophyte endosymbionts as remnants of green algae. *Proc. Natl Acad. Sci. USA* **93**: 7744–7748.

Weeden, N.F. (1981) Genetic and biochemical implications of the endosymbiotic origin of the chloroplast. *J. Mol. Evol.* **17**: 133–139.

Wendt, U.K., Hauschild, R., Lange, C. et al. (1999) Evidence for functional convergence of redox regulation in G6PDH isoforms of cyanobacteria and higher plants. *Plant Mol. Biol.* **40**: 487–494.

Wu, G., Henze, K. and Müller, M. (2001) Evolutionary relationships of the glucokinase from the amitochondriate protist, *Trichomonas vaginalis*. *Gene* **264**: 265–271.

Zauner, S., Fraunholz, M., Wastl, J. et al. (2000) Chloroplast protein and centrosomal genes, a tRNA intron, and odd telomeres in an unusually compact eukaryotic genome, the cryptomonad nucleomorph. *Proc. Natl Acad. Sci.* **97**: 200–205.

Zillig, W., Klenk, H.-P., Palm, P. et al. (1989) Did eukaryotes originate by a fusion event? *Endocyt. Cell Res.* **6**: 1–25.

Dating the Age of the Last Common Ancestor of All Living Organisms with a Protein Clock

Ronald M. Adkins and Wen-Hsiung Li

Using highly-conservative proteins that have evolved in an approximately clock-like manner and taking into consideration the variability of rates among sites, the date of the divergence between eubacteria and eukaryotes/achaebacteria, i.e. the age of the last common ancestor (LCA) of all living organisms, is estimated to be ~2.2 billion years ago. This estimate, which is close to that by Doolittle et al. may not necessarily be incompatible with the appearance of microfossils before 3.5 billion years ago because prokaryotes might have existed long before this divergence occurred. The minimum age for the progenitor lineage may be estimated by calculating the timing of ancient gene duplications that preceded the divergence of the three modern domains of life. Using three pairs of duplicate genes for which reasonable alignments could be obtained, the ages of the duplications are estimated to be roughly 1.3 to 1.6 billion years older than the age of the LCA, or roughly 3.5 to 3.8 billion years old. Therefore, the estimate of 2.2 billion years for the age of the LCA does not appear so recent as to be incompatible with the microfossil record.

INTRODUCTION

The extant forms of life are currently divided into three domains: the Bacteria (eubacteria), the Archaea (archaebacteria), and the Eukarya (eukaryotes). Molecular data strongly indicate that the bacterial and archaebacterial lineages diverged first and later the eukaryotic lineage branched off from the archaebacterial lineage (Iwabe et al., 1989; Brown and Doolittle, 1995; Baldauf et al., 1996; Lawson et al., 1996). Therefore, the date of divergence between the bacterial and the archaebacterial-eukaryotic lineages represents the last time that all living organisms shared a common ancestor, that is, the age of the last common ancestor (LCA) of all extant forms of life, or the age of the cenancestor (Fitch and Upper, 1987). This age is obviously of great interest to evolutionary biologists. Recently, Doolittle et al. (1996) used extensive protein sequence data to obtain an estimate of between 2 and 2.2 billion years for this age, and also an estimate of between 1.8 and 1.9 billion years for the Archaea–Eukarya divergence (i.e. the prokaryote–eukaryote divergence). Although these estimates are within the wide range (from 1.3 to 2.6 billion years) of the previous estimates of the prokaryote–eukaryote divergence based on limited molecular data (Jukes, 1969; McLaughlin and Dayhoff, 1970; Kimura and Ohta, 1973), they strongly contradict the view of some biologists and paleontologists that the prokaryote–eukaryote divergence occurred 3.5 billion years ago (Knoll,

Horizontal Gene Transfer
ISBN: 0-12-680126-6

Copyright © 2002 by Academic Press.
All rights of reproduction in any form reserved.

1992; Martin 1996), the fossil date for the first appearance of cellular life (Schopf, 1993). It was therefore worth obtaining another estimate by restricting the analysis to proteins that are highly conservative and appear to have evolved in a fairly regular (i.e. clock-like) manner, and also by taking into account the rate of variation among amino acid sites, which can have a strong effect on date estimation.

It is worth noting that the estimate that the eukaryote lineage arose only about 2 billion years ago is not necessarily incompatible with the fossil record that prokaryotes emerged before 3.5 billion years ago because prokaryotes could have existed long before the emergence of eukaryotes (Mooers and Redfield, 1996). Considerable insight into this issue may be obtained by estimating the date of a gene duplication that occurred before the Bacteria–Archaea divergence and gave rise to two genes that still exist in both prokaryotes and eukaryotes and can be readily recognized as duplicate genes. Note that for two genes to be readily recognizable as duplicate genes, their structure and function must have been well conserved and should be similar to the structure and function of their common ancestor. In other words, the structure and function of the ancestral gene was already or almost already as

sophisticated as the present-day duplicates. Since it would take a long time for a primitive protein to evolve to a sophisticated one, the duplication date might have been much younger than the emergence of prokaryotes. Therefore, if the duplication date can be shown to be considerably older than the age of the LCA, then there would be no conflict between the estimate of 2 to 2.2 billion years for the age of the LCA (or the estimate of 2 billion years for the age of the eukaryote lineage) and the fossil record that the prokaryotes emerged before 3.5 billion years ago. Three sets of ancient duplications were found to be suitable for this purpose.

Age of the last common ancestor of living organisms

Thirteen protein datasets were used to estimate the eubacteria–eukaryote divergence or the age of the last common ancestor (LCA) of living organisms. The result of this analysis is shown in Table 28.1. The alpha parameter of the gamma distribution indicates the strength of rate heterogeneity in a protein; the values of 0.5, 1 and 2 may be taken as strong, intermediate, and weak, respectively. The majority of these proteins have an alpha value not far from one and

TABLE 28.1 Estimation of the age of the last common ancestor (LCA) of all living organisms

Locus	N^a	Alpha^b	Age of LCA (Myr)
Argininosuccinate lyase	7	0.91	1480
1,4-α-Glucan branching enzyme	8	1.40	3410
Acetyl CoA C-acetyl transferase	8	1.27	1805
Dihydrolipoamide dehydrogenase	9	1.30	1970
Glucose 6-phosphate dehydrogenase	8	1.07	1767
Glutamine fructose 6-phosphate transaminase	4	1.05	2185
Glycine dehydrogenase	4	0.81	1101
Glycine hydroxymethyl transferase	10	0.99	1800
Ribonucleotide reductase (large subunit)	6	1.68	3985
Threonine tRNA ligase	4	1.96	2006
Valine tRNA ligase	5	0.88	1253
Phosphoglycerate kinase	10	1.15	2786
Isoleucine tRNA ligase	5	1.83	2994
Average			2196

[a]Number of sequences used. Sequences were initially aligned by CLUSTALW 1.6 (Thompson et al., 1994), and regions of ambiguous alignment were excluded.
[b]The alpha parameter of the gamma distribution of amino acid substitution rates among sites was estimated from ML trees by the program CODEML in the computer package PAML 1.1 (Yang, 1995).

therefore have intermediate rate heterogeneity. Two of the proteins, threonine tRNA ligase and isoleucine tRNA ligase, have an alpha value close to two and so have only weak rate heterogeneity.

The estimates of the age of the LCA vary greatly among proteins and there are some extreme values. For example, the estimate from glycine dehydrogenase is only 1101 Myr. This could possibly be due to a rate acceleration after the animal-yeast divergence, leading to an overestimate of the substitution rate. Or, it could be due to a horizontal gene transfer either from a prokaryote to a eukaryote or the other way around; however, there is no sequence from archaea available, so it can not be determined whether any horizontal transfer has occurred. On the other extreme, the estimate from ribonucleotide reductase (large subunit) is 3985 Myr; this could possibly be due to a slowdown in rate after the yeast–animal divergence, leading to an underestimate of the substitution rate. Because of these uncertainties, the estimates should be taken with much caution. Surprisingly, however, the average estimate of 2196 Myr is almost identical to Doolittle et al.'s estimate of 2156 Myr (after correction for rate heterogeneity). Therefore, 2200 Myr (i.e. 2.2 billion years) may be taken as an estimate of the age of LCA. At any rate, the main purpose was to see if this estimate is too young as claimed by Martin (1996) and others.

Dates of ancient duplications

Three pairs of ancient duplicate proteins exist for which reliable alignment can be obtained for a substantial part of the sequences. The first one is carbamoylphosphate synthetase. Although this protein was not duplicated, it contains an internal duplication that is present in all living organisms; the internal duplicates are called duplicates 1 and 2 and have been used to root the tree of life (Lawson et al., 1996). Figure 28.1 shows a tree constructed from the two duplicates of animal, yeast, archaeal, and eubacterial carbamoylphosphate synthetase sequences. Although some of the sequences with extreme rates have been eliminated, there are still some sequences that show very different rates; e.g. the

two eubacterial (*Neisseria* and *P. stutzeri*) duplicate-2 sequences show a much slower rate than the other duplicate-2 sequences. As proposed by Li and Tanimura (1987), to estimate divergence dates it is better to use only lineages that have evolved in a relatively regular manner. For example, to estimate the eukaryote–archaea divergence date (T_e) from the duplicate-2 sequences, the animal and yeast lineages will give similar estimates because their branch lengths are nearly equal, whereas the *Neisseria* and *P. stutzeri* lineages are short and will give biased estimates. In the yeast lineage the branch length from the eukaryote–eubacterium node to the animal–yeast node is 0.191 and that from the latter node to the present yeast sequence is 0.358. So, T_e can be estimated as $[(0.191 + 0.358)/0.358] \times 0.96$ billion years = 1.47 billion years. Similarly, the age of the LCA is estimated as $[(0.314 + 0.191 + 0.358)/(0.191 + 0.358)] \times 1.47 = 2.31$ billion years. For duplicate 1, if the Squalus III lineage is used, then estimates of 1.81 and 2.16 billion years are obtained for the two dates. So, the average estimate for the age of LCA is 2.24 billion years, which is very close to the estimate in Table 28.1. What is the age of the internal duplication? Note that although the root of the tree clearly is between the two duplicates, the exact placement of the root is uncertain. However, the total branch length between the two eukaryote–eubacterium divergence nodes inferred from the two duplicates can be computed as $0.290 + 0.567 = 0.857$. From this the duplication is estimated to be 1.32 billion years older than the age of the LCA, or the duplication occurred $2.2 + 1.3 = 3.5$ billion years ago.

Figure 28.2 shows a tree constructed from elongation factors EF-Tu and EF-G. As before, only the sequences that appear to have evolved at similar rates should be used. However, it seems that the rate of amino acid substitution slowed down in the eukaryotic lineages in recent times because the branch lengths from the animal–plant–fungus node to *Hydra*, *Arabidopsis* and *Rhizomyces* are only about one-half of the branch length between their common node and the archaea–eukaryote divergence node. For this reason, if the animal–fungus (or animal–plant) divergence is used to calibrate the substitution rate, one would obtain an estimate of ~3 billion years for the archaea–eukaryote divergence, an estimate of ~4.5 billion years for the age of the LCA, and an estimate of >6 billion years for the gene duplication, a date older than the age of the Earth! In any case, it is clear that the duplication date is considerably older than the age of the LCA. This date can be calculated by assuming that the age of the LCA is 2.2 billion years. For EF-Tu the *Borrelia* lineage, the longest among the EF-Tu lineages, gives the smallest estimate of the duplication date. The total branch length of this lineage since the LCA is 0.302 + 0.297 = 0.599. For the same reason, for EF-G the *Homo* lineage should be used. Its total length since the LCA is 0.157 + 0.242 + 0.117 = 0.516. The sum of the *Homo* and *Borrelia* lineages is 0.599 + 0.516 = 1.115. On the other hand, the total branch length between the two LCA nodes inferred from EF-Tu and EF-G sequences is 0.368 +

0.451 = 0.819. Therefore, the duplication date is estimated to be (0.819/1.115) × 2.2 = 1.62 billion years older than the LCA, or to be 1.6 + 2.2 = 3.8 billion years old.

What estimates arise from the leucyl-tRNA, isoleucyl-tRNA, and valyl-tRNA synthetase genes? In Figure 28.3, these three synthetases are denoted l, i, and v, respectively. In Figure 28.3 the root can simply be chosen as the midpoint between the two longest lineages, which are the *Tetrahymena* and *Neurospora* lineages. This location, which probably does not represent the true root, illustrates the consequences of using unfavorable conditions for inferring the age of the second duplication. Under this assumption, the first gene duplication gave rise to leucyl-tRNA synthetase and the common ancestor of the isoleucyl- and valyl-tRNA synthetases, and the second duplication gave rise to the isoleucyl- and valyl-tRNA synthetases. Note

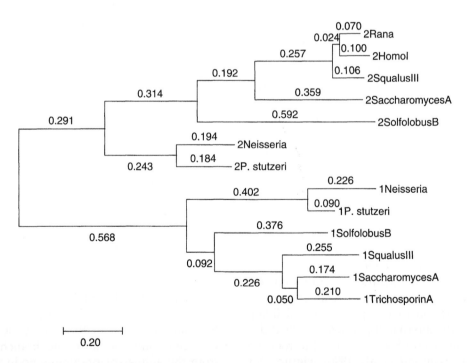

FIGURE 28.1 Neighbor-joining tree (Saitou and Nei, 1987) constructed by MEGA 1.01 (Kumar et al., 1993) using gamma-corrected distances (alpha = 1.14) calculated from the two internal duplicates of the synthetase domain of carbamoylphosphate synthetase. The alpha parameter was estimated from the ML tree for the protein sequences by the program CODEML in the PAML 1.1 package (Yang, 1995). Prefixes 1 and 2 refer to the separate duplicates. Suffixes I, III, A and B refer to the locus designations used by Lawson et al. (1996). Generic names are given for each taxon except for *Pseudomonas stutzeri* (*P. stutzeri*). Branch lengths are in terms of the number of amino acid substitutions per site and are indicated above the branches.

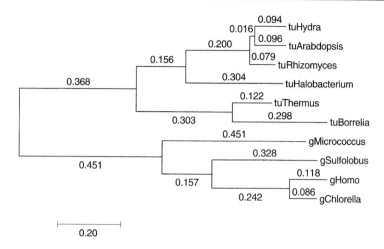

FIGURE 28.2 Neighbor-joining tree constructed from gamma-corrected distances (alpha = 1.22) calculated from elongation factor Tu and G sequences. The two loci are indicated by the prefixes tu or g. Branch lengths are shown above the branches. The sequences were aligned by CLUSTALW 1.6 (Thompson et al., 1994).

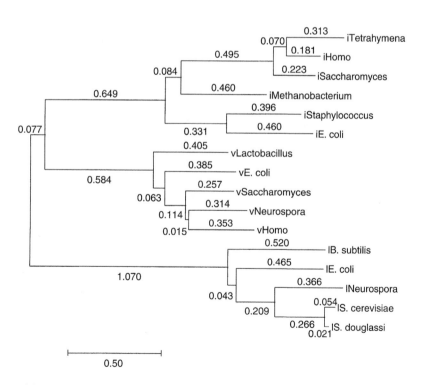

FIGURE 28.3 Neighbor-joining tree constructed from gamma-corrected distances (alpha = 1.26) calculated from leucyl-, valyl-, and isoleucyl-tRNA synthetases. The three loci are identified by the prefixes i (isoleucine), v (valine) or l (leucine). Branch lengths are shown above the branches. Generic names are given for taxa except for *Escherichia coli* (*E. coli*), *Bacillus subtilis* (*B. subtilis*), and *Saccharomyces* (*S. cerevisiae* and *S. douglassi*).

that both of the branches from the latter duplication to the common node of all the isoleucyl-tRNA sequences and the common node of all the valyl-tRNA synthetase sequences are long (0.649 and 0.584), suggesting that the duplication is much older than the LCA. The isoleucyl-tRNA synthetase appears to have an accelerated rate in the branch between the nodes for the eukaryote and the *Saccharomyces* lineages; the branch length is 0.495. So, if it is assumed that the *Saccharomyces* lineage arose 0.96 billion years ago and this lineage is used to calibrate the rate, a date of [(0.495 + 0.22)/0.22] × 0.96 = 3.1 billion years for the age of the eukaryote lineage and an even older date for the age of the LCA are obtained. To use unfavorable conditions for estimating the date of the second duplication, one can assume 2.2 billion years for the age of the LCA and use the *Tetrahymena* lineage (the longest lineage); its branch length from the LCA to the present is 0.083 + 0.495 + 0.070 + 0.312 = 0.960. By this means the duplication date is estimated as [(0.649 + 0.960)/0.960] × 2.2 = 3.7 billion years ago. The valyl-tRNA synthetase sequences can be used to obtain another estimate. Note that the *Lactobacillus* and *E. coli* lineages are shorter than the *Neurospora* and *Homo* lineages, suggesting either a rate slowdown in the former lineages or a rate acceleration in the latter. Either case will lead to an underestimate of the age of the LCA. For example, if the *Homo* lineage (the longest lineage) is used and it is assumed that this lineage arose 0.96 billion years ago, then the age of the LCA is estimated as [(0.063 + 0.114 + 0.014 + 0.352)/0.352] × 0.96 = 1.4 billion years old, which appears to be too young. Using the same assumptions, the duplication date is estimated as (0.584/0.352) × 0.96 = 1.6 billion years older than the LCA. If the age of the LCA is assumed to be 2.2 billion years, then the duplication date is 1.6 + 2.2 = 3.8 billion years old, which is close to the first estimate. Note that the first duplication should be older than the second duplication.

DISCUSSION

The estimate of 2.2 billion years for the age of the LCA involves some uncertainties. First, the number of proteins used is not large and the individual estimates show large variation. To reduce the sampling effect, more proteins should be used. Secondly, the estimation process did not consider the possibility of horizontal gene transfer. Martin (1996) has emphasized the importance of endosymbiotic gene transfer because there is increasing evidence that many genes in the nucleus (e.g. GAPDH, PGK, TPI) might have been acquired from mitochondrial or chloroplast genomes. This type of transfer is unidirectional (i.e. from eubacteria to nucleus) and can cause underestimation of the age of the LCA. This possibility can not be ruled out for many of the proteins because very few sequences from eubacteria and archaebacteria are presently available for a thorough phylogenetic analysis. However, despite these uncertainties, the estimate is very close to that of Doolittle et al. (1996), although only a subset of their data and different methods of analysis were used.

Is the estimate of 2.2 billion for the age of the LCA really incompatible with the fossil record that life emerged more than 3.8 billion years ago (Schidlowski 1988) and that prokaryotes existed before 3.5 billion years ago? To answer this question the dates of three ancient duplications were considered, and estimates of at least 3.5 billion years were obtained. In this estimation a constant rate of amino acid substitution since the duplication was assumed. This assumption may not hold well because duplicate genes may evolve faster following duplication if relaxation in functional requirement or positive selection for functional changes or modifications occurs (Goodman, 1976, 1981; Li, 1985; Ohta, 1994). However, even if the average substitution rate between the duplication and the LCA is two times that of the average rate after the LCA, the estimates are still 2.9, 3.0, and 3.0; also remember, the first duplication for the aminoacyl tRNA synthetases should be even older. As noted above, at the time of duplication each of these proteins had already or almost already evolved the sophisticated function and protein structure we see today. Since it is likely to take much time for a primitive protein to evolve to a protein with a sophisticated function and structure, there is not much room to push back the duplication date. Therefore, these ancient duplication dates, which were estimated on the assumption of 2.2 billion years for the age of the LCA, suggest that there is no severe conflict between the estimated age of the LCA and the fossil record that life started more than 3.8 billion years ago.

Martin (1996) noted that if all present forms of life can be traced to a common ancestor only 2.2 billion years ago, then there would be no members of the cyanobacteria as we know them today that were photosynthesizing 3.5 billion years ago. This difficulty would not exist if all extant forms of life descend via direct filiation from cyanobacteria-like forefathers, but this is highly unlikely. It is also unlikely that the cyanobacteria-like organisms that were apparently very successful and produced the first large amount of oxygen on earth became extinct, whereas the present-day cyanobacteria emerged later from other eubacteria. However, this problem remains unless one assumes the age of the LCA to be nearly as ancient as 3.5 billion years old. In fact, it remains a problem as long as we believe that all extant eubacteria are a monophyletic group and share a common ancestor younger than, say, 3 billion years of age, because we still have to explain what happened to the cyanobacteria-like organisms that existed 3.5 billion years ago.

ACKNOWLEDGMENTS

This study was supported by NIH grants. We thank Russell F. Doolittle, James R. Brown and Fiona S. Lawson for sending us the sequence data in Table 28.1, the alignment of the tRNA synthetases and the alignment of the carbamoylphosphate synthetases, respectively.

REFERENCES

Baldauf, S.L., Palmer, J.D. and Doolittle, W.F. (1996) The root of the universal tree and the origin of eukaryotes based on elongation factor phylogeny. *Proc. Natl Acad. Sci. USA* **93**: 7749–7754.

Brown, J.R. and Doolittle, W.F. (1995) Root of the universal tree of life based on ancient aminoacyl-tRNA synthetase gene duplications. *Proc. Natl Acad. Sci. USA* **92**: 2441–2445.

Doolittle, R.F., Feng, D.-F., Tsang, S. et al. (1996) Determining divergence times of the major kingdoms of living organisms with a protein clock. *Science* **271**: 470–477.

Fitch, W.M. and Upper K. (1987) The phylogeny of tRNA sequences provides evidence for ambiguity reduction in the origin of the genetic code. *Cold Spring Harbor Symp. Quant. Biol.* **52**: 759–767.

Goodman, M. (1976) Protein sequences in phylogeny. In *Molecular Evolution* (ed., F.J. Ayala), pp. 141–159, Sinauer, Sunderland, MA.

Goodman, M. (1981) Decoding the pattern of protein evolution. *Prog. Biophys. Mol. Biol.* **38**: 105–164.

Iwabe, N., Kuma, K.-I., Hasegawa, M. et al. (1989) Evolutionary relationship of archaebacteria, eubacteria, and eukaryotes inferred from phylogenetic trees of duplicated genes. *Proc. Natl Acad. Sci. USA* **86**: 9355–9359.

Jukes T.H. (1969) Recent advances in studies of evolutionary relationships between proteins and nucleic acids. *Space Life Sciences* **1**: 469–490.

Kimura, M. and Ohta, T. (1973) Eukaryotes–prokaryotes divergence estimated by 5S ribosomal RNA sequences. *Nature New Biol.* **243**: 199–200.

Knoll, A.H. (1992) The early evolution of eukaryotes: a geological perspective. *Science* **256**: 622–627.

Kumar, S., Tamura, K. and Nei, M. (1993) *MEGA: Molecular Evolutionary Genetics Analysis, version 1.01*. The Pennsylvania State University, University Park, PA.

Lawson, F.S., Charlebois, R.L. and Dillon, J.-A.R. (1996) Phylogenetic analysis of carbamoylphosphate synthetase genes: complex evolutionary history includes an internal duplication within a gene which can root the tree of life. *Mol. Biol. Evol.* **13**: 970–977.

Li, W.-H. (1985) Accelerated evolution following gene duplication and its implication for the neutralist-selectionist controversy. In *Population Genetics and Molecular Evolution* (eds, T. Ohta and K. Aoki), pp. 333–352, Japan Scientific Societies Press, Tokyo.

Li, W.-H. and Tanimura, M. (1987) The molecular clock runs more slowly in man than in apes and monkeys. *Nature* **326**: 93–96.

Martin, W.F. (1996) Is something wrong with the tree of life? *BioEssays* **18**: 523–527.

McLaughlin, P.J. and Dayhoff, M.D. (1970) Eukaryotes versus prokaryotes: an estimate of evolutionary distance. *Science* **168**: 1469–71.

Mooers, A.Ø. and Redfield, R.J. (1996) Digging up the roots of life. *Nature* **379**: 587–588.

Ohta, T. (1994). Further examples of evolution by gene duplication revealed through DNA sequence comparisons. *Genetics* **138**: 1331–1337.

Saitou, N. and Nei, M. (1987) The neighbor-joining method: a new method for reconstructing phylogenetic trees. *Mol. Biol. Evol* **4**: 406–425.

Schidlowski, M. (1988) A 3,800-million-year isotopic record of life from carbon in sedimentary rocks. *Nature* **333**: 313–318.

Schopf, J.W. (1993) Microfossils of the early Archaen apex chart: New evidence of the antiquity of life. *Science* **260**: 640–646.

Thompson, J.D., Higgins, D.G. and Gibson, T.J. (1994) CLUSTAL W: improving the sensitivity of progressive multiple sequence alignment through sequence weighting, position specific gap penalties and weight matrix choice. *Nucleic Acids Res.* **22**: 4673–4680.

Yang, Z. (1995) *Phylogenetic analysis by maximum likelihood (PAML), version 1.1.* Institute of Molecular Evolutionary Genetics, The Pennsylvania State University, University Park, PA.

Parallelisms and Macroevolutionary Trends

The earlier sections have addressed the following questions:

- "Do the mechanisms of horizontal gene transfer exist?" We have established that these mechanisms exist for transfer from bacteria to bacteria, from fungi to fungi and from bacteria to plants and animals.
- "Do we see evidence of horizontal gene transfer?" The answer, again, is yes. It is observed, most conspicuously, among bacteria in the recent past and also among Archaea, Bacteria and Eukaryota over a billion years ago.

Finally, in this last section, we will address the question "Does horizontal gene transfer play a significant role in evolution, especially in the evolution of the plants and animals that have shown the most dramatic changes during the last 500 million years?" We cannot directly answer this question yet. We simply have not identified those genes responsible for the morphologies that characterize the major geological periods. This section presents a series of more speculative pieces that describe the explanatory power of a theory of horizontal gene transfer as it may affect a general theory of evolution. Let us keep in mind that any new theory must not only predict new phenomena, but must also account for preexisting fact – especially those facts that are not easily explained by standard theory. Horizontal gene transfer theory more easily explains a number of classical observations than does standard theory – especially in the fields of comparative paleontology and comparative morphology.

In Chapter 29, Krassilov presents his idea, first described 25 years ago, that the modern angiosperm evolved from multiple non-angiosperm lineages that can be traced back more than 250 Myr through the fossil record. Syvanen in Chapter 30 shows, using 18S rRNA sequences from plants, that molecular clock estimations are consistent with the divergence times proposed by Krassilov. It is not so much that Chapters 29 and 30 are presenting new ideas. They are not. Rather, these chapters are included because the entire field of botanical systematics has not yet recognized these patterns, nor has offered any sensible explanations for the data upon which they are based.

Parallel and convergent evolution were popular subjects a century ago, especially among the first generation of comparative morphologists and paleontologists who succeeded Darwin. Berry and Hartman, in Chapter 31, describe the nature of parallelisms seen in the fossil record of one group of animals – the graptolites – that summarizes the nature of this problem and offers a theory of horizontal gene transfer that comfortably accommodates this phenomena. In Chapter 32, Williamson presents a review of some of the more dramatic examples of "convergent" evolution to arise from studies of comparative morphology and modern zoological systematics. His ideas are presented in terms of his larval transfer theory, which involves a specific hypothesized mechanism of horizontal gene transfer.

In Chapter 33, Hartman describes a stress-response horizontal gene transfer idea that is tied to the hypothesis of "snowball earth," which caused evolutionary bottlenecks at hydrothermal vents. He organizes his ideas around his kronocyte theory for the origin of the eukaryotic cell. Henze et al., in the introduction to Chapter 27, give a good summary of the various competing ideas on the origin of the eukaryotic cell, of which the kronocyte theory is one.

Olendzenski et al., in Chapter 34, offer the proposal that groups of related species that are recognized by systematists and organized into higher taxonomic units could possibly be related to each other, not only through time but also in the present via the mechanism of horizontal gene transfer.

CHAPTER 29

Character Parallelism and Reticulation in the Origin of Angiosperms

Valentin A. Krassilov

Angiosperms have appeared as assemblages of different life forms in association with the advanced gnetoid and other proangiosperms.

Their origin was not a solitary event but rather a result of parallel evolution. Typical angiosperm characters, such as vessels, areolate venation, enclosed ovules, extraovular pollen germination, double fertilization, etc., appear scattered among different lineages of proangiospermous plants. Their ensembling by horizontal gene transfer seems even more plausible due to the recently obtained direct evidence of interaction between proangiospermous plants and pollinivorous insects, with certain pollen characters, such as taeniate exine or columellate infrastructure, spreading across taxonomic boundaries. Insects might facilitate horizontal gene transfer in plants by transferring microorganisms capable of gene transduction. Major events in angiosperm evolution occurred during widespread environmental crises making plant populations more receptive to extraneous genetic material. Evolutionary significance of horizontal gene transfer is discussed.

INTRODUCTION

Angiosperm origin is still sometimes perceived as a single evolutionary event. However, a wealth of data drawn from different sources seems more reconcilable with the notion of a long and intricate angiospermization process in which a lateral spread of genes between the parallel evolving lineages might be the genetic basis for how characters that evolved in a single group become, as it were, a common heritage (Krassilov, 1977; Syvanen, 1994). These might be characters under simple genetic control, as in pollen grains, haploid structures providing wonderful examples of translineage parallelisms, or the more complex characters arising from developmental accelerations or sequential changes caused by insertions of regulatory elements.

Morphological evidence of horizontal gene transfer is, by necessity, indirect. There are some interesting points, however, pertaining to angiosperm prehistory, angiosperm diversity, and the ecology of angiosperm appearances, including the role played by insects, in which a theory of evolution incorporating horizontal gene transfer has explanatory power.

Seed plant phylogeny

Angiosperm evolution starts with protected ovules that appeared in the mid-Devonian time, about 400 million years ago. Currently, seed plants are thought to be rooted in progymnosperms of which both the heterosporous archaeopterids and homosporous aneurophytes are considered as potential ancestors (Rothwell and Erwin, 1987). Moreover, there were herbaceous plants of simpler axial anatomy but with elaborate cupule-like sporangial clusters as in *Lenlogia* (Krassilov and Zakharova, 1995). The early seed plant diversity

Copyright © 2002 by Academic Press.
All rights of reproduction in any form reserved.

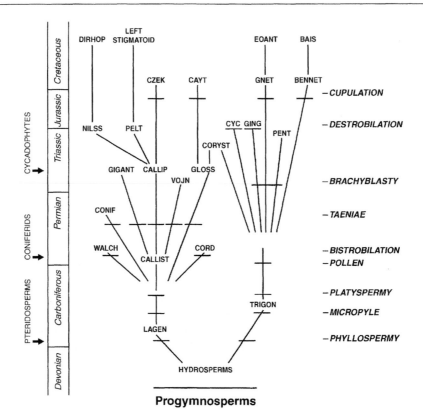

FIGURE 29.1 Grades and phylogeny of gymnosperm and proangiosperm higher taxa based on chronological relations and intermediate or mosaic forms. Horizontal dashes show parallel appearances of characters indicated at the right margin. Abbreviations: Bais, Baisia; Bennet-bennettites; Callist, callistophytes; Callip, callipterids; Cayt, caytonialeans; Conif, conifers; Cord, cordaites; Coryst, corystosperms; Cyc, cycads; Czek, czekanowskias; Dirhop, Dirhopalostachys; Eoant, Eoantha; Gigant, gigantopterids; Ging, ginkgo; Gnet, gnetaleans;Gloss, glossopterids; Lagen, lagenostomaleans; Lept, Leptostrobus (stigmatoideus); Nilss, nilssonialeans; Pelt, peltasperms; Trigon, trigonocarps; Walch, walchians).

is insufficiently known, however, for establishing progenitorial relations other than in a very general form. The possibility of gymnosperm anatomy and seeds originating in different lineages and then becoming combined by horizontal gene transfer is an attractive idea.

Insofar as phylogeny primarily conveys our understanding of homology, it cannot be more objective than the latter. The objective elements in it are chronological relationships as well as morphological continuities furnished by (1) intermediate forms that show character states midway in the morphocline, and (2) mosaic forms that combine typical characters of different taxonomic units thus serving as phylogenetic links or stepping stones. Contemporaneous plant

groups are likely to have independent origins or, if connected by intermediate forms, fraternal rather than progenitorial, relations. On the other hand, members of successive age groups, if connected by intermediate or mosaic forms, are likely to have progenitorial relations. A phylogenetic tree based on the combined chronological information and intermediate/mosaic morphologies of the linking forms (Figure 29.1) shows the following general tendencies:

(1) The increase of the higher rank diversity that, for non-angiosperm seed plants, has peaked in the early Mesozoic, at the time of proangiosperm appearances. The subsequent decrease at the expense of

proangiospermous orders might reflect the latter evolving into conventional angiosperms.

(2) The genetic continuity of seed plant lineages indicates that few disappeared without leaving descendants that in turn contributed genetic material to the next evolutionary stage up to the proangiosperm level. Angiosperms thus inherited most of the genetic information that had accumulated during their prehistory.

(3) New groups of gymnosperms and proangiosperms appear to occur in clusters rather than by a succession of dichotomies.

(4) A notable feature of the nodal arrangement is a widespread homoplasy of fundamental derived characters simultaneously appearing in fraternal lineages, e.g. phyllospermy in lagenostomaleans and trigonocarps; true pollen and platyspermic ovules replacing prepollen and radiospermic ovules in trigonocarps, callistophytes and cordiates; compound strobili in cordiates, walchians, and conifers, radial synangiate and ovuliferous structures in pteridosperms, glossopterids, vojnovskialeans and bennettites; taeniate pollen grains in glossopterids, peltasperms and conifers and their simultaneous transition to asaccate morphotypes; specialized laminar ovuliphores – seed-scales – in peltasperms, nilssonialeans, conifers, etc.; their fusion to subtending bracts assisting in seed dispersal in glossopterids, conifers and gnetaleans; peltate imbricate cone scales in peltasperms, cycads, several groups of conifers; destrobilation in *Ginkgo*, *Cycas*, Taxaceae and other conifers as well as Mesozoic gnetophytes; secondary cupules in caytonialeans, czekanowskias and other proangiosperms; asaccate anasulcate pollen morphologies in Mesozoic nilssonialeans, cycads, ginkgoaleans, etc.

(5) There are cycles of morphological evolution, in particular, the appearance of cupulate ovules in the Late Devonian–Early Carboniferous as the first round of angiospermization followed by gymnospermization in the late Paleozoic in turn followed by a second round of angiospermization in the late Mesozoic. A

morphological distance between cupulate structures of extinct seed plants and angiosperm gynoecia appears much shorter than between the latter and the scaly ovuliphores of extant gymnosperms.

PROANGIOSPERMS AS A MORPHOLOGICAL POOL

Seed plant evolution has proceeded through a series of successional hydrosperm, pteridosperm, cycadophyte and proangiosperm grades (Figure 29.1) before reaching to angiosperm level. Proangiosperm grade became apparent due to a series of palaeobotanical findings (Krassilov, 1975, 1977, 1982, 1984, 1986; Krassilov and Bugdaeva, 1982) that brought to light seed plants chronologically preceding or contemporaneous with the earliest angiosperm records and showing some critical characters but lacking in other critical characters on which conventional recognition of angiosperms is based. Thus vessels appeared in gnetophytes and bennettites, paracytic stomata in bennettites, graminoid leaves in gnetophytes, compound-palmate to lobatopalmate leaves with reticulate venation in *Sagenopteris* – Scoresbya group (Caytoniales and allies), "dendroid" androclades anticipating fasciculate androecia in Caytonanthus, pollen grains with protocolumellar infrastructure and zonal, as well as porous, protoapertures in *Classopollis* (Hirmerellaceae); cupular gynoecia in *Dirhopalostachys* of cycadophyte descent (with solitary anatropous ovule) and *Basia* of bennettitalean descent (with solitary orthotropous ovule), four-membered cupulate gynoecia with bracteate perianth in *Eoantha* (gnetophytes), ascidiform cupule with many ovules in *Caytonia*, syncupulate capsules with stigmatic crests in *Leptostrobus* (Czekanowskiales), etc. Proangiosperm groups are, thus, complimentary in forming a morphological pool containing an almost complete set of typical angiosperm characters.

In angiosperms we find derived states of the characters that appeared in their prototypal states in proangiosperms. But no single proangiosperm lineage could conceivably give rise to the basic angiosperm diversity evidenced by the

TABLE 29.1 Example morphological features

	Gnetophytes	Monocotyledons	Dicotyledons
Flower	−	+	+
Stamen	−	+	+
Carpel	−	+	+
Cupule	+	−	−

mid-Cretaceous fossil record. Rather the diversity of angiosperm morphologies could arise by recombination of the diversity of progymnosperm morphologies. Angiosperms as a whole are "pachyphyletic" in the sense that they descended from proangiosperms as a whole. At the same time, the basic dichotomy of seed plant morpho- clines with receptacular orthotropous ovules versus appendicular anatropous ovules is traceable through the gymnosperm and proangiosperm grades to the early angiosperms giving clue to their progenitorial relations, e.g. linking the gnetophyte-bennettite group of proangiosperms to graminoid monocots and the caytonialean – czekanowskialean proangiosperms to ranunculids and related dicot and monocot orders.

ANGIOSPERM DIVERSITY

Traditional taxonomy represents angiosperms as a diverse but fairly integral group bound up by such critical characters as areolate leaf venation, vessels, flower, carpel, double fertilization, etc., the structures and the process lacking, at least nominally, in other seed plants, thus constituting a unique diagnostic complex that could arise only once in the history of the plant kingdom. The morphological diversity of angiosperms is then reducible to a single ancestral prototype.

Comparative morphology of angiosperms is governed by this creed. The drive for unity makes a taxonomist feel unsatisfied until all the items are brought to one end. In effect, monophyletic systems are much more popular than polyphyletic systems. Cladistics provide a seemingly objective approach to the problem by numerical estimates of intergroup versus outgroup similarities. In the case of angiosperms, the objectivity of the results indicating monophyly must be doubted because monophyly is assumed *a priori*. There are other reasons to doubt, notably the semantic. Strict morphological definitions are wanting for most of the characters used, therefore their presence or absence depends on how we define them. For example, Table 29.1 is senseless unless we know that "flower," "carpel," etc. mean the same thing for all the compared groups while often this is still unknown. Incidentally, flowers, familiar by association with weddings and funerals, escaped precise morphological definition. Historically, flowers were scarcely considered a discriminative feature: Linnaei said that all plant species had flowers and fruits, even if concealed from our eyes (Linnaei, 1751). While flower is used discriminatively at present as a special feature of angiosperms separating them from other seed plants, it seems logical to define flower in relation to reproductive structures of non-angiospermous seed plants. In a majority of angiosperms, flowers consist of sporangiophores and/or ovuliphores formed of a floral meristem that is similar to the apical meristem of vegetative shoots but is mitotically more active and of a less distinct zonal structure. Gymnosperms also have specialized reproductive shoots, but their apices are not fully fertile, bearing sterile scales (in bennettites) or, in *Ginkgo*, fully developed leaves intermingled with ovuliphores. Although differing from typical flowers these structures correspond to some anomalous "flowers" with bracts in the gynoecial zone, e.g., in *Eupomatia* where they form the "inner corolla" between carpels and stamens. Such floral structures occurring in a number of angiosperm families are actually preflowers rather than typical flowers.

The situation with other typical structures is similarly semantically biased. Characters assigned to a certain morphological type are not necessarily homologous and, in fact, are rarely so. For example, follicles, often thought of as basic leaf-like gynoecial structures, are either monomerous ascidiform (Rohweder, 1967) or pseudomonomerous (Vink, 1978). The unitegmic condition can result from fusion or reduction or integumental shifting (Bouman and Callis, 1977) while bitegmic condition can result from splitting or modification of peripheral nucellar tissues in respect to the pollen conducting function of the inner integument (Heslop-Harrison et al., 1985). Even some characteristic biochemical compounds are end-products of dissimilar biosynthetic pathways (Kubitzki, 1973). Pseudohomology is a prolific source of phylogenetic misconceptions, such as morphological integrity of angiosperms.

Comparative morphological analysis shows that none of the critical characters is shared by all the species currently assigned to angiosperms (there are forms lacking distinct stratification of apical meristem, vessels, typical sieve element companion cells, with atypical double fertilization, as in *Onagraceae*, with embryogenesis nuclear up to 64–128 or even 256-nucleate stage, as in *Paeonia*, etc.) On the other hand, such characters are not exclusively confined to angiosperms but occur, though less consistently, in seed plants that are not formally recognized as angiospermous (this includes not only preflowers and carpel-like cupules, but also double fertilization, as in *Ephedra*). The morphological boundary between angiosperms and gymnosperms is thus not absolute.

Notably the anomalous character states occur in angiosperm taxa that are generally considered primitive. An assembly of all such characters may be closer to an ancestral form then the paradygmatic angiosperm. This form may not be conventionally classified as angiosperm thus making angiosperms cladistically paraphyletic with an implication that the typical angiosperm characters, not yet occurring in the common ancestor, appeared in parallel in the descendant lines and, therefore, are not uniquely derived.

Actually shared by all angiosperms is developmental acceleration resulting in highly condensed and/or chimeric structures of great morphological plasticity. But this is an evolutionary trend simultaneously evolving in different seed plant groups.

COLLECTIVE BREAKTHROUGH

Origin of the angiosperm is conceived here as a process that does not have a strictly defined beginning. A semitectate pollen grain witnesses a step in morphological evolution towards angiospermy rather than the existence of angiosperms. Such records go back to the Triassic (Cornet, 1979) and even further. The ensembled record of several angiosperm traits accumulate, after a period of single-trait appearances, in the Barremian–early Aptian, about 115 Mya. Close to this date are the associate records of authentic angiosperms and various proangiosperms, notably the advanced gnetaleans, in central Asia, Middle East, Atlantic coasts, and Australia. They not only testify to the early angiosperm appearance in these areas, but also evidence the on-going process of angiospermization.

Remarkably, the most important localities have yielded not only the first angiosperms and angiosperm-like fossils, but also the remains of advanced proangiosperms. In the Barremian *Baisa* locality in the upper reaches of the Vitim River, Transbaikalia, angiosperm leaves *Dicotylophyllum pusillum* and pollen grains *Asteropollis* and *Tricolpites* (Vakhrameev and Kotova, 1977) are accompanied by the abundant achene-like disseminules of *Baisia*, a one-seeded cupule on persistent bristled receptacles (Krassilov and Bugdaeva, 1982). Recently, intact inflorescences of this plant were studied by Krassilov and Bugdaeva (in press) confirming the previously suspected bennettitalean derivation of this proangiospermous plant. From the same plant-bed came *Eoantha*, a bracteate preflower with a four-lobed gynoecium and with *Ephedripites* pollen grains in the pollen chambers of orthotropous ovules (Krassilov, 1986). A recently found attached flower (Figure 29.2) enabled us to assign to this plant the associated graminoid leaves. New finds have added inflorescences with staminate preflowers of gnetalean type (Figure 29.2).

Even more diverse are angiospermoid and proangiosperm fossils in the roughly contemporaneous localities Manlay, Gurvan-Eren and Bon-Tsagan in the western Gobi, Mongolia (Krassilov, 1982). Angiospermoid fruits *Gurvanella* and *Erenia* are accompanied there by the monocot-like

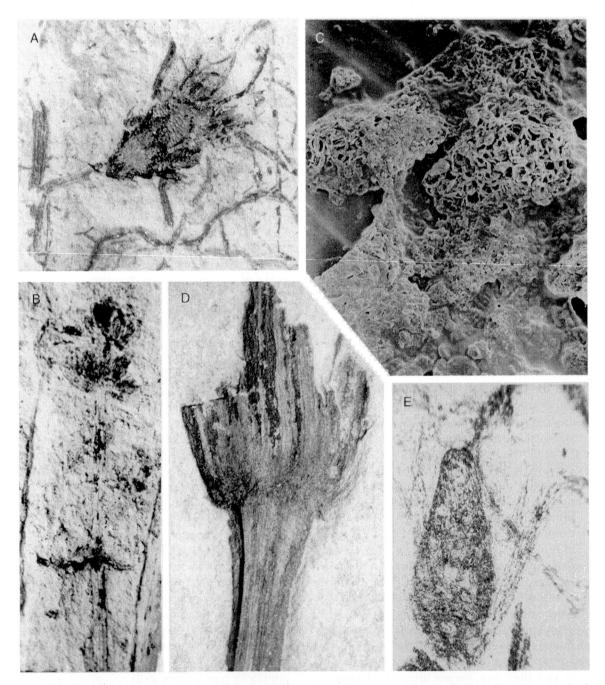

FIGURE 29.2 Proangiosperms of the Baisian assemblage, Early Cretaceous, Transbaikalia: 1, Eoantha, attached. ovulate preflower, ×8; 2, Gnetalean inflorescence with pollen preflowers, ×8; 3, single ovulifore with several sporangia from the same specimen, scanning electronmicrograph, ×170: 4, Graminoid leaf of Eoantha plant, ×8; 5, Basia, a cupule on persistent receptacle with bristles, ×12.

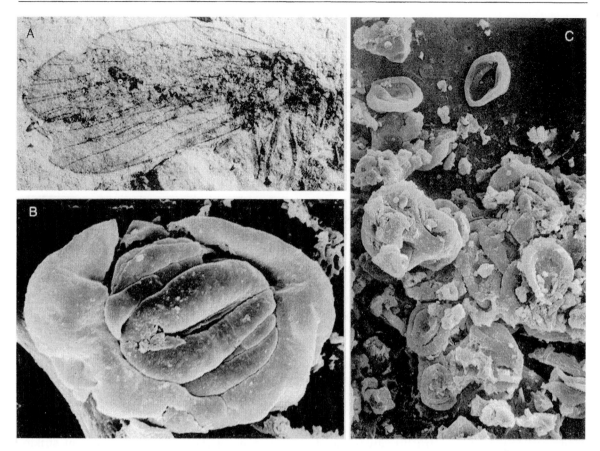

FIGURE 29.3 Pollen in the gut compressions of fossil insects; 1, Idelopsocus, a Permian hypoperlid insect, ×9; 2, taeniate pollen Luntisporites from the intestine of the above specimen, SEM, ×1700; 3, Classopollis form the gut of a Jurassic katydid insect, scanning electronmicrograph, ×700.

Cyperacites, Graminophyllum and unassigned *Sparganium*-like and *Potomageton*-like fruiting axes, as well as a pappose reed-mace-like *Typhaera*. Their preservation is unfavorable for detailed morphological studies, thereby making their angiospermous or proangiospermous status uncertain.

Similar situations are described in other parts of the world. In the Potomac flora of the Atlantic coast, early angiosperms (Doyle and Hickey, 1972) appeared in association with *Drewria*, a herbaceous gnetaiean plant (Crane and Upchurch, 1987). In the English Wealden, the entry of angiosperms in the pollen record was paralleled by the rise of gnetoid pollen (Hughes and McDougall, 1987). In Koonwarra, southern Australia, ceratophyllean fruits (Dilcher et al. 1996) and racemose inflorescences (Taylor and Hickey, 1990) are joined by ephedroids (Douglas, 1969; Krassilov et al., 1996) and perhaps some other gnetaleans represented by ovulate bracts and bracteate pollen cones (Drinnan and Chambers, 1986). In a small collection from the Aptian "amphibian bed" of Makhtesh-Ramon, Israel, angiospermoid fruits are found together with Sagenopteris-type leaves of caytonialean proangiosperms (Krassilov and Dobruskina. 1995).

Thus, the appearance of angiosperms was not a lonely breakthrough against a static background, pushing other plants aside, but was rather a collective breakthrough involving a number of parallel lineages that grew side by side as members of breakthrough plant communities.

INSECT ROLE

The communal association of proangiosperms makes horizontal gene transfer between them at least plausible. *Sporangia* of microscopic endo-parasitic fungi frequently found in fossil pollen grains (Krassilov, 1987) may indicate a potential transducing agent. Interactions with insects and other animal components of breakthrough communities might in turn mediate transfer of such agents.

Direct evidence of plant–insect interaction in biotic communities of the geological past is provided by the pollen preserved in gut compressions of fossil insects first obtained from the Early Cretaceous *Xyeiidae* (Krassilov and Rasnitsyn, 1982) from the *Baisa* proangiosperm–early angiosperm locality (see above). Insects are known to transfer various gene-transducing microorganisms that can confer parallel genetic changes in the target plants. This mechanism might have been already in action in the Permian, as evidenced by the recently described striate pollen in the gut compressions of hypoperlid and gryllobtattid insects of this age (Rasnitsyn and Krassilov, 1996).

Striate (taeniate) exinal structure gives one of the most spectacular examples of a morphological parallelism simultaneously appearing in major gymnosperm groups, glossopterids, peltasperms and conifers, dominating late Paleozoic plant communities of Eurasia and Gondwanaland (Clement-Westerhof, 1974; Zavada, 1985). Our next recent finding of widespread angiospermoid *Classopollis* pollen grains in the intestines of Jurassic grasshopper-like katydid insects (Krassitov et al., in press; Figure 29.3) suggests a similar role of Mesozoic insects in innovation of reproductive morphology occurring in proangiosperm lineages that evolved in parallel in the evolution of the angiosperm.

The katydid example shows that pro-angiosperms and, by implication, their succeeding early angiosperms with their small gregarious flowers could use unconventional pollinators, while beetles and other then existing anthophilous insects were engaged with more conspicuous bennettite preflowers (from the insects perspective, they were flowers regardless of what plant morphologists might think of them).

With the decline of bennettites at the end of Early Cretaceous, their insect retinue passed over to their succeeding angiosperms, perhaps mediating horizontal gene transfer between these groups. Actually, solitary bennettite-like flowers of many parts (Dilcher and Crane, 1984) first appeared in angiosperms at this time.

ECOLOGICAL CRISES AS A GENE TRANSFER SITUATION

It follows from the above discussion that angiosperm origin was a communal event. The above-mentioned localities of early angiosperms and their accompanying proangiosperms reflect xeromorphic brachyphyllous communities widespread in the ecotonal zones of temperate summer-green and subtropical evergreen to winter-green vegetation, at about 50°N and 40°S. Angiospermization might have been conceivably occurring throughout these zones. However, most of the actual records are confined in downfaulted grabens and semigrabens of the Early Cretaceous rift systems. Thus, the Transbaikalian and Mongolian basins are linear depressions of the extensive rift system striking northeast from Mongolia to the Sea of Okhotsk. The lacustrine facies of the rift zone are typical of stratified lakes, with thick finely laminated black shale sequences intervened by psammitic and carbonate interbeds, the latter abounding in fish and aquatic insect larvae remains. Large dragonfly, mayfly and beetle (coptoclavid) larvae pile up on the bedding planes suggesting mass mortality perhaps related to abrupt pH fluctuations caused by volcanic acid rains. The taphonomic data suggest stressful environments as a factor impelling developmental acceleration and condensation of morphological structures characteristic of both proangiosperms and early angiosperms.

Major evolutionary novelties appear after major environmental crises at least for two reasons. One is abbreviation of seral sequences (Krassilov, 1992). Successional species are, as a rule, more "fine-grained" than climax species. Their relatively broad ecological niches are potentially splittable into narrower ecological niches. Elimination of the climax phase thus leaves a community more open for new species entries either by invasion or by speciation. In addition, in perturbed communities, a decrease of stabilizing selection pressure provides opportunities for evolutionary experimentation. While species occupying well-defined ecological niches have to insulate their finely adapted genetic system from invading genes that are likely to decrease their fitness, the opposite is true for post-crisis species that tend to be highly polymorphic. In the pioneer stage of ecological expansion introgression of genetic material is likely to be advantageous as a source of additional genetic variability. Therefore, gene transfer, both vertical, by hybridization, and horizontal, by microorganisms, is promoted by ecological crises. Actually, new groups appearing after ecological crises show not only elaboration, but also recombination of characters occurring in their preceding groups. Thus, recombination of proangiosperm characters in angiosperms can be taken as indirect evidence of interspecies gene transfer.

COMMUNAL GENE POOL

Our data on fossil plant communities that included a number of co-evolving proangiosperm lineages (Krassilov and Bugadeva, 2000) indicate that plant community is more than an assemblage of species growing together. The populational genetic pool concept implies that genetic information stored in soil and transducted by various microorganisms. Therefore, their populational gene pools are included in a higher order system of interspecific – communal – gene pool. A constant flow of genetic material might serve as a factor of integration and coadaptation owing to which biotic community evolves as a unit.

Plant/arthropod interactions are commonly recognized as a leading factor in the origin and evolution of flowering plants. But they are of a more general significance, commencing at the onset of land plant evolution (on evidence of coprolites with spores found in the late Silurian already). The arthropod herbivory is at least partly responsible for the appearance of secondary tissues, as well as for aggregation and protection of sporangia that eventually led to flowers. Coevolution of insect pollinivory increases the probability of viral transduction, because plant viruses are regularly spread by insects.

CONCLUDING REMARKS

The horizontal gene transfer concept is potentially of great importance for leading evolutionary thinking beyond the ossified tenets of "synthetic" theory, first of all, by introducing a long sought – and vigorously denied by traditionalists – mechanism by which macromutations can spread. The ubiquitous evolutionary parallelisms receive a new explanation. And intercommunal interactions between co-evolving organisms appear in a new light not only as competitive, but also cooperative, including at least episodic sharing of a communal gene pool, thus enforcing the idea of community as an evolutionary unit.

Although, at first glance, horizontal gene transfer may seem accidental, the above suggested association with ecological crises means that there could be method in its accidentality: it is effective when actually required as a mechanism of genetic enrichment promoting adaptive innovation and ecological expansion. At least, interpretation of the fossil record seems easier with horizontal gene transfer than without it, which can be taken as indirect evidence in favor of the mechanism.

REFERENCES

Bouman, F. and Callis, J.L.M. (1977) Integumentary shifting – a third way to unitegmy. *Ber. Dt. Bot. Ges.* **90**: 15–28.

Clement-Westerhof, J.A. (1974) *In situ* pollen from gymnospermous cone of the Upper Permian of the Italian Alps – a preliminary account. *Rev. Palaeobot. Palynol.* **17**: 65–73.

Cornet, B. (1979) Angiosperm-like pollen with tectate columellate wall structure from the per Triassic (and Jurassic) of the Newark Supergroup, USA. *Palynology* **3**: 281–282.

Crane, P.R. and Upchurch, R., Jr (1987) *Drewia potomacensis* gen. et sp. nov., an early Cretaceous member of Gnetales from the Potomac Group of Virginia. *Am. J. Bot.* **74**: 1722–1736.

Dilcher, D.L. and Crane, P.R. (1984) *Archaeanthus*: an early angiosperm from the Cenomanian of the western interior of North America. *Ann. Mo. Bot. Gard.* **71**: 351–383.

Dilcher, D.L, Krassilov, V.A. and Douglas, J.G. (1996) Angiosperm evolution: fruits with affinities to Ceratophyllales from the Lower Cretaceous. *Abstr. Fifth Conf. Int. Org. Paleobot. Santa Barbara*: 23.

Douglas, J.G. (1969) The Mesozoic floras of Victoria, 1–2. *Mem. Geot. Surv. Victoria* **28**: 1–310.

Doyle, J.A. and Hickey, L.J. (1972) Coordinated evolution in Potomac Group angiosperm pollen and leaves. *Am. J. Bot.* **59**: 660.

Drinnan, A.N. and Chambers, T.C. (1986) Flora of the Lower Cretaceous Koonwarra fossil bed (Korumburra Group) South Gippsland, Victoria. *Mem. Ass. Austr. Palaeontol.* **3**: 1–77.

Heslop-Harrison, Y., Heslop-Harrison, J. and Reger, B.J. (1985) The pollen-stigma interaction in the grasses. 7. Pollen-tube guidance and the regulation of tube number in *Zea mays* L. *Acta Bot. Neer.* **34**: 193–211.

Hughes, N.F. and McDougall, A.B. (1987) Record of angiospermid pollen entry into succession. *Rev. Palaeobot. and Palynol.* **50**: 255—272.

Krassilov, V.A. (1975) Dirhopalostachyaceae – a new family of proangiosperms and its bearing on the problem of angiosperm ancestry. *Palaeontographica* **153B**: 100–110.

Krassilov, V.A. (1977) The origin of angiosperms. *Bot. Rev.* **43**: 143–176.

Krassilov, V.A. (1982) Early Cretaceous flora of Mongolia. *Palaeontographica* **181B**: 1–43.

Krassilov, V.A. (1984) New paleobotanical data on origin and early evolution of angiospermy. *Ann. Mo. Bot. Gard.* **71**: 577–592.

Krassilov, V.A. (1986) New floral structure from the Lower Cretaceous of Lake Baikal area. *Rev. Palaeobot. Palynol.* **47**: 9–16.

Krassilov, V.A. (1987) Fungi sporangia in the pollen of the Early Cretaceous conifers. Palynology of the Soviet Far East. *Vladivostok, Far East Sci. Center* 6–8.

Krassilov, V.A. (1992) Ecosystem theory of evolution and social ethics. *Riv. Biol.* **87**: 87–104.

Krassilov, V.A. (1995) Scytophyllum and the origin of angiospermous leaf characters. *Paleont. J.* **29**: 110–115.

Krassilov, V.A. and Bugdaeva, E.V. (1982) Achene-like fossils from the Lower Cretaceous of the Lake Baikal area. *Rev. Palaeobot. Palynol.* **36**: 279–295.

Krassilov, V.A. and Bugadeva, E.V. (1999) An angiosperm cradle community and new proangiosperm taxa. *Acta Palaeobot. Suppl.* **2**: 111–127.

Krassilov, V.A., Dilcher, D.L. and Douglas, J.G. (1996) Ephedroid plant from the Lower Cretaceous of Koonwarra, Australia. *Abstr. Fifth Conf. Int. Org. Paleobot. Santa Barbara*: 54.

Krassilov, V.A. and Dobruskina, I.A. (1995) Angiosperm fruit from the Lower Cretaceous of Israel and origins in rift valleys. *Paleont. J.* **29**: 63–74.

Krassilov, V.A. and Rasnitsyn, A.P. (1982) Unique finding: pollen in the guts of the Early Cretaceous xyelotomid insects. *Palaeontol. Zh.* **4**: 83–96.

Krassilov, V.A., Shilin, P.V. and Vachrameev, V.A. (1983) Cretaceous flowers from Kazakhstan. *Rev. Palaeobot. Palynol.* **40**: 91–113.

Krassilov, V.A. and Zakharova, T.V. (1995) Moresnetia-like plants from the Upper Devonian of Minusinsk Basin, Siberia. *Paleont. J.* **29**: 35–43.

Krassilov, V.A., Zherikhin, V.V. and Rasnitsyn, A.P. Classopollis in the gut of Jurassic insects. Unpublished.

Kubitzki, K. (1973) Probleme der Grosssystematik der Blütenpflanzen. *Ber. Dt. bot. Ges.* **85**: 259–277.

Linnaei, C. (1751) *Philosophia Botanica*, Stockholm.

Rasnitsyn, A.P. and Krassilov, V.A. (1996) Pollen in the gut of Early Permian insects: first direct evidence of pollinivory in the PaIeozoic. *Paleoht. Zh.* **3**: 1–6.

Rohweder, O. (1967) Karpellbau und Synkarpie bei Ranunculaceen. *Ber. Schweiz. Bot. Ges.* **77**: 376–425.

Rothwell G.W. and Erwin, D.M. (1987) Origin of seed plants: an aneurophyte/seed fern link elaborated. *Am. J. Bot.* **74**: 970–973.

Salzer, P., Hubner, B., Sirrenbery, A. and Hager, A. (1997) Differential effect of purified spruce chitinases and β-1,3-glucanases on the activity of elicitors from ectomycorrhizal fungi. *Plant Physiol.* **114**: 957–968.

Syvanen, M. (1994) Horizontal gene transfer: evidence and possible consequences. *Annu. Rev. Genet.* **28**: 237–261.

Taylor, D.W. and Hickey, L.J. (1990) An Aptian plant with attached leaves and flowers: implications for angiosperm origin. *Science* **247**: 702–704.

Vakhrameev, V.A. and Kotova, I.Z. (1977) Early angiosperms and their accompanying plants from the Lower Cretaceous of Transbaikalia. *Paleont. Zh.* **4**: 101–109.

Vink, W. (1978) The Winteraceae of the Old World. 3. Notes on the ovary of Takhtajania. *Blumea* **24**: 521–525.

Zavada, M. S. (1991) The ultrastructure of pollen found in the dispersed sporangia of arberilla (Glossopteridaceae). *Bot. Gaz.* **152**: 248–255.

CHAPTER 30

Temporal Patterns of Plant and Metazoan Evolution Suggest Extensive Polyphyly

Michael Syvanen

The current work has determined the divergence times between major eukaryotic clades based on an analysis of 18S ribosomal RNA. A trifurcation rate test is employed which renders it unnecessary to assume that the molecular clocks in the different lineages under comparison are the same. This test suggests divergence times between some of the major clades that are consistently earlier than would be suggested by the fossil record. For example, the trifurcation rate test suggests a molecular divergence time for monocots and dicots at about 175–205 Myr ago, while the fossil record shows that the angiosperm radiation occurred 110 Myr ago. Similar discrepancies are seen between molecular and paleontological estimates of divergence times when the lineages being compared include angiosperms, gymnosperms, bryophytes and some of the metazoan phyla. This suggests that major clades are polyphyletic in that the same modern characters evolved in different lineages (i.e. extensive parallelism has occurred). This discrepancy between molecular time estimates and paleontological estimates is not as extensive with the animals in that most of the major phyla diverged at a time consistent with the Cambrian radiation. There are two exceptions – the Cnidaria and Porifera diverged from the lineage leading to other metazoan phyla about 400 Myr before the Cambrian radiation. A single simple explanation for these widespread parallelisms, ambiguous higher taxonomic categories and polyphyly is that asexual transfer of genetic information between remotely related plants and animals was a major factor in shaping their macroevolutionary patterns.

INTRODUCTION

When Haeckel (1866) took up Darwin's challenge and began a program to unravel metazoan phylogeny, the subsequent effort forced the question whether shared characters were due to descent from common ancestry or whether they arose independently in parallel lineages. Thus, in 1870, Lankester coined the phrase "homoplasy" and formalized the question: Were characters homologous (modern definition) or homoplastic? This problem has bedeviled biologists since. As vexing as the problem was to zoologists (they were, after all, able to find consensus by the 1920s on what constituted the major metazoan phyla), it has proven nearly insoluble in botany. This is because the morphological characters upon which phylogenetic trees are constructed create radically different branching patterns depending upon the weighting of the characters. Botanists have, at different times, come to agree on taxonomic categories, but these periods of agreement have been fleeting (Stevens, 1984). Paleontologists have the potential to record the phylogenetic changes through time and hence, identifying clades as they arise. This requires, however, identifying an ancestral form that shares the

Horizontal Gene Transfer
ISBN: 0-12-680126-6

Copyright © 2002 by Academic Press.
All rights of reproduction in any form reserved.

characters of the resulting clade of descendants. Identifying such ancestral forms has proven problematic. Thus, it has been frequently suggested that major plant clades are polyphyletic, i.e. that they arose from multiple ancestors, none of whom share the clade's defining characters. The difficulty in determining whether clades are monophyletic or polyphyletic is especially problematic when attempting to reconstruct the evolutionary history of the modern flowering plants – the angiosperms – but is also seen in other higher taxonomic groups such as the gymnosperms, ferns and mosses.

The introduction of molecular evolution studies was greeted with great expectations; it was assumed that the taxonomic ambiguities of plant classification would be resolved. These hopes, however, were quickly dashed (Peacock and Boulter, 1975; Boulter and Gilroy, 1992). Phylogenetic tree constructions based on molecular sequences have confronted the same problem that confronts morphological character arrays – namely homoplasy. We documented this in an earlier study of cytochrome *c* sequences and showed, furthermore, that the problem of homoplasy of the plant sequences was significantly different from homoplasy among animal sequences (Syvanen et al., 1989). Molecular sequence information has a potential advantage over morphological character arrays due to the possibility of inferring temporal relationships through the operation of what is commonly called the molecular clock. That is, if we know the rate at which two lineages have diverged from a common ancestor and the extent of divergence, then we can calculate the time of divergence. Molecular clock estimations can therefore be tested against the fossil record. Molecular clock estimations have traditionally assumed that the rate of evolution in two lineages under comparison is constant (Sarich and Wilson, 1973).

Recently, the 18S ribosomal RNA sequences from a large number of different land plants have been determined and deposited in public gene banks; this development was greeted with great expectations, and again it appears there are some difficulties in finding a consensus interpretation. For example, the first 90 plant rRNA sequences that were deposited in GenBank are referenced in the annotations as:

"Darwin's abominable mystery revisited: Ribosomal RNA insights into flowering plant evolution, Unpublished (1991) by Hamby et al." However, no paper ever appeared with only some summaries of these data appearing (Doyle et al., 1994). Published analyses of a larger dataset seems to contain more encouraging results in that both the angiosperm and gymnosperm descended from their own unique ancestor (Chaw et al., 1995, 1997; Soltis and Soltis, 1997; Soltis et al., 1999), though it is clear that topologies based on 18S rRNA are frequently inaccurate. More recently, a number of studies on a variety of proteins have appeared that have been hailed as a major breakthrough in plant classification. Unfortunately the new topologies conflict with 18S rRNA trees and have also resulted in the demise of the monocot–dicot division which used to be one of the more solid higher taxonomic groupings of the angiosperm (Donoghue and Alverson, 2000; Donoghue and Doyle, 2000; Mathews and Donoghue, 1999; Soltis et al., 1999).

So far the 18S rRNA sequences have been used to define phylogenetic topologies, a concerted effort to infer times of divergence using molecular clock assumptions has not been attempted. The major reason for this is that the rate of 18S rRNA evolution appears to be highly variable among different lineages, thereby invalidating the assumption of a constant molecular clock (Romano and Palumbi, 1996; Sogin et al., 1996; Sorhannus, 1996). In the current chapter I describe a procedure for estimating divergence times when there are unequal rates of evolution in the lineages being compared; this is straightforward application of an approach suggested by Fitch and Margoliash (1967) and has found application on numerous occasions (Chapter 28). The different approach I am taking is to average corrected distances for numerous members of individual clades and to use these averages in constructing trees. In addition, taxa that introduce inconsistencies are removed from the analysis. This test will be used to estimate the time of major radiations of the land plants (where the metazoan and fungi are used as outgroups), and the time of metazoan radiation (where plants and fungi are used as outgroups).

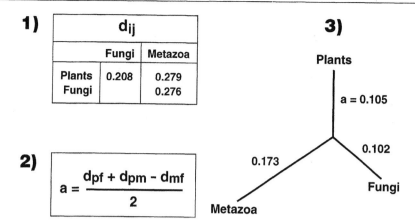

FIGURE 30.1 The trifurcation test. Estimation of evolutionary rates when outgroup taxa are of unequal rates. (1) The distance matrix gives an estimate of the number of nucleotide substitutions that have occurred since the separation of two groups under comparison. Because of the large variation in d_{ij}s among individual taxa, the numbers are averages from many different comparisons. (2) The subscripts p, f, and m refer to plant, fungi and metazoa respectively. Shown is the example for calculating the number of substitutions that occurred in the plant lineage back to the trifurcation. (3) Time to trifurcation in 1.05×10^9 years for fungi and metazoa.

COMPUTATIONS

Computer methods

All computer calculations were performed using GCG software (Genetics Computer Group, Madison, WI) in a VMS or UNIX operating environment.

Sequence alignment

In the current analysis, over 60 sequences were chosen with representatives from angiosperms, gymnosperms, bryophytes, fungi and metazoa from a much larger group. The 18S sequences were recovered from GenBank and aligned using the GCG program Pileup. The alignments were then edited manually to correct alignment errors and to remove uninformative sequences. The 5' and 3' ends of each were trimmed such that the 5' sequence began with the highly conserved TTAAGCCATG and the 3' sequence ended with the relatively conserved GGGCGGTCG 3'. The highly variable regions, as defined from plant, fungi and metazoan comparisons, were deleted from the sequence. This step had the effect of removing nearly 200 characters that were phylogenetically informative among, for example, vascular plants but not

between the different kingdoms. After adjusting the alignment, 1650 nucleotide positions were compared. In a second alignment, with a larger group of metazoans, more extensive editing was performed and only 1350 nucleotide positions were compared.

The trifurcation test

Figure 30.1 illustrates application of the trifurcation test of Fitch and Marguliash (1967) to fungi, f, plants, p, and metazoans, m. If the length of each line in units of the number of nucleotide substitutions since divergence from the common ancestor is given as a, b, c in the figure, then the number of substitutions that separate each taxa, i.e. the number of nucleotide differences divided by the number compared (d_{ij}) is

$$d_{pm} = a + b$$
$$d_{pf} = a + c$$
$$d_{mf} = b + c$$

and a, b and c are easily solved since d_{ij} values are given in the distance matrix calculated from aligned sequences, and is shown in Figure 30.1. These simple relations were first suggested by Fitch and Margoliash (1967) and have been incorporated into programs designed to calculate trees and iterations of this simple calculation are

TABLE 30.1 Taxa used in the present study

GeneBank designation	Domain/phyla	Species
Plants		
AASRG18S	Dicot	*Asarum canadense*
ACURG18S	Monocot	*Acorus calamus*
ADPRG18S	Magnoliophyta	*Antidaphne viscoidea*
AKEERG	Dicot	*Akebia quinata* Houtt
ARU42494	Dicot	*Acer rubrum*
CUORGE	Dicot	*Caulophyllum thalictroides*
D29773	Monocot	*Trachycarpus wagnerianus*
D29776	Dicot	*Magnolia acuminata*
D85299	Gnetophyta	*Welwitschia mirabilis*
D85303	Filicophyta	*Asplenium nidus*
DAU38314	Dicot	*Dillenia alata*
DCU42532	Dicot	*Drosera capensis*
EGLRG18S	Magnoliophyta	*Englerina woodfordioides*
EOARG18S	Magnoliophyta	*Exocarpos bidwillii*
GACRGE	Dicot	*Glaucidium palmatum*
GAU43012	Gymnosperm	*Gnetum africanum*
GDORGE	Monocot	*Gladiolus buckerveldii*
GILRG18S	Magnoliophyta	*Ginalloa*
GSU42541	Dicot	*Geranium* sp.
GUU42417	Gymnosperm	*Gnetum urens*
ISPRGE	Monocot	*Isophysis tasmanica*
LH18SRRNA	Bryophyta	*Lophocolea heterophylla*
LUORGEA	Dicot	*Lepuropetalon spathulatum*
NSU42787	Dicot	*Nepenthes* sp.
OEU42791	Monocot	*Oncidium excavatum*
PIN18SRR	Gymnosperm	*Pinus luchuensis*
RH18SRRNA	Bryophyta	*Riella helicophylla*
RP18SRRNA	Bryophyta	*Rhacocarpus purpurascens*
SBIRGE	Dicot	*Sabia swinhoei*
SNDNA18RR	Filicophyta	*Salvinia natans*
SPIRG18S	Dicot	*Spinacia oleracea*
SRMRG18S	Monocot	*Sparganium eurycarpum*
TUERG18S	Magnoliophyta	*Tupeia antarctica*
ZMU42796	Monocot	*Zea maize*
Choanoflagella		
CFGRGDL	Choanoflagellida	Rosette agent
Fungi		
BBU59062	Basidiomycota	*Bondarzewia berkeleyi*
BRU42477	Ascomycota	*Botryosphaeria ribis*
NEORR18S	Chytridiomycota	*Neocallimastix* sp.
RDU42660	Ascomycota	*Reddellomyces donkii*
UHRRNA18S	Ascomycota	*Urnula hiemalis*
Metazoans		
apu43190	Porifera	*Axinella polypoides*
bgu65223	Mollusca	*Biomphalaria glabrata*
bgu65223	Mollusca	*bloodfluke planorb*
bl18srrna	Echinodermata	*Brissopsis lyrifera*
cc18srrna	Echinodermata	*Centrostephanus coronatus*

TABLE 30.1 Contd

GeneBank designation	Domain/phyla	Species
ccu42452	Porifera	*Clathrina cerebrum*
cpz86107	Cnidaria	*Coryne pusilla*
eb18srrna	Echinodermata	*Echinodiscus bisperforatus*
et18srrna	Echinodermata	*Eucidaris tribuloides*
ewu29492	Arthropoda	*Eusimonia wunderlichi*
funrg18s	Vertebrata	*Fundulus heteroclitus*
humrge	Vertebrata	*Homo sapiens*
ll18rr	Mollusca	*Littorina littorea*
mh18srna	Tardigrada	*Macrobiotus hufelandi*
oeu88709	Mollusca	*Ostrea edulis*
pau42453	Cnidaria	*Parazoanthus axinellae*
peu29494	Nemertea	*Prostoma eilhardi*
pm18srrnx	Echinodermata	Sand urchin
ps18srrn3	Vertebrata	*Polyodon spathula*
ratrge4a	Vertebrata	*Rattus norvegicus*
rp18srr	Vestimentifera	*Ridgeia piscesae*
scu29493	Arthropoda	*Scolopendra cingulata*
sebrg18s	Vertebrata	*Sebastolobus altivelis*
sv18rr	Mollusca	*Scutopus ventrolineatus*
th18srrna	Echinodermata	*Temnopleurus hardwickii*
xelrge14	Vertebrata	*Xenopus laevis*

used in algorithms that determine distance trees of unequal rates. If the time, T, of the trifurcation is known, for some clades, then the clock in each lineage is simply a/T, b/T and c/T, respectively. Hence, it is not necessary to assume that the rate of evolution in each lineage is constant to calibrate the clock.

Clearly, before these rates can be used to determine divergence times of taxa within lineage a, for example, we must assume that the rate of evolution in the lines leading to the taxa under comparison are the same and equal to a. This assumption can be tested using the relative rate test (Sarrich and Wilson, 1973) and deviations can be accommodated (see below).

DETERMINATION OF DIVERGENCE TIMES FOR THE LAND PLANTS

The plant, animal and fungi kingdoms diverged from a common ancestor about 1×1000 Myr ago (Yokoyama and Harry, 1993; Doolittle et al., 1996). Extensive analysis of numerous protein sequences that show clear clock-like behavior

fix this time within an error of about 10%. It appears that the metazoa and fungi are about 5% more closely related to each other than either is to plants (Baldauf and Palmer, 1993; Nikoh et al., 1994; Sidow and Thomas, 1994; Drouin et al., 1995; Kumar and Rzhetsky, 1996; Sogin et al., 1996; VandePeer et al., 1997). Hence, if metazoa and fungi diverged 1.0×10^9 Myr ago, then plants diverged from the trifurcation point 1.05×10^9 Myr ago. This will be one of the calibration points I will use in calculating the divergence times. The trifurcation test will first be used to determine divergences in the major plant clades where we will use fungi and two metazoan phyla as our outgroup. In another section, we will apply this test to metazoan phyla where plants and fungi will be used as outgroups.

Selection of taxa and multisequence alignment

There are over 300 nearly complete 18S rRNA sequences of different land plants deposited in the public databases. About 30 were chosen for the purposes of this study. They were primarily collected by Soltis et al. (1997) (Table 30.1). The three mosses and the fern were grouped together and are collectively referred to as "bryophytes," though they are more accurately called lower archaegoniates.

A group of fungi and metazoan taxa were chosen from a much larger group found in the GenBank. These were analyzed by picking representative samples which were characterized by the fact that, after alignment, the simple d_{ij} among both the fungi and metazoa were greater than 0.10. The amount of divergence among the metazoans is highly variable.

These sequences were aligned and corrected distances were calculated (Table 30.2). The average distance between all members of each assemblage was computed and Table 30.2 presents these averages. On the basis of 18S rRNA sequences alone, the seven groups shown in Table 30.2 define clearly delineated clades. These clades are seen using either parsimony or distance methods. For purposes of presentation I am showing the results where Cnidaria and Porifera are the metazoan outgroup, though the results using a larger number of protostomes and deuterostome did not affect the final calculations of plant divergence times (not shown). However, representatives of these latter taxa were present during sequence alignment and their presence influenced the final alignment.

Sources of error

There are three sources of error that contribute to any estimated divergence times using the data in Table 30.2. The first is the variance in raw d_{ij} values among, for example, different pairs of monocots and dicots. In the present analysis, this variation contributes the least error to the final estimate, since we are using averages from a relatively large number of taxa resulting in a small standard error.

A second source of variation is in the choice of the outgroup taxa. This is illustrated in Figure 30.2. In this example, averages from the same set of plant and fungi sequences are compared against individual metazoan taxa and the average distance to trifurcation for the plants is computed. As can be seen, there are significant differences depending on the choice of the metazoan. Similar variations in plant distance to trifurcation is seen if the metazoan taxa are held constant and individual fungi are used. As with the error due to variation in d_{ij}, this error is minimized by averaging over relatively large numbers of taxa so the ultimate standard error is small.

The third source of error is due to the estimated substitution rate, i.e. have we calibrated the clock properly? This error contributes the greatest uncertainty to our final estimated divergence times. Recall that we are using a 1.05 Byr time to trifurcation to calibrate a clock

TABLE 30.2 Average distances between major plant clades. Just the cnidarian and porifera taxa were used as the metazoan out group but representative of the other phyla were included in determining the final sequence alignment. The numbers outside the parentheses are d_{ij} values. The numbers inside the parentheses are the comparable patristic distances from the tree in Figure 30.3. A statistical correction for multiple substitutions was applied. A gamma distribution of multiple hits was assumed (Jin and Nei, 1990), though using a Poisson distribution (Jukes and Cantor, 1969) gave comparable results

	1	2	3	4	5	6	7
1. Monocots	–	4.32 (4.3)	9.62 (9.6)	10.09 (9.9)	21.50 (21.6)	22.92 (22.6)	19.97 (19.6)
2. Dicots		–	9.44 (9.5)	10.07 (9.8)	21.52 (21.5)	22.46 (22.5)	19.71 (19.5)
3. Gymnosperms			–	9.97 (10.4)	22.52 (22.2)	23.05 (23.2)	19.76 (20.2)
4. Bryophytes				–	21.52 (20.1)	20.20 (21.0)	17.40 (18.0)
5. Fungi					–	19.70 (19.7)	16.71 (16.7)
6. Cnidaria						–	14.80 (14.8)
7. Porifera							–

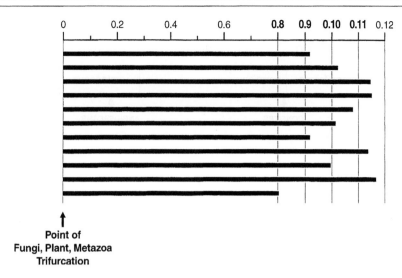

FIGURE 30.2 Variation in the rate due to choice of one outgroup in the trifurcation test.

that will be used to estimate 200–500 Myr divergence times. As mentioned above, in order for the trifurcation test to work, it is best that, for example, the different plant lineages being compared have comparable rates of evolution. To a first approximation, this is the case, as can be seen by applying the relative rate test to the four major plant assemblages; monocots, dicots, gymnosperms and bryophytes in Table 30.2. For example, the distance from the fungi to these four groups are 21.5, 21.5, 22.5 and 21.5 respectively, while the distance from the Porifera is 20.0, 19.7, 19.8 and 17.1 respectively. However, on closer examination, there are significant differences between the rates of evolution among the four plant assemblages since their divergence. This is seen most clearly in the complete distance tree shown in Figure 30.3. (This tree is constructed by a reiteration of the trifurcation test and a weighted least squares distance average for individual branches.) As can be seen, the bryophytes evolve the slowest and the gymnosperms the fastest. This then raises the question: which of the rates is representative of the land plants as a whole? This problem was dealt with by calculating divergence times by two means – in the first, the slowest evolving lineage is considered to be representative of plants as a whole, and in the second, the fastest evolving lineage is. In Figure 30.4, the branching times are given as the average of these two

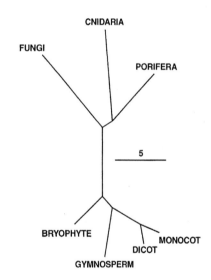

FIGURE 30.3 Distance tree based on data in Table 30.2. The space bar gives a distance of five substitutions per hundred nucleotides.

values and the error bar gives the range between slowest and fastest.

The tree in Figure 30.5 can be used to estimate the overall robustness of the approach I have employed in this analysis. The distances through the finally computed tree (i.e. the patristic distances) are shown in parentheses next to the d_{ij} values in Table 30.2. As can be seen, there is quite good agreement between these two values. If we

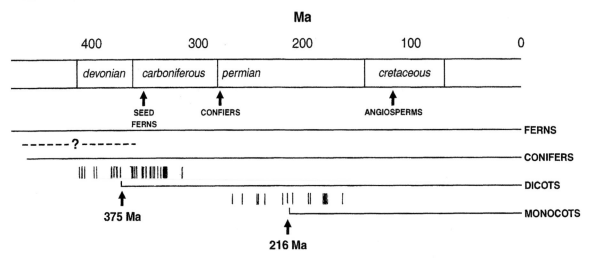

FIGURE 30.4 Times of divergence of major plant clades. The times calculated in this study are compared to the geologic periods. The first appearance of the different groups of plants are shown by the arrows. The scatter of points between the gymnosperm/angiosperm split and the monocot/dicot diversion reflect the magnitude of variation when individual taxa are used to compute patristic distances.

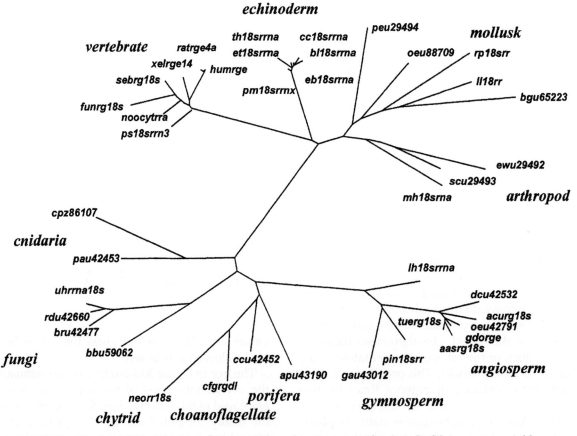

FIGURE 30.5 Untrooted tree showing larger groups of metazoans. This is a Jin/Nei nearest neighbor tree provided in the GCG package and is basically a distance tree that does not assume constant. Arrow denotes trifurcation point to the prokaryotes.

exclude lower archaegoniates, the agreement is 98% or better. In real datasets, patristic distances and d_{ij} values rarely show such close agreement; the reason for the agreement in the present case is because the values in Table 30.2 represent the average from large numbers. Further, all characters that had experienced too many changes, i.e. were approaching or at saturation, were deleted from the alignment.

DETERMINATION OF DIVERGENCE TIMES OF METAZOAN EVOLUTION

Porifera and Cnidaria in the above discussion were included as an outgroup. In this section I will apply the trifurcation rate test to some metazoan phyla. In this case, plants and fungi will serve as the outgroups. Six metazoan phyla are analyzed and include representatives from vertebrate, echinoderms, arthropods and mollusks as well as Porifera and Cnidaria.

The number of metazoan 18S sequences deposited in GenBank is about 4500. A number of criteria were used to pick the 29 used in this study. First, I picked out about 100 full-length 18S sequences somewhat arbitrarily to include equal numbers of protosomes and deuterosomes. An effort was made to align them automatically – many were too highly diverged to align and these were eliminated. Finally, the database was broken up into smaller groups and distance trees were computed. There is huge variation in the rate of evolution in the 18S rRNA sequences. The rates varied from those that were similar to plant and fungi rates to others that were two to three times faster. It was found that the trifurcation rate test completely failed when those fast evolving sequences were compared with the slower evolving sequences.

The failure manifested itself in tree inconsistencies; namely, we could see obvious examples of "long branch attractions" and, more importantly, cases where patristic distances through the minimal tree scattered widely from the d_{ij} values. (These examples are not shown here.) To avoid this problem, all sequences that gave rise to long branches were excluded from the analysis. After identifying long branches, sequences were recompiled, aligned and edited, and new trees

constructed. Figure 30.5 shows that tree. For the purpose of an outgroup comparison, five land plants and five fungi were included. This tree has a number of features that are expected. Arthropods and mollusks define a subclade to the exclusion of vertebratae and echinoderms – this is consistent with known relationships between protosomes and deuterosomes. It is also clear that the major metazoan clade displays a greatly accelerated rate of evolution compared with the rest of the eukaryotes, even though the fastest evolving metazoan taxa were removed from the analysis, as can be seen from branch lengths from the root (noted by the arrow). These patterns are consistent with earlier studies. There are two unexpected details implied by this tree. One is that Cnidaria (e.g. jellyfish) and Porifera (the sponges) diverged from the other metazoans at a time very close to the fungal, plant and metazoan radiation. The second is that Porifera appears in a clade that includes choanoflagellates (such as chlamydomonas and volvox) and chytrids. (The association with plants that is seen is too deep to be of much significance.) The association of Porifera with choanoflagellates and chytrids is more significant. This result suggests that if the sponges are considered metazoan, then the metazoans are at least biphyletic. A clade that includes what was previously thought of as (1) a metazoan, (2) a single-celled plant, and (3) a fungus, is not really totally unexpected. Namely, it has been noted, even prior to molecular evidence, that these three taxa seem to have similar choanocyte flagella (Mohri et al., 1995).

The metazoan phyla shown in Figure 30.6 were submitted to the same trifurcation rate test as was done for the plants above (Table 30.3) and results are summarized in Figure 30.6. As can be seen, when distances of multiple taxa with the various clades are averaged, we can obtain a distance tree whose patristic distances reasonably reconstruct d_{ij} values. As is clear from the tree in Figure 30.7, the Cnideria and Porifera diverged from the other metazoans about 1 Byr ago. In addition, the major diversification of metazoan phyla shows a diversification of 430–500 Myr ago, times not at great odds with the metazoan radiation of 530 Myr ago. This latter fact lends credence to the major finding of this analysis – namely that Cnidaria and

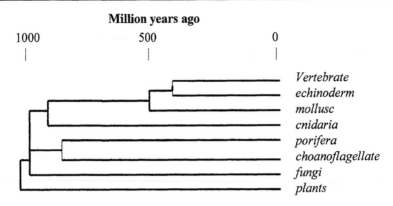

Million years ago

FIGURE 30.6 Time of divergence of major metazoan groups.

TABLE 30.3 Average distances between major metazoan clades. The taxa used in the averages are shown in the tree in Figure 30.5. See Table 30.2 legend for details. Patristic distances are for a minimal distance tree not shown (see Figure 30.6 for topology)

	1	2	3	4	5	6	7
1. Plants		12.31 (12.3)	10.50 (11.1)	12.62 (12.1)	18.28 (18.3)	18.27 (17.6)	17.44 (18.1)
2. Fungi			9.15(9.2)	10.62 (10.26)	16.30 (16.3)	16.53 (15.6)	15.21 (16.1)
3. Porifera				8.43(9.0)	15.86 (15.2)	16.53 (14.6)	15.35 (15.1)
4. Cnidaria					15.76 (16.2)	15.36 (15.7)	14.81 (16.1)
5. Vertebrata						11.01 (10.89	10.06 (10.1)
6. Mollusca							10.31 (10.2)
7. Echinoderm							

Porifera had diverged from the other metazoans 0.85–1.0 Byr ago.

FURTHER DISCUSSION

Evolution of plants

The rates of land plant 18S ribosomal RNA sequence evolution estimated in the current work, coupled with the divergence data, indicate that the higher taxonomic groups such as monocots and dicots, gymnosperms and angiosperms have roots that are much deeper in time than is shown by the fossil record (Figure 30.4). Let us first consider the monocots and dicots within the angiosperm clade. To be sure, taken by itself, the 18S rRNA data support a monophyletic ancestry for the angiosperms because they show that all angiosperms evolved from a common ancestor that excludes other extant plant clades.

However, the angiosperms are defined by multiple morphological characters which are apparent in the fossil record, not only by their 18S rRNA. This record shows that modern flowering plants appeared only 110 Myr ago, though some, but not all, of their defining characters have been found as far back as 140 Myr ago (see Chapter 29). However, the data in this paper, and in earlier molecular clock calculations, place, for example, the monocot–dicot divergence time estimates back greater than 170 Myr ago (Wolfe et al., 1989; Savard et al., 1994). These early divergence times have been seen for cytochrome *c* evolution as well (Syvanen, 1994). Among the dicots, there are also multiple families whose divergence also goes back as far; in the current work, I have seen that all extant Rosidae are descended from a common ancestor but they diverged from the other dicots at this early date (not shown). This leads to the inescapable conclusion that multiple plant lineages existed in Jurassic (and probably the late Permian) that left no fossils recognizable as angiosperms, but evolved to become what today are the flowering plants. Krassilov (1977, 1998; see Chapter 29) describes this as the "angiospermization" of multiple lineages.

The divergence time of 340 Myr ago for the angiosperm and gymnosperm means that the angiosperms had already diverged from the line leading to modern gymnosperm before the appearance of fossil gymnosperms, early after the appearance of the first progymnosperms. This means the tracheophytes are also likely to be polyphyletic. And finally, the results in this paper show that the line leading to tracheophytes diverged from that leading to the modern bryophytes in the early Devonian, a time prior to the appearance of any vascular plants in the fossil record. The conclusion that major clades of vascular plants may be polyphyletic has also been drawn from an analysis of Devonian and Carboniferous fossils (reviewed in Bell, 1992).

One common explanation for this timing problem is that the plant fossil record is incomplete; this is known as the *hidden-forests* hypothesis by critics of paleobotany. This is not an attractive idea; it demands gaps in the fossil record that extend over tens of millions of years and involves multiple lineages. A much more parsimonious explanation is that major taxonomic groups are polyphyletic.

Polyphyly means that major characters that are shared by different angiosperm lineages must have arisen by extensive parallelisms. (E. Mayr has attempted to define this situation as "parallelophyly," and V. Krassilov has coined "pachyphyly.") To use Lankester's terminology, these characters are homoplastic, not homologous. Classically, parallel developments are explained by the general morphogenetic potential found in the various plant lineages and are inherited linearly and differentially expressed by environmental factors. This, in fact, was the hypothesis that Went explicitly addresses in his 1972 review; he argues against the possibility of independent evolution because patterns of homoplastic characters that he described occur throughout the same geographical areas. In different geographical regions, different suites of

homoplastic characters are encountered, that is, different lineages on different continents display a lack of homoplasy even though they contain the same morphogenetic potential. However, when the lineages share a habitat, parallel development occurs. Vavilov (1922) also noticed the tendency of plants to vary in parallel and called the phenomenon "homologous variation" which may be an accurate explanation for variation that is determined by transfer of homologous genes.

Today, given what we know of molecular genetic mechanisms and molecular evolution, I would propose that the simplest explanation of polyphyly at this level is that horizontal gene transfer has played a major role in the evolution of higher plants. I am not claiming any examples of horizontally transferred genes among the 18S rRNAs studied here. Rather, I am claiming that many of the major morphological traits, as revealed in the fossil record, that are shared and therefore characterize higher taxonomic groups, are controlled by genes that have horizontally transferred during the emergence of that group (Syvanen, 1994). This hypothesis will be given a direct test once those genes controlling major developmental patterns that distinguish angiosperm families and genera are identified and phylogenetic analysis of them is possible.

Evolution of the metazoa

The current finding that Porifera, Cnidaria and the other metazoans diverged from each other near the time of the plant, fungi and metazoan trifurcation is strongly supported. If this is the case, then either metazoans are polyphyletic (at least triphyletic) or metazoans do not include Porifera and Cnidaria. Indeed for more than a century, the sponges were not even considered metazoa, but given their own kingdom of parazoa. West and Powers (1993) uncovered molecular evidence that sponges had an unexpected close affinity to some protozoans but they provided a different interpretation to the one given here. In either case, this situation points out a major case of parallel evolution. Sponges appear in the fossil record at the same moment as the other metazoans – during the Cambrian radiation. During this time, we may presume that the ancestor to the modern sponges learned how to utilize $CaCO_3$ or, for the glass sponges, silicates for skeleton construction. This ability appeared simultaneously in sponges, in other metazoans and in numerous protozoan lineages. At this stage, the simplest explanation is that the genes controlling the construction of these hard inorganic skeletons spread like an infection across multiple lineages. The abrupt appearance of hard fossils at the beginning of the Cambrian has been understood for over 100 years as representing an example of massive parallelism and has been cited as supporting a major role for horizontal gene transfer in metazoan evolution (Reanney, 1976; Erwin and Valentine, 1984; Syvanen, 1985; Jeppsson, 1986).

A number of molecular studies, based on protein molecular clocks, have placed the major metazoan radiation (such as divergence of chordates from other in vertebratae or deutrosomes from protostomes) to times back in the Precambrian (Wray et al., 1996; Bromham et al., 1998, 2000; Gu, 1998; Cutler, 2000). The current study does not support these estimates: indeed the protostome/deuterostome divergence estimated here is consistent with the time of the Cambrian radiation. This difference between the rRNA and protein time estimates will have to be resolved in the future.

REFERENCES

Baldauf, S.L. and Palmer, J.D. (1993) Animals and fungi are each other's closest relatives: congruent evidence from multiple proteins. *Proc. Natl Acad. Sci. USA* **90**: 11558–11562.

Bell, P.R. (1992) *Green Plants; Their Origin and Diversity*, Dioscorides Press, Portland, OR.

Boulter, D. and Gilroy, J.S. (1992) Partial sequences of 18s ribosomal RNA of two genera from each of six flowering plant families. *Phytochemistry* **31**: 1243–1246.

Bromham, L., Rambaut, A., Fortey, R. et al. (1998) Testing the Cambrian explosion hypothesis by using a molecular dating technique. *Proc. Natl Acad. Sci. USA* **95**: 12386–12389.

Bromham, L.D. and Hendy, M.D. (2000) Can fast early rates reconcile molecular dates with the Cambrian explosion?. *Proc. Roy. Soc. Biol. Sci. Series B* **267**: 1041–1047.

Chaw, S.M., Sung, H.M., Long, H. et al. (1995) The phylogenetic positions of the conifer genera *Amentotaxus*, *Phyllocladus*, and *Nageia* inferred from 18S rRNA sequences *J. Mol. Evol.* **41**: 224–230

Chaw, S.M., Zharkikh, A., Sung, H.M. et al. (1997) Molecular phylogeny of extant gymnosperms and seed plant evolution: analysis of nuclear 18S rRNA sequences. *Mol. Biol. Evol.* **14**: 56–68

Cutler, D.J. (2000) Estimating divergence times in the presence of an over-dispersed molecular clock. *Molec. Biol. Evol.* **17**: 1647–1660.

Donoghue, M.J. and Alverson, W.S. (2000) A new age of discovery. *Ann. Missouri Bot. Garden* **87**: 110–126.

Donoghue, M.J. and Doyle, J.A. (2000) Seed plant phylogeny: Demise of the anthophyte hypothesis? *Curr. Biol.* **10**: R106–R109.

Doolittle, R.F., Feng, D.F., Tsang, S. et al. (1996) Determining divergence times of the major kingdoms of living organisms with a protein clock. *Science* **271**: 470–477.

Doyle, J.A., Donoghue, M.J. and Zimmer, E.A. (1994) Integration of morphologic and ribosomal RNA data on the origin of angiosperms. *Ann. Missouri Bot. Garden* **81**: 419–450.

Drouin, G., Moniz de Sa, M. and Zuker, M. (1995) The *Giardia lamblia* actin gene and the phylogeny of eukaryotes. *J. Molec. Evol.* **41**: 841–849.

Erwin, D.H. and Valentine, J.W. (1984) "Hopeful monsters", transposons, and metazoan radiation. *Proc. Natl Acad. Sci. USA* **81**: 5482–5483.

Fitch, W.M. and Margoliash, E. (1967) A method for estimating the number of invariant amino acid coding positions in a gene using cytochrome c as a model case. *Biochem. Genetics* **1**: 65–71.

Gu, X. (1998) Early metazoan divergence was about 830 million years ago. *J. Molec. Evol.* **47**: 369–371.

Haeckel, E. (1866) *General Morphology of Organisms*, reprinted Walter de Gruyter, Berlin, 1988.

Jeppsson, L. (1986) A possible mechanism in convergent evolution. *Paleobiology* **12**: 337–344.

Jin, L. and Nei, M. (1990) Limitations of the evolutionary parsimony method of phylogenetic analysis. *Molec. Biol. Evol.* **7**: 82–102.

Jukes, T.H. and Cantor, C.R. (1969) Evolution of protein molecules. In *Mammalian Protein Metabolism* (ed., H.N. Munro), vol. III, pp. 21–132, Academic Press, New York.

Krassilov, V.A. (1977) The origin of angiosperms. *Bot. Rev.* **43**: 143–176.

Krassilov, V.A. (1998) Character parallelism and reticulation in the origin of the Angiosperms. In *Horizontal Gene Transfer* (eds, M. Syvanen. and C. Kado), pp. 409–424, Chapman & Hall, London.

Kumar, S. and Rzhetsky, A. (1996) Evolutionary relationships of eukaryotic kingdoms. *J. Molec. Evol.* **42**: 183–193.

Lankester, E.R. (1870) On the use of the term homology in modern zoology and the distinction between homogenetic and homoplastic agreements. *Ann. Mag. Natl Hist.* **6**: 34–43.

Lee, M.S.Y. (1999) Molecular clock calibrations and metazoan divergence dates. *J. Molec. Evol.* **49**: 385–391.

Mathews, S. and Donoghue, M.J. (1999) The root of angiosperm phylogeny inferred from duplicate phytochrome genes. *Science (Washington DC)* **286**: 947–950.

Mohri, H., Kubo-Irie, M. and Irie, M. (1995) Outer arm dynein of sperm flagella and cilia in the animal kingdom. *Mem. Mus. Natl d'Hist. Naturelle* **166**: 15–22.

Nikoh, N., Hayase, N., Iwabe, N. et al. (1994) Phylogenetic relationship of the kingdoms Animalia, Plantae, and Fungi, inferred from 23 different protein species. *Molec. Biol. Evol.* **11**: 762–768.

Peacock, D. and Boulter, D. (1975) Use of amino acid sequence data in phylogeny and evaluation of methods using computer simulation. *J. Mol. Biol.* **95**: 513–527.

Reanney, D. (1976) Extrachromosomal elements as possible agents of adaptation and development. *Bacteriolog. Rev.* **40**: 552–590.

Romano, S.L. and Palumbi, S.R. (1996) Evolution of scleractinian corals inferred from molecular systematics. *Science (Washington DC)* **271**: 640–642.

Sarich, V.M. and Wilson, A.C. (1973) Generation time and genomic evolution in primates. *Science* **179**: 1144–1147.

Savard, L., Li, P., Strauss, S.H. et al. (1994) Chloroplast and nuclear gene sequences indicate late Pennsylvanian time for the last common ancestor of extant seed plants. *Proc. Natl Acad. Sci. USA* **91**: 5163–5167.

Sidow, A. and Thomas, W.K. (1994) A molecular evolutionary framework for eukaryotic model organisms. *Curr. Biol.* **4**: 596–603.

Sogin, M.L., Morrison, H.G., Hinkle, G. and Silberman, J.D. (1996) Ancestral relationships of the major eukaryotic lineages. *Microbiologia* **12**: 17–28.

Soltis, P.S., Soltis, D.E., Wolf, P.G. et al. (1999) The phylogeny of land plants inferred from 18S rDNA se-quences: Pushing the limits of rDNA signal? *Molec. Biol. Evol.* **16**: 1774–1784.

Soltis, P.S. and Soltis, D.E. (1997) Phylogenetic analysis of large molecular data sets. *Bol. Soc. Bot. Mexico* **59**: 99–113.

Soltis, D.E., Soltis, P.S., Nickrent, D.L. et al. (1997) Angiosperm phylogeny inferred from 18S ribosomal DNA sequences. *Ann. Missouri Bot. Garden* **84**: 1–49.

Soltis, P.S., Soltis, D.E. and Chase, M.W. (1999) Angiosperm phylogeny inferred from multiple genes as a tool for comparative biology. *Nature (London)* **402**: 402–404.

Sorhannus, U. (1996) Higher ribosomal RNA substitution rates in Bacillariophyceae and Dasycladales than in Mollusca, Echinodermata, and Actinistia-Tetrapoda. *Molec. Biol. Evol.* **13**: 1032–1038.

Stevens, P.F. (1984) Metaphors and typology in the development of botanical systematics 1690–1960 or the art of putting new wine in old bottles. *Taxon* **33**: 169–211.

Syvanen, M. (1985) Cross-species gene transfer: implications for a new theory of evolution. *J. Theoret. Biol.* **112**: 333–343.

Syvanen, M., Hartman, H. and Stevens, P.F. (1989) Classical plant taxonomic ambiguities extend to the molecular level. *J. Molec. Evol.* **28**: 536–544.

Syvanen, M. (1994) Horizontal gene transfer: Evidence and possible consequences. *Annu. Rev. Genetics* **28**: 237–261.

Valentine, J.W., Jablonski, D. and Erwin, D.H. (1999) Fossils, molecules and embryos: New perspectives on the Cambrian explosion. *Development (Cambridge)* **126**: 851–859.

Vavilov, N. (1922) Law of homologous series in variation. *J. Genet.* **12**: 47–89.

Van De Peer, Y.A.D. and De Wachter, R. (1997) Evolutionary relationships among the eukaryotic crown taxa taking into account site-to-site rate variation in 18S rRNA. *J. Molec. Evol.* **45**: 619–630.

Went, F.W. (1972) Parallel evolution. *Taxon* **20**: 197–226.

West, L. and Powers, D. (1993) Molecular phylogenetic position of hexactinellid sponges in relation to the Protista and Demospongiae. *Mol. Marine Biol. Biotechnol.* **2**: 71–75.

Wolfe, K.H., Gouy, M., Yang, Y.W. et al. (1989) Date of the monocot-dicot divergence estimated from chloroplast DNA sequence data. *Proc. Natl Acad. Sci. USA* **86**: 6201–6205.

Wray, G.A., Levinton, J.S. and Shapiro, L.H. (1996) Molecular evidence for deep Precambrian divergences among metazoan phyla. *Science (Washington DC)* **274**: 568–573.

Yokoyama, S. and Harry, D.E. (1993) Molecular phylogeny and evolutionary rates of alcohol dehydrogenases in vertebratas and plants. *Mol. Biol. Evol.* **10**: 1215–1226.

Graptolite Parallel Evolution and Lateral Gene Transfer

William B.N. Berry and Hyman Hartman

The horizontal transfer of genes by infectious agents such as viruses or viroids is suggested to have been a mechanism for the parallel evolution of morphological characters in the paleontological record. In particular, the case of stunting in the initial developmental phase of microaerophilic graptolite colonies which took place in a number of genera after near-extinction of most of these graptolites about 437 million years ago is considered. The microaerophilic graptolites that nearly became extinct lived in nitrate and nitrite-reducing, hypoxic environments on the margins of oceanic oxygen minimum zones. When those environments were markedly diminished, the microaerophilic graptolites suffered oxygenic stress and/or loss of the nitrate and nitrite-reducing bacteria upon which they presumably fed. An infectious viroid model for the development of stunting is proposed as a possible horizontal transfer of a control gene.

INTRODUCTION

A symposium held at the University of Chicago in 1959 celebrated the centenary of publication of Darwin's *Origin of Species*. The three volumes of papers that resulted from that symposium provide ample evidence of the vitality of Darwinian natural selection. The major focus of the symposium was the synthesis of Mendelian genetics with Darwinian theory of natural selection. The so-called synthetic theory of evolution,

as discussed at the Chicago symposium, was applied to several fields of biology, such as taxonomy and paleontology. The centerpiece of the synthetic theory was the problem of the formation of species, which was defined as a population of sexually reproducing organisms that shared a common gene pool. Speciation would result when this population was split into two or more separate gene pools by geographic isolation for an extended period of time. No genes could flow from the gene pool of one species to that of other species. A species was an isolated gene pool.

In the field of paleontology, however, the fossil record posed certain problems to the synthetic theory. Among those discussed at the symposium was the observation of parallel evolution in the paleontological record. Parallel evolution was defined at this symposium by Olson (1960) as:

> parallelism in major morphologic structures, and especially in suites of major structures, in evolving lines of populations related only at high categorical levels, and with remote common ancestors in which common structures did not exist.

This is a phenomenon observed frequently in the fossil record.

It should be noted that parallel evolution as defined by Olson poses certain difficulties for the synthetic theory. For example, there is no common ancestor detectable in the fossil record, and also the complex morphological character

Horizontal Gene Transfer
ISBN: 0-12-680126-6

Copyright © 2002 by Academic Press.
All rights of reproduction in any form reserved.

appears in populations which in taxonomic terms are above the level of species and hence not capable of exchanging genes. The only possible explanation for parallel evolution, using this sense of species and speciation, is similarity in selective forces operating on the genetically isolated populations.

In recent years, due to advances in molecular biology, possibilities for transfer of genes by means of viral and other agents have been demonstrated (Syvanen and Kado, 1998). Fuhrman (1999) discussed genetic exchange among marine microorganisms, indicating that marine viruses have been identified as having a role in horizontal gene transfer in the marine environment. Jiang and Paul (1998) suggested that transduction, in which a virus picks up DNA from one host species and transfers it to another, may take place in marine settings. Chiura (1997) discussed conditions for gene transfer by virus-like elements from marine bacteria, and Paul et al. (1993) indicated that viral lysis caused release of DNA from one host which could be transferred to other host organisms through natural transformation in marine environments. Based upon current understanding of mechanisms available to facilitate lateral transfer of genes, horizontal gene transfer of genes across species or higher taxonomic levels may be a viable explanation for some instances of parallel evolution observed in the fossil record. This chapter examines a case of parallel evolution in graptolites, an extinct fossil group.

GRAPTOLITES

Graptolites, because they have an extensive paleontological record resulting from a great many collections of large numbers of individuals from closely spaced rock layer intervals, are a favorable group in which to see evolutionary patterns. Graptolite morphologies, systematics and habitats have been discussed by Berry (1987), Finney and Berry (1997), and Berry and Finney (1998, 1999).

The most commonly-found graptolites in the fossil record are the remains of colonial marine plankton. They lived from early in the Ordovician into the Early Devonian (a time span of about 500 to 390 million years ago). The majority of graptolites are found as silhouettes on rock surfaces. Rare finds of little chemically altered, uncrushed specimens that may be freed from the rock matrix using acids provide most of the known morphological information concerning graptolites. That information was used to interpret what is seen in silhouettes.

The majority of planktonic graptolites lived in tropical oceans, where they inhabited waters over the outer parts of continental shelves and upper parts of adjacent continental slopes of the time. The greatest number of planktonic graptolite colonies lived within or close to sites of oceanic upwelling. Some graptolites inhabited surface and near-surface waters that were relatively oxic. Most graptolites, however, appear to have lived at some depth, inhabiting hypoxic waters along the margins of the oxygen minimum zone. They encountered high concentrations of denitrifying bacteria that reduced nitrous oxides in those waters (Berry and Finney, 1998, 1999). Graptolite zooids are presumed to have fed on these denitrifying bacteria (Finney and Berry, 1997; Berry and Finney, 1999).

Based on a few impressions and on diameters of zooidal tubes that comprise the graptolite colony shell, or test, zooids ranged in size from about 0.15 to 0.8 mm in diameter. The zooids seem to have possessed a lophophore (tentacle-bearing arm-like structure) and they appear to have budded serially from the preceding zooid. The budding pattern is considered to have been similar to that of modern rhabdopleurans (Berry, 1987; Rigby, 1994; Urbanek, 1994). Study of growth of colonies of the pterobranch hemichordate *Cephalodiscus* led Dilly (1993) to suggest that living cephalodiscids and graptolites are so similar that cephalodiscids are living graptolites. Urbanek (1994), however, concluded from an analysis of budding patterns seen in certain graptolites and in the living pterobranch *Rhabdopleura* that graptolites were more closely similar to rhabdopleurans and that *Rhabdopleura* is a living graptolite and, therefore, a "living fossil."

The first-formed part of the shell, or test, of the colony (the sicula) house a presumably sexually produced individual. That part of the test differs in morphological details from the test of

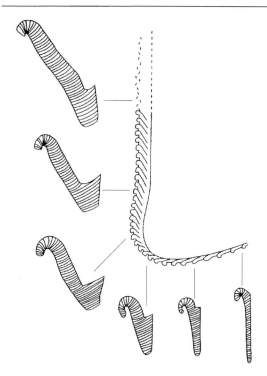

FIGURE 31.1 Monograptid graptolite showing change in size and shape of zooidal cups from immature to mature part of the colony test (adapted from Bulman, 1970, Figure 43).

the remainder of the colony. Test material appears to have been proteinaceous in composition for collagen-like fibers have been identified in it.

Development of the graptolite colony records changes in zooid size and shape during passage from the immature to mature state (Figure 31.1). In many colonies, zooidal tubes in the immature part of the colony are small-diameter, slender cones. Later in colony development, as the colony became more mature, zooidal tubes achieved greater diameters. Such changes from slender to more robust zooidal tubes gave the whole colony test a tapered appearance. That appearance reflects the change from small to relatively larger zooids produced as the colony became more mature. The example in Figure 31.1 is a monograptid (it has uniserial colony form) and it is used to illustrate how changes in zooidal cup size in the development of the colony result in distinctive colony shape. That shape is described hereafter as tapering.

CONDITIONS AT A NEAR-EXTINCTION AMONG GRAPTOLITES

Graptolites nearly became extinct within about a 100 000–200 000 year interval during the onset of massive continental glaciation in the Late Ordovician (about 436 million years ago) (Berry and Finney, 1999; Finney and Berry, 1999). Continental ice covered most of the land mass of a large southern hemisphere continent, Gondwanaland, at that time. During maximum glacial development, sea level was lowered by 70–80 m. As a consequence, upwelling conditions that had prevailed along many continental shelf margins were either diminished markedly or suppressed entirely. Because the upwelling conditions were either lost or diminished, oxygen minimum zones that had existed under them were also either lost or markedly contracted. The record of the Late Ordovician glacial interval graptolites as well as the geochemistry of the rocks bearing them indicate that the hypoxic, nitrous oxides-rich waters inhabited by most graptolites were wiped out entirely or confined to a number of small, relatively local centers (Berry et al., 1995; Berry and Finney, 1999; Finney and Berry, 1999; Finney et al., 1999).

Another consequence of glacio-eustatic sea level fall was marked reduction of shelf sea areas and of any sites of hypoxic waters that had been present during high stands of sea level. Accordingly, continental glaciation destroyed a great many of the habitats inhabited by the majority of graptolites. Graptolites that lived in oxic waters continued to find available habitats. The microaerophilic graptolites, in contrast, were stressed in the Late Ordovician glacial interval both by the loss of oxygen-poor environments and the spread of oxygenic waters. It is these conditions that resulted in the near-extinction of the microaerophilic graptolites (Berry et al., 1995; Berry and Finney, 1999; Finney and Berry, 1999).

Stunted or dwarfed colonies typify the majority of microaerophilic graptolite colonies that survived near-extinction at a number of localities. At certain sites, large numbers of such colonies occur crowded together on rock surfaces that accumulated during and immediately after the

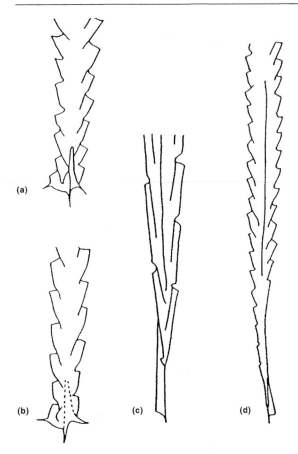

FIGURE 31.2 (a, b) Pre- and (c, d) post-Late Ordovician near-extinction graptolites. Both (a) *Orthograptus* of the *O. calcaratus* group and (b) *Glyptograptus tenuissimus* have relative "square" initial regions of the colony and bell-shaped siculae. Compare these characters with the long, slender siculae and elongate initial parts of (c) *Akidograptus ascensus* and (d) *Dimorphograptus confertus swanstoni*. The appearances of both *Akidograprus* and *Dimorphograptus* are indicated in Figure 31.3.

interval of near-extinction. Stunting is expressed not only in small colonies composed of relatively few zooidal tubes, but also as slender, markedly tapered colonies. Certain of the tapered colonies are essentially clusters of a few long, thin zooidal tubes. Some of these colonies are uniserial (tubes on only one side of the colony axis) initially and then they become biserial. Colonies that existed both before and after near-extinction are illustrated in Figure 31.2.

RE-RADIATION AFTER NEAR EXTINCTION

Studies of localities with rock sequences bearing graptolites that survived the extinction and were involved in radiations after it indicate that most of these localities were isolated from one another, leading to marked provincialism of immediate post-extinction graptolites (Koren and Melchin, 2000). Analyses of the graptolites obtained at these localities reveals that nearly all of the survivor graptolites were those that lived in relatively oxic waters. In certain areas, these survivor graptolites (they are primarily normalograptids and diplograptids commonly included within *Glyptograptus*) radiated somewhat during that interval in which the microaerophilic graptolites nearly became extinct (Storch and Serpagli, 1993; Melchin, 1998; Finney and Berry, 1999). Each area appears to have been populated with mostly endemic species (see, for example, taxa described from the southern Urals in western Kazakhstan by Koren and Rickards (1996), southwestern Sardinia by Storch and Serpagli (1993) and eastern Uzbekistan by Koren and Melchin (2000)). Certain graptolites with tapered colony form occur at most localities at which immediate post-extinction taxa have been found. During postglaciation sea-level rise, some of these taxa with tapered colony form that seem to have been restricted to one site initially were transported to other sites.

Near-extinction of microaerophilic graptolites occurred during glacio-eustatic lowstand and markedly diminished hypoxic, nitrous-oxides rich habitats preferred by these graptolites. A few taxa did persist in these diminished habitats. When climates warmed and glaciers melted, sea level rose worldwide. Upwelling oceanic conditions and oxygen minimum zones were re-established in many continental shelf seas. These developments led to expansion of the hypoxic nitrous-oxide-rich habitats. As these environments expanded, microaerophilic taxa radiated and new lineages developed (Rickards et al., 1977; Melchin and Mitchell, 1991; Storch, 1996; Berry and Finney, 1999). Many of the initial representatives that occur at the genus or family taxonomic level in these new lineages of microaerophilic graptolites have markedly tapered colony form. Rickards (1988) stated specifically in this regard that colonies of species within the new re-radiation genera are characterized by a "drawn-out, thorn-like" first-formed part of the colony and that the initial zooidal cups in these colonies are long slender cones. These conical zooidal cups are significantly thinner and more tapering than zooidal cups in the remainder of the colony. Rickards et al. (1977) discussed the re-radiation lineages, indicating that colonies in the genera *Akidograptus, Atavograptus, Cystograptus, Dimorphograptus, Glyptograptus, Paraorthograptus* and *Rhaphidograptus* are typified by markedly tapered colonies. Other genera with similar morphologies include *Parakidograptus* and *Pseudorthograptus*. Examples of two of these new genera are illustrated in Figures 31.2c and 31.2d, andthe pattern of appearances of the new, reradiation genera is shown in Figure 31.3. Tapered colony morphologies include those in which the zooidal tubes are arranged biserially and colonies in which zooidal tubes have uniserial arrangement in the immature part of the colony but become biserial in the mature part. The tapered colony morphology characterizes most monograptid graptolites (Figure 31.1) – those graptolites in which zooidal cups are arranged uniserially throughout the colony. The size of the zooidal tubes changes from the immature to mature portion of the colony (Figure 31.1). The uniserial morphologies of monograptids parallel the uniserial immature parts of those colonies that become biserial in the mature parts of the colony, such as the colony shown in Figure 31.2d. Slender immature region zooidal tubes (Figures 31.1, 31.2c and 31.2d) characterize most re-radiation taxa and post-extinction lineages of microaerophilic graptolites.

The common explanation for the relatively sudden appearance of the new genera is that they developed from undetected ancestral populations that were small and relatively isolated. Despite carefully conducted searches for potential ancestors of most new, post near-extinction genera (Rickards et al., 1977; Melchin and Mitchell, 1991), none has been identified.

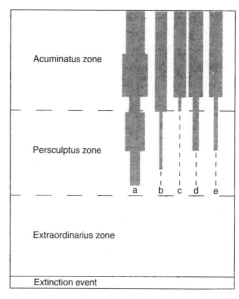

FIGURE 31.3 Chart illustrating parallelism among new, post-near extinction, re-radiation genera (adapted from Melchin and Mitchell, 1991, Figure 5). Appearances where known in the fossil record are indicated by shading. Widths of the shaded areas are indicative of the relative number of species. The narrowest shaded area is one species. The time interval indicated from the Extinction Event through the Accumulation Zone is somewhat more than a million years. The widths indicated for the zones do not reflect relative time durations: the Extraordinus Zone may have been only a few hundreds of thousands of years in duration and fossils indicative of this zone are rarely preserved. New appearances seen in the rock record of the Persculptus Zone are controlled by preservation of graptolite bearing rocks. (a) *Pseudorthograptus*; (b) *Akidograptus*; (c) *Parakidograptus*; (d) *Dimorphograptus*; (e) *Atavograptus*.

GENE EXCHANGES

Eukaryotic cells have the ability to form endosymbioses with bacteria. This ability resulted in the formation of mitochondria and chloroplasts. This, of course, is a way to transmit genes above the species level.

Molecular biologists also have discovered an extensive system of gene exchanges mediated by plasmids, episomes and viruses among bacteria, especially *E. coli*. These observations led

Anderson (1970) to suggest that viral mechanisms existing in sexual populations may transport genes across species barriers. In particular, Anderson (1970) indicated that parallel evolution is due to continuous gene flow among populations at higher levels than that of species. Went (1971) independently noted that parallel evolution is the rule rather than the exception in plant evolution, and he suggested that a viral mechanism may transfer genes between plant populations representative of different families. Reanny (1974) reviewed ideas on the potential significance of viral mechanisms for lateral gene transfer.

Syvanen (1985, 1987, 1994) discussed gene transfer by viruses, bacteria and eukaryotes. He interpreted certain instances of parallel evolution among fossil invertebrates to be the result of cross-species gene transfer (Syvanen, 1985) and included an analysis of parallel evolution among graptolites, which had been discussed originally by Bulman (1933). In another analysis of fossil taxa, Jeppsson (1986) suggested cross-species gene transfer by viruses as a mechanism for convergence in five lineages of the extinct fossils, conodonts. Erwin and Valentine (1984) commented on the possibility of linking horizontal transfer of genetic information by RNA-based viruses with the Early Cambrian radiation among metazoans. The fossil record appears to include evidence suggesting that lateral gene transfer has been a significant element in certain radiations. Among extant organisms, Beneviste (1985) proposed retroviruses as a mechanism for gene transfer among mammalian species. His study indicates that lateral gene transfer occurs in multicellular, eukaryotic, sexually reproducing organisms.

VIROIDS

An interesting discovery in recent years suggests that viroids may be a mechanism for lateral gene transfer. Viroids are infective agents of plants. They are single-stranded RNAs of low molecular weight. The number of nucleotides in viroids ranges from 245 to 370. They do not code for any protein and they are replicated by enzymes found in the plant host. Semancik and Conjero-Tomas (1987) reviewed the properties of viroids and suggested that viroids could be escaped regulatory agents. Sequence homology between small nuclear RNAs and viroids has been observed (Diener, 1979). Furthermore, numerous sequence homologies between Group I introns and viroids have been demonstrated (Dinter-Gottlieb, 1987). Diener (1979) indicated that viroids are recent in origin, but the assumption is made that a viroid originated in the wild and spread from "a wild carrier species to a susceptible cultivated plant species." The major observable effect of viroid infection is stunting in the infected host plant.

The feature of stunting of infected hosts may explain the described example of parallel evolution observed in the extinct fossils, the micro-aerophilic graptolites. Parallel evolution appears to be prominent in the post near-extinction re-radiation among these latest Ordovician-earliest Silurian graptolites. Stunting is one characteristic of certain survivor and re-radiation taxa.

DISCUSSION

It is clear that a number of possible scenarios may explain the slender tapered colony morphologies observed among the new post near-extinction, re-radiation microaerophilic graptolite colonies that represent new genera. The explanation of the observed phenomenon as having been the result of lateral gene transfer by viruses or viroids has the merit of directing the experimentalist to search for such agents in present-day evolution. Viroid infection by plants suggests a mechanism for what could have taken place among the microaerophilic graptolites in the latest Ordovician near-extinction and subsequent re-radiation. Potentially, a viroid-like RNA passed among environmentally-stressed graptolites, causing stunting and delayed budding (the long slender zooidal tubes in immature parts of colonies reveal that budding must have been delayed when compared with budding seen among pre-near-extinction colonies) among infected colonies. The RNA was incorporated into the host genome, leading to permanent delayed budding from the asexually-produced zooids in infected colonies. In many re-radiation colonies, frequency of budding did increase eventually. In certain of them, the monograptids, frequency of budding remained delayed throughout maturation of the colony. The origins of monograptids have been an unresolvable issue, thus the possibility of monograptid origination as a consequence of lateral gene transfer is plausible and reasonable. The restricted, nutrient-poor environments in which survivor microaerophilic graptolites lived during the interval of near-extinction would have facilitated transfer of a viroid-like RNA. Fuhrman (1999) drew attention to the enormous number of viruses (approximately 10 billion per liter) in hypoxic marine waters. He also noted (Fuhrman, 1999) that viral abundance is closely correlated with bacterial abundance in the oceans. Based upon fossil occurrences of microaerophilic graptolites in the near-extinction rock record, individual colonies were crowded together in the hypoxic environments they preferred. The enormous abundance of viruses as well as of DNA and RNA released from bacteria by lysis in these waters, as noted by Furhman (1999), and the crowding of graptolite colonies in the same environment would have created ample opportunities for passage of infecting RNA from colony to colony. The signal for the horizontal transfer of viroids was the oxygenic stress that the microaerophilic graptolites experienced at that time. Horizontal transfer of genes may be an adaptive response set off by a stressed population of organisms. Interestingly, that response to environmental stress was induced by climate change (Berry and Finney, 1999).

To date, much of the discussion of lateral gene transfer has centered on structural genes. The latest Ordovician near-extinction and subsequent re-radiation of the microaerophilic graptolites may be an example of lateral transfer of genes that control development. The evidence from the latest Ordovician microaerophilic graptolites suggests that lateral transfer of genes did take place.

In a discussion of the origin of the eukaryotic cell, it was postulated that small RNAs were used to control complex patterns of differentiation in multicellular eukaryotes (Hartman, 1986). These RNAs could possibly be remnants of an ancient RNA-based cell. It is ironic that the study of parallel evolution in the paleontological record may lead to the search for molecular mechanisms of development in multicellular organisms: ontogeny may capitulate to phylogeny (or paleontology).

REFERENCES

Anderson, N.S. (1970) Evolutionary significance of virus infection. *Nature* **227**: 1346–1347.

Beneviste, R.E. (1985) The contribution of retroviruses to the study of mammalian evolution. In *Molecular Evolutionary Genetics* (ed., R.J. MacIntyre), pp. 359–385, Plenum Press, New York.

Berry, W.B.N. (1987) Phylum Hemichordata (including Graptolithina). In *Fossil Invertebrates* (eds, R.S. Boardman, A.H. Cheetham and R.J. Rowell), pp. 612–635, Blackwell Scientific Publications, Palo Alto, CA.

Berry, W.B.N. and Finney, S.C. (1998) Significance of oceanic denitrification zones for graptolite occurrence. *Proceedings of the 6th International Graptolite Conference.* (eds, J.C. Gutierrez-Marco and I. Rabano), Temas Geologico-Mineros ITGE **23**: 152–153.

Berry, W.B.N. and Finney, S.C. (1999) New insights into Late Ordovician graptolite extinctions. *Acta Universit. Carolinae Geol.* **43**: 191–193.

Berry, W.B.N., Quinby-Hunt, M.S. and Wilde, P. (1995) Impact of Late Ordovician glaciation-deglaciation on marine life. In *Effects of Past Global Change on Life* pp. 34–46, Studies in Geophysics Series, NRC Board on Earth Sciences and Resources, National Academy Press, Washington, DC.

Bulman. O.M.B. (1933) Programme-evolution in the graptolites. *Biol. Rev. Cambridge Philos. Soc.* **8**: 311–334.

Bulman, O.M.B. (1970) *Graptolithina with Sections on Enteropneusta and Pterobranchia. Treatise on Invertebrate Paleontology,* Part V, The Geological Society of America, Boulder, CO.

Chiura, H.X. (1997) Generalized gene transfer by virus-like particles from marine bacteria. *Aquatic Microb. Ecol.* **13**: 75–83.

Diener, T.O. (1979) *Viroids and Viroid Diseases,* Plenum Press, New York.

Dilly, P.N. (1993) *Cepalodiscus graptoloides* sp. Nov., a probable extant graptolite. *J. Zool.* **229**: 69–78.

Dinter-Gottlieb, G. (1987) Possible viroid origin: viroids, virusoids and group I introns. In The Viroids (ed., T.O. Diener), pp. 189–204, Plenum Press, New York.

Erwin, D.H. and Valentine, J.W. (1984) "Hopeful Monstors," transposons, and Metazoan radiation. *Proc. Natl Acad. Sci. USA* **81**: 5482–5483.

Finney, S.C. and Berry, W.B.N. (1997) New perspectives on graptolite distributions and their use as indicators of platform magin dynamics. *Geology* **25**: 919–922.

Finney, S.C. and Berry, W.B.N. (1999) Late Ordovician graptolite extinction: the record from continental margin sections in central Nevada, USA. *Acta Universit. Carolinae Geol.* **43**: 195–198.

Finney, S.C., Berry, W.B.N., Cooper, J.D. et al. (1999) Late Ordovician mass extinction: A new perspective from stratigraphic sections in central Nevada. *Geology* **27**: 215–218.

Fuhrman, J.A. (1999) Marine viruses and their biogeochemical and ecological effects. *Nature* **399**: 541–548.

Hartman, H. (1986) The origin of the eukaryotic cell. *Spec. Sci. Technol.* **7**: 77–81.

Jeppsson, L. (1986) A possible mechanism in convergent evolution. *Paleobiology* **12**: 80–88.

Jiang, S.C. and Paul, J.H. (1998) Gene transfer by transduction in the marine environment. *Appl. Environ. Microbiol.* **64**: 22780–2787.

Koren, T.N. and Rickards, R.B. (1996) Taxonomy and evolution of Llandovery biserial graptoloids from the southern Urals, western Kazakhstan. *Palaeontol. Assoc. Special Papers Palaeont.* **54**: 1–103.

Koren, T.N. and Melchin, M.J. (2000) Lowermost Silurian graptolites from the Kurama Range, eastern Uzbekistan. *J. Paleont.* **74**: 1093–1113.

Melchin, M.J. (1998) Morphology and phylogeny of some early Silurian "diplograptid" genera from Cornwallis Island, Arctic Canada. *Palaeontology* **41**: 263–315.

Melchin, M.J. and Mitchell, C.E. (1991) Late Ordovician extinction in the Graptoloidea, In *Advances in Ordovician Geology* (eds, C.R. Barnes and S.H. Williams), Geological Survey of Canada Paper 90-9, pp. 143–156.

Olson, E.C. (1960) Morphology, paleontology, and evolution. In *Evolution after Darwin* (ed., S. Tax), pp. 523–545, University of Chicago Press, Chicago, IL.

Paul, J.H., Rose, J.B. and Jiang, S.C. (1993) Distribution of viral abundance in the reef environment of Key Largo, Florida. *Appl. Environ. Microbiol.* **59**: 718–724.

Reanny, D. (1974) Viruses and evolution. *Int. Rev. Cytology* **37**: 21–52.

Rickards, R.B. (1988) Graptolite faunas at the base of the Silurian. *Brit. Mus. Nat. Hist. (Geol.) Bull.* **43**: 345–349.

Rickards, R.B., Hutt, J.E. and Berry, W.B.N. (1977) Evolution of Silurian and Devonian graptoloids. *Brit. Mus. Nat. Hist. (Geol.) Bill.* **28**: 1–120.

Rigby, S. (1994) Hemichordate skeletal growth: shared patterns in *Rhabdopleura* and Graptoloids. *Lethaia* **27**: 317–324.

Semancik, D.S. and Conjero-Tomas, V. (1987) Viroid pathogenesis and expression of biological activity. In *Viroids and Viroid-like* (ed., J.S. Semancik), pp. 71–126, Pathogen CRC Press, Boca Raton, FL.

Storch, P. (1996) The basal Silurian *Akidograptus ascensus-Parakidograptus acuminatus* Biozone in peri-Gondwanan Europe: graptolite assemblages, stratigraphical ranges and palaeobiography. *Vestnik Ceskeho Geologickeho Ustavu* **71**: 177–188.

Storch, P. and Serpagli, E. (1993) Lower Silurian graptolites from the southwestern Sardinia. *Boll. Della Soc. Paleontologica Italiana* **32**: 3–57.

Syvanen, M. (1985) Cross-species gene transfer: implications for a new theory of evolution. *J. Theoret. Biol.* **112**: 333–343.

Syvanen, M. (1987) Molecular clocks and evolutionary relationships: possible distortions due to horizontal gene flow. *J. Molec. Evol.* **26**: 15–23.

Syvanen, M. (1994) Horizontal gene transfer: evidence and possible consequences. *Annu. Rev. Genet.* **28**: 237–261.

Syvanen, M. and Kado, C.I. (eds) (1998) *Horizontal Gene Transfer*, Chapman & Hall, London.

Urbanek, A. (1994) When is a pterobranch a graptolite? *Lethaiia* **27**: 324.

Went, F.W. (1971) Parallel evolution. *Taxon* **21**: 197–226.

CHAPTER 32

Larval Transfer in Evolution

Donald I. Williamson

Larval transfer is the hypothesis that genes specifying larval forms have been transferred between taxa at infrequent intervals during the course of animal evolution. It accounts for the occurrence of similar larval forms in the life histories of some remotely related animals and of radically different larval forms in the life histories of some closely related animals. Hybridization provides the probable mechanism. It is now proposed that the basic forms of all embryos and larvae were transferred from other taxa and all originated as non-larval forms.

Transfers by hybridization have probably taken place in most metazoan phyla; sponge crabs (Dromiidae), echinoderms and hemichordates are here used as examples. Adult Dromiidae are usually classified as true crabs (Brachyura), but their larvae resemble those of hermit crabs (Anomura). In the starfish *Luidia sarsi*, the bilateral larva and the radial juvenile of the same individual can co-exist for three months after metamorphosis. Other echinoderms furnish less spectacular examples of overlap between larva and juvenile, as do polychaetes and nemerteans with trochophore-like larvae and urochordates with tadpole larvae. Co-existence of the larva and juvenile of the same individual, previously unexplained, is consistent with the suggestion that the two phases originated as separate genomes: echinoderm larvae and hemichordate tornaria larvae originated as adult plancto- sphaeroids, trochophore larvae as adult rotifers, and urochordate tadpole larvae as adult appendicularians.

Experimental hybrids between ascidian eggs and sea urchin sperm are described. More information will come from more hybridizations, including those between less distantly related species.

Some molecular phylogenies are incompatible with evolution within separate lineages but are consistent with larval transfer. Classifications and cladograms of the animal kingdom that ignore larval transfer are unsound.

INTRODUCTION

The larval transfer hypothesis challenges the assumption that larvae and corresponding adults have always evolved together as part of the same life history; it proposes that genes specifying the basic forms of all larvae have been transferred from other taxa at infrequent intervals during the course of animal evolution. According to the hypothesis, the first echinoderm larva was a transferred hemichordate tornaria, and the first urochordate tadpole larva was a transferred appendicularian. Hybridization is the probable method of transfer. To Darwin (1859), evolution was an essentially gradual process which occurred within separate lines of descent, and he assumed that all stages in the life history of an animal had always evolved together, as part of the same lineage. Larval transfer and endosymbiosis (Margulis, 1993) are, however, saltational evolutionary processes, involving fusion of lineages. Larval transfer, endosymbiosis, and the accumulation

Horizontal Gene Transfer
ISBN: 0-12-680126-6

Copyright © 2002 by Academic Press.
All rights of reproduction in any form reserved.

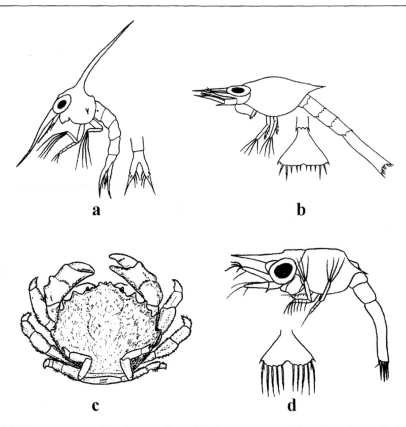

FIGURE 32.1 (a) *Cancer pagurus* (Brachyura, Cancridae): zoea I (lateral), telson (dorsal). (b) *Pagurus carneus* (Anomura, Paguridae): zoea I (lateral), telson (dorsal). (c, d) *Dromia personata* (Brachyura, Dromiidae): (c) adult (dorsal), (d) zoea I (lateral), telson (dorsal). (Adapted from Williamson, 1992.)

of adaptive mutations are complementary methods of evolution, not mutually exclusive alternatives. Lineages have occasionally fused, producing saltations in the otherwise gradual process of lineal evolution. Lineal and synlineal evolution were discussed by Williamson (1996). In *Larvae and Evolution* Williamson (1992) presented evidence for larval transfer in representatives of eight animal phyla, and Williamson (2001) proposes that the basic forms of all embryos and larvae were transferred from foreign taxa and that all have their origins in adult forms. The present chapter illustrates the principles of larval transfer and updates and expands parts of *Larvae and Evolution*, to which readers are referred for additional illustrations and references.

The larval transfer hypothesis presupposes that genes prescribing embryonic and larval features act largely independently of those that prescribe juvenile and adult features. The suggestion that genes specifying embryos and larvae can all be traced back to the genomes of non-larval forms goes some way to explaining their postulated independence. The hypothesis also assumes that, while hybrids between closely related species may show a simultaneous mixture of characters of both parents, those between distantly related species seldom do. The result may be a sequential chimera, in which the larval form of one parent may be followed by the juvenile phase of the other, with no mixing of characters within each phase (Williamson, 1991).

LARVAE

All the larvae considered in this chapter are marine and planktonic, and convergent evolution of larvae has been suggested as the explanation of several of the anomalies dicussed below. Convergent evolution results from organisms that are not necessarily closely related adopting similar shapes as adaptations to life in the same enviromnent or to similar patterns of behavior. There are undeniable cases of convergent evolution in many environments, and the streamlined shapes of many aquatic, fast moving animals, both living and extinct, provide well-known examples. Streamlining is an adaptation to swimming quickly, but there is no one adaptation to swimming slowly or drifting. The vast variety of shapes found in marine plankton implies that this environment allows many different forms. There is not one planktonic shape but thousands, and, judging by their success, they are all adapted to the environment. In consequence, while it would be rash to say that there are no cases of convergent evolution in marine larvae, it is difficult to imagine that such cases are common, so the concept of convergent evolution does not explain away all unexpected larval similarities, particularly when these go far beyond external shape. Also, adaptation to the planktonic environment should not be used to account for all unexpected dissimilarities between marine larvae, particularly when the forms occur together, eat a similar range of foods, and exhibit similar behavior.

It is generally accepted that larvae and adults have both evolved, but some larvae appear to suggest different classifications, and hence different phylogenies, from corresponding adults. Under conventional evolutionary theory, which assumes lineal evolution in which all parts of a life history evolved together, such incongruities are attributed to misclassification of the larvae or the adults or both. Where it can be shown that there is no misclassification, conventional theory must be questioned.

Anomalous crustacean larvae

In most cases, the larvae of decapod crustaceans can be classified into the same groups as the adults. This is what we should expect if adults and their larvae had evolved together, irrespective of the origin of the larval forms. In the sponge crabs (Dromiidae), however, classification based on larval characters would lead to a very different grouping from that based on adult characters. The great majority of brachyuran crabs have very distinctive larvae, and a typical example is shown in Figure 32.1a. Such larvae are quite unlike those of hermit crabs (Figure 32.1b) and other anomurans. An adult dromiid (Figure 32.1c) resembles a brachyuran rather than an anomuran in external shape, in the form of the internal (endophragmal) skeleton, and (in females) in the presence of seminal receptacles. Some dromiids have no zoea larvae, but others have larvae resembling those of hermit crabs (Paguroidea) (Figure 32.1d). Adult characters, therefore, place the Dromiidae in the Brachyura, but larval characters place the group in the Anomura. Those seeking to resolve this paradox along conventional lines have suggested that:

(1) adult dromiids are only superficially like brachyuran crabs but are really crab-like hermit crabs; or
(2) dromiid larvae are only superficially like those of hermit crabs, some suggesting that their anomuran features arose by convergent evolution in response to life in the plankton; or
(3) brachyuran crabs evolved from hermit crabs and that the dromiids have retained the ancestral larval condition; or
(4) brachyurans, anomurans, and dromiids all evolved from former dromioids with anomuran-type larvae.

Over thirty years of study have convinced me that the brachyuran characters of adult dromioids and the anomuran characters of the larvae are not superficial, that no other evidence supports the evolution of brachyurans from hermit crabs, and that there are great difficulties in deriving adult anomurans and larval brachyurans from dromioid ancestors.

Some of the evidence that seems at variance with suggestions (3) and (4) comes from homoloid larvae. The Homoloidea and the Dromioidea are both widely regarded as primitive crabs and are often grouped together as the Dromiacea. Some skeletal characters of adult homoloids appear to be more primitive than those of dromioids, and the oldest homoloids predate those of the oldest dromioids in the fossil record. Homoloid larvae, however, are not anomuran or pre-anomuran, but they show many features from which brachyuran larvae may be derived. Larval and adult characters are thus consistent with the derivation of brachyurans from homoloids, but the larval characters of dromioids seem to preclude their close relationship to either homoloids or brachyurans (Williamson, 1988a,b). The dromioid problem seemed to defy explanations in terms of accepted evolutionary theory, but it could be resolved if one accepted the hypothesis that an animal could acquire a larval form from another taxon and that this larval form would be inherited. In this case it is postulated that a dromioid with no zoea larva acquired this larval form of a hermit crab. Serious consideration of the larval transfer hypothesis began as an attempt to explain the unexpected larvae of dromioids, although it gradually emerged that many comparable or even more bizarre developmental anomalies in many phyla also lend themselves to similar solutions. Now there is a molecular study relevant to the phylogeny of the dromioids and to the hypothesis of larval transfer.

Spears et al. (1992) published cladograms based on 18S ribosomal RNA showing the relationships of species representing several families of crabs, two species of dromiids, a hermit crab, and a caridean shrimp. The study was later extended to cover several more species of Dromiidae (L.G. Abele, personal communication). The 18S rRNA sequences show a considerable gap between the brachyurans (*sensu stricto*) and the only anomuran investigated, the diogenid hermit crab *Clibanarius vittatus*. Most of the dromiids showed sufficient affinity to brachyurans to merit inclusion in the Brachyura *sensu lato*, but the two species of *Hypoconcha* (including one investigated after the published report) showed remarkable similarity to *Clibanarius*. No-one had previously questioned the classification of *Hypoconcha* as a dromiid, and it certainly bears little morphological resemblance to a hermit crab. The 18S gene, of course, tells us nothing of morphology; all it tells us is the ancestry of the 18S gene. Relationships inferred by parsimony analysis (Spears et al., 1992) do not entirely preclude the suggestion that the last common ancestor of the

Brachyura and the Anomura was a dromiid, but such a conclusion does not explain the brachyuran affinity of most dromiids and the diogenid affinity of *Hypoconcha*. The ancestral form in question could equally well have been an early brachyuran, an early anomuran, or (as I favor) a more generalized reptant decapod. Schram (1982) interpreted the fossil evidence as supporting the latter view.

The results of the 18S rRNA investigation might have been expected to show whether the Dromiidae were brachyurans (*sensu lato*), as their adult characters suggest, or anomurans, as their larval characters suggest, or a mixture, consistent with the suggestion that modern dromiids are descended from a hybrid. In the event, they showed that the Dromiidae as a whole show both brachyuran and anomuran genic affinities, but no one species showed a mixture. The only conventional explanation that has been offered implies that *Hypoconcha* is a diogenid in disguise, but no-one who has examined adult specimens accepts this. A solution in terms of larval transfer was proposed by Williamson and Rice (1996), who suggested that the dromiids acquired their larvae by two or more separate hybridizations. The larvae of *Dromia* and *Dromidia* are similar to those of *Pagurus* (Figure 32.1b,c), but they show some characters not found present-day hermit crabs but which probably occurred in ancestors of the modern Paguridae (Pike and Williamson, 1960). All the larval characters of *Hypoconcha*, on the other hand, correspond to those of present-day Diogenidae. I suggest that an ancestor of *Dromia* and *Dromidia* acquired its larval form from an early pagurid, but an ancestor of the extant species of *Hypoconcha* acquired its larval form much later, from a recent diogenid. Williamson and Rice further suggested that the cross that produced the first *Hypoconcha* with larvae might have been the reciprocal of that which gave rise to dromiids with *Pagurus*-type larvae. Such an explanation presupposes that the genes of hybrids resemble those of one parent only. Evidence from a laboratory hybrid is consistent with this (see below). Presumably, after the hybridizations, descendants inherited only one of the two possible 18S rRNA genes.

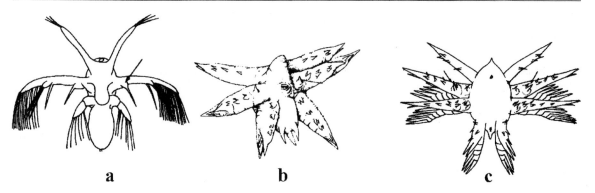

FIGURE 32.2 A recent nauplius and two Cambrian forms resembling nauplii, all in ventral view. (a) Nauplius of the copepod *Temora longicornis* (Recent). (b) "Larva A" (Müller and Walossek) (Cambrian). (c) paranauplius of *Martinssonia elongata* (Cambrian). (Adapted from Müller and Walossek, 1986a.)

The hypothesis that the Dromiidae acquired their larvae by hybridization between members of the Brachyura and Anomura, originally based purely on morphology, is certainly strengthened by the molecular evidence linking some members of this supposed family to the Brachyura and others to the Anomura. Obviously more investigations are needed into the phylogeny of these and other crab-like animals, including molecular studies looking at more genes and more species. I hope these studies will include members of the Homolidae and the spider crab *Dorhynchus thomsoni*. The zoeal carapace of this latter species has 14 spines arranged in a pattern described previously only in homolid larvae, but the adult characters and most larval characters apart from the carapace place it in the Majidae, subfamily Inachinae. A hybrid origin for this species was proposed by Williamson (1992).

Williamson & Rice (1996) proposed that most crustacean larvae, other than megalopas and postlarvae, evolved from transferred forms, and I now believe that all larvae, including megalopas and postlarvae, originated in other taxa. The crustacean examples attributed to larval transfer by Williamson (1992) are now seen as comparatively recent examples of this process, which, I claim, has occurred at intervals throughout the evolution of the group.

Nauplius, protozoea and zoea/mysis larvae provide earlier examples affecting a wide range of crustaceans, all illustrating my thesis that the basic types of all larvae were transferred from foreign taxa and all can be traced back to adult forms. These will now be considered briefly.

Crustaceans are arthropods in which the foremost appendages are two pairs of pre-oral antennae and one pair of post-oral mandibles. Nauplius larvae, however, differ significantly from this definition in having one pair of uniramous, pre-oral appendages and two pairs of similar, biramous, post-oral appendages (Figure 32.2a). "Larva A" (Figure 32.2b) was the name given to 67 specimens of a nauplius-like fossil species from Swedish Upper Cambrian strata by Müller and Walossek (1986a). The material came from 16 different samples in three different zones, spanning some 30 million years, but, in spite of the range in place and time, the specimens were remarkably similar. There were no metanauplii or later developmental stages. This is consistent with my view that "larva A" was not a larva but an adult member of the Naupliomorpha, a phylum of non-crustacean arthropods (Williamson, 2001). I suggest that representatives of several groups of crustaceans acquired nauplius larvae by hybridizing with different species of naupliomorphs, each with three pairs of appendages. The non-crustacean arthropod *Martinssonia*, described from other Swedish Cambrian material by Müller and Walossek (1986b), develops from a "paranauplius" larva with four pairs of appendages (Figure 32.2c). I suggest that some naupliomorphs had an additional pair of appendages and that an ancestor of *Martinssonia* acquired a paranauplius larva by hybridizing with such a naupliomorph. The Cambrian arthropod *Plenocaris* has a carapace, appendages on the head and anterior thoracic somites, uropods at the posterior end of the abdomen, but no other appendages (Briggs, 1983). It thus resembles a protozoea larva of a penaeid shrimp, and I suggest that an early penaeid acquired such a larva by hybridizing with a plenocarid. The resemblance between mysidaceans and the zoea larvae of many decapod crustaceans is reflected in the fact that such larvae are also known as mysis larvae. I suggest that early decapods acquired zoea larvae by hybridizing with mysidaceans (Williamson and Rice, 1996; Williamson, 2001).

Symmetry conflicts in echinoderms

Echinoderm larvae present an obvious incongruity in having an entirely different form of symmetry from corresponding adults. The conventional explanation, repeated in numerous textbooks, is that the original echinoderms were bilaterally symmetrical throughout life and the larvae have remained so; the adults, however, adopted radial symmetry as an adaptation to sedentary life. This solution seems commendably simple, but it is entirely lacking in supporting evidence. The oldest known echinoderms were not bilaterally symmetrical, and they were not all sessile (Paul, 1979). Proponents of the conventional explanation usually put forward the homolozoans (including carpoids and cinctans) as bilaterally symmetrical, Cambrian echinoderms. Palaeontologists debate whether the homolozoans were bilaterally symmetrical or asymmetrical, and there is no general agreement on the relationship of this group to the echinoderms, but there seems to be no doubt that radially symmetrical echinoderms existed before, during and after the occurrence of homolozoans. The ancestral, bilateral echinoderm remains purely hypothetical.

The assumption that a group of sedentary, bilaterally symmetrical animals could have evolved radial symmetry should also not go unchallenged. There is no precedent for this in any taxon. All known members of the Bryozoa and the Brachiopoda are sessile and bilateral, and their long fossil records suggest that they have always been so.

I postulate that adult echinoderms have always been radially symmetrical. Some early echinoderms might have had ciliated blastulas and gastrulas, but otherwise they had no larva until one acquired a bilateral larva from a hemichordate by hybridization. Further hybridizations within the echinoderms, together with continuing gradual evolution of both adults and larvae, led to the existing situation.

All echinoderm larvae are enterocoelous deuterostomes (the coelom develops from offshoots of the enteron, and the blastopore does not become the mouth), and this includes those with abbreviated larval development, often with much modified larvae. Other echinoderms have secondarily adopted direct development, with vestigial larvae persisting as embryos, and these too develop as enterocoelous deuterostomes. The phylum, however, also includes a number of brittle stars and the sea daisies (Concentricyclomorpha) that show no trace of larvae and are radially symmetrical throughout life. The brittle stars in question are schizocoelous protostomes (the coelom develops from splits in the mesenchyme, and the blastopore becomes the mouth) (Fell, 1941, 1968). The sea daisies probably develop in a similar way, although some features are undescribed (Rowe et al., 1988). For over a century the two types of coelom and mouth formation have been widely regarded as marking a fundamental cleavage in metazoan evolution. If this is so, either the enterocoelous, deuterostomatous echinoderms evolved from quite different ancestors from the schizocoelous, protostomatous forms, or (as I believe) the larvae were added later. This postulated later addition of bilateral larvae could have resulted from the fertilization of eggs of a Carboniferous holothurian with sperm from a hemichordate. Resemblances between the holothurian auricularia larva and the hemichordate tornaria have long been widely accepted (MacBride, 1914).

The Hemichordata includes (1) the Enteropneusta, some of which have tornaria larvae, (2) *Planctosphaera*, a pelagic animal that resembles a giant tornaria, and (3) the Pterobranchiata, some of which have larvae resembling trochophores. Trochophore larvae are protostomes and are discussed below. Specimens of *Planctosphaera*

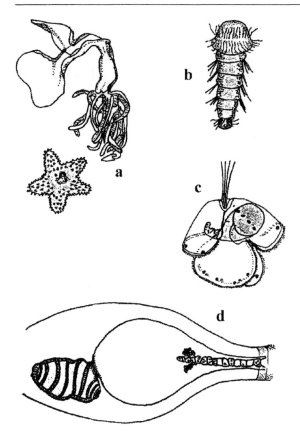

FIGURE 32.3 Co-existence of two body forms of the same individual. (a) *Luidia sarsi* (Echinodermata, Asteromorpha): juvenile detached from bipinnaria larva. (b) *Armandia brevis* (Annelida, Polychaeta): segmented larva protruding from trochophore larva. (c) *Cerebratulus* sp. (Nemertea, Anopla): juvenile within pilidium larva. (d) *Doliolum mulleri* (Urochordata, Thaliacea): juvenile within tadpole larva. Juvenile or second larva shaded in each case. (After Williamson, 2001.)

pelagica ranging in diameter from 8 to 25 mm showed little variation apart from size (Hart et al., 1994). Even in the largest, there was no development of a waist or elongation of the body such as occurs in developing tornaria larvae, and I suggest that *Planctosphaera* is not a larva but a form that matures without metamorphosis. I regard the Planctosphaeromorpha as a distinct phylum that provided the original source of larvae for both enteropneusts and echinoderms (Williamson, 2001). Enteropneusts with tornaria larvae probably include descendants of the original hybrid between an enteropneust and a planctosphaeromorph and possibly descendants of subsequent hybridizations between enteropneusts with larvae and others without. The observations that some enteropneusts lack larvae and some pterobranch larvae resemble trochophores are consistent with the suggestion that the tornaria was not an ancestral hemichordate larva, and that the original enteropneust tornaria larva was acquired by transfer from a planctosphaeromorph. I believe that a hybrid between an echinoderm an enteropneust with tornaria larvae produced the first echinoderm with bilateral larvae, and subsequent hybridizations within the echinoderms led to the spread of such larvae throughout most of the phylum. I also claim that early echinoderms had no larvae and developed as schizocoelous protostomes, and that this method of direct development has been retained by modern sea daisies and some brittle stars.

The foregoing examples that I attribute to larval transfer are limited to crustaceans, echinoderms and hemichordates. Examples in many other taxa are given in Williamson (1992), and I now believe that the basic forms of all embryos and larvae, from blastulas to caterpillars, had their origins in adult forms in foreign taxa (Williamson, 2001).

METAMORPHOSIS

If all stages in development have evolved, methods of metamorphosis must also have evolved. The metamorphoses of many animals, however, are difficult or impossible to explain in terms of lineal evolution but suggest fusion of lineages.

Figure 32.3a shows the swimming, bilaterally symmetrical larva and the crawling, radially symmetrical juvenile of the starfish *Luidia sarsi* shortly after their separation (Tattersall and Sheppard, 1934). These two body forms developed from the same egg, and, in this instance, they continued their independent existence for a further three months before the larva eventually died. Clearly the larval *L. sarsi* does not "develop into" the juvenile, but how did this situation arise? What was the evolutionary process that produced an egg with the genetic recipes for two different co-existing body forms? The only explanation that has been put forward is mine. I claim that *Luidia*, like all echinoderms with larvae, is descended from a hybrid between an echinoderm and a hemichordate. Genes specifying the bilateral hemichordate larva and the radial echinoderm juvenile have continued to function largely independently ever since the hybridization, although both developmental phases must have continued to evolve gradually (Williamson, 1992).

The development of *Luidia sarsi* is a particularly clear case of a very widespread phenomenon. The unusual feature is the length of overlap between the larval and juvenile phases, but metamorphosis in echinoderms always involves some overlap. No echinoderm larva "develops into" the juvenile, but, unlike *L. sarsi*, the larva usually dies when the juvenile separates. In all echinoderms with larvae, the radial rudiment of the juvenile first appears in a group of undifferentiated cells lining the left mesocoel of the bilateral larva. It incorporates some larval ectoderm as it grows like a parasite within the larva, eventually migrating to the outside before separating. The amount of larval tissue assimilated by the juvenile varies between different groups of echinoderms, and it is greatest in sea cucumbers, sea lillies, and feather stars, all of which go through a doliolaria larval stage. The juvenile grows within the barrel-shaped doliolaria and eventually incorporates virtually all of it. In these cases there is no larva left to die, but the radial juvenile nevertheless grew from a coelomic sac of the bilateral larva. Even when the juvenile uses larval nervous tissue, it is employed to make an entirely different nervous system from that of the larva, and the two systems function independently. An egg that develops into two distinct bodies, overlapping in time and space, seems quite inexplicable in terms of evolution within one lineage. Such a situation, however, is consistent with the suggestion that early echinoderms developed radially from the undifferentiated cells of the gastrula, just as sea daisies and some brittle stars do today. Most modern echinoderms have bilateral larvae, acquired, I claim, by transfer. The radial juvenile of such species develops from cells lining a larval coelomic sac, in much the same way that ancestral echinoderms developed from cells lining the blastocoel.

I have yet to see an explanation of the evolution of the metamorphosis of echinoderms in terms of natural selection or any other form of lineal evolution. We can be certain, however, that the echinoderms could not have evolved in the way suggested by their ontogeny. If the original echinoderms were bilateral, natural selection would have acted against the development of a radial quasiparasite. Natural selection operates on living organisms and functioning organs, and, as Darwin insisted, it is gradual. It cannot look ahead in evolutionary time and foster a gradually evolving second body form which, initially, was incapable of independent existence. A radial echinoderm can develop from undifferentiated cells in a bilateral larva. This, the only known method of changing the type of animal symmetry, clearly works during one life cycle, but a comparable process spread over numerous life cycles would defy natural selection.

Echinoderms are the only animals to change from bilateral to radial symmetry during development, but they are far from being the only animals in which the juvenile nervous system is quite distinct from that of the larva. Some annelids, echiurans, sipunculans, and mollusks hatch as trochophore larvae, which are planktonic protostomes with a characteristic pattern of cilia. The pilidium larva of nemerteans is similar. The wriggling, segmented, juvenile annelid can, in some cases, be seen protruding from the late trochophore (Figure 32.3b), and the juvenile nemertean worm can be seen flexing within the late pilidium (Figure 32.3c). In these cases, the independence of the larval and juvenile body forms is self-evident. Even when the distinction between the larva and the juvenile is less obvious, the nervous systems of the trochophore and the subsequent phase are quite distinct and overlap in time.

Ascidians and doliolids are urochordates with tadpole larvae. In ascidians the larval and juvenile nervous systems are quite distinct and overlap, and in doliolids the complete juvenile zooid can be seen at the anterior end of the late tadpole larva (Figure 32.3d). In all the foregoing taxa, the egg gives rise to two distinct and overlapping nervous systems, a condition that seems quite inexplicable in terms of evolution within one lineage.

Bryozoans cover a considerable range in larval form. Some have only a short planktonic life, and metamorphosis to the adult is by gradual modification of the larva. Others spend weeks in the plankton either as a trochophore or as a shelled larva called a cyphonautes, and, in these cases, the subsequent metamorphosis is drastic. The larva, whether trochophore or cyphonautes, has well developed nervous, locomotory and digestive systems, but it eventually settles and undergoes histolysis to produce two undifferentiated vesicles. No larval organs or tissues are preserved, and the juvenile grows from these vesicles. Here again, the juvenile does not "develop from" the larva. Such a life history seems to contradict most of von Baer's laws of development, particularly those which state that, during development, an animal progresses from general to specialized characters, and departs more and more from the form of other animals. Also, the evolution of animals that have two very different body forms connected only by undifferentiated cells, is almost as difficult to explain in conventional terms as that of animals with two overlapping body forms.

It is possible to indicate the probable non-larval sources of both trochophore and tadpole larvae.

A larval polyclad turbellarian, known as Müller's larva, resembles a trochophore, but its metamorphosis is quite different. As it grows, it gradually changes shape, retaining and modifying the larval organs. Garstang (1966), in a poem published posthumously but written before 1922, said:

> Johannes Müller's larva is the primal Trochophore
> That shows how early worms grew up from fry in days of yore:
> No drastic metamorphosis! – each youngster keeps her skin:
> Her larval frills are not thrown off, but eaten from within.

Until recently, I agreed with Garstang that Müller's larva was "the primal trochophore," which, I postulated, had been transferred to other groups by hybridization. If, however, this larval form had evolved in ancestors of the Turbellaria, why is it today restricted to polyclads? I now propose that the primal trochophore was not Müller's larva but a trochophore-like member of the Rotifera. *Trochosphaera* and other rotifers show a marked resemblance to trochophore larvae, and this led Hatschek to suggest that rotifers evolved from pedomorphic trochophores (Hyman, 1951). I believe that rotifers antedated trochophores and that a polyclad flatworm acquired trochophore larvae by hybridizing with a rotifer. There is "no drastic metamorphosis" because none is required to transform a rotifer into a flatworm. Other groups, I claim, acquired trochophores either from rotifers or from groups that had acquired their larvae from rotifers. The transient segmentation in the development of echiurans suggests that members of this phylum acquired their trochophores from polychaetes (Williamson, 1992).

I suggest that the first tadpole larvae of ascidians and doliolids were transferred appendicularians, a group of animals that are tadpoles throughout postembryonal life.

In this and the previous sections, I have postulated that tornaria larvae and echinoderm larvae were derived from planctosphaeroids, urochordate tadpole larvae from appendicularians, and trochophore larvae fom rotifers. In each case, the original source of the larval form was an animal that matured without metamorphosis, and I propose that all larval and embryonic forms originated as non-larvae (Williamson, 2001).

HYBRIDIZATION

Larvae are, by definition, different from adults, and, whatever their origins, there must be mechanisms to prevent the larval and adult forms mixing and to regulate the order and timing of development. In postulating that larval transfer can result from hybridization, it is assumed that these same mechanisms will continue to operate in hybrids.

The urochordate *Ascidia mentula* normally hatches as a tadpole larva, and the sea urchin *Echinus esculentus* normally hatches as a blastula which develops into a pluteus, with arms supported by calcareous rods. These species are not only very different as adults and larvae but their eggs are also very different. An *Ascidia* egg has a firm but nodulose outer membrane, and the inner egg is surrounded by follicle cells. An *Echinus* egg is surrounded by a transparent, sticky, jelly layer, and there are no follicle cells. These differences are very obvious even at low magnifications.

Eggs of *A. mentula* were cross-fertilized with sperm of *E. esculentus* several times in 1988–90. These experiments were briefly described in Williamson (1992), but I here give more details of experimental procedure and the history of the hybrids. Specimens of *A. mentula* were obtained from an old, covered, seawater storage tank at the Marine Laboratory, Port Erin, Isle of Man, and specimens of *E. esculentus* were collected from the open sea nearby. Ripe eggs were gently squeezed from large ascidians, and sperm was obtained from *Echinus* inverted over beakers for up to two hours. Similar procedures were followed in both experiments that produced many hybrid larvae (see below). *Echinus* sperm was obtained one morning and one drop used to fertilize *Echinus* eggs. This culture of developing *Echinus* eggs and all unused eggs were removed from the laboratory. *Ascidia* eggs were obtained the following morning, washed on a mesh in filtered seawater, then counted. Ascidians are hermaphrodites, and eggs obtained by squeezing would sometimes be self-fertilized. Only batches that remained undivided for at least three hours were used further. In these cases, the water was filtered from the eggs, and undiluted, viscid *Echinus* sperm was added to the eggs on the mesh and left for 20–30 minutes. This was followed by repeated washing to remove excess sperm. The eggs were then washed off the mesh, counted again, and kept under observation for cleavage.

In most experiments, no *Ascidia* eggs divided. In four cases, however, some hatched next day as ciliated blastulas which developed into plutei, indistinguishable from those of *Echinus*. The first three experiments yielded one pluteus, over 200 plutei and one pluteus, respectively, but none metamorphosed. In an experiment in 1990, however, over 3000 hybrid pluteus larvae

were obtained, and some of these metamorphosed. Sea urchin rudiments developed in the wall of the left larval mesocoel in over 70 cases. Some died at various stages during the growth of the rudiment, but, 37–50 days after fertilization, 20 free-living sea urchins were counted. Of the four that survived for a year, the two largest were pentaradial, like normal *Echinus*, and the other two were tetraradial. The smallest reached a diameter of 9 mm by the end of the second year, and, without growing further, died at 3 years 3 months. The other three all produced eggs at 3 years 6 months and at intervals over the next 9 months. Some eggs from each were fertilized with wild *Echinus* sperm and produced healthy pluteus larvae. These three hybrid urchins died at the age of 4 years 3 months when the seawater circulation failed. Their respective diameters were 65, 52 and 43 mm.

Of the 3000 hybrid pluteus larvae, those that developed *Echinus* rudiments were only a minority. Many larvae died, but several hundred, after attaining full development as plutei and a length of about 1.0 mm, gradually resorbed their arms and rods to become ciliated spheroids, each with a small protuberance. Each spheroid was about 0.25 mm in diameter, and it could attach itself by the protuberance, release, and re-attach. This stage was reached in 57–77 days from fertilization, and all died without developing further. We can only speculate whether, under different conditions, they would have been capable of metamorphosing and to what. Several older publications describe the development of *Echinus* and other sea urchins in detail, but they do not mention spheroids developing from plutei. These may be regarded tentatively as the type of larvae that ascidians had before they acquired tadpoles, but this must remain conjecture until more specimens are obtained.

Hart (1996) investigated nucleotide sequences for the COI mitochondrial gene and the 28S ribosomal gene extracted from tube feet of three hybrid urchins, and, in each case, found near identity with wild *Echinus esculentus*, with no ascidian components. Unfortunately this work was carried out some three years after the death of the spheroidal larvae. Hart concluded that, because the COI and 28S sequences agreed with those of *Echinus*, the putative hybrid urchins could not have hatched fron ascidian eggs, and he suggested that "a hermaphrodite [*Echinus*] used in the cross-fertilization experiments to provide sperm may have provided eggs as well". *Echinus* eggs are very difficult to filter because they stick to any mesh, but the hermaphrodite hypothesis implies that the eggs I recovered from the mesh were all *Echinus*, that their number tallied with my earlier count of *Ascidia* eggs (>200 in 1989; >3000 in 1990), that they were still undivided when counted, in spite of having been in undiluted *Echinus* sperm for more than 24 hours, and that they subsequently hatched. In practice, *Echinus* eggs exposed to this concentration of sperm for a minute do not hatch, probably as a result of polyspermy. Obviously I reject the hermaphrodite supposal and also Hart's alternative of contaminated cultures. Contamination could not explain the spheroids into which the majority of the hybrid plutei developed. There is no doubt that the urchins in question came from *Ascidia* eggs.

Clearly nucleotide sequences of more hybrids should be investigated, but Hart's findings on my hybrid urchins are relevant to the anomalies of dromiids (mentioned earlier). To summarize: 18S rRNA showed marked differences between brachyuran crabs and the only hermit crab investigated, and while the affinities of dromiids of several genera were with the brachyurans, those of members of one genus were with the hermit crab (Spears et al., 1992). Incongruities between adult and larval morphologies in the Dromiidae had earlier prompted the suggestion that dromiids with zoea larvae are descended from dromiids without such larvae that hybridized with hermit crabs (Williamson, 1988a,b, 1992). I suggest that each gene of a hybrid is maternal or paternal rather than a mixture of both. Such a proposal agrees with the results on dromiids and on my *Ascidia* × *Echinus* crosses. It is also borne out by analysis of 11 genes in interspecific hybrids between the sea urchins *Heliocidaris erythrogramma* and *H. tuberculata* (Nielsen et al., 2000). Some genes of these hybrids were maternal and some paternal, but none were mixtures.

Hatching occurred in about one in four attempts to fertilize eggs of *Ascidia mentula* with sperm of *Echinus esculentus* in 1988–90. In two experiments all the eggs hatched as tadpoles. In another two, some hatched as blastulas that quickly died and others as tadpoles. The *A. mentula* were all from the old storage tank, but this was leaking so badly that it had to be drained in 1990. In over 100 subsequent attempts to repeat this hybridization with *A. mentula* from other sources, no eggs hatched as blastulas. In some experiments some eggs hatched as tadpoles, and in others some eggs divided once or twice but then re-fused. Occasional experiments with other species of ascidian gave similar results. In one experiment with *Ciona intestinalis*, two eggs were rapidly changing shape and repeatedly dividing and refusing 24 hours after exposure to *Echinus* sperm.

These experiments show that, in spite of many failures, species in different phyla may occasionally hybridize in the laboratory. The eggs may hatch as maternal or paternal larvae, and there is no mixing of maternal and paternal features. Generalizations based on a few hybridizations may be misleading, and the source of specimens may be important. The only paternal hybrid larvae were from eggs of specimens of *Ascidia mentula* from the old storage tank. This population was growing on a substrate of old, crumbling concrete, and the specimens may well have been genetically similar, originating from one or two tadpole larvae, but the importance of these factors is unknown.

An ascidian and an echinoderm were chosen for these experiments because their larvae are very different and could not be confused, but there is no suggestion that representatives of these groups have hybridized in nature. Natural intergroup hybridizations are probably extremely rare, and I suggest that only about 20 interphyletic hybrids and rather more hybridizations between species within the same phylum or the same class have taken place during the course of animal evolution. There is evidence that cross-fertilizations between less remotely related species are easier to achieve in the laboratory, and Giudice (1973) listed many within the Echinodermata. The regular sea urchin *Strogylocentrotus purpuratus* and the sand dollar *Dendraster excentricus* have been hybridized on three occasions (Flickinger, 1957; Moore, 1957; Brookbank, 1970). Unfortunately none of these hybrid larvae were kept long enough to see whether they would metamorphose, and all the experiments were conducted before the age of gene sequence analysis. Now is the time to repeat them. Experimental cross-fertilizations between species at all levels of relationship are needed before we can generalize on the morphology and genetics of the larval and adult hybrids.

MOLECULAR AND MORPHOLOGICAL CLASSIFICATION

Today many zoologists are trying to reconcile molecular phylogenies of the animal kingdom with classifications based on morphology (references in Raff, 1996). Most morphological classifications rely heavily on the characters of embryos and larvae, and they assume that all stages in development have always evolved together. If, however, genomes specifying the forms of embryos and larvae have occasionally been transferred between phyla, the whole basis of these classifications must be questioned. For example, all echinoderm larvae are enterocoelous deuterostomes, and this has been regarded as justification for claiming affinity between echinoderms and chordates, some of which also develop as enterocoelous deuterostomes. If, however, the bilateral larval form of echinoderms was a later addition to the radial form of the adults, it forms no basis for claiming affinity between adult echinoderms and chordates. Even if this argument is rejected, the observations that some echinoderms develop directly as schizocoelous protostomes should not be ignored. Comparable incongruities occur in other phyla, including the gastropod mollusk that develops directly as a deuterostome, while mollusk larvae are protostomes (Williamson, 1992).

It has been claimed (Raff, 1996) that molecular systematics, based on 18S rRNA, upholds the grouping of echinoderms and chordates in the deuterostome superphylum, with the associated characters of radial cleavage and enterocoely. I contend that all echinoderms with larvae are descended from a hybrid between a protostome and a deuterstome but some echinoderms develop directly as schizocoelous protostomes. Most brittle stars of the families Amphiuridae, Ophiomyxidae and Gorgonocephalidae come in this latter category. Their molecular affinities should be investigated, but I do not know of any such study that includes brittle stars. Michael Syvanen's 18S rRNA phylogram, however, includes several other echinoderms and a hemichordate (Figure 32.4). This previously unpublished tree indicates a number of groupings totally at variance with morphological classifications, such as the association of chaetognaths with a tick and of an insect larva with a clam. Clearly the 18S gene has not always evolved in association with genes coded for metazoan morphology. On the other hand, the 18S phylogeny of the hemichordate and echinoderms investigated can be explained in terms of larval transfer, if it is assumed that the 18S ribosomal gene was transferred with genes specifying larval form. The 18S gene of *Balanoglossus* (Hemichordata) shows the greatest affinity to that of *Cucumaria* (Holothuromorpha), followed, in order, by *Asterias* (Asteromorpha), and *Strongylocentrotus* and *Psammechinus* (Echinomorpha). Interpreted in a conventional way, Syvanen's 18S tree supports the evolution of echinoderms and hemichordates from a common ancestor, but it also implies that hemichordates are more closely related to echinoderms than to urochordates, and that holothurians, starfish and sea urchins evolved in that order. This sequence of echinoderm evolution, however, does not agree with the fossil evidence (Smith, 1988). My alternative explanation proposes that the 18S gene of hemichordates and echinoderms was acquired with their larvae, which were transferred from group to group, and the degree of modification of the larval form indicates the sequence of acquisition. Thus the first hemichordate tornaria larva was acquired from a planctosphaeromorph, the first holothurian auricularia larva was acquired from a hemichordate, the first starfish bipinnaria larva was acquired from a holothurian, and the first sea urchin pluteus larva was acquired from a starfish. 18S rRNA tells us nothing about the

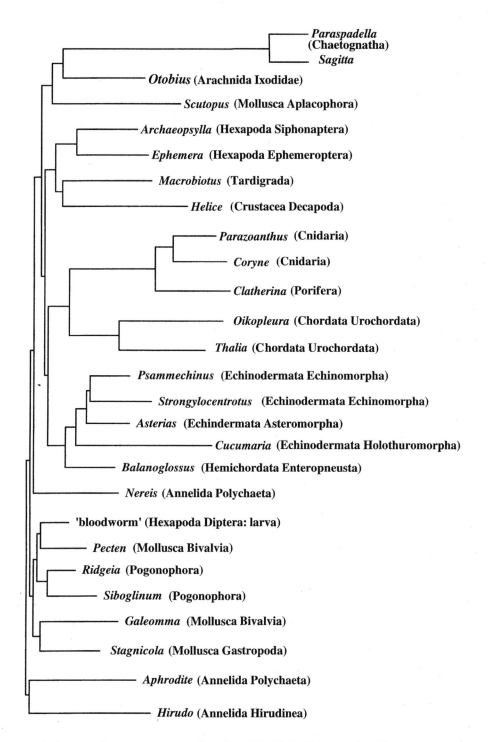

FIGURE 32.4 A phylogram of some metazoans, based on 18S rRNA. (After Michael Syvanen, unpublished.)

relationship between adult hemichordates and echinoderms.

Raff (1996) admits that some groups pose particular problems for molecular phylogeny. One such group is the chaetognaths, which develop directly by radial cleavage, deuterostomy, and enterocoely, and are therefore usually classified in the chordate superphylum. Their 18S rRNA, however, apparently excludes them from the "deuterostomes" (Telford and Holland, cited in Raff, 1996), and in Syvanen's tree their closest relative is a tick. I claim that the phylogeny of virtually all metazoans is complicated by one or more fused lineages in their ancestry. The phylogeny of the 18S gene in those hemichordates and echinoderms included in Figure 32.2 is explicable in terms of larval transfer. I suggest that many other unexpected affinities, indicated in this and other molecular trees, resulted from comparable transfers of genomes by hybridization and possibly by other methods.

The molecular phylogeny of groups traditionally regarded as chordates or near-chordates is also unexpected. Turbeville et al. (cited in Raff, 1996) found that maximum parsimony analysis linked ascidians (Urochordata) with hemichordates rather than with chordates. Raff also doubted that the 18S gene reflects adult phylogeny, saying "the morphological features … that link ascidians with chordates are so persuasive that the 18S rRNA-based inference is unacceptable without strong independent support." The main link between chordates and urochordates is through the tadpole larvae of the latter group, and I interpret the overlapping nervous systems of a tadpole larva metamorphosing to a juvenile ascidian as a result of fusion of lineages. I believe that the first ascidian tadpole larva was a transferred appendicularian and that the 18S gene of ascidians was transferred with the larval form.

The molecular findings on the Dromiidae and on my hybrid urchins are consistent with the suggestion that the genes of hybrids resemble those of one parent only, but, at present, we cannot predict which. If this is so, it will further complicate molecular phylogeny. This is not to advocate abandonment of this branch of science, but classifications based on one or two genes should be treated with caution. The evolutionary theories of Lamarck and Darwin sought to explain the distribution of morphologies, but animal phylogenies based on adult morphology, larval morphology, and 18S rRNA are frequently in conflict. I regard these conflicts as evidence of lineage fusion. Adults, larvae, embryos, and ribosomes have all evolved, but not always along the same phylogenetic paths.

I (Williamson, 1992) reproduced the phylogenetic tree of the animal kingdom in Barnes (1980), and I criticized it as a typical example of a dendrogram based largely on larval features without considering the origins of the larvae. Two commentators on Williamson (1992) considered Barnes' tree outdated, and suggested that a computer-generated cladogram based on Hennigian systematics would have been beyond reproach. Figure 32.5 gives an example of just such a cladogram from Brusca and Brusca (1990). It uses the same larval criteria as Barnes (1980) to separate the phyla (node "d" denotes deuterostomy, "e" denotes protostomy), and it arrives at much the same groupings. One important exception is that Brusca and Brusca placed the lophophorates in the deuterostome superphylum, a grouping that apparently ignores the protostomatous trochophore larvae of some bryozoans. Neither Barnes nor the Bruscas took account of those hemichordates with larvae resembling trochophores, those echinoderms that develop as schizocoelous protostomes, and those molluscs and annelids that do not have trochophore larvae. Hennig accepted Darwin's assumption that all organisms have evolved by lineal descent, but I do not. I agree with both Barnes (1980) and Brusca and Brusca (1990) that methods of mouth formation and coelom formation mark a fundamental cleavage in animal evolution, but the main divisions in Barnes' tree and Bruscas' cladogram reflect differences in larval development. I am convinced that all embryos and larvae originated in foreign groups and all can be traced back to adult forms (Williamson, 2001). The basic forms of embryos and larvae tell us nothing about the evolution of the adults in the groups to which they now belong.

The phylogenetic origins of animals should be re-investigated.

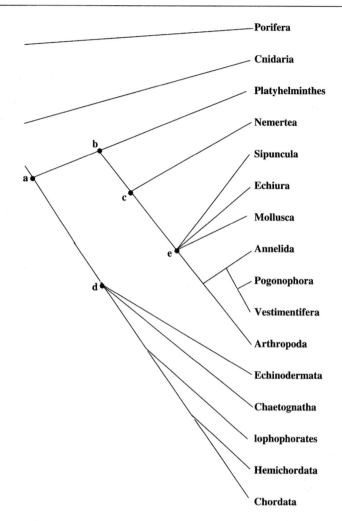

FIGURE 32.5 A cladogram depicting the relationships of the major animal phyla. a–e, nodes marking major evolutionary events, other evolutionary changes not marked. (Adapted from Brusca and Brusca, 1990.)

ACKNOWLEDGMENTS

I am very grateful to Michael Syvanen for allowing me to include his 18S rRNA tree in this chapter and to Jim Ludgate for preparing the illustrations.

REFERENCES

Barnes, R.D. (1980) *Invertebrate Zoology*, 4th edn, Saunders, Philadelphia.

Briggs, D.E.G. (1983) Affinities and early evolution of the Crustacea: evidence from Cambrian fossils. In *Crustacean Phylogeny* (ed., F.R. Schram), pp. 1–28. Balkema, Rotterdam.

Brookbank, J.W. (1970) DNA synthesis and development in reciprocal interordinal hybrids of a sea urchin and a sand dollar. *Dev. Biol.* **21**: 29–47.

Brusca, R.C. and Brusca, C.G. (1990) *Invertebrates* pp. 922, Sinauer Associates, Sunderland, MA.

Darwin, C. (1859) *The Origin of Species by Means of Natural Selection or the Preservation of Favoured Races in the Struggle for Life*, John Murray, London.

Fell, H.B. (1941) The direct development of a New Zealand ophiuroid. *Quart. J. Microsc. Sci.* **82**: 377–441.

Fell, H.B. (1968) Echinoderm ontogeny. In *Treatise on Invertebrate Paleontology, Part 5, Echinodermata* (ed., R.C. Moore), pp. S 60–S 85, Geological Society of America and University of Kansas Press, Lawrence, KS.

Flickinger, R.A. (1957) Evidence from sea urchin sand dollar hybrid embryos for a nuclear control of alkaline phosphatase activity. *Biol. Bull.* **112**: 21–27.

Garstang, W. (1966) *Larval Forms and other Zoological Verses*, Blackwell, Oxford.

Giudice, G. (1973) *Experimental Biology of the Sea Urchin Embryo*, Academic Press, New York.

Hart, M.W. (1996) Testing cold fusion of phyla: maternity in a tunicate × sea urchin hybrid determined from DNA comparisons. *Evolution* 50(4): 1713–1718.

Hart, M.W., Miller, R.N. and Madin, L.P. (1994) Form and feeding mechanism of a living *Planctosphaera pelagica* (phylum Hemichordata). *Mar. Biol. (Berlin)* 120: 521–533.

Hyman, L.H. (1951) *The Invertebrates: Acanthocephala, Aschelminthes, and Entoprocta. The Pseudocoelomate Bilateria*, vol. III, McGraw-Hill, New York.

MacBride, E.W. (1914) *Text-Book of Embryology*, vol. 1. *Invertebrata*, Macmillan, London.

Margulis L. (1993) *Symbiosis in Cell Evolution: Microbial Communities in the Archean and Proterozoic Eons*, W.H. Freeman, New York.

Moore, A.R. (1957) Biparental inheritance in the interordinal cross of sea urchin and sand dollar. *J. Experim. Zool.* 135: 75–83.

Müller, K.J. and Walossek, D. (1986a) Arthropod larvae from the Upper Cambrian of Sweden. *Trans. R. Soc. Edinburgh, Earth Sci.* 77: 157–179.

Müller, K.J. and Walossek, D. (1986b) *Martinssonia elongata* gen. et sp. n., a crustacean-like euarthropod from the Upper Cambrian Orsten of Sweden. *Zool. Scripta* 15: 73–92.

Nielsen, M.G., Wilson, K.A., Raff, E.C. and Raff, R.A. (2000) Novel gene expression patterns in hybrid embryos between species with different modes of development. *Evol. Devel.* 2(3): 133–144.

Paul, C.R.C. (1979) Early echinoderm radiation In *The Origin of Major Invertebrate Groups* (ed., M.R. House), pp. 443–481, Systematics Association Special Volume No. 12, Academic Press, London.

Pike, R.B. and Williamson, D.I. (1960) Larvae of decapod Crustacea of the families Dromiidae and Homolidae from the Bay of Naples. *Pubbl. Staz. Zool. Napoli* 31: 553–563.

Raff, R.A. (1996) *The Shape of Life. Genes, Development, and the Evolution of Animal Form*, University of Chicago Press, Chicago.

Rowe, F.W.E., Baker, A.N. and Clark, H.E.S. (1988) The morphology, development and taxonomic status of *Xyloplax* Baker, Rowe and Clark (1986) (Echinodermata, Concentricycloidea), with description of a new species. *Proc. Royal Soc. Lond. (Ser. B)* 233: 431–459.

Schram, F.R. (1982) The fossil record and evolution of Crustacea. In *The Biology of Crustacea, Vol. 1, Systematics, the Fossil Record, and Biogeography* (ed., L.G. Abele), pp. 93–147, Academic Press, New York.

Smith, A.B. (1988) Fossil evidence for the relationships of extant echinoderm classes and their times of divergence. In *Echinoderm Phylogeny and Evolutionary Biology* (eds, C.R.C. Paul and A.B. Smith), pp. 85–97, Clarendon Press, Oxford.

Spears, T., Abele, L.G. and Kim, W. (1992) The monophyly of brachyuran crabs: a phylogenetic study based on 18S rRNA. *Syst. Biol.* 41: 446–461.

Tattersall, W.M. and Sheppard, E.M. (1934) Observations on the bipinnaria of the asteroid genus *Luidia*. In *James Johnstone Memorial Volume*, pp. 25–61, Liverpool University Press, Liverpool.

Williamson, D.I. (1988a) Incongruous larvae and the origin of some invertebrate life-histories. *Prog. Oceanogr.* 19: 87–116.

Williamson, D.I. (1988b) Evolutionary trends in larval form. In *Aspects of Decapod Crustacean Biology* (eds, A.A. Fincham and P.S. Rainbow), pp. 11–15, Symposia of the Zoological Society of London No. 59.

Williamson, D.I. (1991) Sequential chimeras. In *Organism and the Origin of Self* (ed., A.I. Tauber), pp. 299–336, Kluwer, Dordrecht.

Williamson, D.I. (1992) *Larvae and Evolution: Toward a New Zoology*, Chapman and Hall, New York.

Williamson, D.I. (1996) Types of evolution. *J. Natl Hist.* 30: 1111–1112.

Williamson, D.I. (2001) Larval transfer and the origins of larvae. *J. Linn. Soc. (Zool.)* 131: 111–122.

Williamson, D.I. and Rice, A.L. (1996) Larval evolution in the Crustacea. *Crustaceana* 69(3): 267–287.

Macroevolution, Catastrophe and Horizontal Transfer

Hyman Hartman

Macroevolution occurs when there is extensive horizontal exchange of genes between taxa at levels higher than at the species level. This exchange of genes is triggered by catastrophes in the biosphere and allows the organism to adapt to the radical alteration of the environment due to the catastrophe.

In bacteria the major mechanism for gene exchange is by the transfer of transposons and plasmids. In the single-celled eukaryotes, the major mechanism is the endosymbiosis of bacteria. In the plants, fungi and animals (the multicellular eukaryotes) the major mechanisms for gene exchange are the transposons, retroposons and retroviruses.

Macroevolution is not microevolution writ large.

INTRODUCTION

There are two major themes in this chapter: (1) macroevolution is mediated by the horizontal transfer of genes, and (2) macroevolution occurs in a biosphere that has undergone a global catastrophe. These two themes are then specialized to the two types of cellular organization (i.e. the bacterial and the eukaryotic cells). After we have dealt with the simpler cases of horizontal transfer of genes (plasmids) in bacteria and the formation of the eukaryotic cell by

means of endosymbiosis, we will speculate on the possible mechanisms for horizontal gene transfer in multicellular eukaryotes – plants, fungi and animals.

BACTERIA

There are two bacterial kingdoms, eubacteria, and archaebacteria. There are now about 30–40 eubacteria and archaebacteria whose genomes have been sequenced. The surprise is that there has been extensive lateral transfer of genes between the eubacteria and the archaebacteria. In a recent review of bacterial genomics, W.F. Doolittle concluded "that lateral transfer is not just a molecular phylogenetic nuisance supported in evidence by a few anecdotal cases. Instead it is a major force, at least in prokaryotic evolution. Our trees are gene trees, not organism trees" (Doolittle and Logsdon, 1998).

In response to the global application of antibiotics, bacteria have responded by putting multiple resistance genes on a plasmid that is then sent out with the address "to whom it may concern." This is Macroevolution: a response to a crisis in the eubacterial biosphere by means of the lateral or horizontal transfer of genes.

In the area where our knowledge is greatest due to the complete sequences of the genomes of archaebacteria and eubacteria, there is no

Copyright © 2002 by Academic Press.
All rights of reproduction in any form reserved.

doubt that a mechanism for horizontal transfer of genes (i.e. the plasmid or transposon) exists. This mechanism is used in response to stresses in the biosphere (i.e. antibiotics). The result is a movement of genes among otherwise unrelated taxa of bacteria (i.e. the Gram-positive and Gram-negative eubacteria). In the past the genes moved between the kingdoms of the archaebacteria and eubacteria.

THERMOPHILIC BACTERIA

The thermophilic archaebacteria and eubacteria are the most deeply branching groups based on the small ribosomal sequences (Pace, 1997). This has usually been interpreted to imply that the last common ancestor of the archaebacteria and the eubacteria was a thermophilic cell. The recent sequence of the genome of *Thermotoga maritima*, a deeply branching thermophilic eubacterium, strongly suggests that there had been extensive lateral transfer of genes between the thermophilic eubacteria and archaebacteria (Nelson et al., 1999).

I would like to offer a new speculation for how the thermophilic bacteria came into being based on the lateral transfer of genes. This hypothesis assumes that the eubacteria and archaebacteria went through a thermophilic bottleneck due to a snowball earth – an earth covered in ice where hydrothermal vents were the only place where prokaryotic autotrophs could thrive. The horizontal transfer of genes allowed the cells to adapt to the new and challenging environment.

This speculation is based on the accumulating evidence that about 2.4 billion years ago there was a catastrophic cooling of the earth resulting in the freezing of the oceans. This resulted in "isolation of the oceans from sunlight, the interruption of the hydrological cycle, and the absence of liquid water on continents for tens of millions of years" (Kirschvink et al., 2000). This would have limited the cellular microorganisms to a limited number of viable ecological niches. The hydrothermal vents at the oceanic ridges, where the archaebacteria and the eubacteria are to be found today, would be where the horizontal transfers of thermophilic genes occurred.

In summary, the catastrophe was the snowball earth. The timing was about 2.4–2.2 billion years ago. The response by the microbial world was the lateral transfer of genes between the eubacteria and archaebacteria resulting in bacteria and archaebacteria that could survive and thrive in the hydrothermal environments of the snowball earth. Another ecological niche that stands out is the hydrothermal springs that were "one of the few places where liquid water was continuously maintained in the presence of sunlight" (Kirschvink et al., 2000). This is the likely location where the photosynthetic bacteria found a refuge.

PHOTOSYNTHETIC BACTERIA

Woese observed that photosynthetic eubacteria were the dominant lineage in the evolution of eubacteria (Woese, 1987). The five phyla of the eubacteria are the purple sulfur and non-sulfur eubacteria, the green sulfur eubacteria, filamentous green gliding eubacteria, the gram-positive eubacteria and the cyanobacteria. What is of great interest is that using a protein clock, it was inferred that the divergences between the cyanobacteria, Gram-positive bacteria and Gram-negative bacteria occurred between 2.2 and 2.5 billion years ago (Feng et al., 1997). This means that the modern photosynthetic eubacteria came into being at the same time as the catastrophic snow ball earth that occurred 2.4 billion years ago.

The gene trees for the proteins of the photosynthetic reaction centers are not congruent with the ribosomal trees of the photosynthetic bacteria. This has led to the hypothesis that the reaction centers have been horizontally transferred in some of the photosynthetic bacteria (Blankenship, 1992). The most significant of such transfers was the horizontal transfer of a reaction center from a green sulfur photosynthetic bacterium into a purple photosynthetic bacterium. This transfer resulted in the formation of the cyanobacteria and oxygenic photosynthesis (Hartman, 1998). This led to the accumulation of oxygen in the atmosphere. It would appear based on the protein clock that there was a burst of horizontal transfer of photosynthetic reaction centers at the time of the snowball earth of 2.2 billion years ago.

The accumulation of oxygen in the atmosphere would cause a new catastrophe in the biosphere that would lead to the formation of the eukaryotic cell.

THE EUKARYOTIC CELL

One of the proposals for the origin of the eukaryotic cell is that the eukaryotic cell is a combination or fusion of eubacteria and archaebacteria. Under the simple fusion hypothesis the last common ancestor of the archaebacteria, eubacteria and the eukaryotic cell would be the last common ancestor of the eubacteria and the archaebacteria.

The split between the archaebacteria and the eubacteria has been dated, using a protein clock, at about 3.5 billion years ago. "The differences between archaebacterial and eubacterial sequences indicate these two groups may have diverged between 3 and 4 billion years ago" (Feng et al., 1997). Therefore the simple fusion model for the formation of the eukaryotic cell should reflect this split. However a recent finding using protein sequences to date the last common ancestor of the eukaryotic cell and the prokaryotic cells contradicts this prediction. The prokaryotes and eukaryotes last shared a common ancestor about 2 billion years ago (Adkins and Li, 1998; Feng et al., 1997). The simplest explanation for these results is that of a horizontal transfer of bacteria into a non-bacterial host cell (Feng et al., 1997). The hypothesis then is that there was a massive intrusion of prokaryotes (eubacteria and archaebacteria) into a host cell, which was not a prokaryote and which resulted in the formation of the nucleus. This horizontal transfer of prokaryotic genes into a non-prokaryotic cell would reset the evolutionary clock for the divergence of the eukaryotic cell from the prokaryotes. It is speculated that the nucleus formed about 2–2.2 billion years ago.

What event in the biosphere might have set off this massive endosymbiosis? I would suggest the answer to this puzzle is the arrival of oxygen. It is at this time that oxygen first appears in the atmosphere (1–10% of present atmospheric level). The geological record is quite clear that dramatic changes in the oxidation state of the sedimentary rocks occurred about 2 billion years ago (Des Marais et al., 1992). This meant that oxygenic photosynthesis due to the modern cyanobacteria became the dominant form of photosynthesis resulting in the production of oxygen.

There are a number of other later events that seem to be triggered by oxygen, as the appearance of oxygen in the atmosphere seems not to have arisen gradually but in a pulsed fashion. There is evidence that the next pulse of oxygen occurred 1 billion years ago (Des Marais et al., 1992), which can then perhaps be correlated with the appearance of multicellular eukaryotes. It is speculated that the mitochondria may also have appeared at this time. Finally the Cambrian explosion may have been set off by another and final pulse of oxygen which can be correlated with the complex multicellular bauplans. The chloroplasts may have appeared at this time. The correlation of the molecular record with the geological record is one that is still in its infancy but can give rise to interesting and testable hypotheses.

The origin of the nucleus, mitochondria and chloroplasts can now perhaps be dated approximately by both molecular and geological methods. This is an exciting development giving us the means to test the various theories concerning the origin and evolution of the eukaryotic cell.

THE NUCLEUS AS ENDOSYMBIONT AND THE RETROVIRUSES

If the nucleus was an endosymbiont, then the most important question concerns the nature of the host cell. It is unlikely to have been an archaebacterium or a eubacterium as neither is capable of phagocytosis. In 1984, the host cell was postulated to be an RNA-based cell called a kronocyte. "The earliest cell to dominate the biosphere was a cell based on RNA. Surrounded by a membrane, it eked out its anaerobic existence. Its genome was not united but dispersed throughout the cell attached to the membrane. The cytoskeleton composed of proteins and RNA kept the cell in expanded form and allowed it to move about. From this ancient monster evolved the bacteria (eubacteria and archaebacteria) whose genome was based on DNA. There then ensued a set of events, very much like that catalogued in classical mythology. The kronocyte (named after the Greek god Kronos) began to swallow its progeny. After many endosymbiotic events a nucleus was formed" (Hartman, 1984). The bacteria involved in the endosymbiosis were both eubacteria and archaebacteria. The RNA genome of the kronocyte was eventually reverse transcribed into the DNA of the nucleus by means of a reverse transcriptase. This resulted in the eukaryotic cell retaining primitive RNA mechanisms such as splicing and reverse transcriptases (Hartman, 1992a,b).

According to this model, the RNA viruses with their RNA-directed RNA polymerases are remnants of the kronocyte. The retroviruses which reverse transcribe their RNA genome into DNA are then evolved from the kronocyte's genome, which was transferred into the DNA of the nucleus. The reverse transcriptase by which the retroviruses read their RNA genome into a DNA is evolved from an RNA-directed RNA polymerases (Xiong and Eickbush, 1990). This added to the capabilities of the eukaryotic cell for horizontal gene transfer as mediated by retroposons and retroviruses.

MULTICELLULAR EUKARYOTIC ORGANISMS

The geological record of the last 600 million years is filled with evidence of global catastrophes. There are claims that the earth was frozen over just before the Cambrian explosion 600 million years ago (Hoffman et al., 1998). There is evidence in the fossil record for other catastrophic events. An example is that which occurred 65 million years ago at the K-T boundary. An incoming comet or asteroid caused the demise of the dinosaurs and the rise of the mammals (Lewis et al., 1982). Associated with each of these and other catastrophes is a radical change in the biosphere.

The hypothesis then is that the response to these global catastrophes is a massive horizontal exchange of genes resulting in new and more complex phenotypes. One should note that these mechanisms are also available to local biota when there are local stresses.

The amount of horizontal transfer of genes in plants, animals and fungi is only now beginning to be understood. What is clear is that the genomes of these phyla are extremely fluid due to the hopping or transposition of genes along and between chromosomes. Barbara McClintock pointed out that, "In the future, attention undoubtedly will be centered on the genome, with greater appreciation of its significance as a highly sensitive organ of the cell that monitors genomic activities and corrects common errors, senses unusual and unexpected events, and responds to them often by restructuring the genome" (McClintock, 1984). What Barbara McClintock did not consider was the possibility of hopping genes not only transposing between chromosomes but also jumping between organisms and across species boundaries.

For example, the mariner transposon is small (1300 bp) stretch of DNA with short inverted terminal repeats of 30 bp. This small stretch of DNA codes for an enzyme involved in transposition called a transposase. It is widely spread throughout the insects, other invertebrates and vertebrates (including man). This distribution is evidence for extensive horizontal transmission of the mariner transposon across phyla. The "evidence points to repeated episodes of horizontal transfer to new hosts as the primary means by which these transposons perpetuate themselves" (Robertson and Lampe, 1995; Robertson et al., 1998). The transmission vectors are not known but the following vectors have been proposed: "External parasites, infectious agents, intracellular parasites and symbionts (especially those in the germline), DNA viruses, RNA viruses, retroviruses, even hitchhiking in other transposable elements" (Hartl et al., 1997). This collection of possible vectors is evidence of the great variety of agents that can mediate the horizontal transmission of genes in multicellular eukaryotes.

Then there are the large terminal repeats retroposons (LTR) such as gypsy-Ty3 and copia-Ty1. These LTR retroposons are found in plants, insects and yeast. Related LTRs are found in vertebrates. These elements are called the retroviral like elements (RLE) as they have all the genes of a retrovirus but lack the gene for the envelope protein. In his article "Macroevolution and retroviral elements" J.F. McDonald points out that under stress these retroposons become active and that as a result "RLEs are frequently associated with major morphological and developmental mutants in eukaryotic species" (McDonald, 1990).

There is also evidence for the horizontal transfer of LTR retroposons or RLES presumably through an infective retrovirus (Flavell, 1999). This means that the RLEs have picked up a gene coding for the envelope protein and have become infective. The RLEs are not only agents of gene transfer within an organism but they can also be infective and hence transfer genes between organisms. Thus under stress not only can an organism reshape its genome but it can also infect other organisms to do likewise. This may explain the amount of parallel evolution seen in the fossil record (Reanney, 1974; Syvanen, 1985, 1994; Berry and Hartman, 1998).

This is the great challenge for the future. How much of the evolution of the multicellular eukaryotes is based on the horizontal transfer of genes stimulated by stresses in the biosphere? The answer lies in the DNA of their genomes. Horizontal transfer of genes has played large role in the evolution of bacteria and single-celled eukaryotes. The speculation is that it played an equally large role in the evolution of the plants, fungi and animals.

REFERENCES

Adkins, R.M. and Li. W. (1998) "Dating the age of the last common ancestor of all living organisms with a protein clock". In *Horizontal Gene Transfer* (eds, M. Syvanen and C.I. Kado), Chapman & Hall, London.

Blankenship, R.E. (1992) Origin and early evolution of photosynthesis. *Photosyn. Res.* **33**: 91–111.

Berry, W.B.N. and Hartman. H. (1998) "Graptolite parallel evolution and lateral gene transfer". In *Horizontal Gene Transfer* (eds, M. Syvanen and C.I. Kado), Chapman & Hall, London.

Des Marais, D.J., Strauss, H., Summons, R.E., and Hayes J.M. (1992) Carbon isotope evidence for the stepwise oxidation of the Proterozoic enviroment. *Nature* **359**: 605–609.

Doolittle, W.F. and Logsdon, J.M. (1998) Do Archaea have a Mixed Heritage? *Current Biol.* **8**: R209-R211.

Feng, D., Cho, G. and Doolittle, R.F. (1997) Determining divergence times with a protein clock: update and reevaluation. *PNAS* **94**: 13028–13033.

Flavell, A.J (1999) Long terminal repeat retrotransposons jump between species. *PNAS* **96**: 12211–12212.

Hartl, D.L., Lohe, A.R. and Lozovskaya, E.R (1997) Modern Thoughts on an Ancyent Marinere: Function, Evolution, Regulation. *Annu. Rev. Genet.* **31**: 337–358.

Hartman, H., (1984) The origin of the eukaryotic cell. *Speculations Sci. Technol.* **7**: 77–81.

Hartman, H. (1992a) The eukaryotic cell and the RNA–protein world. In *Frontiers of Life* (eds, J. and K. Tran Thanh Van, J.C. Mounolou, J. Schneider and C. Mckay), Editions Frontieres, Paris.

Hartman, H. (1992b) The eukaryotic cell: evolution and development. In *The Origin and Evolution of the Cell* (eds, H. Hartman and K. Matsuno), World Scientific Press, Boca Raton, FL.

Hartman, H. (1998) Photosynthesis and the origin of life. *Origins of Life* **28**: 515–521.

Hoffman, P.F., Kaufman, A.J., Halverson, G.P. and Schrag, D.P. (1998) A Neoproterozoic Snowball Earth. *Science* **281**:1342–1346.

Kirschvink, J.L., Gaidos, E.J., Bertani, L.E. et al. (2000) Paleoproterozoic snowball Earth: Extreme climatic and geochemical global change and its biological consequences. *PNAS* **97**: 1400–1405.

Lewis, J.S., Watkins, G.H., Hartman, H. and Prinn, R.G. (1982) Chemical consequences of major impact events on Earth. Geological Society of America, Special Paper 190.

McClintock, B. (1984) The significance of responses of the genome to challenge. *Science* **226**: 792–801.

McDonald, J.F. (1990) Macroevolution and retroviral elements. *BioScience* **40**: 183–191.

Nelson, K.E., Clayton, R.A., Gill, S.R. et al. (1999) Evidence for lateral gene transfer between Archaea and bacteria from genome sequence of *Thermotoga maritima*. *Nature* **399**: 323–329.

Pace, N.R. (1997) A molecular view of microbial diversity and the biosphere. *Science* **276**: 734–740.

Reanney, D. (1974) Viruses and evolution. *Int. Rev. Cytol.* **37**: 21–52.

Robertson, H.M. and Lampe, D.J. (1995) Distribution of transposable elements in arthropods. *Annu. Rev. Entomol.* **40**: 333–357.

Robertson, R.M., Soto-Adames, F.N., Walden, K.O. et al. (1998) "The mariner transposons: horizontally jumping genes". In *Horizontal Gene Transfer* (eds, M. Syvanen and C.I. Kado), Chapman & Hall, London.

Syvanen, M. (1985) Cross-species gene transfer: implications for a new theory of evolution. *J. Theoret. Biol.* **112**: 333–343.

Syvanen, M. (1994) Horizontal gene transfer: evidence and possible consequences. *Annu. Rev. Genet.* **28**: 237–261.

Woese, C. (1987) Bacterial evolution. *Microb. Rev.* **51**: 221–271.

Xiong, Y. and Eickbush, T.H. (1990) Origin and evolution of retroelements based upon their reverse transcriptase sequences. *EMBO J.* **9**: 3353–3362.

CHAPTER 34

Horizontal Gene Transfer: A New Taxonomic Principle?

Lorraine Olendzenski, Olga Zhaxybayeva and J. Peter Gogarten

Although the occurrence of horizontal gene transfer (HGT) among prokaryotes is well established, microbial taxonomy relies on the assumption that evolution in microbes is dominated by vertical inheritance. However, an alternative explanation also appears to be compatible with the molecular data: if HGT-frequency occurs on a shallow declining gradient from more closely related species to less closely related species, the present taxonomic categories (including the domain level) could be exclusively determined by horizontal transfer frequency, i.e. Bacteria contain mainly bacterial genes because they more frequently exchange genes with other Bacteria than with Archaea. Levels of transfer would decrease among more distantly related lineages, with fewest occurring between the domains. Under this theory, organisms now considered to represent early branching lineages would appear as early branching only because these lineages partook less in intra-domain horizontal transfers than more shallow branching lineages. If one takes intramolecular recombination into consideration, the surprising congruence between genome content trees and 16S rRNA phylogeny, which has been used to discount the importance of HGT, is in perfect agreement with the theory that prokaryotic taxonomic units exclusively reflect HGT frequencies.

INTRODUCTION

The important role of horizontal gene transfer (HGT) in the spread of antibiotic resistance, the formation of pathogenicity islands, and selfish genes has long been established (Lawrence, 1999). HGT also has been suggested as a major force in structuring prokaryotic genomes, e.g. Lawrence and Roth's selfish operon theory (Lawrence and Roth, 1996; see also Chapters 9 and 24). In addition to genes under sporadic selection, HGT also affects house keeping genes, e.g. the archaeal-type ATP synthase transferred to Deinococcaceae and Spirochetes (Hilario and Gogarten, 1993), and functions involved in information processing, e.g. the archaeal type lysyl-tRNA synthase in rickettsia and spirochetes (Ibba et al., 1997) and archaeal type prolyl tRNA synthase in *Thermus* and *Deinococcus*, representatives of the Bacteria (Gogarten et al., 1999; Olendzenski et al., 2000). The widespread occurrence of HGT has been amply documented through the analyses of microbial genomes that have been completely sequenced (Bult et al., 1996; Koonin et al., 1997; Deckert et al., 1998; Nelson et al., 1999). Nevertheless, the sheer amount of HGT suggested from comparative genome analyses came as a surprise to most microbiologists.

Today, the fact that HGT occurred is no longer disputed; however, the impact of HGT

Copyright © 2002 by Academic Press.
All rights of reproduction in any form reserved.

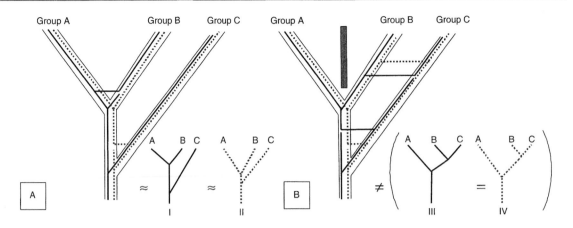

FIGURE 34.1 Depiction of two extremes regarding the influence of horizontal gene transfer (HGT) on our ability to reconstruct organismal evolution. The gray and the dotted line represent two genes. The solid black lines represent the evolution of the species (or taxonomic unit). In panel A the HGT frequency drops off steeply when considering less related species. As a result most of the transfers occur close to bifurcation events of the organismal tree, and the individual molecular phylogenies (I and II) are similar to the organismal tree (far left). Panel B represents the opposite extreme. Taxonomy is based on HGT frequency. In this illustration Group A is a more recently branching taxonomic unit; however, after its separation from group B, it occupies a niche, or develops a physiology that drastically reduces the frequency with which genes are shared with group B and C. In contrast group B and C continue to exchange genes. As a result the molecular phylogenies (III and IV) differ from the organismal phylogeny. In particular the deep branching lineage in the gene trees (III and IV) is the one that less frequently participates in HGT.

on microbial evolution, in particular on our ability to reconstruct organismal evolutionary history, remains controversial. The question addressed here is: how much does horizontal gene transfer impact our ability to reconstruct the evolutionary history of organisms from molecular data? With regard to the role of HGT in shaping bacterial phylogeny, we propose the possibility that HGT frequency is the main determinant of taxonomic units in prokaryotes.

Since the acceptance of Darwin's idea of natural selection, the goal of taxonomy has been to develop a natural system for the classification of organisms, i.e. one that is based on shared ancestry. A simplified view of the evolutionary process considers the generation of offspring with variation and selection as the main ingredients of biological evolution. The selection process is often described as survival of the fittest. While there are many well studied instances of fitness increase through collaboration (e.g. lichen, kefir, and the nucleated cell of euaryotes are all the result of symbiosis (Margulis, 1993)) the traditional view of evolution is based on

vertical inheritance, i.e. the passing on of genetic information from parent to offspring, and competition/selection of individuals within a population.

The ability to study the evolution of molecules, and the enormous increase of available data has dramatically changed our view of microbial evolution. Comparison of 16S rRNA sequence successfully provides a method for relating all organisms to each other, and especially has aided in microbial taxonomy. Using animal and plant evolution as a paradigm, microbial evolution has been assumed to be a process determined primarily by vertical inheritance. This view of systematics is based on gene trees constructed from markers thought to have never or rarely undergone horizontal transfer. Interspecies horizontal gene transfer is regarded as an important but rare exception. Under this vertical inheritance model of evolution, a natural bacterial classification reflecting shared ancestry and based on the majority of vertically inherited genes would emerge after weeding out the few horizontal transfers.

The present points of view form a continuum (Doolittle, 1999; Woese, 2000) whose two extremes are described in Figure 34.1. Figure 34.1A summarizes the present, conservative model in which taxonomy is based on vertical inheritance. HGT occurs under the vertical inheritance model, however, most of the HGT events occur between closely related species. As a result most molecular phylogenies will reflect the organismal phylogeny. There is some HGT between less related species, but these events are easily detected as conflicts with the majority consensus of "conserved" genes. If reality is close to this model dominated by vertical inheritance, then interdomain and interkingdom HGT events provide an excellent means of correlating the evolution in the different parts of the tree of life. However, the other extreme (Figure 34.1B) is equally compatible with the available data. Under this model HGT-frequency determines "relatedness," i.e. bacteria are more similar to other bacteria, and are recognized as bacteria, because they more frequently exchange genes with other bacteria than with less related groups (e.g. Archaea). The same statement would be true for other taxonomic categories: e.g. alpha-purple bacteria are more similar to other alpha-purple bacteria because they more frequently exchange genes with other alpha-purple bacteria than with beta-, gamma- or delta-purple bacteria. Under this extreme interpretation the recognized taxonomic categories would be created exclusively through a gradient in the frequency of HGT.

A crucial parameter is the steepness of the gradient with which HGT-frequency drops when considering less "related" species. If HGT-frequency occurs on a shallow declining gradient from more closely related species to less closely related species, the present taxonomic categories (including the domain level) could be exclusively determined by horizontal transfer frequency, i.e. Bacteria contain mainly bacterial genes because they more frequently exchange genes with other Bacteria than with Archaea. Levels of transfer would decrease among more distantly related lineages, with fewest occurring between the domains. The degree of relatedness could be defined as the degree of similarity between two genomes, as the distance in a genome content tree, or as the distance in a 16S rRNA phylogeny – however, under the assumptions of a HGT frequency model (Figure 34.1B), more "relatedness" does not correlate with more recently shared ancestry, but with higher HGT frequency. At first sight this definition might seem circular; however, the point is that relatedness as defined by the usual operational measures might be due exclusively to higher rates of HGT.

Surprisingly, the rather heretical point of view depicted in panel B cannot be easily discounted given the current data. To the contrary, it makes some interesting predictions that appear to agree with microbial taxonomy: Figure 34.1. panel B details two successive speciation events, the more recent one, between groups A and B, giving rise to a lineage (A) that adapts to an ecological niche that prevents frequent HGT with the other lineages. As a result of HGT ongoing among the other lineages, these lineages will become more similar to one another, and the more isolated lineage will be recovered as the deepest branching lineage in most molecular phylogenies (e.g. gene trees III and IV). This isolation with respect to HGT could either be caused by an environment that is less conducive to HGT, or by physiological or genetic isolating mechanisms that evolve within the organisms. An example of the former could be seen in the adaptation to extreme thermophily: in a hot environment naked DNA would be less stable and therefore transformation in this environment would be much less frequent. Is it a coincidence that the deep branching lineages in both the bacterial and the archaeal domain are occupied by extreme thermopiles? Could the Thermotogales be Gram positives that adapted to an extremely hot environment, where the more recent inventions made in the bacterial domain were not available to them by HGT and thus did not reach them, and where the only genes available were mostly from extremely thermophilic Archaea? HGT could still occur via transfection and conjugation under hot conditions, but these mechanisms appear to be most effective only for close relatives. An example for an internal barrier toward HGT would be the invention, or adaptation, of transcription or translation features that make the machinery of the cell incompatible with incoming DNA. Could it be that the Archaea form a distinct, and in molecular phylogenies, deep branching lineage because they invented a different approach to transcription? With the discovery of the many differences between Archaea and Bacteria the idea that Archaea might only be a derived group of Bacteria had been rejected (Woese and Gupta, 1981), however, the idea that HGT-frequency might be the cause of the recognizable features used in taxonomy resurrects this possibility. The evolution of the RNA polymerase functioning in mitochondria provides an example that drastic replacements in the transcription machinery can occur (Cermakian et al., 1997; Rousvoal et al., 1998; Schinkel and Tabak, 1989).

The hypothesis that taxonomy might be based exclusively on HGT-frequency and not on vertical inheritance makes predictions about the physiology and ecology of deep branching lineages. Clearly, the same characteristics are also compatible with the interpretation that microbial taxonomy is a natural systematic system, i.e. one based on shared ancestry. Under a vertical inheritance model of taxonomy, particular traits, e.g. living in isolated environments, having a different transcription machinery, are assumed to reflect historical facts: the last common ancestor was an extreme thermophile (Pace, 1991) or the environments on early earth selected for thermophilic organisms (Gogarten-Boekels et al., 1995). The last common ancestor is also assumed to have had a less well established transcription machinery than modern day prokaryotes; therefore the two lineages descending from the common ancestor (Bacteria and Archaea/Eukaryotes) evolved a different machinery for regulation of DNA directed RNA synthesis (Marsh et al., 1994). However, while both taxonomic models are compatible with the distribution of phenotypes on the tree of life, it is noteworthy that the model based on vertical inheritance would also be compatible with the converse observations (e.g. if deep branching lineages were found to be living in the same environment as the majority of other microorganisms), whereas the HGT based taxonomic model, which predicts that deep branches contain organisms that live in environments that

isolate them from HGT, would be falsified through finding different properties of the deep branching lineages (e.g. that they did not live in environments with inherent DNA instability or that they did not have incompatible transcription machineries). However, this slight predictive advantage is clearly insufficient to decide where between the two extreme scenarios actual microbial evolution occurs and has occurred in the past.

A strong argument against an important role of HGT in determining microbial taxonomy was the finding that phylogeny based on gene content were surprisingly similar to the phylogeny of rRNA (Fitz-Gibbon and House, 1999; Snel et al., 1999; Tekaia et al., 1999). Could it be that some genes, and in particular ribosomal RNA operons, were less prone to HGT? However, an alternative explanation for the congruence between gene content and ribosomal RNA phylogenies is that both the genome and the ribosomal RNA are mosaics which are formed through HGT. In both cases the similarity of both types of trees might mainly reflect HGT frequency and not vertical inheritance.

MOSAICISM IN GENOMES AND WITHIN GENES

The horizontal transfer of genes and their complete uninterrupted incorporation into the genome of the recipient, lead to a mosaic genome where different parts of the genome reflect different histories. The potential difference between a molecular or gene tree and the organismal evolution has been widely recognized, the latter is often assumed to be net-like or reticulate (Hilario and Gogarten, 1993; Gogarten, 1995). However, genes are not immutable units of inheritance, and even at the gene level, evolution is not always treelike. It is known that if multiple identical or similar copies of a gene are present in the same cell, gene conversion events tend to make these two copies of the genes more similar to one another (Gogarten and Olendzenski, 1999). These conversion tracts, i.e. the stretches of genomic information that are copied from one gene to another, are usually in the range of only a few hundred nucleotides (Sweetser et al., 1994; Betran et al., 1997; Yang and Waldman, 1997) – much smaller than the lengths of a typical gene. The result is that genes do not evolve as a whole. Rather different parts of genes can have different histories.

Many instances of apparently independent and parallel gene duplication events (for example in interferon and chaperonin evolution) might be due to gene conversion events that result in copies within one lineage being more similar to one another than to the orthologues in the different lines of descent (Gogarten et al., 1992; Archibald et al., 1999; Gogarten and Olendzenski, 1999). Only in a few cases does other evidence provide additional support for these conversion events. For example, the less conserved promoter sequences of the different interferon genes retain their functional specificity even though the protein coding parts of the genes are largely homogenized (Sick et al., 1998). The enolase of green algae and land plants shares a short deletion with one group of unrelated protists, whereas most of the gene reflects the proper systematic position of green algae and plants (Keeling and Palmer, personal communication). In Archaea a partial gene conversion event was detected for one of the chaperonin subunits (Archibald et al., 1999). Additionally, Goodman and collaborators have mapped conversion tracts in globin evolution (e.g. Fitch and Goodman, 1991).

The frequency of recombination between a gene obtained by HGT and a homologue already present in the genome depends on several factors: the presence of at least one sequence that allows homologous recombination (the presence of a second sequence with sufficient similarity increases recombination by several orders of magnitude, but recombination events relying on a single homologous sequence occur so frequently that they can be measured on a laboratory timescale, Wackernagel, personal communication); the time over which the homologues coexist in the same cell; and the number of copies of each of the genes in the genome. Transfer between closely related organisms will often result only in a couple of hundred nucleotides being replaced in the genome of the recipient:

restriction enzymes chop down the incoming DNA, sequences between donor and recipient organisms are so similar as to provide many possible recombination points (McKane and Milkman, 1995). Intra-gene mosaicism can be expected to be the rule rather than the exception when considering populations within a species or closely related species. The impact of this recombination between nearly identical gene versions on the reconstruction of gene trees, however, will be rather limited because the parts that form the mosaic have nearly identical histories. The recombination between genes from different genera, families or phyla will have a more dramatic impact on phylogenetic reconstruction.

Protein coding genes are mainly under selection for the amino acid sequence they encode. Because of the redundancy of the genetic code, two sequences can diverge in their nucleotide sequence but retain the same encoded function. If a protein coding gene is transferred between divergent organisms, the new copy will either replace the old gene, or both might be retained because the two homologues have acquired some different functionality. Recombination between the two copies, while detected in a few instances, is expected to be a rare event for divergent protein encoding genes.

In contrast, ribosomal RNAs function on the RNA level. Some parts of the rRNA sequences are universally conserved. This high level of conservation at the nucleotide sequence level is the reason that the environmental polymerase chain reaction is so successful for rRNA sequences. These conserved regions also might allow for homologous recombination between different copies of the rRNA coding genes. Other factors that are expected to favor recombination of rRNA genes are the presence of multiple copies per genome; the compatibility *in vitro* and *in vivo* of rRNA genes from one organism with the cellular machinery of another; and the frequency with which the translation machinery is a target of antibiotics. All of these factors suggest that the agreement between whole genome phylogenies and rRNA phylogenies might be due to the fact that both the genome and the rRNA genes are mosaics that were formed through HGT and recombination.

Gene conversion tends to homogenize the different members of a gene family; however, several cases of divergent rRNA encoding operons present in the same genome have been reported. For the present discussion the most striking is the case of *Thermomonospora chromogena* (Yap et al., 1999). The rRNA operons present in this organism provide a snapshot of homogenization in progress. Of the five operons sequenced in this actinomycete, four are very similar to each other, whereas the fifth is more similar to the rRNA operons from *Microbispora bispora* (Wang et al., 1997). Analyses by Yap et al. indicate that this operon is still expressed, and that while overall more similar to the *Microbispora bispora* operons, gene conversion events between the other *Thermomonospora chromogena* operons and the one transferred from *Microbispora bispora* have resulted in a

mosaic rRNA operon (Figure 34.2). With additional gene conversion events occurring over evolutionary time, this mosaicism can be expected to become more and more fine-grained until it is no longer recognizable.

Whole genome phylogenies agree with traditional rRNA phylogenies to a surprising degree (Fitz-Gibbon and House, 1999; Snel et al., 1999; Tekaia et al., 1999). These genome phylogenies appear to be very robust and independent from the particular measures and algorithms used. Examples for whole genome phylogenies are given in Figure 34.3. Note that the deepest split within the Archaea separates Cren- and Euryarcheotes, and that the two deepest branching Bacteria are *Thermotoga* and *Aquifex*.

Due to their central role in translation (and the complexity of the ribosome) genes for rRNAs are commonly assumed non-transferable. However,

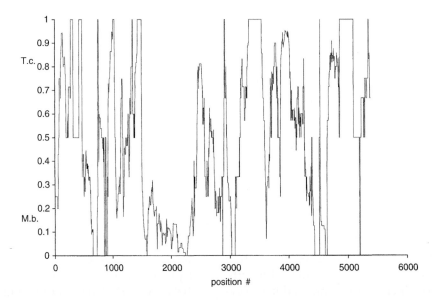

FIGURE 34.2 Comparison of rRNA operons form *Thermomonospora chromogena* with those from *Microbispora bispora*. The sequences analyzed are from Wang et al. (1997) and Yap et al. (1999). The graph depicts the relative identity between the one "atypical" *T. chromogena* rRNA operon with the remaining five *T. chromogena* rRNA operons and with the four from *M. bispora*. The ordinate gives the relative similarity of the atypical rRNA operon for a sliding window of 100 nucleotides. The relative similarity is calculated as a proportion of nucleotide differences between the atypical operon and other *T. chromogena* operons to the total number of observed differences between the atypical operon and all operons in the dataset. Relative similarity values above 0.5 indicate that the atypical rRNA operon is more similar to the remaining *T. chromogena* rRNA operons. Values below 0.5 indicate that the atypical rRNA operon is more similar to *M. bispora* operons. For identical regions the value of the relative similarity is set to 0.5. While on the whole the atypical *T. chromogena* operon is more similar to the *M. bispora* operon, this atypical operon clearly represents a mosaic formed from frequent recombination events between the different rRNA operons present in *T. chromogena*.

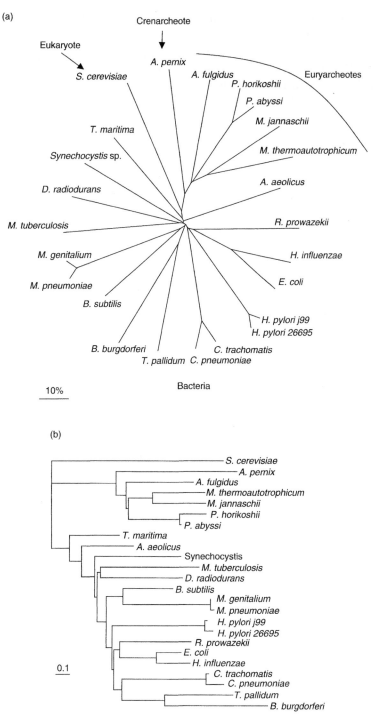

FIGURE 34.3 Genome content trees. (a) Tree calculated from a distance matrix based on the asymmetric distance measure (100 minus the percent of shared genes). Genes were considered as shared, when their BLAST score had an E-value below 10^{-8}. The distance matrix was calculated using Robert L. Charlebois' NeuroGadgets site at http://ngi.bio.uottawa.ca/. (b) Tree calculated using a cut-off value of 10^{-20}, and a distance measure that takes saturation into account: distance = $-\ln(1 - (\% \text{ shared genes})/100)$. Compare also the analyses of Fitz-Gibbon and House (1999), Snel et al. (1999), and Tekaia et al. (1999).

recent studies have shown that rRNA genes *can be* transferred (replacing the native complement) in the laboratory, and *have been* transferred in nature (Asai et al., 1999; Yap et al., 1999). Furthermore, Asai et al. (1999) managed to substitute a domain of yeast 28S rRNA for a domain of *E. coli* 23S rRNA without drastic effects on viability. These findings show that rRNA encoding genes are not immune to HGT, and that after HGT has occurred, gene conversion events can produce mosaic rRNA genes. Therefore the congruence between gene content based phylogenies and rRNA phylogenies could be due to the mosaic nature of both the genome and the rRNA. If this mosaicism is generated by the same process, HGT with a graded frequency distribution, the resulting phylogenies are expected to be similar.

It is clear that both vertical inheritance and horizontal transfer have played a role in microbial evolution. However, a quantitative assessment of HGT frequencies is needed to determine the relative contributions of horizontal and vertical inheritance in the formation of what we define as microbial taxonomic units. The information available at present is compatible with both views, that trees of microbial taxonomy reflect phylogeny dominated by vertical inheritance, or the other more heretical view, that these trees reflect relationships shaped predominantly by gradients of horizontal gene transfer. In either case, it is important to understand what the rRNA based view of microbial taxonomy is really telling us, and to keep an open mind regarding possible alternative interpretations of these trees.

REFERENCES

Archibald, J.M., Logsdon, J.M. and Doolittle, W.F. (1999) Recurrent paralogy in the evolution of archaeal chaperonins. *Curr. Biol.* **9**(18): 1053–1056.

Asai, T., Zaporojets, D., Squires, C. and Squires, C.L. (1999) An *Escherichia coli* strain with all chromosomal rRNA operons inactivated: complete exchange of rRNA genes between bacteria [see comments]. *Proc. Natl Acad. Sci. USA* **96**(5): 1971–1976.

Betran, E., Rozas, J., Navarro, A. and Barbadilla, A. (1997) The estimation of the number and the length distribution of gene conversion tracts from population DNA sequence data. *Genetics* **146**(1): 89–99.

Bult, C.J., White, O., Olsen, G.J. et al. (1996) Complete genome sequence of the methanogenic archaeon, *Methanococcus jannaschii. Science* **273**(5278): 1058–1073

Cermakian, N., Ikeda, T.M., Miramontes, P. et al. (1997) On the evolution of the single-subunit RNA polymerases. *J. Mol. Evol.* **45**(6): 671–681.

Deckert, G., Warren, P.V., Gaasterland, T. et al. (1998) The complete genome of the hyperthermophilic bacterium *Aquifex aeolicus. Nature* **392**(6674): 353–358.

Doolittle, W.F. (1999) Phylogenetic classification and the universal tree. *Science* **284**(5423): 2124–2129.

Fitch, D.H. and Goodman, M. (1991) Phylogenetic scanning: a computer-assisted algorithm for mapping gene conversions and other recombinational events. *Comput. Appl. Biosci.* **7**(2): 207–215.

Fitz-Gibbon, S.T. and House, C.H. (1999) Whole genome-based phylogenetic analysis of free-living microorganisms. *Nucleic Acids Res.* **27**(21): 4218–4222.

Gogarten, J.P. (1995) The early evolution of cellular life. *Trends Ecol. Evol.* **10**: 147–151.

Gogarten, J.P. and Olendzenski, L. (1999) Orthologs, paralogs and genome comparisons. *Curr. Opin. Genet. Dev.* **9**(6): 630–636.

Gogarten, J.P., Starke, T., Kibak, H. et al. (1992) Evolution and isoforms of V-ATPase subunits. *J. Exp. Biol.* **172**: 137–147.

Gogarten, J.P., Murphey, R.D. and Olendzenski, L. (1999) Horizontal gene transfer: pitfalls and promises. *Biol. Bull.* **196**(3): 359–361; discussion 361–362.

Gogarten-Boekels, M., Hilario, E. and Gogarten, J.P. (1995) The effects of heavy meteorite bombardment on the early evolution – the emergence of the three domains of life. *Orig. Life Evol. Biosph.* **25**(1–3): 251–264.

Hilario, E. and Gogarten, J.P. (1993) Horizontal transfer of ATPase genes – the tree of life becomes a net of life. *Biosystems* **31**(2–3): 111–119.

Ibba, M., Bono, J.L., Rosa, P.A. and Soll, D. (1997) Archaeal-type lysyl-tRNA synthetase in the Lyme disease spirochete *Borrelia burgdorferi. Proc. Natl Acad. Sci. USA* **94**(26): 14383–14388.

Koonin, E.V., Mushegian, A.R., Galperin, M.Y. and Walker, D.R. (1997) Comparison of archaeal and bacterial genomes: computer analysis of protein sequences predicts novel functions and suggests a chimeric origin for the archaea. *Mol. Microbiol.* **25**(4): 619–637.

Lawrence, J.G. (1999) Gene transfer, speciation, and the evolution of bacterial genomes. *Curr. Opin. Microbiol.* **2**(5): 519–523.

Lawrence, J.G. and Roth, J.R. (1996) Selfish operons: horizontal transfer may drive the evolution of gene clusters. *Genetics* **143**: 1843–1860.

Margulis, L. (1993) *Symbiosis in Cell Evolution*, W.H. Freeman, New York.

Marsh, T.L., Reich, C.I., Whitelock, R.B. and Olsen, G.J. (1994) Transcription factor IID in the Archaea: sequences in the *Thermococcus celer* genome would encode a product closely related to the TATA-binding protein of eukaryotes. *Proc. Natl Acad. Sci. USA* **91**(10): 4180–4184.

McKane, M. and Milkman, R. (1995) Transduction, restriction and recombination patterns in *Escherichia coli. Genetics* **139**(1): 35–43.

Nelson, K.E., Clayton, R.A., Gill, S.R. et al. (1999) Evidence for lateral gene transfer between Archaea and bacteria from genome sequence of *Thermotoga maritima*. *Nature* **399**(6734): 323–329.

Olendzenski, L., Liu, L., Zhaxybayeva, O. et al. (2000) Horizontal transfer of archaeal genes into the Deinococcaceae: detection by molecular and computer based approaches. *J. Mol. Evol.* **51**: 587–599.

Pace, N.R. (1991) Origin of life – facing up to the physical setting. *Cell* **65**(4): 531–533.

Rousvoal, S., Oudot, M., Fontaine, J. et al. (1998) Witnessing the evolution of transcription in mitochondria: the mitochondrial genome of the primitive brown alga *Pylaiella littoralis* (L.) Kjellm. Encodes a T7-like RNA polymerase. *J. Mol. Biol.* **277**(5): 1047–1057.

Schinkel, A.H. and Tabak, H.F. (1989) Mitochondrial RNA polymerase: dual role in transcription and replication. *Trends Genet.* **5**(5): 149–154.

Sick, C., Schultz, U., Munster, U. et al. (1998) Promoter structures and differential responses to viral and nonviral inducers of chicken type I interferon genes. *J. Biol. Chem.* **273**(16): 9749–9754.

Snel, B., Bork, P. and Huynen, M.A. (1999) Genome phylogeny based on gene content. *Nat. Genet.* **21**(1): 108–110.

Sweetser, D.B., Hough, H., Whelden, J.F. et al. (1994) Fine-resolution mapping of spontaneous and double-strand break-induced gene conversion tracts in *Saccharomyces cerevisiae* reveals reversible mitotic conversion polarity. *Mol. Cell. Biol.* **14**(6): 3863–3875.

Tekaia, F., Lazcano, A. and Dujon, B. (1999) The genomic tree as revealed from whole proteome comparisons. *Genome Res.* **9**(6): 550–557.

Wang, Y., Zhang, Z. and Ramanan, N. (1997) The actinomycete *Thermobispora bispora* contains two distinct types of transcriptionally active 16S rRNA genes. *J. Bacteriol.* **179**(10): 3270–3276.

Woese, C.R. (2000) Interpreting the universal phylogenetic tree. *Proc. Natl Acad. Sci. USA* **97**(15): 8392–8396.

Woese, C.R. and Gupta, R. (1981) Are archaebacteria merely derived "prokaryotes"? *Nature* **289**(5793): 95–96.

Yang, D. and Waldman, A.S. (1997) Fine-resolution analysis of products of intrachromosomal homeologous recombination in mammalian cells. *Mol. Cell. Biol.* **17**(7): 3614–3628.

Yap, W.H., Zhang, Z. and Wang, Y. (1999) Distinct types of rRNA operons exist in the genome of the actinomycete *Thermomonospora chromogena* and evidence for horizontal transfer of an entire rRNA operon. *J. Bacteriol.* **181**(17): 5201–5209.

Index

Printed and bound by CPI Group (UK) Ltd, Croydon, CR0 4YY

08/05/2025

01864993-0001